Preparative Methods in Solid State Chemistry

CONTRIBUTORS

W. Albers
A. M. Anthony
S. S. Batsanov
David Bellavance
R. Bougon
R. Collongues
Michel Dugue
J. Ehretsmann
J. B. Goodenough
Jean François Goullin
J. A. Kafalas
W. Kwestroo
J. M. Longo
Pierre Merenda
Michel Moulin
R. Naslain
H. Nowotny
M. Paulus
J. Portier
C. J. M. Rooymans
Harald Schäfer
A. Tressaud
H. J. A. Van Dijk
S. Windisch
Aaron Wold

Preparative Methods in
SOLID STATE CHEMISTRY

Edited by PAUL HAGENMULLER

University of Bordeaux
Talence, France

ACADEMIC PRESS New York and London 1972

ACADEMIC PRESS, INC.
111 Fifth Avenue, New York, New York 10003

United Kingdom Edition published by
ACADEMIC PRESS, INC. (LONDON) LTD.
24/28 Oval Road, London NW1

LIBRARY OF CONGRESS CATALOG CARD NUMBER: 74-182666

PRINTED IN THE UNITED STATES OF AMERICA

Contents

Modern Methods of Growing Single Crystals of High-Melting-Point Oxides

A. M. Anthony and R. Collongues

Chemical Transport as a Preparative Procedure

Harald Schäfer

Preparation of Transition Metal Compounds by Electrolytic Reduction of Fused Salts

Aaron Wold and David Bellavance

Preparation of Single Crystals of III–V Compounds

Michel Dugue, Jean François Goullin, Pierre Merenda, and Michel Moulin

Preparation of CdS Single Crystals by Epitaxial Growth on Germanium Substrates

H. J. A. Van Dijk

Oriented Eutectic Crystallization

W. Albers

Preparative Methods for Solid Fluorine Compounds

R. Bougon, J. Ehretsmann, J. Portier, and A. Tressaud

Boron Syntheses

R. Naslain

Preparation Conditions of the Ferrites

M. Paulus

Preparation of Transition-Metal Carbides and Related Compounds

S. Windisch and H. Nowotny

Preparation of Chalcogenides and Pnyctides at Low Temperature

W. Kwestroo

List of Contributors

Numbers in parentheses indicate the pages on which the authors' contributions begin.

W. Albers, Philips Research Laboratories, N. V. Philips' Gloeilampenfabrieken, Eindhoven, The Netherlands (367)

A. M. Anthony, Centre National de la Recherche Scientifique, Orleans, France (147)

S. S. Batsanov,* Institute of Thermophysics, Academy of Science, Novosibrisk, U.S.S.R. (133)

David Bellavance, Department of Chemistry and Division of Engineering, Brown University, Providence, Rhode Island (279)

R. Bougon, Centre d'Études Nucléaires de Saclay, Gif-sur-Yvette, France (401)

R. Collongues, University of Paris, Paris, France (147)

Michel Dugue, Thomson-CSF Laboratories, Corbeville, France (309)

J. Ehretsmann, Centre d'Études Nucléaires de Saclay, Gif-sur-Yvette, France (401)

J. B. Goodenough, Lincoln Laboratory, Massachusetts Institute of Technology, Lexington, Massachusetts (1)

Jean François Goullin, Thomson-CSF Laboratories, Corbeville, France (309)

J. A. Kafalas, Lincoln Laboratory, Massachusetts Institute of Technology, Lexington, Massachusetts (1)

W. Kwestroo, Philips Research Laboratories, N. V. Philips' Gloeilampenfabrieken, Eindhoven, The Netherlands (563)

*Present address: State Committee for Standards, Moscow, U.S.S.R.

J. M. Longo,* Lincoln Laboratory, Massachusetts Institute of Technology, Lexington, Massachusetts (1)

Pierre Merenda, Thomson-CSF Laboratories, Corbeville, France (309)

Michel Moulin, Thomson-CSF Laboratories, Corbeville, France (309)

R. Naslain, Service de Chimie Minérale Structurale associé au CNRS, Université de Bordeaux, Talence, France (439)

H. Nowotny, University of Vienna, Vienna, Austria (533)

M. Paulus, Groupe d'Étude et de Synthèse des Microstructures, Centre National de la Recherche Scientifique, Meudon-Bellevue, France (487)

J. Portier, Service de Chimie Minérale Structurale associé au CNRS, Faculté des Sciences de Bordeaux, Talence, France (401)

C. J. M. Rooymans, Philips Research Laboratories, N. V. Philips' Gloeilampenfabrieken, Eindhoven, The Netherlands (71)

Harald Schäfer, Institute of Inorganic Chemistry, Münster University, Münster, Germany (251)

A. Tressaud, Service de Chimie Minérale Structurale associé au CNRS, Faculté des Sciences de Bordeaux, Talence, France (401)

H. J. A. Van Dijk, Philips Research Laboratories, N. V. Philips' Gloeilampenfabrieken, Eindhoven, The Netherlands (361)

S. Windisch, University of Vienna, Vienna, Austria (533)

Aaron Wold, Department of Chemistry and Division of Engineering, Brown University, Providence, Rhode Island (279)

* Present address: Esso Research and Engineering, Corporate Research Laboratories, Linden, New Jersey.

Preface

During the past several years, considerable theoretical development has taken place in the field of solid state chemistry. Solid state specialists have been able to specify the localization and energy levels of electrons, through more systematic and quantitative determination of electrical, magnetic, and optical properties, and are now able to formulate general rules which, in turn, allow the prediction of those properties. These are closely related to structures of the long- or short-range type, which depend on the nature and also the thermal history of the elements in question. The time is coming when it will be possible to prepare, on demand, the materials corresponding to a given set of physical properties. The detailed study of properties, as well as the problem of modifying them, requires a complete mastery of preparative methods.

It has been known for a long time that temperature has a definite influence on crystalline structure. Usually at higher temperatures there is a tendency toward formation of structures less ordered, richer in thermal defects, and more symmetric. The influence of temperature on symmetry is sometimes combined with that of composition, whereby the presence of vacancies or "floating cations" is the cause of dissymmetry. For example the phases Ca_xWO_3 and Al_xWO_3 have an increasing symmetry as a function of the extent of insertion x, as well as of temperature; the monoclinic distortion of Na_3FeF_6 disappears at 630°C, and also when sodium is replaced by the larger potassium atom. The use of high pressures is a more recent event because of the technical problems which accompany it. An important part of this volume is devoted to this important technique, because of the role which it will play in inorganic synthesis in the coming years. Crystal growth and other modern aspects of preparative solid state chemistry are also extensively developed.

This work concerns certain modern methods of solid state synthesis, and also compares the various preparative possibilities of some advanced fields. The structural aspects are well developed and in certain cases the questions of stoichiometry and purity.

This volume had its origin at a EUCHEM conference held in 1968, organized under the very relevant initiative of Dr. Rabenau at Aachen, which dealt with the preparative methods of solid state chemistry. This conference brought to a focus the importance of the chemist's role in preparing materials of desired quality, and in obtaining new materials according to the requirements of the user, the physicist for example. The articles represent only part of the lectures which were given on that occasion; the framework has been enlarged, and an equilibrium corresponding to actual needs has been sought.

The editor would like to thank all those who contributed to this work, and hopes furthermore, that this volume will be useful not only for the high quality of information which it will furnish to the solid state chemist, but also in bringing about new international cooperations of this type.

Introduction

The article by Goodenough, Kafalas, and Longo describes for the first time in a detailed manner the new and numerous possibilities for syntheses using high pressures. Certainly one can expect an increase in coordination, and also changes in structure related to growing cationic repulsions, stabilization of high oxidation states (Fe^{4+}, Ni^{3+}, or Cu^{3+} for example) as well as intermediate states such as those of Cr^{4+} or Cr^{5+}, extension of homogeneity ranges of nonstoichiometric phases, and formation of phases stabilized by metallic bonds. The solid state chemist will find in high pressure synthesis a new dimension, which is at the same time an extension of possibilities and a refinement.

The complement to such an article is the detailed description of methods which allow the chemist to obtain high pressures. Rooymans describes with great accuracy the processes for attaining high pressures through transmission by gaseous or liquid fluids or in the solid state. The applications to certain types of measurements under pressure are also considered.

Batsanov deals with the method of shock waves both from the point of view of thermodynamics and thermoelasticity, and of the possibility of using superpressures for evidently revolutionary applications.

A very important topic covered in this book is the synthesis of single crystals. Crystal growth occupies an increasing role in the activities of the chemist for reasons which are becoming more and more urgent—for example, the study of tensorial physical properties, the distribution of lattice defects, investigations into the kinetics of surface reactions, and applications in the field of electronics. The preparation of single crystals remains a difficult problem insofar as it requires a change of state.

These difficulties are particularly important in the growth of single crystals of high melting point materials. The detailed chapter by Anthony and Collongues is devoted to the synthesis of single crystals of refractory oxides either at high temperatures (essentially liquid–solid transformations) or at lower temperatures in the presence of a solvent or a chemical reagent.

The results obtained by the various methods proposed, in particular those which concern the nature of defects, are subjected to a critical study.

Numerous examples of vapor transport reactions in a temperature gradient are presented in the chapter by H. Schäfer. He discusses the thermodynamic fundamentals, techniques used, importance of impurities, and influence on solid state reactions.

A less-known but nonetheless delicate method is the production of single crystals by electrolytic reduction in molten salts. Wold and Bellavance show how this method can be used with competence.

Crystal growth of III–V compounds is discussed by Dugue, Goullin, Merenda, and Moulin, both from the point of view of the variety of methods utilized and of their industrial importance in the field of electronic components. The problems of ternary phases and thin films are considered.

The chapter by Van Dijk is devoted to epitaxial deposits of cadmium sulfide.

A chapter is devoted to composite materials of the whiskers type because of their mechanical and, more recently, magnetic applications. One of the more productive methods for the formation of whiskers, presented by Albers, is oriented growth starting from eutectics, monotectics, or eutectoids.

Some chapters on fields of inorganic chemistry now actually in development and of interest to researchers wishing to specialize in these domains are included in this volume. The study of preparative methods is considered in a critical manner so as to be of interest to those concerned with the choice of a method as a function of a given problem.

Bougon, Ehretsmann, Portier, and Tressaud have collaborated here to describe the various ionic fluoride preparative methods. Reactions in solution, in the solid state, solid–gas reactions, and reactions in the gaseous phase are presented. The use of fluorine under pressure and the problem of the conditions required are considered.

The chapter by Naslain is devoted to boron preparative methods which deal with the degree of purity and the crystallographic nature of the various samples obtained, as well as with the operational precautions required.

The article by Paulus concerns ferrite preparative methods, the importance of which is great from the point of view of applications. These methods present delicate problems in terms of property dependences on cationic distribution, anionic vacancies, and texture. The thermodynamic aspects of the problem are considered. Although this chapter is limited to ferrites, it will interest all those who prepare oxides with definite properties.

Windisch and Nowotny in their chapter on carbides of transition elements, describe the various preparative methods (metallurgy of powders, procedures utilizing the molten metal or volatile compounds) and give precise information on the varying parameters for each particular group. The ternary carbides are also considered.

Kwestroo presents a recent method for the preparation of chalcogenides and pnyctides of high purity. These are important semiconducting and luminescing materials, whose direct synthesis sometimes leads to inhomogeneous materials or to unexpected compositions. The effectiveness of this method which uses a halogenide of the same metal is described and justified.

High-Pressure Synthesis†

J. B. GOODENOUGH, J. A. KAFALAS, and J. M. LONGO‡

Lincoln Laboratory, Massachusetts Institute of Technology
Lexington, Massachusetts

† This work was sponsored by the Department of the Air Force.

‡ Present address: Esso Research and Engineering, Corporate Research Laboratories, Linden, New Jersey.

I. Introduction

High-pressure synthesis has a practical as well as a scientific interest, since many high-pressure products are either stable or metastable at atmospheric pressure to temperatures well above 300°K. To illustrate the strategy and present-day techniques of high-pressure synthesis, we have chosen to review the high-pressure studies, through June 1970, performed on compounds having the chemical formula ABX_3 or $(AX)_n ABX_3$, where cation A is always larger than cation B.

Several other reviews provide either general summaries of synthesis under pressure (Kleber and Wilke, 1969; Klement and Jayaraman, 1967; Rooymans, 1967, 1969; Morimoto and Akimoto, 1969; Syono, 1969; Prewitt and Shannon, 1969; Popova, 1964; Marezio, 1969) or specialized summaries of works on spinels (Ringwood and Reid, 1969; Reid and Ringwood, 1970; Liebertz and Rooymans, 1967), AX_2 compounds (Seifert, 1967; Kabalkina *et al.*, 1969; Austin, 1969), and sulfides (Range *et al.*, 1968a–d, 1969a–i). Several review papers also appear in the Proceedings of the Symposium on Crystal Structure at High Pressure (*Trans. Amer. Crystallogr. Ass.* **5,** 1969).

At atmospheric pressure, the ABX_3 or $(AX)_n ABX_3$ compounds crystallize in a variety of different structures, or do not form at all. High pressures stabilize preferentially the more dense phase. The most dense ABX_3 phase has the cubic-perovskite structure, and if the A cation is not too small, pressures obtainable in the laboratory are generally sufficient to stabilize it. In the case of oxides, fluorides, and chlorides, these high-pressure perovskites, as well as the intermediate-pressure polymorphs, are generally stable at atmospheric pressure and room temperature.

After a brief description in Section II of general preparative techniques, several ABX_3 and $(AX)_n ABX_3$ structures are compared in Section III, special attention being given to cation-site coordinations and relative densities. The experimental summary is divided into three operational groupings:

1. High-pressure transformations of compositions prepared at atmospheric pressure.
2. Composition formations requiring elevated pressures.
3. Syntheses greatly facilitated by pressure.

Within the first grouping, the relative stabilities of five hexagonal-perovskite polytypes are determined by the relative ionic sizes and the pressure. These polytypes are distinguished from the other polymorphs, since extensive data on the relationships between them are now available. The relative stabilities of other ABX_3 polymorphs depend upon the outer-electron configurations at the cations as well as on relative ionic sizes and pressure. Some general features of these interrelationships are presented and used to classify the second grouping into six distinguishable categories, each representing a different set of conditions, modifiable by high pressure, that inhibit composition formation at atmospheric pressure. This classification is not intended to be exhaustive, but to illustrate, from available data on ABX_3 compounds, considerations pertinent to the choice of high pressure as a tool for chemical synthesis. The final section merely calls attention to the fact that high pressure may also alter the kinetics of a reaction so as to greatly facilitate synthesis of a composition stable at atmospheric pressure.

II. General Techniques

Several apparatuses have been used for high-pressure syntheses: the cylindrical die and piston, the "belt," the "girdle," the tetrahedral anvil, the cube, the Bridgman anvil, and various hydrothermal units. These systems and their use are adequately described in other chapters of this volume.

We have used a high-pressure "belt" apparatus similar to that illustrated in Fig. 1. It is capable of developing pressures to 90 kbar and temperatures

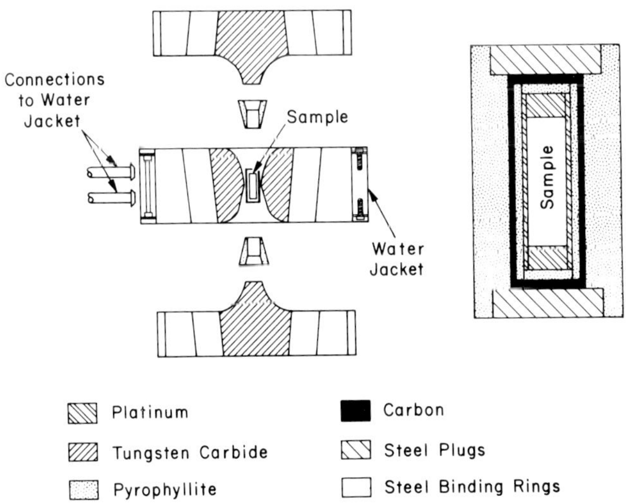

FIG. 1. High-pressure belt apparatus and expanded view of typical sample cell.

to 2000°C. The specimen cell uses pyrophillite as the pressure-transmitting medium and a graphite sleeve as an internal heater. The specimen is encapsulated in a nonreactive, malleable material. For oxides, the capsules we used consisted of 0.1-mm-thick noble-metal sleeves with 2-mm-thick noble-metal end plugs. For temperatures above 1000°C, platinum has proven satisfactory. At lower temperatures, gold is preferable, since it is more economical, is easier to handle, and makes better high-pressure seals. Some workers (Sugawara *et al.*, 1968a; Syono *et al.*, 1969a; Tomashpol'skii *et al.*, 1967) have used graphite specimen capsules, the capsule also serving as a cell heater. Neither gold nor platinum is sufficiently inert for use in the synthesis of many compounds containing sulfur, selenium, or tellurium. In these cases, it is necessary to encapsulate the specimens in BN, MgO, or some other inert material.

In a typical synthesis of a quenchable high-pressure phase, the encapsulated specimen is subjected to pressure before the temperature is raised. After holding at temperature (30 min is generally ample time), the sample is quenched under pressure at about 400°C sec^{-1} by turning off the power to the cell. The pressure is then released and the specimen recovered for examination. Since the high-pressure phase is retained at room temperature and atmospheric pressure, the high-pressure product may be identified by normal x-ray powder diffraction. This technique is independent of the pressure unit used—belt, girdle, tetrahedral anvil, cube, or Bridgman anvil.

The synthesis of some materials, such as $CaFeO_3$ and $CaMnO_3$, requires a high oxygen pressure within the specimen capsule. One solution is to divide the capsule into two compartments: one containing the specimen and the other filled with CrO_3. For separators, we have used disks of BN or platinum with a small center hole; Kanamaru *et al.* (1970) have used sintered disks of ZrO_2. At elevated temperatures, the decomposition of CrO_3 to $CrO_2 + \frac{1}{2}O_2\uparrow$ provides several kbar of oxygen pressure at the specimen. Such a split-cell arrangement could also take up decomposition products, such as O_2 or CO_2, if CrO_3 were replaced by a getter.

For synthesis at pressures to 200 kbar and 1000°C, Ringwood and Major (1968) have used a pair of 0.5-in.-faced Bridgman anvils operating on a cell with an internal nickel heater. Their specimens were isolated from the heater by a gold foil 0.0002 in. thick. They report that a 3–5 min pressure treatment at 1000°C in this device is sufficient to drive most high-pressure transformations to completion. While their device yields scarcely enough specimen for x-ray identification, its enormous pressure range and low cost make it an excellent tool for exploratory work and feasibility studies.

In many instances, polymorphic transformations can be made to occur at lower pressures and to give a higher yield or more crystalline product in shorter time by the addition of a suitable flux. Water is the most common flux used. Ringwood and Major (1967b, 1968) use 2% water in most of their research on reactions of geophysical significance. We have found that the addition of a small amount of water to samples of $Ln_2Ge_2O_7$, which transforms to the pyrochlore structure (Shannon and Sleight, 1968), leads to a pure and very crystalline product. In many cases, an excess of one of the reactants can serve as a flux. This excess may usually be removed by washing with a suitable solvent.

Post and Banks (1968) have reported crystal growth of $Na_{0.8}WO_3$ by an electrolytic reaction within a high-pressure cell of boron nitride that was fitted with platinum electrodes to prevent leakage of any molten material. Mixtures of Na_2WO_4 and WO_3 were heated to 700°C at 40 kbar, and the $Na_{0.8}WO_3$ crystals grew at the cathode. This technique should be applicable to a variety of metallic oxides.

III. Structural Considerations

A. ABX_3 Perovskite Polytypes

Oxides and fluorides with the chemical formula ABX_3 tend to form a large class of structurally related compounds provided the larger A cation has a radius $0.9 < r_A < 1.9$ Å and the smaller B cation is stable in an octahedral interstice ($0.5 < r_B \lesssim 0.9$ Å). These hexagonal-perovskite polytypes are also found where X is a heavier anion, such as sulfur or chlorine, but the cubic-perovskite structure is more difficult to obtain in this case.

This class of compounds is characterized by a stacking of close-packed

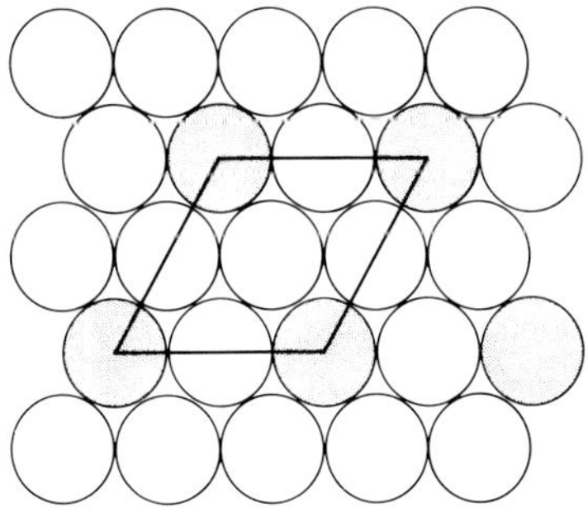

FIG. 2. An AX_3 close-packed layer showing order of A cations (shaded circles) and X anions (open circles).

AX_3 layers that are ordered as illustrated in Fig. 2, the B cations occupying all the interlayer X-ion octahedra. As with any sequence of close-packed layers, successive layers may be stacked in either of two ways. Therefore the stacking of any AX_3 layer may be cubic or hexagonal with respect to its two neighboring layers, depending upon whether it is in the middle of an *aba* or of an *abc* sequence. If the stacking is entirely cubic, the B-filled octahedra share only corners in three dimensions to form the cubic-perovskite structure of Fig. 3. Figure 4a shows the cubic network of B-filled octahedra as viewed perpendicular to the cubic [111] axis, which forms the *c* axis of a hexagonal cell. If all the stacking is entirely hexagonal, as in $CsNiCl_3$, the B-filled octahedra form isolated *c*-axis chains of face-shared octahedra. Figure 4b illustrates such a chain.

Polytypes are different structural modifications of a given composition having similar unit cells in two dimensions, but different stacking of layers along the third axis to make its variable length an integral multiple of some common unit. In the hexagonal-perovskite polytypes discussed here, the integral unit is the spacing between close-packed AX_3 layers. The cubic-perovskite and hexagonal $CsNiCl_3$ structures have the same unit-cell dimensions within the AX_3 layers and repeat distances of three and two layers, respectively. Therefore they will be referred to as the 3C and 2H structures, in accordance with the usual polytype nomenclature.

In the 3C structure of Fig. 3, the corner-shared octahedra form an A-site cage of finite size; this constraint places a definite restriction on the range of A-cation sizes that can be accommodated in this structure. Because there is an optimum A–X bond length, the presence of an A atom generally distorts the BX_3 array so as to optimize the A–X bonding. If this distortion

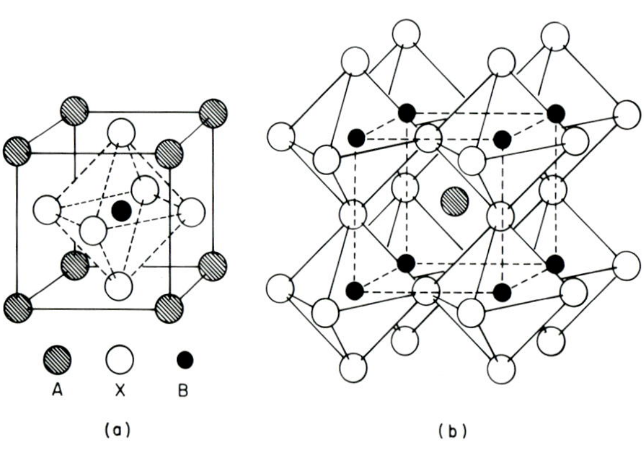

FIG. 3. Ideal perovskite structure: (a) A cation at origin, (b) B cation at origin.

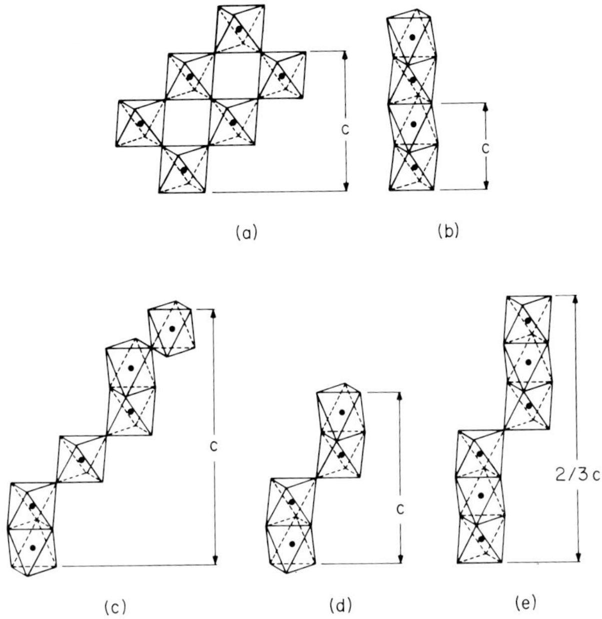

FIG. 4. Stable structures intermediate between (a) cubic-perovskite and (b) 2H $CsNiCl_3$ structures: (c) 6H $BaTiO_3$ structure, (d) 4H $SrMnO_3$ structure, (e) 9R $BaRuO_3$ structure.

is too large, other structures become competitive. Goldschmidt (1926) defined the tolerable limits on the radius r_A of the A cation relative to r_X and r_B of the anion and B cation via a tolerance factor

$$t = (r_A + r_X)/2^{1/2}(r_B + r_X) \quad (1)$$

By geometry, the ideal cubic structure should have $t = 1$. The perovskite structure occurs at atmospheric pressure only within the range $0.75 < t < t_c \approx 1.0$, where the value of t_c obtained depends upon the choice of ionic radii. We use the radii of Shannon and Prewitt (1969). However, to have t within the range $0.75 < t < t_c$ is not a sufficient condition, since the A and B cations must, in themselves, be stable in twelvefold (12 or 8 + 4 or 6 + 6) and sixfold coordinations, respectively. This sets lower bounds for the cationic radii. In oxides, these bounds are $r_A > 0.90$ and $r_B > 0.51$ Å. In addition, Megaw (1946) noted that, if $0.75 < t < 0.9$, a cooperative buckling of the corner-shared octahedra to optimize the A–X bond lengths introduces a superstructure to the unit cell. Geller (1956) showed that in $GdFeO_3$, this buckled structure has the orthorhombic Pbnm symmetry of

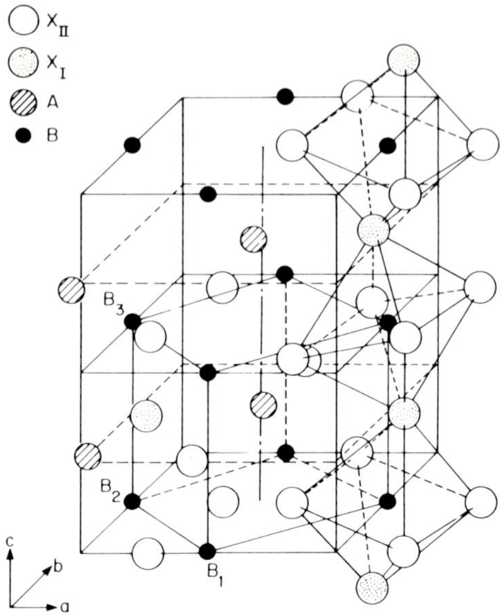

FIG. 5. The orthorhombic perovskite with space group Pbnm.

Fig. 5. Although small distortions to other orthorhombic space groups may occur, this appears to be the most common. On the other hand, if $0.9 < t < t_c \approx 1$, such buckling is not found, although small distortions to rhombohedral (usually $R\bar{3}c$) symmetry may occur. These distortions of the cubic-perovskite structure must be distinguished from those induced by various types of electron ordering (Goodenough and Longo, 1970). However, what is relevant for high-pressure synthesis is not the possible distortions within the perovskite-phase field, but the character of the competitive phases at the edges of this field.

Where $t > t_c$ because r_A is too large, hexagonal stacking of the AX_3 layers is stabilized. As Babel (1967) has pointed out, this generates isolated chains of face-shared octahedra that can, by adjusting the spacing between chains, accommodate the larger A cations without altering the B–X bond length. However, formation of hexagonal stacking occurs at the expense of the electrostatic Madelung energy, since there is a larger B–B repulsion across a shared octahedral-site face. This loss in electrostatic binding energy can be reduced by confining the length of any face-shared chain to two or, at most, three units. This permits a displacement of the cations along the c axis, with a corresponding contraction of the anion triangle composing the shared octahedral-site face. The three most probable stacking sequences for achieving this are illustrated in Figs. 4c–e. The 6H hexagonal structure of Fig. 4c corresponds to the two-to-one cubic-to-hexagonal stacking sequence (cchcch), the 4H hexagonal structure of Fig. 4d to a one-to-one stacking (chch), and the 9R rhombohedral structure of Fig. 4e to the one-to-two cubic-to-hexagonal sequence (chhchhchh). (A more complete discussion of these structures is given by Katz and Ward, 1964.) Although an infinite number of ordered structures is theoretically possible, each polytype having a different ratio of cubic to hexagonal stackings, only these three intermediate phases have been identified in stoichiometric ABX_3 compounds. They represent an optimization of the total energy

$$E = E_M + E_B + E_p \tag{2}$$

where E_M is the Madelung energy, E_B is the core-repulsion energy that determines the optimum M–X bond length, and E_p is the energy required to alternate the stacking sequence. Clearly, E_p is 1.5 times as large for the 4H structure as for the 6H and 9R structures, so the range of phase stability for the 4H structure will be relatively small. In fact, this structure has only been identified in oxides, and there, over a narrow phase field.

Van Veen (1971) has reported an ordered 4H structure for the compound $RbCsMn_2Cl_6$. Alternate AX_3 layers contain 3:1 and 1:3 Rb/Cs ratios, and the alkali-metal atoms are ordered within these layers. It is unusual to find an ordering among alkali-metal ions of different size, and this ordering as well as the 4H polytype are undoubtedly stabilized by a mutual interaction between these two structural features.

In general, the optimum A–X bond length decreases more rapidly with pressure than does the optimum B–X bond length. This means that pressure decreases the tolerance factor ($dt/dP < 0$) and therefore stabilizes cubic versus hexagonal stacking. This is also consistent with pressure stabilizing preferentially the more dense phase, since in the 3C polytype, an A cation has twelvefold coordination, whereas in the hexagonal polytypes, some or all of the A cations have this coordination reduced from 12 to (6 + 6) by deviations of the c/a ratio from its ideal value.

The intermediate phases occur only in a narrow transitional interval of the tolerance factor. The width Δt of this interval is larger the greater the electrostatic contribution E_M to Eq. (2) : $\Delta t \approx 0.08$ in the fluorides (Babel, 1967) and is even smaller in the chlorides and sulfides. Because Δt is small, it is possible to find examples where the various polytypes are stabilized successively with increasing pressure. Some of these structures have physical properties of potential technical significance. For example, the $A^+Ni^{2+}X_3^-$ compositions having the 6H structure are ferrimagnetic and optically transparent.

In the $A^{2+}B^{4+}O_3$ oxides, and to an even greater extent in the $A^{+}B^{5+}O_3$ oxides, the electrostatic contribution E_M is relatively large, and a $t_c > 1.0$ is found. The magnitude of $(t_c - 1)$ increases with the effective charge on the B cation, but the stability of the perovskite 3C phase having a $(t - 1) > 0$ is decreased by the associated stretching of the B–X bond. Stretching of the B–X bond also enhances any tendency for a ferroelectric-type displacement of the B cation from the center of symmetry of its octahedral interstice to form shorter and longer B–X bonds. Thus $BaTiO_3$, which has a $t = 1.06$, is cubic at high temperatures and becomes ferroelectric below a $T_c = 120°C$. However, slight reduction of $BaTiO_3$ transforms it to the 6H polytype, showing that $t_c \approx 1.06$ in this compound. By contrast, $CaTiO_3$ ($t = 0.97$) exhibits no ferroelectric-type distortion and $SrTiO_3$ ($t = 0.99$) has a more complex transformation only below $-163°C$. The compounds $RbNbO_3$ ($t = 1.085$) and $RbTaO_3$ ($t = 1.085$) cannot be synthesized at atmospheric pressure. However, Kafalas (1971) has synthesized them at pressures $P > 65$ kbar and $P > 90$ kbar, respectively. Significantly, heating of these metastable compounds leads to their decomposition rather than to transformation to one of the hexagonal polytypes. This indicates that the electrostatic force between B^{5+} ions sharing common octahedral-site faces is too large to permit stabilization of the hexagonal polytypes. On the other hand, a $t = 1.085$ is too large to permit stabilization of the cubic phase at atmospheric pressure. In addition, room-temperature $RbNbO_3$ is ferroelectric, having the structure of the orthorhombic ferroelectric phase of $BaTiO_3$ stable in the interval $-90 < T < 5°C$, whereas room-temperature $RbTaO_3$ remains cubic. Thus the condition for a ferroelectric-type phase transformation induced by B cations appears to consist of at least two parts: stretching of the B–X bond and nearly equivalent tetrahedral and octahedral site energies (Goodenough and Longo, 1970).

B. ABX_3 Polymorphs: Small A Cations

Polymorphism is the existence of the same composition in more than one crystalline form. Polytypism is a particular kind of one-dimensional polymorphism. We refer to polymorphism as distinguishable from polytypism.

If the size of the A cation falls below about 0.9 Å, then at atmospheric pressure, other ABX_3 structures become competitive with that of perovskite. Where the B cation is well stabilized in an octahedral site, the competitive structures are corundum or corundum-related, provided an ABX_3 composition is stable (see Sections V.A, G). Where the B cation is more stable in tetrahedral coordination, a pyroxene structure is generally the more stable ABX_3 structure at atmospheric pressure, but the garnet structure is also found. However, if the B cation is as stable in the fivefold coordination of a trigonal bipyramid as in either tetrahedral or octahedral coordination and if the A cation is near the lower limit of r_A for formation of the perovskite structure, then the $YAlO_3$ or the related $YMnO_3$ structure may be found at atmospheric pressure. If the A cation has a tetrahedral-site preference energy, an ABX_3 compound may not form at atmospheric pressure.

1. *Sixfold-coordinated B cations*

If the A and B cations are both stable in octahedral sites, then $A^{3+}B^{3+}O_3$ compounds generally crystallize in the corundum structure of Fig. 6. It consists of a close-packed anion array in which the chains of face-shared octahedra parallel to the *c* axis are each two-thirds occupied, occupied pairs alternating with vacancies. This ordering permits ionic displacements that optimize the electrostatic Madelung energy. Here, the A and B cations are randomly distributed over the occupied sites.

In the case of $A^{2+}B^{4+}O_3$ compounds, electrostatic and covalent forces order the A^{2+} and B^{4+} cations into alternate basal planes to give the ilmenite structure. At high pressures, these may transform directly to the perovskite structure, or to a disordered corundum phase of smaller volume.

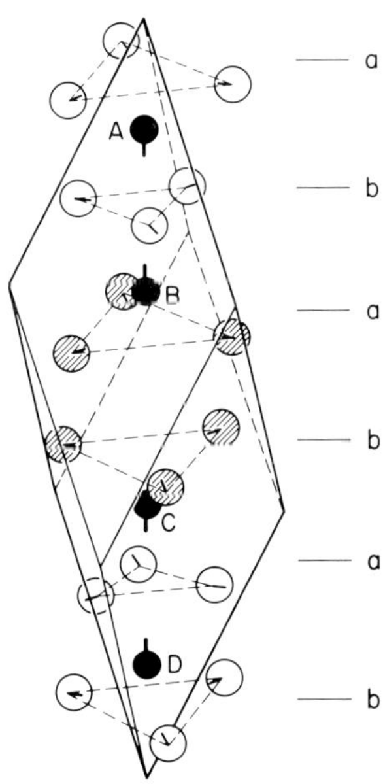

FIG. 6. Rhombohedral unit cell of the corundum structure.

The compositions $A^{+}B^{5+}O_3$ generally favor the $LiNbO_3$ or $LiSbO_3$ structures if r_A is small. In $LiNbO_3$, the *c*-axis chains of face-shared octahedra contain Nb^{5+} ions in every third octahedral site and the Li^{+} ions, instead of occupying near-neighbor octahedra to make Li^{+}–Nb^{5+} pairs as would be the case in the ilmenite structure, are displaced to the octahedral-site face midway between its two neighboring Nb^{5+} ions in the chain. Only below the ferroelectric Curie temperature $T_c = 1210°C$ are the Li^{+} ions displaced cooperatively along the *c*-axis to one or the other side of the (001) oxygen planes. In $LiSbO_3$, the Sb^{5+} ions occupy alternate octahedra of two-thirds of the *c*-axis chains and the Li^{+} ions fill all the octahedra of the remaining *c*-axis chains.

The corundum and ilmenite structures have the space groups $R\bar{3}c$ and $R\bar{3}$, respectively, and can therefore be related to the rhombohedral perovskite structure having the same space group (Moreau *et al.*, 1970; Michel *et al.*, 1971). Furthermore, the rhombohedral perovskite structure has the greater anion coordination about the A cation, $(6 + 6)$ instead of 6, so that it is more dense. Therefore pressure tends to stabilize the perovskite relative to the ilmenite structure; and where A is not too small, a phase transformation can be induced by pressures available in the laboratory. In fact, even α-Fe_2O_3, which has two identical cations and the corundum structure at low pressure, may undergo disproportionation ($2Fe^{3+} \rightarrow Fe^{2+} + Fe^{4+}$) as it transforms at very high pressures to the perovskite structure (Reid and Ringwood, 1969; Clark, 1966). The sesquioxide Co_2O_3, which has been synthesized at high pressures by Marezio *et al.* (1971), has two corundum phases: an atmospheric-pressure phase consisting of high-spin Co^{3+} ions and a high-pressure phase of smaller volume containing low-spin Co^{III} ions.

2. *Fivefold-coordinated B cations*

Structures having an anion coordination number $Z_B < 6$ about each B cation ($r_B < 0.51$ Å) may also transform directly to the perovskite structure. Hexagonal $YAlO_3$, for example, contains Al^{3+} ions in fivefold coordination (trigonal-bipyramid sites) and Y^{3+} ions in eightfold $(6 + 2)$ coordination, as illustrated in Fig. 7 (Bertaut and Mareschal, 1963). This structure seems to occur where both the A cations and the B cations are at the lower limit for tolerable cationic size: $r_B = 0.51$, $r_A = 0.90$ Å. The antiferromagnetic, ferroelectric compound $YMnO_3$ has a similar structure, but with an *a* axis $(3)^{1/2}$ larger than that of $YAlO_3$, as is also illustrated in Fig. 7 (Yakel *et al.*, 1963). The Mn^{3+} ion, though of larger radius, can be stabilized in a trigonal-bipyramid site because it has four, parallel-spin outer d electrons. In this site, the d-electron configuration is $e_{g\sigma}^2 e_{g\pi}^2 a_{1g}^0$,

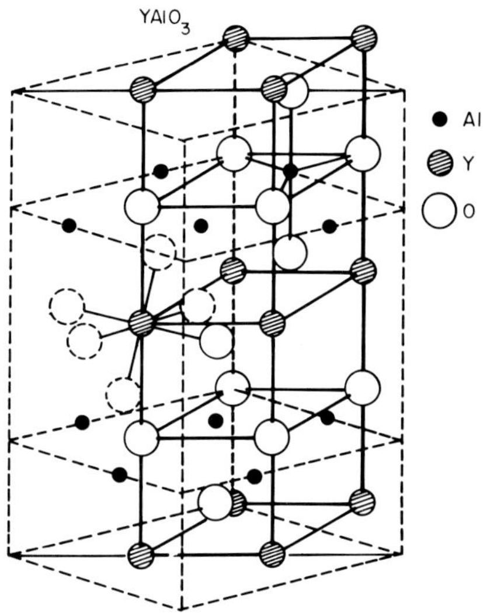

FIG. 7. Comparison of the unit cells of $YAlO_3$ (solid lines) and $YMnO_3$ (dashed lines). After Bertaut and Maraschel (1963), *C. R. Acad. Sci. Paris Ser. C*, **257**, 867; $a = 3.678$, $c = 10.52$ Å

where the empty a_1 orbital is directed along the c axis to bond covalently with the two collinear oxygen ions. The ferroelectric transition that occurs below 600°C is apparently due to the relatively large size of the Mn^{3+} ion, which creates a large enough eightfold interstice for a Y^{3+}-ion c-axis displacement that lowers the anion coordination from eight toward seven. These two structures transform under pressure directly to the perovskite structure.

3. *Fourfold-coordinated B cations*

The pyroxenes contain simple chains of BO_4 tetrahedra that may be arranged to allow sixfold or eightfold coordination about the A cation, depending on its size. Ringwood and Seabrook (1962a) have reported the high-pressure transformation of several germanates $A^{2+}GeO_3$ from a pyroxene to the ilmenite structure, showing that ilmenite may be but an intermediate polymorph between a pyroxene and the perovskite or disordered structure. This transformation again illustrates the principle that high pressure stabilizes preferentially the more dense phase, which is characterized by a higher anion coordination about the cations. However, not all germanates with the pyroxene structure transform under pressure to the ilmenite form. Some disproportionate into $Ge[A_2]O_4$ (spinel) and GeO_2 (rutile), only half of the Ge^{4+} ions becoming sixfold-coordinated. Presumably at still higher pressures, these two products would rereact to form an ilmenite or corundum phase, if not a perovskite.

The mineral garnet, $Mn_3[Al_2](Si_3)O_{12}$, contains the largest Mn^{2+} ion in eightfold coordination, and the smaller Al^{3+} and Si^{4+} ions in octahedral and tetrahedral sites, respectively. Interestingly, Ringwood and Major (1967b) report that $MnSiO_3$ glass transforms to the pyroxene structure at 60–90 kbar and to the garnet structure $Mn_3[MnSi](Si_3)O_{12}$ between 120 and 170 kbar. In this case, accommodation of Mn^{2+} ions in eightfold coordination requires less pressure than a change of all the Si^{4+} ions from fourfold to sixfold coordination. ABO_3 compounds having the garnet structure $A_3[AB]B_3O_{12}$ at atmospheric pressure have been transformed at high pressures directly to a perovskite phase. Presumably $MnSiO_3$ has a perovskite polymorph at highest pressures.

The $YAlO_3$ structure may, like ilmenite, also be an intermediate polymorph between perovskite and a structure having fourfold coordination of the B cations. $InGaO_3$ prepared at atmospheric pressure has the β-Ga_2O_3 structure, in which the Ga^{3+} ions are in tetrahedral sites and the In^{3+} ions have sixfold coordination. Shannon and Prewitt (1968) have transformed $InGaO_3$ at 1200°C and 65 kbar to the $YAlO_3$ structure. This observation suggests that at even higher pressures, $InGaO_3$ would transform to a perovskite form.

C. ABX_3 Polymorphs: Normal A Cations

1. *Small B cations*

With a larger A cation and a B cation small enough to preferentially occupy tetrahedral sites at atmospheric pressure, the wollastonite structure is normally stable. This structure occurs in $CaGeO_3$ and transforms via a tetragonal garnet intermediary to the perovskite phase at elevated pressures. So, too, does the more complex structure of $CdGeO_3$, the Cd^{2+}-ion radius being intermediate to that of In^{3+} in $InGaO_3$ and that of Ca^{2+} in $CaGeO_3$ (Ringwood and Seabrook, 1963; Ringwood and Major, 1967b; Prewitt and Sleight 1969).

2. *Large B cations*

If the B cation is too large to be well stabilized in an octahedral site, one of the rare-earth sesquioxide phases (in the order C, B, A with increasing r_B) may be stabilized. In the C structure, but not in the B and A structures, $Z_A = Z_B = 6$, as in ilmenite or corundum, but the anions occupy six corners of a cubic site rather than forming an octahedral interstice. Again perovskite is the more dense phase, since it contains $Z_A = 12$ and $Z_B = 6$, so that high pressures may force the larger B cation into an octahedral coordination in order to gain the energy associated with twelvefold coordination about the larger A cation.

D. The Defect-Pyrochlore and the Cubic $KSbO_3$ Structures

The cubic-pyrochlore structure illustrated in Fig. 8a corresponds to the chemical formula $A_2B_2X_6X'$. As in perovskite, the B cations occupy corner-shared octahedra. The A cations, on the other hand, occupy an eightfold, hexagonal-bipyramid interstice. Two types of tetrahedrally coordinated anions may be distinguished: X anions having two A-cation and two B-cation near-neighbors; and X′ anions having four A-cation near-neighbors. The X′ anions are located at the top and bottom of the apex-shared hexagonal bipyramids. Removal of the X′ ion leaves the octahedral sites unaltered; but it creates a highly anisotropic A site, and it introduces a large electrostatic repulsion between the A cations facing each other across the anion vacancy. If the tolerance factor were $0.75 < t < 1.0$, the defect pyrochlore structure $A_2B_2X_6\square$ would not appear to be competitive with the perovskite structure. Nevertheless, several ABX_3 compounds with the defect-pyrochlore structure are known. All but one of them contain Tl^{1+}, Pb^{2+}, or Bi^{3+} ions, which have highly polarizable $6s^2$ cores (Longo *et al.*, 1969). The exception is $AgSbO_3$, which contains an acidic Ag^+ ion and strongly covalent Sb–O linkages (Sleight, 1969a).

It is necessary to distinguish two types of strongly covalent B–X linkages: those that involve d orbitals at the cations and those that do not. If the B ion has no d orbitals available for bonding, as in the case B = Sb or Bi, then only the σ-bonding orbitals contribute significantly to the covalent

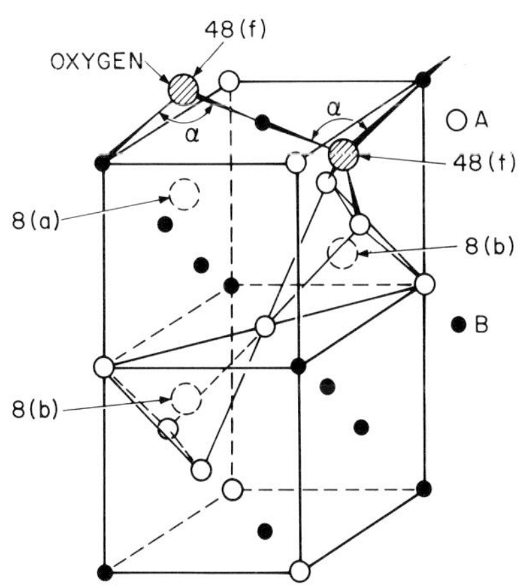

Fig. 8a. Cation sublattice for one-quarter unit cell for cubic-pyrochlore structure. Two of the 48(f) anions are also shown. The B cations share common octahedral site corners making the B–O^{2-}–B angle α. The 8(a) positions are empty and the 8(b) anion positions are also empty in the defect pyrochlore. After Longo *et al.* (1969), reprinted with permission of Pergamon Press.

FIG. 8b. The $Sb_{12}O_{36}$ array of the cubic unit cell for high temperature $KSbO_3$. Reprinted from Longo and Sleight *Inorg. Chem.* 7, 10 (1968).

bonding. In the perovskite structure, where two B cations are on opposite sides of each anion, covalent bonding by the single anion p_σ orbital must be shared by the two cations. In the defect-pyrochlore structure, on the other hand, the B–X–B angle approaches 120°, and two different hybrid orbitals at the anion may be used in bond formation. If the additional covalent energy gained by reducing the angle of the Sb–O–Sb linkage from 180° is larger than the associated loss in Madelung energy, then the defect-pyrochlore structure becomes more stable than the perovskite structure. Support for this reasoning comes from failure to stabilize $AgSbO_3$ in the perovskite structure at pressures $P < 100$ kbar. Instead, at pressures $P > 90$ kbar the defect-pyrochlore structure transforms to a body-centered-cubic structure with space group Im3 (Kafalas, 1971). This cubic structure is closely related to the high-temperature form of $KSbO_3$, space group Pn3, obtained by Spiegelberg (1940). In both of these structures edge-shared octahedral pairs Sb_2O_{10} are linked by common corners to form the cubic matrix illustrated in Fig. 8b. This subarray contains only 90° and ~130° Sb–O–Sb linkages, and the Im3 and Pn3 structures are apparently stabilized relative to the cubic perovskite structure for the same reasons as is the defect-pyrochlore structure. The Pn3 structure contains ordered A^+ ions in the three-dimensional tunnels of the subarray of Fig. 8b. Clustered about the origin, located at the center of the front face of Fig. 8b, are four A^+ ions forming a tetrahedron, each at $\sim(0.15\sqrt{3})a$ from the origin, and four on the remaining body diagonals a distance $(0.25\sqrt{3})a$ from the origin. This arrangement results in pairs of A^+ ions between the origin and half the body-center positions and single A^+ ions midway between the origin and the remaining four body-center positions. In the Im3 structure, the A^+ ions are disordered. It is difficult to predict whether a particular A^+SbO_3 or A^+BiO_3 compound will have the defect-pyrochlore or the cubic $KSbO_3$ structure. In fact, $AgSbO_3$ can be prepared in the $KSbO_3$ structure at atmospheric pressure by ion exchange (Hong, 1972).

Where B is a transition-metal atom, covalent mixing involves cation-d as well as cation-s and cation-p orbitals. Furthermore, if the d orbitals are empty, as in $A^{2+}TiO_3$ or A^+NbO_3 and A^+TaO_3, then the anion p_π orbitals participate in covalent bonding in the perovskite structure. In fact, this participation is strong enough in the oxides to break down the localized character of the d orbitals (Goodenough, 1971). Where all the anion p orbitals share in covalent bonding in the perovskite phase, a significantly smaller covalent energy is gained by reducing the B–X–B angle from 180°, and the defect-pyrochlore structure only competes with the perovskite structure where $A = Tl^+$, or the associated loss in Madelung energy is minimal. However, if the π-bonding d orbitals are occupied, as at a low-spin Ru^{IV} ion, then relatively little energy is gained by π-bond covalent mixing, and the defect-pyrochlore or cubic $KSbO_3$ structure may become competitive if $A = Tl^+$, Pb^{2+}, or Bi^{3+}. For these A cations, the loss in Madelung energy associated with the defect-pyrochlore or cubic $KSbO_3$ structure is minimized by a spontaneous polarization of the $6s^2$ core electrons.

An x-ray electron-density difference map for $PbRuO_3$, which has the defect-pyrochlore structure if prepared at atmospheric pressure, shows that the $6s^2$ core electrons at the Pb^{2+} ions are strongly perturbed by the crystalline fields. In fact, the crystalline fields stabilize 6s6p hybrid orbitals that interact via the anion vacancies to form itinerant-electron states having a large electron density in the vacancy site (Longo *et al.*, 1969). This extreme core polarization is compatible with a highly anisotropic A site, and electron transfer to the anion vacancies effectively screens the positive A cations from one another. Thus the defect-pyrochlore structure may be competitive with the perovskite structure even in the presence of higher effective charges on the A cations, provided $6s^2$ core polarization transfers electronic charge to the anion vacancies so as to screen the electrostatic repulsion between the A cations. However, $6s^2$ core polarization is not, by itself, sufficient to stabilize the defect-pyrochlore structure; there must also be a large σ and small π contribution to the B–X bonding. Thus $PbTiO_3$ is a tetragonal perovskite at atmospheric pressure, and the composition $PbSnO_3$ is only synthesized under pressure in the perovskite structure (Sugawara *et al.*, 1968a).

Since the perovskite structure is the more dense, hydrostatic pressure should stabilize the perovskite versus the defect-pyrochlore structure, and indeed $PbRuO_3$ transforms to the perovskite structure above 90 kbar (Kafalas and Longo, 1970a).

E. INTERLAYER COMPOUNDS $(AX)_nABX_3$

A number of compositions $(AX)_nABX_3$ crystallize in an interlayer structure, perovskite ABX_3 slabs alternating with rock salt AX slabs along a common c axis to give the body-centered tetragonal unit cells of Fig. 9. The relative widths of the slabs depend upon $n = 0, \frac{1}{3}, \frac{1}{2}, 1$. One end member ($n = 0$) is just the ABX_3 composition of perovskite, Fig. 9a. The

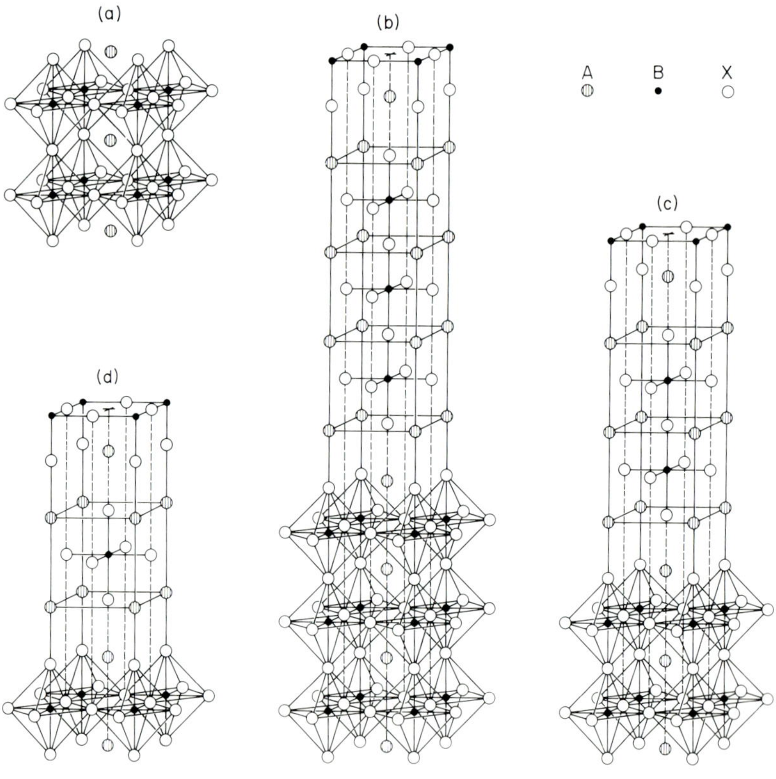

FIG. 9. Comparison of (a) the perovskite ABX_3 with the interlayer compounds $(AX)_nABX_3$, where (b) $n = \frac{1}{3}$, (c) $n = \frac{1}{2}$, and (d) $n = 1$. After Longo and Kafalas (1969c).

other end member ($n = 1$) consists of the alternating perovskite and rock salt units found in K_2NiF_4, Fig. 9d. The $A_4B_3X_{10}$ ($n = \frac{1}{3}$) structure of Fig. 9b consists of a rock salt unit interleaved between every three perovskite units, and $A_3B_2X_7$ ($n = \frac{1}{2}$) has two perovskite layers for each rock salt layer, Fig. 9c.

Like perovskite itself, these structures represent dense ionic packing. Therefore we may anticipate that these are the high-pressure forms of compositions $A_{1+n}BX_{3+n}$, $n = 0, \frac{1}{3}, \frac{1}{2}, 1$.

IV. High-Pressure Transformations

A. Prologue

The most effective use of high pressures in solid state synthesis has been the inducing of structural transformations in compositions that can be prepared at atmospheric pressure. Most high-pressure transformations are reversible and cannot be retained by quenching to room temperature. Such transitions are normally detected by monitoring at pressure some physical property, the most convenient being electrical resistivity. However, thermal, optical, magnetic, and structural properties have all been monitored under pressure with varying degrees of success. In a few cases, the high-pressure phase could only be retained by quenching to very low temperatures, but its properties could then be measured at 1 atm while held in the cold state (Kafalas *et al.*, 1962). Fortunately, there are, however, a great many compounds whose high-pressure forms may be retained at 1 atm and room temperature. The high-pressure perovskites and perovskite-related structures discussed here belong to this group. These phases are metastable, and they retransform at 1 atm. Some retransform at room temperature, while others must be heated to over 1200°C for many hours before retransformation is complete.

B. ABX_3 Polytypes in Oxides

1. *Ruthenates*

One of the largest known classes of materials having quenchable high-pressure phases is the series of polytypes illustrated in Fig. 4. The known compositions exhibiting transitions from one of these polytypes to another are given in Table I. From the table, it is immediately apparent that in all cases studied, the effect of pressure is to increase the proportion of cubic

TABLE I

ABX_3 POLYTYPES

Compound	t	2H	9R	4H	6H	Perovskite	Equip.	Container	Temp. (° C)	Time (hr)	Remarks	Ref.
$BaTiO_3$	1.06				*	Atm					(*)When oxygen deficient	1
$BaCrO_3$	1.08		*	*	*		T. A.†	Pt	750-1400	1-4	(*)All polytypes found as mixed phase at 60-65 kbar	2
$BaMnO_3$	1.09	Atm-3	30-75	90			T. A.†	Pt	400-1200			3, 4
$SrMnO_3$	1.02			Atm-50	50-95		T. A.†	Pt	700-1300			3, 4
$BaFeO_3$	1.07				Atm	*					(*)When oxygen deficient	5
$BaRuO_3$	1.05		Atm-15	15-30	30-90	$\gtrsim$120*	Belt	Pt	1000	0.5	(*)Estimated	6
$SrRuO_3$	.98					Atm*					(*)See $Ba_{1-x}Sr_xRuO_3$ P vs x in Fig. 6	6
$BaIrO_3$	1.04		Atm	30	65*		Belt	Pt	1000	0.5	(*)Monoclinic distortion	7
$SrIrO_3$	.97				Atm-40*	40-65	Belt	Pt	1000	0.5	See P-T diagram in Fig. 7; (*)Monoclinic distortion	7, 8
$CsNiF_3$	1.12	Atm-3	3-48	---	48-90		Belt	Au	700	0.5	See Fig. 9(a)	9, 10
$TℓNiF_3$	1.08				Atm	30	T. A.†	C	700			3
$RbNiF_3$	1.07				Atm-20	20-65	Belt	Au	700	0.5	See Fig. 9(b)	3, 11
$CsMgF_3$	1.11	*	30-40	---	40-65		Belt	Au	700	0.5	See Fig. 8, (*)No stable 1:1 compound to 30 kbar	10

$CsZnF_3$	1.09	*			30-80		Belt	Au	700	0.5	See Fig. 8, (*)No stable 1:1 compound to 30 kbar	10
$CsCoF_3$	1.10		9-20	---	20-80		Belt	Au	700	0.5	See Fig. 8	10
$CsFeF_3$	1.08				Atm-70	70-80	Belt	Au	700	0.5	See Fig. 8	10
$RbFeCl_3$	.97	Atm-15	15-20	---	20-35	*	Belt	Au	700	0.5	(*)Perovskite retransforms at room temperature	12
$CsMnF_3$	1.05				Atm-30	30-65	Belt	Au	700	0.5	See Fig. 8	10
$CsMnCl_3$	.99		Atm-10	---	10-25	25-65	Belt	Au	700	0.5		12
$RbMnCl_3$	.95				Atm-7	7-65	Belt	Au	700	0.5		12
$CsCdCl_3$	.95				Atm	*					(*)Perovskite from water solution	13
Ba_2FeSbO_6	1.04				Atm	65	Belt	Pt	1000	0.5		14
$BaVS_3$	1.00	Atm-90**					Belt	Au	1000	1	(**)See below	14
$BaTaS_3$	.97	Atm-90**					Belt	Au	1000	1	(**)See below	14

† Tetrahedral Anvil

* See the Remarks column.

** Any transformation would have been undetected if retransformation occurs at room temperature.

References:

1. Arend and Kihlborg, 1969
2. Chamberland, 1969a
3. Syono *et al.*, 1969a
4. Chamberland *et al.*, 1970
5. Mori, 1965
6. Longo and Kafalas, 1968
7. Longo *et al.*, 1971
8. Longo and Kafalas, 1969c
9. Longo and Kafalas, 1969a
10. Longo and Kafalas, 1969b
11. Kafalas and Longo, 1968
12. Kafalas *et al.*, 1970
13. Siegel and Gebart, 1964
14. This chapter

to hexagonal stacking of the AX_3 layers. The first demonstration of this relationship (Longo and Kafalas, 1968) was in the $Ba_{1-x}Sr_xRuO_3$ system, which has the composition–pressure ($P < 90$ kbar) phase diagram shown in Fig. 10.

$BaRuO_3$ was originally prepared by Randall and Ward (1959) and shown by Donohue *et al.* (1965) to be the first example of the 9R structure. This compound may be synthesized at atmospheric pressure by the solid state reaction of $BaCO_3$ or BaO_2 and Ru or RuO_2 at 1000°C in oxygen. The resulting 9R form transforms at about 15 kbar to the 4H structure of Fig. 4d. This 4H phase is stable over a relatively narrow pressure range, transforming at 30 kbar to the 6H polytype of Fig. 4c. The 6H phase is stable to at least 90 kbar, but from Fig. 10, it is reasonable to anticipate a further transformation to the 3C perovskite structure near 120 kbar.

$SrRuO_3$ has the perovskite 3C structure, with a small distortion to orthorhombic symmetry (Randall and Ward, 1959). Donohue *et al.* (1966) studied the system $Ba_{1-x}Sr_xRuO_3$ at atmospheric pressure; they found some solid solution in each of the end members and a third 4H phase in the interval $0.1 < x < 0.3$ at 1100°C. Reexamination of this system (Longo and Kafalas, 1968) revealed, in addition, a narrow region ($0.6 < x < 0.7$)

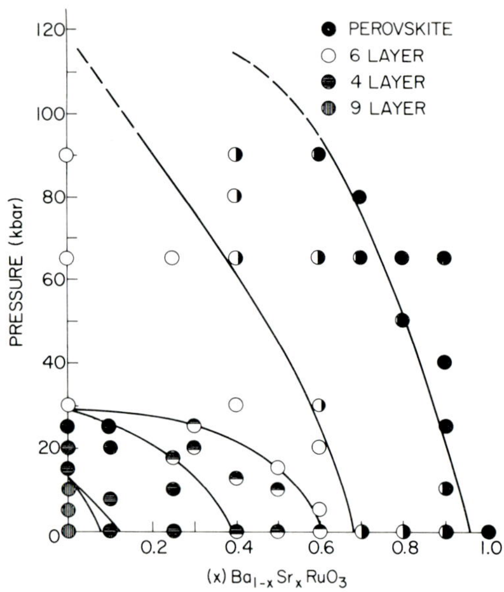

FIG. 10. Pressure–composition phase diagram for $Ba_{1-x}Sr_xRuO_3$ at 1000°C. After Longo and Kafalas (1968), used with permission of Pergamon Press.

TABLE II

CRYSTALLOGRAPHIC DATA FOR $Ba_{1-x}Sr_xRuO_3$[a]

Composition	Structure type	a (Å)	b (Å)	c (Å)	Volume per $ARuO_3$ (Å^3)
$BaRuO_3$	9R	5.75	—	21.60	68.7
$BaRuO_3$	4H	5.73	—	9.50	67.5
$BaRuO_3$	6H	5.71	—	14.00	65.9
$Ba_{0.75}Sr_{0.25}RuO_3$	4H	5.70	—	9.47	66.6
$Ba_{0.75}Sr_{0.25}RuO_3$	6H	5.69	—	13.95	65.2
$Ba_{0.20}Sr_{0.80}RuO_3$	Perovskite	5.56	5.59	7.88	61.1
$SrRuO_3$	Perovskite	5.53	5.57	7.85	60.4

[a] Longo and Kafalas (1968), used with permission of Pergamon Press.

of stability for the 6H structure and a somewhat smaller solubility of $BaRuO_3$ in $SrRuO_3$.

The effect of pressure (to 90 kbar) on this system is essentially the same as an increase in the Sr concentration, thereby demonstrating that the tolerance factor t is the critical parameter determining the polytype structure and that $dt/dP < 0$. This study also shows that the intermediate polytypes 9R, 4H, and 6H occur in oxides only within a narrow range of t.

The high-pressure phases found in this study were stable to 1100°C at atmospheric pressure and only slowly retransformed if heated above 1200°C. This stability permits study and use of the physical properties of the metastable phases over a wide temperature range, and high-pressure synthesis provides access to a given polytype over a much wider range of compositions than is available to synthesis at atmospheric pressure. For example, the 6H form is constrained to the interval $0.6 < x < 0.7$ at atmospheric pressure, whereas at 30 kbar, it may be obtained over the entire interval $0 \leqslant x < 0.6$. The stability region of the perovskite phase is increased from $0.95 < x \leqslant 1.0$ to $0.7 < x < 1.0$ at 90 kbar and, presumably, to the whole range of x for $P > 120$ kbar. A summary of crystallographic data for this system is given in Table II.

2. *Manganates*

Establishing these principles has helped to clarify several other systems. For example, the compounds $AMnO_3$ (A = Ba, Sr, Ca) all crystallize in different polytypes at atmospheric pressure: $BaMnO_3$ is 2H (Hardy, 1962), $SrMnO_3$ is 4H (Syono *et al.*, 1969a), and $CaMnO_3$ is 3C with a small distortion to orthorhombic symmetry (MacChesney *et al.*, 1967). The decomposition of $BaMnO_4$ at 450°C (Hardy, 1961) or the reaction of $BaCO_3$ and MnO_2 in O_2 at 900–1000°C (Gushee *et al.*, 1957) yields the 2H structure of $BaMnO_3$. Treatment of this phase at 3 kbar (700°C) or 65 kbar (1000°C) gives the 9R structure (Chamberland *et al.*, 1970a). Syono *et al.* (1969a) have also reported that heating $BaMnO_3$ to 1200°C at $P = 30$ and 75 kbar gives the 9R structure; but in addition, they found transformation to the 4H polytype at 90 kbar. However, it should be noted that heating the 2H form above 1000°C at atmospheric pressure gives oxygen-deficient $BaMnO_{2.85}$ (Chamberland *et al.*, 1970a), which has a structure resembling the 4H polytype (Hardy, 1962). $SrMnO_3$ has the 4H structure to 50 kbar, where it transforms ($850 < T < 1300$°C) to the 6H form, which is stable to at least 95 kbar (Syono *et al.*, 1969a; Chamberland *et al.*, 1970a).

3. *Iridates*

The compound $SrIrO_3$, first reported by Rodi (1963), may be prepared at atmospheric pressure by the reaction of $SrCO_3$ and IrO_2 (900°C, 20 hr).

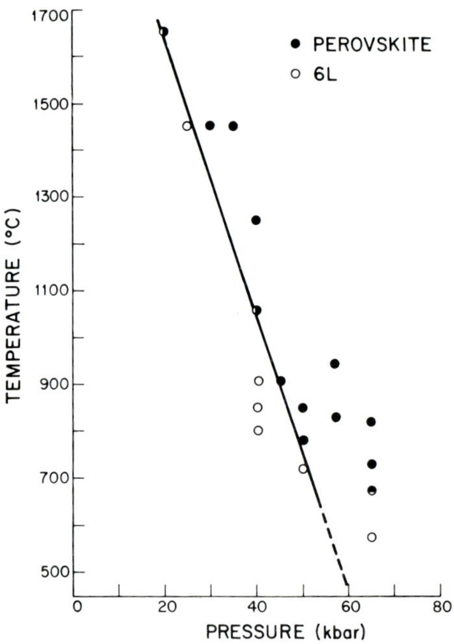

FIG. 11. Pressure–temperature phase diagram for $SrIrO_3$. After Longo *et al.* (1971).

Longo *et al.* (1971) have indexed the powder pattern on a monoclinic unit cell ($a_M = 5.60$, $b_M = 9.62$, $c_M = 14.17$ Å, $\beta = 93.26°$) that is closely related to a pseudohexagonal 6H polytype cell through the relationships $a_M \approx a_H$, $b_M \approx (3)^{1/2}a_H$, and $c_M \approx c_H$. Pressure–temperature treatments of this phase in a belt apparatus have given the phase diagram of Fig. 11. The high-pressure 3C phase, which is distorted to orthorhombic symmetry, is about 3% more dense than the monoclinic form, and it is stable at atmospheric pressure, retransforming only slowly at 1200°C.

It is perhaps significant that $SrIrO_3$ and $SrMnO_3$ are the only $SrBO_3$ compounds (B = Ti, Zr, Hf, Mo, Cr, Tc, Ru, Fe, Sn, Pb) that do not have the perovskite structure at atmospheric pressure. Since $t = 0.97$ for $SrIrO_3$ is well within the stability range for the perovskite structure, it would appear that the hexagonal polytypes are stabilized by the cubic-field outer-electron configurations $t_{2g}^5e_g^0$ and $t_{2g}^3e_g^0$ at the low-spin Ir^{IV} and high-spin Mn^{4+} ions, respectively. In the trigonal fields of these polytypes, the t_{2g}^5 and t_{2g}^3 orbitals of the *c*-axis pairs become $e_{g\sigma}^0e_{g\pi}^4a_{1g}^1$ and $e_{g\sigma}^0e_{g\pi}^2a_{1g}^1$, respectively, which allows for metal–metal bonding along the *c* axis via the half-filled a_{1g} orbitals. Simultaneously, the 90° cation–anion–cation π–σ-bond interactions are stronger than the 180° π–π-bond interactions. Such metal–metal and enhanced metal–anion–metal bonding would tend to favor the relative stability of a hexagonal versus a 3C polytype. Deformation of the $SrIrO_3$ 6H polytype apparently reflects an antiferroelectric displacement of the Sr^{2+} ions, which occupy larger sites in this structure than in the 3C (perovskite) polytype.

At atmospheric pressure, $BaIrO_3$ has essentially the 9R structure (Donahue *et al.*, 1966; Rodi and Babel, 1965). At 65 kbar and 1000°C, this composition is transformed to the monoclinic 6H polytype, but the distortion from hexagonal symmetry is much smaller than that occurring in $SrIrO_3$—consistent with the larger size of the Ba^{2+} ion.

4. $A_2BB'O_6$ *compounds*

The more complex $A_2BB'O_6$ compounds may also show this polytypism with pressure. Blasse (1965) has found that Ba_2FeSbO_6, if prepared at atmospheric pressure, has the 6H structure. At 65 kbar and 1000°C we find that it transforms to the 3C structure ($a = 8.09$ Å), the B and B′ atoms ordering within the simple-cubic array of octahedral sites so as to have unlike nearest neighbors. It is reasonable to anticipate that other mixed-B-site compounds that crystallize in an intermediate 6H polytype, such as some $Ba_2B^{3+}B^{5+}O_6$ and $Ba_3B^{2+}B_2^{5+}O_9$, would transform at high pressure and temperature to an ordered perovskite structure.

5. *Problems with nonstoichiometry*

$BaTiO_3$ prepared in air at atmospheric pressure has the 3C perovskite structure, but it is distorted from cubic symmetry by a series of ferroelectric transitions as the temperature is lowered below $T_c = 120°C$. However, if prepared in H_2 above 1500°C, the oxygen-deficient $BaTiO_{3-\delta}$ phase has the 6H structure (Arend and Kihlborg, 1969). The 6H structure may also be stabilized in the stoichiometric oxide by reducing the average size of the B cation, hence increasing t, by B^{4+}-ion substitutions: $BaTi_{1-x}B_x^{4+}O_3$ (Dickinson *et al.*, 1959, 1961).

Oxygen deficiencies have the opposite effect on the structure of $BaFeO_{3-\delta}$ (Mori, 1965; Negas and Roth, 1969), the 6H structure of $BaFeO_3$ transforming to perovskite with increasing δ.

In both these cases, the preparation or treatment of the 6H forms at high pressure should stabilize the perovskite structure, thereby extending the compositional range over which the properties of the perovskite phase can be studied.

C. ABX_3 Polytypes in Halides

1. *Fluorides*

The high-pressure forms of the ABF_3 compounds may be prepared either by transforming the atmospheric-pressure form or by solid state reaction under pressure of the component fluorides. Syono *et al.* (1969a) transformed $CsMnF_3$, $RbNiF_3$, and $TlNiF_3$ that had been synthesized in an inert atmosphere at 300–500°C for 12 hr via the solid state reaction

$$AF + BF_2 \rightarrow ABF_3$$

They used a tetrahedral-anvil apparatus and encapsulated the sample in a graphite tube that also served as a furnace. Longo and Kafalas (1969b) reacted, at 700°C for 30 min, an intimately mixed, stoichiometric ratio of AF and BF_2 under pressure in a belt apparatus. The samples were loaded into gold capsules within a glove bag filled with dry nitrogen. Use of the glove bag was necessary to prevent hydration of the fluorides, and failure to follow this precaution always led to multiphase products.

The series of polytypes found for $A^{2+}B^{4+}O_3$ compounds is also found for $A^+B^{2+}F_3$ compositions, except that the narrow 4H phase field appears to be completely suppressed. The series of compounds $CsBF_3$, B = Mn, Fe, Co, Ni, Zn, Mg, illustrates well the effect of B-cation size and pressure on the relative stabilities of the various polytypes. The x-ray analysis of the speci-

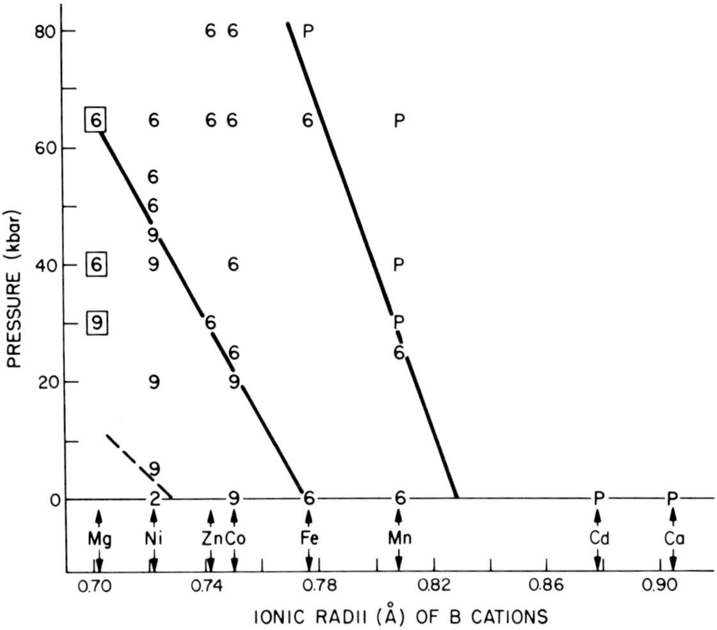

FIG. 12. Pressure–r_B phase diagram for $CsBF_3$ compounds. After Longo and Kafalas (1969b). Structures: P = Perovskite, 9 = 9 layer, 6 = 6 layer, 2 = 2 layer; T = 700°C.

mens prepared at pressure gave the phase diagram shown in Fig. 12 (Longo and Kafalas, 1969b). Careful examination of the 9R to 6H phase boundaries gave no evidence of the 4H polytype. The corresponding cell dimensions and volumes per formula unit are in Table III. The density clearly increases with the proportion of cubic stacking of AX_3 layers. From Eq. (1), t decreases with increasing r_B, and again a $dt/dP < 0$ must be anticipated. Therefore t is largest in the lower left corner of Fig. 12, which is where the 2H structure is encountered, and the proportion of cubic to hexagonal stacking of AX_3 layers increases regularly with decreasing t. The only modification of what was anticipated from our analysis of the oxide data is the absence of a 4H phase, and it was pointed out in Section III why suppression of this phase is not surprising. Inclusion of the full range of polytypes, from 2H to 3C, on a single diagram illustrates how narrow is the transitional range Δt over which the intermediate structures are stable.

The observed transformation pressures, which are given in Fig. 12 for 700°C, were found to be rather insensitive to temperature over the range

TABLE III

UNIT CELL DIMENSIONS (Å) AND VOLUME PER $CsBF_3$ (Å^3) FOR $CsBF_3$ COMPOUNDS[a]

Compound	Structure type 2H	9R	6H	3C
$CsMgF_3$	—	6.16 × 22.13 80.8	6.04 × 14.45 76.1	—
$CsNiF_3$	6.24 × 5.22 88.0	6.15 × 22.32 81.2	6.05 × 14.55 76.9	—
$CsZnF_3$	—	—	6.09 × 14.67 78.5	—
$CsCoF_3$	—	6.20 × 22.36 82.7	6.09 × 14.67 78.5	—
$CsFeF_3$	—	—	6.16 × 14.86 81.3	4.283 78.6
$CsMnF_3$	—	—	6.21 × 15.07 84.0	4.328 81.1

[a] Longo and Kafalas (1969b). For each entry, the first line gives the unit cell dimensions and the second line, the volume per $CsBF_3$.

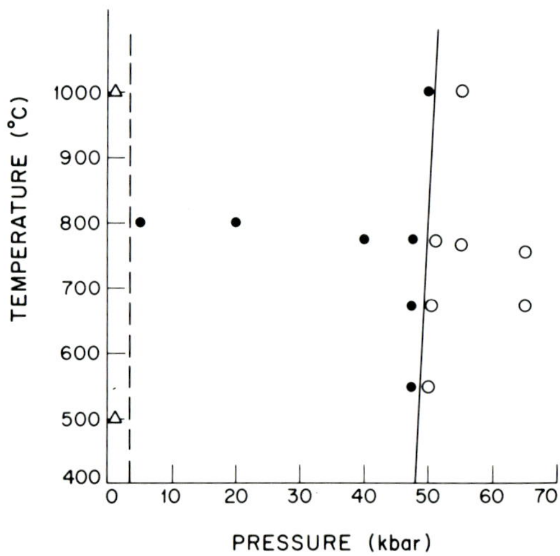

FIG. 13a. Pressure–temperature phase diagram for $CsNiF_3$. After Longo and Kafalas (1969a). △ = 2 layer, ● = 9 layer, ○ = 6 layer.

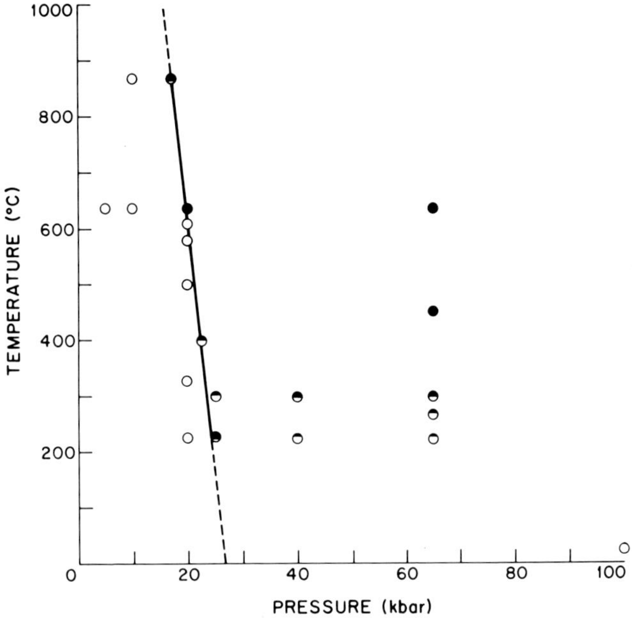

FIG. 13b. Pressure–temperature phase diagram for $RbNiF_3$. After Kafalas and Longo (1968). ○ = hexagonal phase, ● = cubic phase. Used with permission of Pergamon Press.

$400 < T < 900°C$, as is illustrated by the characteristic T–P phase diagram for $CsNiF_3$ shown in Fig. 13a (Longo and Kafalas, 1969a). The ionic radii used in Fig. 12 were calculated from the related series of KBF_3 compounds, all of which have the perovskite structure. This set of radii is consistent with both the cell volumes of the $CsBF_3$ compounds reported here and with the BF_2 compounds, which have the rutile structure.

As shown in Fig. 13b, the 6H atmospheric-pressure form of ferrimagnetic $RbNiF_3$ transforms at 20 kbar to the 3C structure, which has $a = 4.07$ Å at atmospheric pressure (Kafalas and Longo, 1968; Syono *et al.*, 1969a). This diagram again illustrates the relative insensitivity of the transformation pressure to temperature in the ABF_3 compounds.

The 6H form of $TlNiF_3$ has also been found to transform to perovskite (3C) at 30 kbar (Kohn *et al.*, 1967; Syono *et al.*, 1969a).

2. *Chlorides*

The chlorides $CsMnCl_3$, $RbMnCl_3$, and $RbFeCl_3$ have also been shown to exhibit polytypic transformations under pressure (Kafalas *et al.*, 1970). These compounds were prepared by reacting stoichiometric quantities of

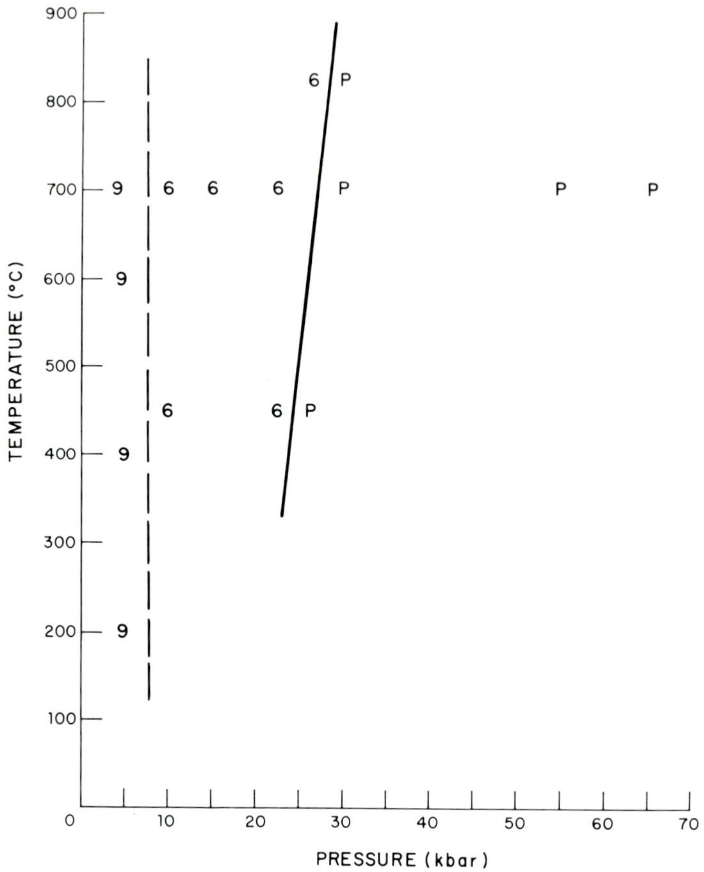

FIG. 14. Pressure–temperature phase diagram for $CsMnCl_3$. After Kafalas *et al.* (1970). Structures: P = perovskite, 6 = 6H, 9 = 9R.

ACl and $BCl_2 \cdot 4H_2O$ in a stream of dry HCl. The components were first heated to 200°C for a few hours to remove all the water of hydration, and then reacted at 500°C for another few hours.

In good agreement with Seifert and Koknat (1965) and with Kestigian *et al.* (1967), the x-ray diffraction data showed $CsMnCl_3$ to have the 9R structure. The temperature–pressure phase diagram, which was obtained by reacting at pressure as in the case of the $CsBF_3$ compounds, shows a transformation from the 9R to the 6H phase at about 10 kbar and from the 6H to the 3C (perovskite) phase at about 25 kbar (see Fig. 14). As in the fluorides, the 4H phase is completely suppressed, and the transition pressures are not very sensitive to temperature. Both high-pressure forms of $CsMnCl_3$ are moderately stable at atmospheric pressure, no evidence appearing of retransformation to the 9R structure after a week at room temperature. However, complete retransformation occurred after only a few hours at 300°C.

The x-ray data for the atmospheric-pressure 6H form of $RbMnCl_3$ are also in good agreement with those reported by Seifert and Koknat (1965) and by Kestigian *et al.* (1967). At 700°C and 7 kbar, this structure transforms to perovskite (3C). The 3C form retransforms at atmospheric pressure if annealed at 400°C for a few hours.

Examination of the tolerance factors t suggested that of all the $ABCl_3$ compounds having the 2H structure, $RbFeCl_3$ was most likely to yield high-pressure polytypes. This expectation was fulfilled, transformations occurring successively with pressure at 700°C from 2H to 9R to 6H to (by inference) 3C (perovskite). Two-phase diffraction patterns were obtained for samples prepared in the range $15 < P < 20$ kbar; those at lower pressures within this range contained 2H and 9R lines, those at higher pressures contained 9R and 6H lines. Apparently the 9R form has so narrow a stability range that pressure inhomogeneities along the specimen always yield two phases. Between 20 and about 35 kbar, single-phase 6H patterns were obtained. After pressure in the range $35 < \mathrm{P} < 80$ kbar, the x-ray patterns of the quenched specimens were broadened, and the 3C phase may be inferred for higher pressures since, after application of 90 kbar, the quenched sample has the 2H structure at room temperature. This observation indicates that the 3C phase is not retained, the structure retransforming all the way back to the 2H form, which is the stable phase at atmospheric pressure. We have observed that the 6H phase retransforms at room temperature to the 2H form with a half-life, as determined from magnetization data, of about one day. After one week, there was no x-ray evidence of the 6H phase.

All of these chlorides are very hygroscopic. However, if prepared in a stream of Ti-scrubbed N_2 instead of dry HCl, they are less hygroscopic and the P–T phase diagram for $RbFeCl_3$, but not the $A^{I}MnCl_3$ compounds, was altered. For example, $RbFeCl_3$ prepared under N_2 transformed to the 3C phase above 45 kbar, and this cubic phase was retained at room temperature and atmospheric pressure, requiring 400°C for retransformation. Furthermore, the 6H phase was not retainable at room temperature and had the same Curie temperature $T_c \approx 109°K$ as that prepared in HCl. These observations indicate that nitrogen substitutes for chlorine (at least at the surface) where, as in the case of Fe^{2+} ions, the B cation can be easily oxidized.

D. ABX_3 Polytypes in Sulfides

The sulfides $A^{2+}B^{4+}S_3$ have thus far been found in only two forms: 2H and 3C (Gardner *et al.*, 1969a, b; Gardner, 1969; Aslanov and Kovba, 1964; Clearfield, 1963; Hahn and Mutschke, 1956). However, Aslanov and Kovba (1964) noted that grinding of $BaTaS_3$ with Pyrex broadens the *hk*1 lines while the *hk*0 lines remain sharp. Gardner *et al.* (1969a, b; Gardner, 1969) found that hot pressing of $BaTaS_3$ and $BaVS_3$ at ~1 kbar led to the same type of line broadening and suggested that it was due to some disorder in the close packing of BaS_3 layers. We have found that treatment at 1000°C and 90 kbar led to further broadening of non-*hk*0 lines, but no apparent transformation to the 9R structures. It appears that either higher pressure is needed for this transformation, or the polytype retransforms to 2H at room temperature as was indicated for $RbFeCl_3$. This point can be settled by quenching to lower temperatures. Other ABS_3 compounds with 2H structure and a smaller t should also be examined.

Although gold was used as the encapsulating material in these studies of $BaTaS_3$ and $BaVS_3$, chalcogenides generally require BN containers because they react with the noble metals (Eatough *et al.*, 1969; Sleight and Bither, 1969).

E. Other Polytypes

The original and most celebrated substance exhibiting polytypism is SiC, which crystallizes into over 40 known modifications (Verma and Krishna, 1966). In each of these, the unit cell is hexagonal: $a = 3.078$ and $c = 2.158m$ Å, where m is an integer. The stable low-temperature modification, β-SiC, forms preferentially below 2000°C. It has all-cubic 3C stacking of the alternating Si and C layers. Fourfold atomic coordination gives this cubic phase the zincblende (sphalerite) structure. Above 2000°C, it transforms to the hexagonal polytypes, generally referred to as α-SiC. The most common of these is 6H, the second most common is 15R (a 9R is not formed), and the third most common is 4H. The all-hexagonal-stacking 2H phase is not found commercially, but may be grown by special methods of gaseous cracking. Sokhor *et al.* (1967) investigated the α–β transformation of SiC placed in graphite capsules and subjected to 30–70 kbar for several minutes at 1200–1400°C. They observed a definite increase with pressure treatment in the proportion of the 3C phase.

Hexagonal polytypes are also found in the rare-earth metals, where the face-centered-cubic cell corresponds to 3C, the hexagonal-close-packed cell to 2H, the Sm structure to 9R, and the double hexagonal-close-packed cell to 4H. Jayaraman (1965) investigated the *P–T* phase diagrams of La, Pr, Nd, Gd, and Tb. The atmospheric-pressure phases progress from 3C to 4H to 9R to 2H on passing from the largest atom La to the smallest Tb. Pressure favors the reverse order: the 3C phase is stable to 40 kbar in La, a 4H $\rightarrow$ 3C transformation occurs in Pr at $P \gtrsim 8$ kbar for $T \lesssim 810°C$ and in Nd at $P \gtrsim 21$ kbar for $T \lesssim 890°C$, a 9R $\rightarrow$ 4H transformation occurs in Sm at $P \gtrsim 18$ kbar for $T \lesssim 915°C$, and in Gd a 2H $\rightarrow$ 9R at $P \gtrsim 8$ kbar for $T \lesssim 1250°C$. Jayaraman and Sherwood (1964) had found that the high-pressure 4H form of Sm can be metastably retained at atmospheric pressure and room temperature. The hexagonal structures have an axial ratio c/a somewhat smaller than ideal for close packing of spheres, which is apparently why the density increases with the proportion of cubic stacking. In this series, the $5p^6$ core–core repulsive forces appear to stabilize cubic stacking. Gschneider and Valletta (1968) have suggested that increasing overlap of the 4f wave functions on neighboring atoms stabilizes the cubic stacking, but this speculation is improbable, even though the energies involved are not too large, since the magnetic ordering is determined by interactions via the intermediary conduction electrons and not by the 4f–4f interactions, which are extremely weak. Furthermore, there is no apparent reason for 4f–4f interactions to favor cubic stacking irrespective of the number of 4f electrons present. On the other hand, the radial extension of the filled $5p^6$ core electrons is no more altered by pressure than that of the 4f orbitals, and overlap of filled core orbitals introduces a strong, repulsive central force that stabilizes the more dense, cubic stacking.

Van Vucht and Buschow (1965) have noted that the $LnAl_3$ phases also exhibit a series of polytypes for stacking of the $LnAl_3$ close-packed layers: $LaAl_3$ to $GdAl_3$ have the 2H structure, YAl_3 has both the 2H and 9R structures, $TbAl_3$ has the 9R and 6H structures, $DyAl_3$ the 4H and 15R $(hchcc)_3$ structures, $HoAl_3$ has the 15R structure, and $ErAl_3$, $TmAl_3$, $LuAl_3$ and $ScAl_3$ have the 3C structure. Although this variation of structure types seems to be just opposite to that found for the elements, it is the same if the lanthanide atomic volume is considered in place of the lanthanide atomic number (Pearson, 1967). Although Gschneider and Pearson (1968) attribute this to the influence of the 4f electrons, it seems that Ln–Al core–core repulsive forces are again the more probable factor determining stabilization of the cubic stacking. Pressure should favor cubic stacking in this series also.

The influence of core–core repulsive forces should also be greater in $CuAu_3$, which has the 3C cubic structure, than in Ni_3Sn, which has the 2H hexagonal stacking. $BaPb_3$ has the 9R stacking and VCo_3 the 6H stacking, and it is reasonable to anticipate core–core interactions of inter-

TABLE IV

ABX_3 POLYMORPHS

Compound	Atm. Press. Structure	High Press., Structure	P (kbar)	T (° C)	Remarks	Ref.
$LnMnO_3$	Hex. $YMnO_3$	Perovskite	45	700-900	Ln = Ho, Er, Tm, Yb, Lu, Y; See Fig. 7	1, 2
$LnA\ell O_3$	Hex. $YA\ell O_3$	Perovskite			Perovskite at 1 atm if prepared T > 900°; See Fig. 7	3
$LnInO_3$	Hex. $YMnO_3$	Perovskite	65	1450	Ln = Dy, Gd, Eu; See Fig. 7	4
$YInO_3$	C-rare earth	Perovskite	65	1250		4
$InGaO_3$	β-Ga_2O_3	$YA\ell O_3$	65	1200		5
$PbRuO_3$	Defect Pyrochlore	Perovskite	90	1400	See Fig. 11 for $Sr_{1-x}Pb_xRuO_3$ P-X diagram	6
$PbIrO_3$	Defect Pyrochlore	---			No transformation to 90 kbar	6
$BiScO_3$	"Defect Pyrochlore"	Perovskite	60-70	700	Possibly I2m3 at atm. press. (applies to $BiScO_3$, $BiNiO_3$, $BiCoO_3$, $BiYO_3$)	7
$BiNiO_3$	"Defect Pyrochlore"	Perovskite	60-70	700		7
$BiCoO_3$	"Defect Pyrochlore"	Perovskite	60-70	700		7
$BiYO_3$	"Defect Pyrochlore"	Perovskite	60-70	700		7
$PbGa_{0.5}Nb_{0.5}O_3$	Defect Pyrochlore	Perovskite	60-70	700		7
$AgSbO_3$	Defect Pyrochlore	Im3	90	1000		8, 24
$T\ell NbO_3$	Defect Pyrochlore	---			No transformation to 90 kbar	9
$CaGeO_3$	Wollastonite	Perovskite	130	900	Transforms to garnet at 40 kbar and 700° C	10, 11
$CdGeO_3$	Complex	Perovskite	130	900	Transforms to garnet at 10 kbar and 700° C	10, 11
$CaGe_{0.7}Si_{0.3}O_3$	Wollastonite	Perovskite	170	900		11

$Y_3Fe_5O_{12}$	Garnet	Perovskite + Fe_2O_3	10	1000		12
$Y_3A\ell_5O_{12}$	Garnet	Perovskite + $A\ell_2O_3$	38	1000		12
$Y_3Ga_5O_{12}$	Garnet	Perovskite + Ga_2O_3	70	1000		12, 13
$Ln_3Ga_5O_{12}$	Garnet	Perovskite + Ga_2O_3	45	1000	Ln = Sm - Lu	13, 14
$MnVO_3$	Distorted Ilmenite	Perovskite	45	1200	See Fig. 24	15
$CdTiO_3$	Ilmenite	Perovskite	20	500	See Fig. 17	16
$MnTiO_3$	Ilmenite	Corundum	60	1200	See Fig. 18	17
$ATiO_3$	Ilmenite				A = Fe, Co, Ni, Mg; No transformation to 90 kbar & 1300°C	17
$MgGeO_3$	Pyroxene	Ilmenite	28	700		18
$MnGeO_3$	Pyroxene	Ilmenite	25	700		18
$AGeO_3$	Pyroxene	$Ce[A_2]O_4 + GeO_2$			A = Fe, Co, $Ni_{0.25}Mg_{0.75}$; Disproportionation between 10-22 kbar	18
$MgSiO_3$	Pyroxene	(Ilmenite)			Shock-wave data	19, 20
Fe_2O_3	Corundum	(Perovskite)			Shock-wave data	19, 21
Rh_2O_3	Corundum	New structure	65	1200		22
Ca_2GeO_4	Olivine	K_2NiF_4	110	1100		23
Sr_2CrO_4	*	K_2NiF_4	65	1000	(*)CrO_4 tetrahedra (Wilhelmi, 1967)	9

References:

1. Waintail and Chenavas, 1967a,b
2. Waintail *et al.*, 1966
3. Bertaut and Mareschal, 1963
4. Shannon, 1967
5. Shannon and Prewitt, 1968
6. Kafalas and Longo, 1970a
7. Tomashpol'skii *et al.*, 1967, 1968
8. Sleight, 1969a
9. Kafalas and Longo, 1970b
10. Ringwood and Seabrook, 1963
11. Ringwood and Major, 1967
12. Marezio *et al.*, 1966a
13. Marezio *et al.*, 1966b
14. Marezio *et al.*, 1968
15. Syono *et al.*, 1970
16. Liebertz and Rooymans, 1965
17. Syono *et al.*, 1969c
18. Ringwood and Seabrook, 1962a
19. Reid and Ringwood, 1969
20. McQueen *et al.*, 1967
21. Clark, 1966
22. Shannon and Prewitt, 1970
23. Ringwood and Reid, 1968
24. Kafalas, 1971

mediate strength in these ordered alloys. In these cases also, pressure can be expected to stabilize a greater proportion of cubic stacking.

Zinc sulfide, which has a structure similar to SiC, has at least ten different polytypes, the 3C (zincblende) phase being stable below 1024°C and the 2H (wurtzite) phase above. The transformation proceeds through intermediary polytypes formed by a differential expansion of a [111] axis of the 3C form. This observation suggests that in this compound also, pressure should favor cubic versus hexagonal stacking such that 2H → 9R → 4H → 6H → 3C, other polytypes possibly appearing as appropriately positioned intermediaries.

The compound CdI_2 exhibits 64 different polytypes (Verma and Krishna, 1966), including 2H, 4H, and 6H. No rhombohedral forms have been identified, in accord with expectations from the dislocation theory of polytypism. Although the structures of the $CdBr_2$ layers are similar to those of CdI_2, they permit the formation of rhombohedral polytypes, and in this compound, 2H, 4H, 6H, and 6R forms, among others, have been identified. Similarly PbI_2 exhibits 2H, 4H, 6H, and 6R polytypes. In all of these, pressure probably favors a higher proportion of cubic stacking. The same is true of other polytypic substances, such as the disulfides and diselenides of Nb, Ta, Mo, and W.

F. Polymorphs

Table IV contains the presently known pressure-induced transformations from a three-dimensional polymorph to the perovskite structure. The structures of each of these polymorphs and their relation to the perovskite structure were discussed in Section III.

1. *Defect pyrochlore*

The only perovskite to defect-pyrochlore system that has been systematically studied is $Sr_{1-x}Pb_xRuO_3$ (Kafalas and Longo, 1970a). Orthorhombic $SrRuO_3$, which has a slightly distorted 3C structure, was prepared by the solid state reaction

$$SrCO_3 + RuO_2 \xrightarrow{1000°C} SrRuO_3 + CO_2 \uparrow$$

Members of the system $Sr_{1-x}Pb_xRuO_3$ were then prepared by placing $(1 - x)SrRuO_3 + xPbO + xRuO_2$ in platinum capsules and subjecting the charge to 15 min of high pressure at 1400°C. A summary of the crystallographic data obtained from quenched specimens is given in Fig. 15 and Table V. The solubility of $PbRuO_3$ in the perovskite $SrRuO_3$ phase in-

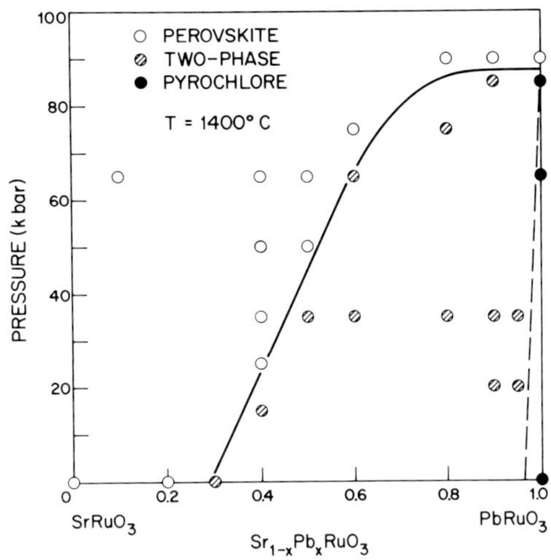

FIG. 15. Pressure–composition phase diagram for $Sr_{1-x}Pb_xRuO_3$. After Kafalas and Longo (1970a), used with permission of Pergamon Press.

creases linearly with pressure from $x = 0.3$ to about $x = 0.6$ at 70 kbar, but increases much more rapidly at higher pressures. The $PbRuO_3$ is itself transformed to (or prepared from a $PbO:RuO_2$ mixture directly in) the 3C structure near 90 kbar. On the other hand, the solubility of $SrRuO_3$ in the defect-pyrochlore structure of $PbRuO_3$ was not detectable. This very limited solubility of Sr^{2+} ions is consistent with the suggestion that stabilization of the defect-pyrochlore structure requires hybridization of the

TABLE V

CRYSTALLOGRAPHIC DATA FOR $Sr_{1-x}Pb_xRuO_3$[a]

Composition	a (Å)	b (Å)	c (Å)	Volume (Å^3)	$c/\sqrt{2}$
$SrRuO_3$	5.530	5.570	7.847	241.7	5.550
$x = 0.2$	5.537	5.579	7.851	242.6	5.552
$x = 0.4$	5.545	5.588	7.856	243.4	5.556
$x = 0.6$	5.549	5.596	7.864	244.2	5.562
$x = 0.8$	5.555	5.603	7.863	244.7	5.561
$PbRuO_3$	5.562	5.610	7.862	245.3	5.560

[a] Kafalas and Longo (1970a), used with permission of Pergamon Press.

Pb^{2+}-ion $6s^2$ core orbitals into anisotropic, itinerant-electron 6sp orbitals that place a large concentration of core–electron charge density in the anion-vacancy positions. The outer $4s^24p^6$ core orbitals of a Sr^{2+} ion can sustain little hybridization, since the p^6 orbitals are filled and the empty 3d orbitals are of considerably higher energy.

The defect pyrochlore $AgSbO_3$ transforms at $P > 90$ kbar only to the Im3 structure of disordered, cubic $KSbO_3$ (Kafalas, 1971). Attempts to transform the defect pyrochlore $TlNbO_3$ (Belyaev *et al.*, 1969) have shown that there is no structural change below 90 kbar (Kafalas and Longo, 1970b).

Tomashpol'skii *et al.* (1967, 1968), on the other hand, report transformations from a defect-pyrochlore to a perovskite structure in $PbGa_{0.5}Nb_{0.5}O_3$ at 70 kbar and 700°C, and in $BiBO_3$, where B = Sc, Ni, Co, and Y, at 60–70 kbar and 600–750°C. They also report that $PbNi_{0.5}W_{0.5}O_3$ can be easily prepared with the perovskite structure at 60 kbar and 750°C.

2. *Garnets*

Several ABO_3 compounds having not too small an A cation and tetrahedral Ge and/or Si ions are transformed by pressure first to $A_3[AB](B_3)O_{12}$ garnets and then to ABO_3 perovskites. However, transformation of the

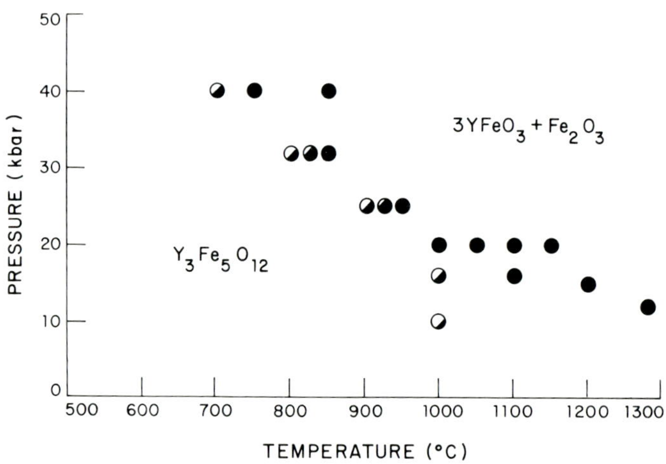

FIG. 16. Pressure–temperature phase diagram for $Y_3Fe_5O_{12}$. After Marezio *et al.* (1966a).

normal garnets is generally accompanied by disproportionation:

$$Ln_3[B_2](B_3)O_{12} \rightarrow 3LnBO_3 \text{ (perovskite)} + B_2O_3 \text{ (corundum)}$$

since the B ions are too small for twelvefold coordination (Marezio *et al.*, 1966a). The transformation pressure at 1000°C for these normal garnets decreases from 70 kbar for B = Ga^{3+} to 38 kbar for B = Al^{3+} to only 20 kbar for B = Fe^{3+}. Figure 16 shows a typical P–T diagram and illustrates the strong temperature dependence of the transformation pressure. Although Shimada *et al.* (1968) report that the garnet $Y_3Fe_5O_{12}$ transforms entirely to the perovskite structure at 20 kbar and 900°C with a 6% increase in density, retransformation occurring after heating for 4 hr at 900°C and atmospheric pressure, the existence of a $(Y_3Fe)Fe_4O_{12}$ perovskite is inconsistent with the phase diagram of Marezio *et al.* (1966a). In addition, Geller *et al.* (1970) have shown that $Y_3Fe_5O_{12}$ disproportionates to $3YFeO_3 + \alpha\text{-}Fe_2O_3$ in an inert atmosphere and to $9YFeO_3 + 2Fe_3O_4 + \frac{1}{2}O_2\uparrow$ in a reducing environment such as that furnished by a tantalum capsule without a platinum liner.

3. *Pyroxene*

Although $MnSiO_3$ transforms from the pyroxene to the garnet structure—and probably to perovskite at still higher pressures—Ringwood and Seabrook (1962a,b) found that none of the germanium pyroxenes that they subjected to pressure transform directly to the garnet form. $MgGeO_3$ and $MnGeO_3$ have the ilmenite structure above 28 and 25 kbar, respectively, the density increasing by 15% in each case, whereas the other $AGeO_3$ compounds (A = Fe, Co, and $Ni_{0.25}Mg_{0.75}$) disproportionate between 10 and 22 kbar into $Ge[M_2]O_4$ (spinel) + GeO_2 (rutile) with a 10% increase in density. Reid and Ringwood (1969) have suggested that the shock-wave data of McQueen *et al.* (1967) indicates a pyroxene → ilmenite transformation at very high pressure in $MgSiO_3$, but there is no information on the structure of the high-pressure phase.

4. *Ilmenite*

The ilmenite structure, whether occurring at atmospheric pressure or as a high-pressure polymorph, has been transformed at higher pressures to the perovskite structure. At 825°C and 1 atm, $CdTiO_3$ shows an ilmenite → perovskite transformation. Liebertz and Rooymans (1965), using the Bridgman anvil and hydrothermal techniques to 40 kbar, have shown that the high-temperature perovskite form is stabilized by high

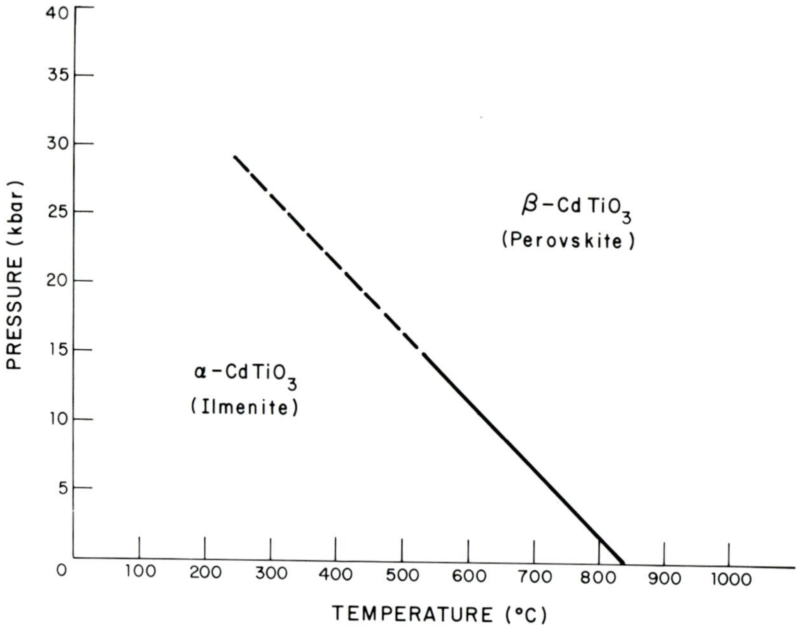

FIG. 17. Pressure–temperature phase diagram for $CdTiO_3$. After Liebertz and Rooymans (1965).

pressure (Fig. 17). $CdSnO_3$, which at 950°C has an ilmenite → perovskite transformation at 1 atm (Morgenstern-Badarau *et al.*, 1963, 1964), would probably show a similar stabilization with pressure of the perovskite phase.

5. *Fivefold-coordinated B cations*

The several compounds $A^{3+}MnO_3$ have the orthorhombic perovskite structure if $r_A \geq 1.03$ Å, the radius of the Dy^{3+} ion (Bertaut and Forrat, 1956). Bertaut and Mareschal (1963) found that for smaller r_A, they crystallize in the hexagonal $LuMnO_3$ structure of Fig. 7 (Yakel *et al.*, 1963). Waintal and Chenavas (1967a,b) and Waintal *et al.* (1966) transformed all of these hexagonal structures to the orthorhombic perovskite in 2 hr at 40–45 kbar and 700–950°C. The reduction in volume at atmospheric pressure was 8–9%.

The heavier $LnAlO_3$ compounds (Eu–Lu) have the hexagonal $YAlO_3$ structure when prepared below 900°C, but have the orthorhombic perovskite structure when prepared above 900°C (Bertaut and Mareschal, 1963). The transformation to perovskite should also be accomplished with high pressure.

This same type of transformation is found for the indates $LnInO_3$ (Ln = Dy, Gd, Eu), which are transformed from the $LuMnO_3$ structure (Schneider, 1961) to the orthorhombic perovskite structure at 65 kbar and 1450°C (Shannon, 1967). At 65 kbar and 1250–1500°C (Shannon, 1967), $YInO_3$ is also transformed to the orthorhombic perovskite structure, although it has the C rare-earth structure at atmospheric pressure (Schneider *et al.*, 1961). The quenched phase is stable at atmospheric pressure to greater than 1000°C.

6. *Limits on A-cation size*

There may be a lower limit to the A-cation size for high-pressure transformation to the perovskite structure. Syono *et al.* (1970) report that above ~45 kbar at 1100°C, the distorted ilmenite $MnVO_3$ transforms to an orthorhombic perovskite with a 6.6% increase in density. If the reported high-pressure structure is correct, it contains the smallest A cation of all the known perovskites. However, without single-crystal data, this structure cannot be considered established, since Shannon and Prewitt (1970) have shown from single-crystal data that the high-pressure form (65 kbar, 1200–1500°C) of Rh_2O_3, which has a powder x-ray pattern similar to

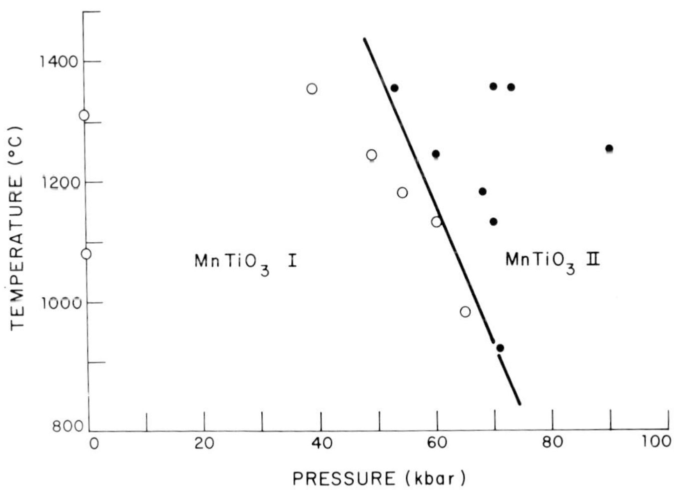

FIG. 18. Pressure–temperature phase diagram for $MnTiO_3$. After Syono *et al.* (1969c), used with permission of Pergamon Press.

perovskite, is actually closely related to the low-pressure corundum structure: pairs of Rh^{III}-occupied octahedra share common faces, but each octahedron shares only two edges rather than three as in corundum. Furthermore, Syono *et al.* (1969c) transformed the ilmenite $MnTiO_3$ to a disordered corundum structure at 60 kbar and 1100°C (Fig. 18), with a 1.6% reduction in volume, and they found no transformation below 90 kbar and 1300°C in other ilmenites $MTiO_3$, M = Fe, Co, Ni, Mg.

In this connection, $CuVO_3$ prepared at atmospheric pressure (Raveau, 1967; Deschanvres and Raveau, 1964) has an as yet unknown structure. Chamberland (1970a) prepared single crystals by hydrothermal methods. He also reported that at 65 kbar and 600–1000°C, $CuVO_3$ transforms to a distorted (triclinic symmetry) ilmenite structure. High-pressure reaction of either $Cu_2O + V_2O_5$ or $CuO + VO_2$ gave V-deficient products with either the triclinic phase or an undistorted ilmenite structure, the latter being furthest from stoichiometry (Cu:V = 1.0:0.9). Both of these high-pressure phases retransform if heated above 340°C at atmospheric pressure. Rea *et al.* (1972) have refined the structure of triclinic $CuVO_3$. It appears to contain Cu^{2+} and V^{4+} ions with strong V^{4+}–V^{4+} interactions.

There also appears to be a minimum A-cation size for transformation to the K_2NiF_4 structure of Fig. 12d. Ringwood and Reid (1968) have shown that Ca_2GeO_4 transforms from the olivine to the K_2NiF_4 structure (a = 3.70, c = 11.88 Å) at 110 kbar, that $CaMgGeO_4$ disproportionates to $CaGeO_3$ (wollastonite) + MgO (rock salt), and that Cd_2GeO_4, Ca_2SiO_4, and Mn_2SiO_4 transform, below 170 kbar, to new structures having complex powder x-ray patterns.

On the other hand, we have found that Sr_2CrO_4, which has tetrahedral Cr^{4+} at low pressures (Wilhelmi, 1967), transforms at 65 kbar and 700°C to the K_2NiF_4 structure (a = 3.82, c = 12.4 Å). This high-pressure phase can also be synthesized directly at high pressure by the reaction of SrO and CrO_2, but a cleaner product is obtained by transforming Sr_2CrO_4, because of difficulties associated with handling SrO. Pistorius (1966) used differential thermal analysis to detect the formation of high-pressure phases of Na_2WO_4 and Na_2MoO_4. From size considerations, the K_2NiF_4 structure may be anticipated for the high-pressure phase, but the structures have not been determined, as they are not retained at room temperature. We failed to find any new high-pressure phase of K_2WO_4 retained at room temperature and 1 atm after a quench from 65 kbar and 1000°C.

At the extreme of large A-cation size, Chamberland (1969a) found that Ba_2CrO_4 treated for 2 hr at 65 kbar and 1000°C disproportionates to $BaCrO_3$ (predominantly 4H polytype) + BaO (rock salt).

V. Syntheses Requiring High Pressure

A. Prologue

Table VI lists the presently known syntheses of ABX_3 compositions requiring high pressure. Reaction between two components to form a compound requires that the new composition have a lower free energy than the sum of the components. Where reaction does not take place at atmospheric pressure, high pressure may induce synthesis by lowering the free energy of the compound relative to that of the components. It is useful to identify those cases where this situation may apply. At least seven may be distinguished:

1. Outer d electrons at transition-metal cations may be either localized or itinerant, depending upon the magnitude of the coupling energy between d electrons on neighboring cations (Goodenough, 1965, 1966, 1967). Where the coupling is of intermediate strength, the d-electron energies are relatively unstable. In this case, application of pressure, which increases the coupling energy to make the d electrons definitely itinerant, may stabilize the compound sufficiently for reaction to occur. Octahedral-site Cr^{4+} ions in oxides appear to illustrate this situation.
2. Stabilization of a cation in a higher valence state may prove difficult at atmospheric pressure if the lower-valence-state ion is stable in a lower anion coordination. In this case, synthesis may require high anion pressure, which not only favors the more dense phase, but also provides the oxidizing power to raise the cation valence. Even if the higher-valence state can be stabilized at atmospheric pressure, higher anion pressures may be required to achieve stoichiometry.
3. The heavy cations Tl^+, Pb^{2+}, and Bi^{3+} have outer-electron cores $6s^2$ that inhibit bonding via empty 6p orbitals unless hybridization occurs either to polarize the core, giving rise to a ferroelectric type of cation displacement, or to make it highly anisotropic as in the defect-pyrochlore structure of $Pb_2Ru_2O_6\square$ discussed in Section III.F (Fig. 8). Unless the B cations exhibit a cooperative ferroelectric displacement, as in $PbTiO_3$, or a cooperative increase in covalent-bond stabilization, as in $Pb_2Ru_2O_6\square$, PbO is more stable than a composition in which the Pb^{2+}-ion concentration is diluted. Since pressure inhibits lattice distortions, it reduces the relative energies of those phases (e.g., PbO and $Pb_2Ru_2O\square$) containing a strong polarization of the $6s^2$ core, while simultaneously stabilizing preferentially the more dense phase. However, the ABX_3 structure synthesized under pressure may be cooperatively distorted at atmospheric pressure.

TABLE VI

HIGH-PRESSURE ABX_3 COMPOUNDS RETAINABLE AT 1 ATMOSPHERE AND ROOM TEMPERATURE

Compound	Atm. Press. Structure	High Press. Structure	P(kbar)	T(°C)	Remarks	Ref.
$CsMgF_3$	$Cs_4Mg_3F_{10} + MgF_2$	9R polytype	30	700	6H polytype stabilized above 35 kbar. See Fig. 12	1
$CsZnF_3$	$Cs_4Zn_3F_{10} + ZnF_2$	6H polytype	30	700	See Fig. 12	1
$BaCrO_3$	$Ba_2CrO_4 + CrO_2$	Hexagonal polytype	65	1000	Multiple polytypes present in preparation	2
$SrCrO_3$	$Sr_2CrO_4 + CrO_2$	Perovskite	65	800		3
$CaCrO_3$	$CaO + CrO_2$	Perovskite	65	700		4
$PbCrO_3$	$PbO + CrO_2$	Perovskite	65	900	See Fig. 20	5, 6
$PbSnO_3$	$PbO + SnO_2$	Perovskite	65	500		7
$PbZn_{1/3}Nb_{2/3}O_3$	$3PbO + ZnO + Nb_2O_5$	Perovskite	25	800	See Fig. 21	8
$BiMnO_3$	$Bi_2O_3 + Mn_2O_3$	Perovskite	40	700		9, 10
$BiCrO_3$	$Bi_2O_3 + Cr_2O_3$	Perovskite	40	700		9, 10
$InCrO_3$	$In_2O_3 + Cr_2O_3$	Perovskite	65	1350	Stable at 1 atm to 1000° C	11
$InRhO_3$	$In_2O_3 + Rh_2O_3$	Perovskite	65	1350	Stable at 1 atm to 1000° C	11
$T\ell CrO_3$	$T\ell_2O_3 + Cr_2O_3$	Perovskite	65	850	Stable at 1 atm to 500° C	11
$T\ell FeO_3$	$T\ell_2O_3 + Fe_2O_3$	Perovskite	65	850	Stable at 1 atm to 650° C	11
$LnGaO_3$	$Ln_2O_3 + Ga_2O_3$	Perovskite	45-70	1000	Ln = Sm, Eu, Gd, Tb, Dy, Ho, Er, Tm, Yb, Lu, Y	12
$T\ell T\ell OF_2$	$T\ell OF + T\ell F$	Perovskite	5	500		13
$RbNbO_3$		Perovskite	65	800		24
$RbTaO_3$		Perovskite	90	800		24
$RbSbO_3$		Cubic $KSbO_3$	20	800		24
$T\ell SbO_3$		Distorted cubic $KSbO_3$	20	800		24
Na_xMoO_3		Cubic bronze	65	700	$0.90 \leq x \leq 0.97$	14

K_xMoO_3		Cubic bronze	65	700	$0.89 \leq x \leq 0.93$	14
		Tetragonal bronze	65	700	$x \approx 0.5$, See Fig. 22	
Rb_xMoO_3		Hexagonal bronze	65	700	$x \approx 0.27$, See Fig. 22	14
		Orthorhombic	65	700	$0.41 \leq x \leq 0.44$ - distortion of hexagonal bronze	
K_xWO_3		Cubic bronze	65	700	$x \approx 0.9$	14
$A_{0.67}CrO_3$		New structure	25	800	A = Na, K, Rb	14, 15
$Li_{0.3}WO_3$		Hexagonal bronze	3	700	See Fig. 22	16
$Li_{0.5}WO_3$		Unknown structure	45	1200		17
$(NH_4)_{0.33}WO_3$		Hexagonal bronze	3	700	See Fig. 22	16
$Sn_{0.24}WO_3$		Hexagonal bronze	3	700	See Fig. 22	16
$Sn_{0.21}WO_3$		Hexagonal bronze	3	400		16
$Na_{0.6}ReO_3$		Distorted cubic bronze	65	1000		18, 19
$K_{0.3}ReO_3$		Hexagonal bronze	65	1000	See Fig. 22	18, 19
$K_{0.6}ReO_3$		Cubic bronze	65	1000		18, 19
$W_{1-x}Re_xO_3$	$WO_3 + ReO_3$	Cubic bronze	65	1000	$x \geq 0.25$, Monoclinic distortion of WO_3 is gradually reduced as x increases to 0.25	18
$WO_{3-x}F_x$		Cubic bronze	3	700	$0.17 \leq x \leq 0.66$	20
$MoO_{3-x}F_x$		Cubic bronze	3	700	$0.74 \leq x\ 0.97$	20, 21
$Sr_3Ir_2O_7$	$Sr_2IrO_4 + SrIrO_3$	K_2NiF_4 related	10	1000	See Fig. 9	22
$Sr_4Ir_3O_{10}$	$Sr_2IrO_4 + 2SrIrO_3$	K_2NiF_4 related	35	1000	See Fig. 9	22
Ba_2RuO_4	$BaO + BaRuO_3$	K_2NiF_4	65	1200	See Fig. 9	23

References:

1. Longo and Kafalas, 1969b
2. Chamberland, 1969b
3. Chamberland, 1967
4. Goodenough *et al.*, 1968
5. DeVries and Roth, 1968
6. Roth and DeVries, 1967
7. Sugawara *et al.*, 1968a
8. Matsuo *et al.*, 1969
9. Sugawara *et al.*, 1965, 1968b
10. Tomashpol'skii *et al.*, 1967
11. Shannon, 1967
12. Marezio *et al.*, 1966b, 1968
13. Demazeau *et al.*, 1969
14. Bither *et al.*, 1966
15. Wilhelmi *et al.*, 1969
16. Gier *et al.*, 1968
17. Post and Banks, 1968
18. Sleight and Gillson, 1966
19. Sleight *et al.*, 1969
20. Sleight, 1969b
21. Pierce *et al.*, 1970
22. Longo and Kafalas, 1969c
23. Kafalas and Longo, 1970b
24. Kafalas, 1971

4. If an ABO_3 or a $\square BO_3$ compound contains a transition-metal B cation in its highest valence state (empty d orbitals), the B cations are generally small enough (Zr^{4+} and Hf^{4+} excepted) either to occupy a tetrahedral interstice or to undergo a ferroelectric-type displacement within an octahedral interstice. These configurations destabilize the empty d orbitals relative to their energy if the cation is in the center of an octahedral interstice, and this makes it difficult to introduce d electrons, whether by A-cation substitution or insertion, by B-cation substitution, or by anion substitution. However, pressure favors octahedral-site occupancy and inhibits ferroelectric-type displacements, thereby stabilizing compositions requiring occupancy of the d orbitals.

5. Insertion-ion, metallic oxides such as the perovskite-related bronzes Na_xWO_3 ($x < 1.0$) contain partially filled, narrow d bands. The range of x that can be synthesized depends upon the density of states $N(\epsilon)$ versus energy ϵ profile of these d bands, optimum stability occurring where the Fermi energy falls at a peak or a valley where $dN(\epsilon)/d\epsilon$ is small. Although a wide compositional range of donor ions may exist, more commonly the compositional range is narrow, or at least limited. Pressure both inhibits formation of distorted competitive phases and reduces extrema created by electron correlations. It simultaneously favors introduction of interstitial cations to increase the density of the product. Therefore, pressure may extend both limits of the phase field.

6. If the A or the B cations have pronounced site-preference energies, either because of their size or their outer-electron configuration, then this places specific constraints on the site-preference energy of the other cation if a stable ABX_3 composition is to be synthesized at atmospheric pressure.

a. Octahedral-site B cations. If the B cation has a strong octahedral-site preference energy, this inhibits formation of ABX_3 compounds having B cations in fivefold or fourfold coordination. In this case formation of a stable ABX_3 compound requires either an $r_A > 0.90$ Å for stabilization of a perovskite polytype or an octahedral-site preference energy for the A cation as well for stabilization of a corundum-related structure. However, if r_A is so large that $t > 1.07$, then an $A^+B^{5+}O_3$ compound may not be stable at atmospheric pressure because large B^{5+}–B^{5+} repulsive forces inhibit any hexagonal polytype and size considerations render the cubic phase unstable.

b. Metal–metal bonding. Octahedral-site B cations having partially filled t_{2g} orbitals bond with neighboring cations sharing a common octahedral-site edge or face. Where these interactions are relatively weak, they give rise to magnetic interactions between localized atomic spins. However, in many cases these interactions are strong enough to delocalize the d electrons and quench any spontaneous magnetism (Goodenough, 1971). Where the interactions are of intermediate strength, there the compound is relatively unstable and tends either to form a distorted structure having stronger metal–metal bonds within cation clusters or to disproportionate, at least one phase having stronger metal–metal bonding. The other may have empty d orbitals. However, strong metal–metal bonding is generally achieved at the expense of anion close packing, and cations with empty d orbitals generally occupy tetrahedral sites or undergo a ferroelectric-type displacement within an octahedral site. Since pressure stabilizes close packing and higher anion coordination, it inhibits disproportionation. Simultaneously, it increases any metal–metal bonding within a more dense product: ABX_3 or A_xBX_3 with $x < 1.0$.

c. *Tetrahedral-site B cations.* If the B cation has a strong tetrahedral-site preference energy, then an ABX_3 composition stable at room temperature could have a pyroxene or $InGaO_3$ structure, provided the A cation does not also have a tetrahedral-site preference energy. But these structures must compete with $B[A_2]O_4$ (spinel) + BO_2 (silica) where the A cation has an octahedral-site preference energy, and with $A_3B_5O_{12}$ (garnet) + A_2O_3 where the A cation is a little too large for octahedral coordination.

7. Where the A cation is large ($t > t_c$), so that the stable ABX_3 compound is one of the hexagonal polytypes, there formation of the interlayer compounds $(AX)_n(ABX_3)$ of Fig. 9 may require a pressure $1 < P \lesssim P_{3C}$, where P_{3C} is the critical pressure required to transform the hexagonal polytype to the 3C (perovskite) structure. A $P \lesssim P_{3C}$ reflects the effective internal pressure present due to the constraints implicit in matching the AX and ABX_3 layers.

B. Stabilization of Cr^{4+} Ions in Oxygen Octahedra

Although chromium ions may be stabilized with formal valences 2+, 3+, 4+, 5+, and 6+, by far the most common are octahedral-site Cr^{3+} and tetrahedral-site Cr^{6+} ions. From a comparison of the physical properties of oxides with the perovskite structure (Goodenough and Longo, 1970), it is apparent that the d electrons in $CaVO_3$ are itinerant, in $CaMnO_3$ are localized, and in $CaCrO_3$ are transitional. In all these perovskites, the B–O–B interactions via π-bond covalent mixing are modulated by the A cation, a more basic A cation competing less for the O^{2-}:2p electrons and therefore allowing stronger interactions. This fact is manifest in a change from metallic conductivity and Pauli paramagnetism in $SrCrO_3$ (Chamberland, 1967) to semiconduction and antiferromagnetism ($T_B = 240°K$) in

$PbCrO_3$ (Roth and De Vries, 1967). Metallic $CaCrO_3$ is also antiferromagnetic, exhibiting a weak ferromagnetic component below 90°K (Goodenough *et al.*, 1968; Weiher *et al.*, 1971). The rutile structure also provides excellent geometry for π-bond formation (Goodenough, 1965), and CrO_2 is a metallic ferromagnet, which again demonstrates the transitional character of the d^2 electrons at the Cr^{4+} ions. Since the occurrence of spontaneous itinerant-electron magnetism is a relatively rare phenomenon in nature, such a transitional state seems to be less stable than either the localized-electron state or the nonmagnetic itinerant-electron state. Furthermore, although Cr^{4+} ions with localized d electrons have been stabilized in isolated oxygen tetrahedra in the presence of basic A cations, synthesis of compounds containing Cr^{4+} ions in oxygen octahedra seems to require high pressure. We interpret this fact to be a manifestation of the relative instability of the 3d electrons in an intermediate state. In the case of chromium, this is enhanced by the particular stability of octahedral-site Cr^{3+} ions, which have localized d^3 electrons stabilized by the crystal fields and intra-atomic exchange.

Roth and DeVries (1967) first synthesized cubic $PbCrO_3$ (a = 4.00 Å)

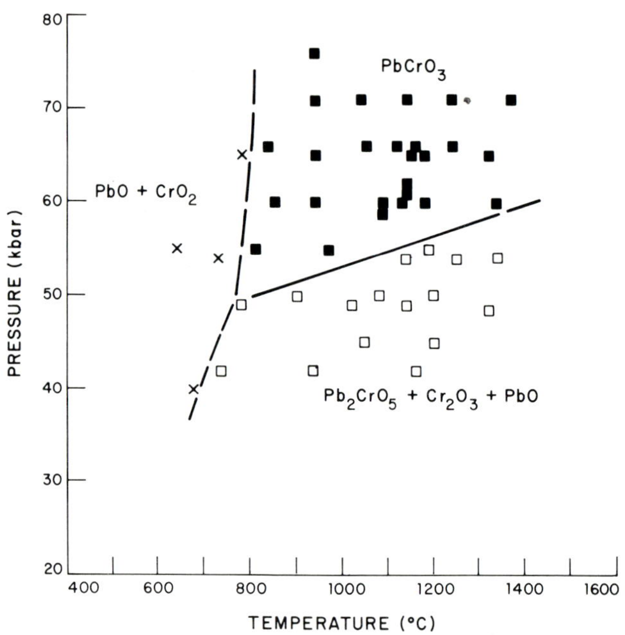

FIG. 19. Pressure–temperature phase diagram for $PbCrO_3$. After DeVries and Roth (1968).

by the reaction of PbO and CrO_2 for 20–30 min in a high-pressure belt apparatus. Their CrO_2 had been prepared by decomposing CrO_3 at 425°C and 0.25 kbar. The mixed powders were encapsulated in 0.001-in. tubular platinum foil crimped at both ends. Figure 19 summarizes the pressure–temperature phase diagram obtained from x-ray analysis of the quenched specimens (DeVries and Roth, 1968). Material quenched from the $PbCrO_3$ stable region contained an unknown impurity of the order of 5%. The x-ray powder patterns obtained by them and by ourselves have always given rather broad lines. The origin of this line broadening is not understood. It does not change appreciably on crossing the Néel temperature $T_N = 240°K$.

Goodenough *et al.* (1968) reported synthesis of the orthorhombic ($a = 5.287$, $b = 5.316$, $c = 7.486$ Å) perovskite $CaCrO_3$ by the solid state reaction of $CaO + CrO_2$ held for 30 min at 65 kbar and 700°C. The CaO was prepared by decomposition of $CaCO_3$, the CrO_2 by decomposition of CrO_3 at 400°C and 20 kbar in a belt apparatus using a platinum-lined cell. CrO_2 has also been prepared by the reaction of $CrO_3 + Cr_2O_3$ at 400°C and 20 kbar (Chapin *et al.*, 1965). Weighing, mixing, and encapsulating the dried CaO–CrO_2 mixture in a nitrogen-filled glove bag proved to be a necessary precaution. A small excess of CaO was required to eliminate a varying amount of Cr_2O_3 impurity that was otherwise present; the residual CaO could easily be removed from the product by washing with dilute acid. In contrast to $PbCrO_4$, which is mostly reduced at 1300°C and 60 kbar to $PbCrO_3$ (DeVries and Roth, 1968), the yellow product $CaCrO_4$ obtained by reacting CaO with CrO_3 did not lose oxygen below 1200°C. Weiher *et al.* (1971) isolated black crystals of $CaCrO_3$ from a product of $CaO + CrO_2$ reacted in platinum at 65 kbar and 1000°C for 1 hr. Their product also contained some acid-soluble β-$CaCr_2O_4$.

The cubic perovskite $SrCrO_3$ ($a - 3.8180$ Å) was first prepared by Chamberland (1967) in a tetrahedral-anvil pressure apparatus via the reaction $SrO + CrO_2$ or $Sr_2CrO_4 + CrO_2$ at 800°C and 65 kbar. The latter reaction is preferable, since it eliminates handling of the very reactive SrO. He also reported that pressures greater than 40 kbar are necessary to prevent reduction of CrO_2 at temperatures of 800–1000°C.

Chamberland (1969a) prepared $BaCrO_3$ by similar reactions at 65 kbar and 750–1400°C. Here also use of Ba_2CrO_4 rather than BaO leads to purer products and eliminates the need to weigh, mix, and encapsulate the compounds in a dry box. The reaction of $BaCO_3 + CrO_2$ at 1000°C and 65 kbar for 2 hr gave no reaction, since the dissociation pressure of $BaCO_3$ is less than 1 atm at 1000°C. The CrO_2 used in this preparation was obtained by hydrothermal reaction of $CrO_3 + Cr_2O_3$ (Swoboda *et al.*, 1961). Chamberland (1969a) also found that the $BaCrO_3$ samples prepared at 60–65 kbar contained mixed polytypes. The reactants $BaO + CrO_2$ gave the 4H, 6H, 9R, and 12 R polytypes, whereas the $Ba_2CrO_4 + CrO_2$ product included the 4H, 6H, 14H, and 27R polytypes. The use of highly reactive BaO or Ba_2CrO_4 plus the large temperature gradients present in many reactions (up to 500°C across the specimen in some cases) probably led to the multiphase products. Although these experiments indicate that $BaCrO_3$ is certainly one of the ABX_3 polytypes—which one, depending upon the reaction pressure and temperature—the appearance of polytypes other than 2H, 9R, 4H, 6H, and 3C is surprising and may indicate departures from stoichiometry.

Our attempts to prepare $CdCrO_3$ by the reaction of $CdO + CrO_2$ at 65 kbar and temperatures to 1000°C were unsuccessful. No reaction occurred below 500°C, whereas loss of oxygen at 650°C gave the spinel $Cd[Cr_2]O_4$. Similar attempts to prepare Pb_2CrO_4 by the reaction of $2PbO + CrO_2$ were also unsuccessful. At temperatures near 500°C, there was little or no reaction, while at temperatures near 1000°C, there was appreciable oxidation of PbO to give a mixture of PbO, Pb_3O_4, and Cr_2O_3.

Chamberland and Cloud (1969) needed high pressure to synthesize $NiCrO_3$. Reaction for 2 hr of NiO and CrO_2 at 65 kbar and 1200°C gave the corundum, rather than the ilmenite, structure. The disordered cation arrangement indicates the presence of Ni^{3+} and Cr^{3+} rather than Ni^{2+} and Cr^{4+} ions, an inference that is supported by magnetic measurements. Interestingly, this is the only known case of high-spin Ni^{3+} in an oxygen octahedral site, and here pressure and its cohabitation with Cr^{3+} ions are probably required to stabilize this ion.

C. Problems with Oxygen Deficiency

Moderate to high oxygen pressures may be required to synthesize stoichiometric ABO_3 compounds, especially if one of the cations is readily stabilized in a lower valence state with lower anion coordination.

This situation is well illustrated by the competition between the cubic perovskite $CaFeO_3$ ($V = 53.58$ Å^3) and brownmillerite $CaFeO_{2.5}$ ($V = 55.78$ Å^3), which has a perovskite structure in which the anion vacancies are so ordered that small anion displacements create local tetrahedral versus octahedral coordination about half of the Fe^{3+} cations, as illustrated in Fig. 20. The relative stability of tetrahedral-site Fe^{3+} ions favors brownmillerite at atmospheric pressure. High oxygen pressure favors the more dense, perovskite phase.

The Mn^{3+} ion, with its d^4 outer-electron configuration, is stable in fivefold

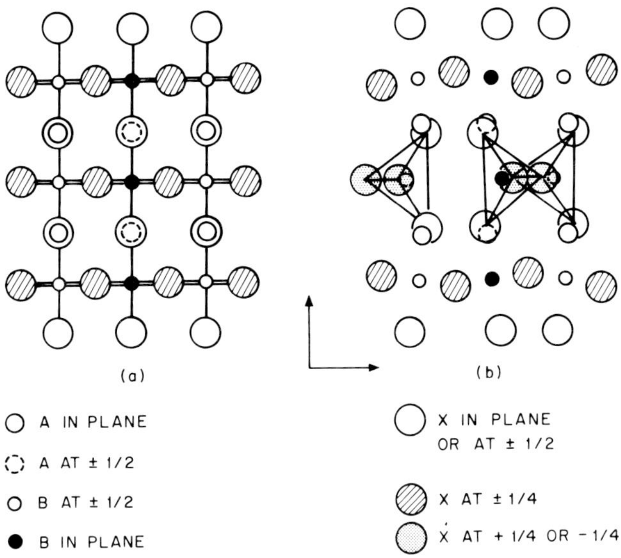

FIG. 20. Projection onto (110) planes of (a) cubic perovskite, and (b) brownmillerite structure, which is formed by removing alternate [110] strings of oxygen from central row of (a) and regrouping remaining oxygen into the tetrahedra shown in (b). After Wadsley (1964).

coordination, which is why $YMnO_3$ has the hexagonal structure of Fig. 7 at atmospheric pressure. Therefore it is not surprising to find that $CaMnO_{3-\delta}$ tends to be anion-deficient unless prepared under high oxygen pressure.

High oxygen pressure may be conveniently supplied by use of CrO_3 in a split cell, as described in Section II, or by hydrothermal techniques. Kanamaru *et al.* (1970) used a two-compartment gold capsule with a ZrO_2 partition to convert brownmillerite $Ca_2Fe_2O_5$ to the cubic (a = 3.770 Å) perovskite $CaFeO_3$. Their capsule was pressurized to 20 kbar and held at 1100°C for 1 hr before quenching.

MacChesney *et al.* (1965a) used hydrothermal techniques to prepare the cubic perovskite $SrFeO_{3.0}$ at 500°C and 340 atm of oxygen. The highest oxygen concentration obtained without oxygen pressure corresponded to $SrFeO_{2.86}$. The product of pure $Fe_2O_3 + 2SrC_2O_4$ air-fired at 1000°C was ground and pressed into disks, which were annealed in air at 1400°C for 16 hr. These disks were then equilibrated at various temperatures and about 340 atm of oxygen in a hydrothermal vessel made of Stellite or René alloy. An internally heated pressure vessel was used for $T > 800°C$. As the temperature increased in the interval $760 < T < 1400°C$, the degree of oxygen deficiency increased. This change was manifest in an increasing cell size, and finally in a tetragonal distortion appearing first in $SrFeO_{2.84}$.

Panyushkin *et al.* (1969) report that some of the Fe^{IV} ions in $SrFeO_{2.86}$ are reduced to Fe^{3+} ions at very high pressures ($\sim$150 kbar). This was determined by monitoring the Mössbauer data as a function of pressure. It is a reversible process that reflects charge transfer from the anion array to the cation array as pressure broadens both the "cation" d bands and the "anion" p bands until the top of the anion p bands overlaps the Fermi energy. This phenomenon is caused primarily by the increased anion–anion interactions and has little to do with compound synthesis.

MacChesney *et al.* (1965b) also prepared oxygen-deficient $BaFeO_{3-\delta}$ hydrothermally, achieving the highest oxygen content $BaFeO_{2.95}$ at 400°C and 1.3 kbar oxygen pressure. This compound had the 6H polytype structure of Fig. 4c.

Nearly stoichiometric $CaMnO_3$, Ca_2MnO_4, $Ca_3Mn_2O_7$, and $Ca_4Mn_3O_{10}$ were prepared by MacChesney *et al.* (1967) at 400°C and hydrothermal oxygen pressures near 1 kbar. We have prepared $CaMnO_{3.0}$ (determined by TGA) at high oxygen pressures using a two-compartment platinum cell in a belt apparatus. One compartment contained the product of CaO and MnO fired in air at 1300°C, and the other contained CrO_3. The partition between them consisted of a boron nitride disk with a center hole. The cell was pressurized to about 20 kbar and held at about 500°C for 2 hr. Magnetic measurements on the product (Menyuk *et al.*, 1969) were in good agreement with those reported by MacChesney *et al.* (1967).

D. Heavy Cations Having $6s^2$ Cores

The ABO_3, $A_2BB'O_6$, and $A_3B_2B'O_9$ compounds having $A = Pb^{2+}$ may crystallize in either the perovskite or the defect-pyrochlore structure, the latter phase being stabilized if the B cations have d electrons. Where the defect-pyrochlore structure is stable at atmospheric pressure, the perovskite phase is probably stable at high pressures, as was shown to be the case for $PbRuO_3$ above 90 kbar. However, there are a few cases where no compound is formed at atmospheric pressure, the conditions not being favorable for formation of either the defect-pyrochlore or the perovskite structure. In these cases, high-pressure synthesis can be expected to yield the perovskite phase.

Sugawara *et al.* (1968a) have synthesized $PbSnO_3$ by reacting reagent-grade PbO and SnO_2 in a tetrahedral-anvil apparatus at 60–70 kbar and 500°C for 40 min. The specimen was encapsulated in a graphite tube that also served as a heating element. The nearly single-phase product had a monoclinically distorted perovskite structure ($a = c = 4.076$, $b = 40.43$ Å, $\beta = 89°45'$) that became cubic on heating above 400°K. The perovskite phase could not be obtained at pressures below 50 kbar or at 60 kbar and 750°C, indicating that the structure is stabilized by high pressure and low temperature.

Since a Sn^{4+} ion has no empty, energetically accessible d orbitals for covalent bonding, the stable atmospheric-pressure form of $PbSnO_3$ would presumably have the defect-pyrochlore structure. However, the disproportion products PbO and SnO_2 are sufficiently stable that this compound is not competitive. At high pressures, a more dense compound must be formed. Since pressure stabilizes the perovskite phase relative to the defect-pyrochlore structure, it is not surprising to find that the compound formed at high pressure is a perovskite.

Bokov and Myl'nikova (1960) have reported growth of small $Pb_3ZnNb_2O_9$ crystals from a PbO flux, but synthesis of ceramic samples by normal solid state reaction was unsuccessful. Matsuo *et al.* (1969) and Matsuo (1970) have synthesized this nearly cubic perovskite by reacting

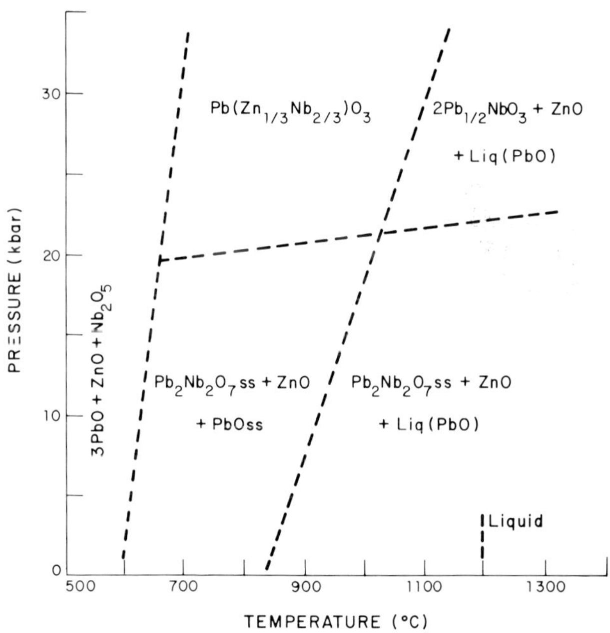

FIG. 21. Pressure–temperature phase diagram for $Pb_3ZnNb_2O_9$. After Matsuo *et al.* (1969).

mixed powders at 15–25 kbar and 600–1100°C for periods of 10 min to 2 hr. Platinum capsules were placed in a lava cell fitted with a graphite heater, and pressure was applied in a piston-cylinder apparatus. The P–T phase diagram obtained is shown in Fig. 21. It shows that high pressure stabilizes the compound $Pb_3ZnNb_2O_9$, which appears to crystallize only in the perovskite structure. Heating above 700°C at 1 atm decomposes it into $Pb_2Nb_2O_7$ (pyrochlore), PbO, and ZnO.

With the exception of $BiFeO_3$, the $BiB^{3+}O_3$ compounds (B = Sc, Cr, Mn, Fe, Co, Ni, and Y) have not been prepared as perovskites at atmospheric pressure with normal solid state reactions. Bokov *et al.* (1965) were able to separate a small amount of $BiMnO_3$ from a Bi_2O_3–Mn_2O_3 flux after slow cooling from 700 to 1000°C at 1 atm, but 40 kbar and 700°C were required for a direct solid state reaction, the product decomposing at 1 atm above 300°C (Sugawara *et al.*, 1968b). In fact, the compositions $BiMnO_3$ and $BiCrO_3$ both seem to require high pressure for solid-state synthesis. Sugawara *et al.* (1965, 1968b) and Tomashpol'skii *et al.* (1967) placed the mixed oxides $Bi_2O_3 + B_2O_3$ in a graphite capsule, which also acted as a heater, and reacted them at about 40 kbar and 700°C for 30–60 min before quenching. Both $BiMnO_3$ and $BiCrO_3$ are triclinically distorted perovskites. So also is $BiScO_3$, whereas $BiFeO_3$ is rhombohedral (Michel *et al.*, 1969) and the others are cubic (Tomashpol'skii *et al.*, 1968). Although magnetic measurements of flux-grown and high-pressure $BiMnO_3$ were in good agreement, neither technique seemed to yield a pure compound. The two Russian groups each indexed $BiMnO_3$ on a doubled unit cell.

Even though the fluoride and oxide ions have similar radii, not many oxyfluorides are known. An interesting example, however, is the cubic (a = 4.59 Å) perovskite $Tl^{1+}Tl^{3+}OF_2$, which was prepared by Demazeau *et al.* (1969) at 4.5 kbar and 500°C by the reaction of either TlF and O_2 or TlOF and TlF. Our attempts to prepare $BaNiOF_2$ from atmospheric pressure to 90 kbar at temperatures up to 1500°C were unsuccessful. The reactants BaF_2 and NiO were inert, and a $BaO:NiF_2$ mixture converted to BaF_2 and NiO.

E. Donating Electrons to Empty d Orbitals

1. *Insertion-ion or A-cation donors*

Octahedral-site Ti^{4+}, Nb^{5+}, Mo^{6+}, and W^{6+} ions are commonly stabilized by a ferroelectric-type displacement at lower temperatures that gives stronger bonds with one, two, or three of their nearest-neighbor anions. On the other hand, Ti_2O_3, NbO_2, MoO_2, and WO_2 are stabilized by metal–

metal bonding. Therefore any attempt to donate d electrons to a BO_3 array via A-cation donors tends to result in disproportionation. High pressure inhibits the ferroelectric-type distortions that destabilize the empty, π-bonding d orbitals. It also favors the higher density associated with incorporation of the donor A cation.

The WO_3 structure consists of the BO_3 array of the cubic perovskite, but it is distorted at room temperature by antiferroelectric displacements of the W^{6+} ions. Thus the A-cation interstices are vacant, and it is not surprising to find that A cations may be inserted into these interstices to form the system A_xWO_3, where the range of x varies with the size and valence of the A ion (Goodenough and Longo, 1970). MoO_3 has a more complex structure in which the ferroelectric-type distortions are stable to the melting point, which suggests that this phase is more strongly competitive with any A_xMoO_3 phase, and indeed no A_xMoO_3 perovskite bronze has been prepared at atmospheric pressure. However, the existence of A_xWO_3 defect perovskites indicates that similar A_xMoO_3 phases may be stabilized sufficiently by high pressure to make synthesis possible.

Bither *et al.* (1966) found that at 65 kbar, the cubic-perovskite system Na_xMoO_3 exists in the limited composition range $0.90 \leqslant x \leqslant 0.97$. At 20 kbar and 700°C, the perovskite phase was stable over a smaller range of x, and hydrothermal reaction at 3 kbar gave only MoO_2 as the reduced phase. The red Na_xMoO_3 crystals slowly decompose at atmospheric pressure and room temperature by forming a coating of MoO_2. A smaller stability range ($0.89 \leqslant x \leqslant 0.93$) was found for K_xMoO_3 at 65 kbar; at about $x = 0.5$, single crystals with the tetragonal bronze structure of Fig. 22a

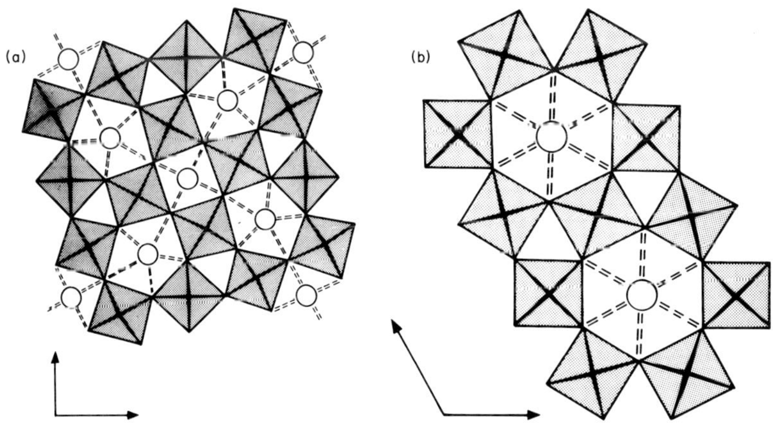

FIG. 22. Bronze structures found in A_xBX_3 systems: (a) tetragonal structure, (b) hexagonal structure. After Wadsley (1964).

were also recovered. This tetragonal structure, which is also found in the system K_xWO_3, contains corner-shared BO_3 octahedra that are buckled in the (001) plane so as to more easily accommodate the larger K^+ ion. Both the cubic and the tetragonal K_xMoO_3 bronzes decompose to MoO_2 at atmospheric pressure. Two structure types were also found at 65 kbar for Rb_xMoO_3: the hexagonal structure of Fig. 22b at $x = 0.37$ and an orthorhombic structure of this same general type in the range $0.41 \leqslant x \leqslant 0.44$.

At atmospheric pressure, the K_xWO_3 system is stabilized in either the tetragonal or the hexagonal structure of Fig. 22, but not the cubic-perovskite phase. However, at 65 kbar, Bither *et al.* (1966) found yellow crystals having the cubic-perovskite structure at $x \approx 0.9$. Gier *et al.* (1968), using a hydrothermal reaction for about 9 hr at 3 kbar and 700°C in gold capsules, obtained three new tungsten bronzes with the hexagonal structure of Fig. 22: $Li_{0.30}WO_3$, where the Li^+ ions occupy the small tunnels, $(NH_4)_{0.33}WO_3$, and $Sn_{0.24}WO_3$. They also recovered hexagonal $Sn_{0.21}WO_3$ crystals after 8 hr at 400°C. At 45 kbar and 1200°C, Post and Banks (1968) obtained $Li_{0.5}WO_3$, which has an undetermined tetragonal structure. None of these compounds has been prepared at atmospheric pressure.

The compound $NaNbO_3$ is a distorted perovskite. $Ca_xNa_{1-2x}NbO_3$ can be prepared at atmospheric pressure, and annealing in Na vapor creates a metallic $Na_{1-x}Ca_xNbO_3$ skin on the crystal (Kunnmann, private communication). It may be possible to synthesize metallic $Na_{1-x}Ca_xNbO_3$ directly under pressure.

2. *B-cation donors*

Metallic ReO_3 has the cubic-perovskite BO_3 array, as does high-temperature WO_3, but the solid solutions $W_{1-x}Re_xO_3$ are not easily synthesized at atmospheric pressure. This is probably due to a segregation of the WO_3 phase at temperatures below which static ferroelectric-type displacements of the W^{6+} ions occur, since such segregation would permit cooperative distortions that minimize the associated elastic energy. Furthermore, the rhenium d orbitals are more stable than the tungsten d orbitals, and segregation of the ReO_3 phase permits stronger Re–O–Re metallic interactions for stabilization of the occupied d orbitals. At high pressures, on the other hand, the static ferroelectric-type distortions may be suppressed to temperatures below which ionic diffusion is limited, and solid solutions can be formed. Sleight and Gillson (1966) obtained solid solutions over the entire range $0 \leqslant x \leqslant 1$ at 65 kbar. The room-temperature monoclinic distortion of WO_3 diminishes with increasing x, disappearing at $x \approx 0.25$. Re_2O_7 was used as a flux in all runs to provide an oxidizing environment and a medium for crystal growth.

3. *Anion donors*

Electrons may also be introduced into the WO_3 or the MoO_3 array by substituting fluorine for oxygen. Sleight (1969b) has prepared $WO_{3-x}F_x$ and $MoO_{3-x}F_x$ at 700°C and 3 kbar in either 48% aqueous HF or anhydrous HF. The cubic ReO_3 structure is found in the ranges $0.17 \leqslant x \leqslant 0.66$ and $0.74 \leqslant x \leqslant 0.97$, respectively. Pierce *et al.* (1970), working hydrothermally at 2 kbar, found a cubic $MoO_{3-x}F_x$ compound at $x = 0.6$. In the narrow range $0.20 \lesssim x \lesssim 0.25$, both Sleight (1969) and Pierce *et al.* (1970) found a compound isostructural with $Mo_4O_{10}(OH)_2$, ignoring the protons. The structure is similar to MoO_3 except for a more symmetric displacement of the molybdenum ion from the center of its octahedral interstice.

4. *Stabilization of Cr^{5+} ions*

The Cr^{6+} ion is generally stabilized in a tetrahedral site, since it is smaller than either Mo^{6+} or W^{6+}, and the formation of A^+CrO_3 phases has never been accomplished at atmospheric pressure. However, Bither *et al.* (1966) did prepare $Na_{0.67}CrO_3$ at 65 kbar. Wilhelmi *et al.* (1969) also found isostructural $K_{0.67}CrO_3$ and $Rb_{0.67}CrO_3$ at 25 kbar and showed that its monoclinic structure, which is unrelated to perovskite, consists of a network of octahedral-site and tetrahedral-site chromium ions. Presumably Cr^{6+} ions occupy the tetrahedral sites, the outer d electrons being shared among the Cr-occupied octahedral sites. It is reasonable to anticipate that at still higher pressures, a cubic, metallic perovskite Na_xCrO_3 would be formed, probably most easily at $x \approx 0.9$.

F. Shifting the Fermi Energy

Correlations among narrow band electrons tend to introduce a deep minimum, if not an energy gap, in the density-of-states versus energy profile where the occupied states correspond to an integral number per atom. For a nondegenerate band, for example, correlation splitting may occur where it is half-filled (Hubbard, 1964). In the localized-electron limit, this phenomenon gives a finite energy difference between different ionic states of the same atom. Displacement of states to form a minimum creates maxima in the density-of-states profile on either side of it. Thus in the system Na_xWO_3, a maximum may be anticipated near $x \approx 0.9$, a minimum at $x = 1.0$. This reasoning is consistent with cubic-perovskite phases in the vicinity of $x = 0.9$ for K_xWO_3, K_xMoO_3, and Na_xMoO_3 prepared under pressure, and with an upper limit of $x \approx 0.9$ for Na_xWO_3

prepared at atmospheric pressure. Since pressure favors higher density, it is possible to prepare at high pressures the end member $x = 1.0$, where the density-of-states profile should have a minimum.

Chamberland (1969b) has described a general method for synthesizing under pressure transition-metal oxides containing alkali-metal insertion ions:

$$x\text{AN}_3 + \text{BO}_3 \rightarrow \text{A}_x\text{BO}_3 + (3x/2)\text{N}_2\uparrow$$

Using this method, he has prepared the end member $NaWO_3$ at 65 kbar and 800°C. He has also used it to prepare K_xReO_3, where here the problem is to raise the Fermi energy from its minimum at $x = 0$.

Three alkali-metal bronzes A_xReO_3 have been prepared by Sleight and Gillson (1966) and by Sleight *et al.* (1969) at 65 kbar and 800–1100°C: $Na_{0.6}ReO_3$ (tetragonal, $a = 3.825$, $c = 3.841$ Å), $K_{0.6}ReO_3$ (cubic, $a = 3.8952$ Å), and $K_{0.3}ReO_3$ (hexagonal of Fig. 22, with $a = 7.335$, $c = 7.480$ Å). Reduction with rhenium metal was controlled so that some alkali perrhenate, $AReO_4$, always remained in the platinum container. This excess perrhenate served as a readily soluble flux.

The lower limit of x in the $A_x^{m+}BO_3$ defect perovskites also seems to be determined by the number of donated electrons mx, where $m = 1$, 2, or 3. This is consistent with a maximum or a minimum in the density-of-states profile at the corresponding energy. Pressure can be expected to extend this limit somewhat by quenching the ferroelectric-type distortions associated with the phase at $x = 0$.

G. Site Preference and Compatible Cation Sizes

1. *Octahedral-site B cations*

Cr^{3+} and Rh^{III} ions have notoriously strong octahedral-site preference energies. The ions In^{3+} and Tl^{3+} have an $r_A < 0.90$ Å, but they are too large to be readily stabilized together with smaller Cr^{3+} or Rh^{III} ions on the same octahedral sites of a disordered corundum structure. Therefore no ABO_3 composition is formed at atmospheric pressure. However, Shannon (1967) prepared $InCrO_3$ and $InRhO_3$ at 65 kbar and 1350°C, thus extending the lower limit on the size of the A cation found in $ACrO_3$ and $ARhO_3$ perovskites. The two compounds were stable at atmospheric pressure to over 1000°C. He also prepared $TlCrO_3$ and $TlFeO_3$ at 65 kbar and 850°C; these perovskites are stable to 500°C and 650°C, respectively.

The Mn^{4+} ion, which is isoelectronic with the Cr^{3+} ion, and the Sn^{4+} ion also have strong octahedral-site preference energies, whereas Mn^{2+}, Zn^{2+}, Co^{2+}, and Mg^{2+} tend to have tetrahedral-site preference energies. This

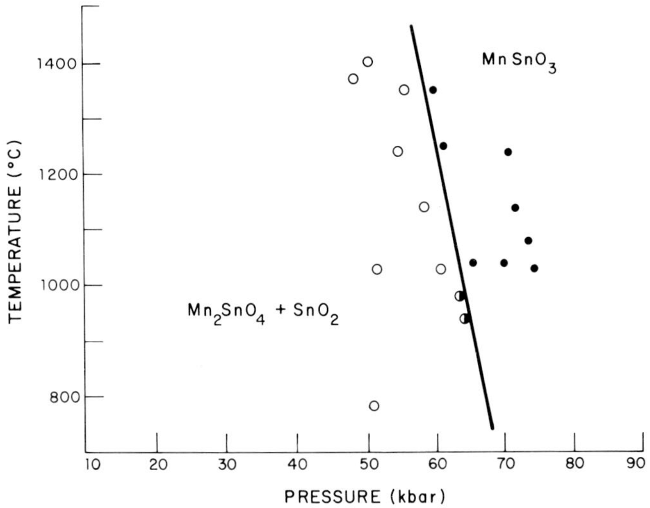

FIG. 23. Pressure–temperature phase diagram for $MnSnO_3$. After Syono *et al.* (1969b), used with permission of Pergamon Press.

makes A[AMn]O_4 (spinel) + MnO_2 (rutile) more stable at atmospheric pressure than the ilmenites $A^{2+}Mn^{4+}O_3$, where A = Zn, Co, Mg. However, Swoboda *et al.* (1958) prepared $CoMnO_3$ hydrothermally in platinum at 3 kbar and 625°C for 3 hr; Chamberland *et al.* (1970b) prepared $MgMnO_3$ and $ZnMnO_3$ in gold capsules at 3 kbar and 700°K for 6 hr. Similarly, no reaction occurs between Mn[MnSn]O_4 and SnO_2 below 65 kbar and 1000°C, whereas above these conditions (Fig. 23), $MnSnO_3$ forms with a disordered corundum structure (Syono *et al.*, 1969b).

The Cu^+ ion has a strong tetrahedral-site preference energy, and high pressures are required to synthesize the $Cu^+B^{5+}O_3$ compounds having B = Nb or Ta. Sleight and Prewitt (1970) found that $CuTaO_3$ formed at 65 kbar and 1000–1200°C has the $LiNbO_3$ structure discussed in Section III.B.1; the $CuNbO_3$ formed under the same conditions has monoclinic symmetry, probably resulting from a distortion of the $LiNbO_3$ structure.

The Cd^{2+} ion also exhibits a tetrahedral-site preference energy, and we found no reaction between CdO and RuO_2 at 65 kbar below 1000°C.

The compounds $RbNbO_3$ and $RbTaO_3$ do not form at atmospheric pressure. With a tolerance factor $t = 1.085$, the Rb^+ ions are too big to be easily accommodated in the cubic perovskite structure, and the effective charge on the B^{5+} ions is too large for formation of the hexagonal perovskites. However, a $t = 1.085$ is not too large for the cubic phase to be accessible with high pressure. Kafalas (1971) prepared $RbNbO_3$ at 65 kbar and 900°C.

At atmospheric pressure, the quenched phase was an orthorhombic 3C perovskite isostructural with the orthorhombic form of ferroelectric $BaTiO_3$. Preliminary DTA measurements suggest transformation at lower temperature to the rhombohedral $BaTiO_3$ phase and at higher temperature to the tetragonal $BaTiO_3$ phase. However, heating above 500°C leads to disproportionation without the formation of any hexagonal-perovskite polytype. Synthesis of $RbTaO_3$ required 90 kbar at 900°C, and the quenched phase at atmospheric pressure and room temperature appeared to be cubic.

The Rb^+ ion is too large to be stabilized in the sixfold coordination of the ilmenite structure, or even to be easily accommodated in the three-dimensional tunnels of the cubic subarray of Fig. 8b. Furthermore, the hexagonal perovskite polytypes are not stabilized with a B^{5+} ion, and covalency considerations inhibit the formation of a cubic 3C perovskite phase. Nevertheless Kafalas (1971) has synthesized $RbSbO_3$ at 20 kbar and 1000°C. Synthesis in the pressure interval $20 < P < 80$ kbar gives the Pn3 phase having ordered Rb^+ ions in the tunnels of the SbO_3 matrix of Fig. 8b. At pressures $P > 80$ kbar, the Rb^+ ions are disordered in the tunnels, resulting in the body-centered structure with space group Im3.

2. *Metal–metal bonding*

Although octahedral-site V^{4+} ions may form metal–metal bonds across a shared octahedral-site face or edge, as in VO_2, disproportionation into V^{5+} and V^{3+} allows stronger metal–metal bonding by octahedral-site V^{3+} ions, since the radial extension of the d wave functions increases with smaller formal valence at the cation. Furthermore, tetrahedral-site V^{5+} ions are very stable. It should not be too surprising, therefore, that $NiVO_3$ is not stable at atmospheric pressure even though the Ni^{2+} ion has a strong octahedral-site preference energy. Chamberland (1970b) was able to prepare $NiVO_3$ at 60–65 kbar and 1400°C for 2 hr in platinum. Heating at atmospheric pressure caused the compound to disproportionate according to the reaction

$$4NiVO_3 \rightarrow Ni_3V_2O_8 + NiO + V_2O_3$$

Here, V_2O_3 contains strong metal–metal bonding (Goodenough, 1971).

Although the structure of $NiVO_3$ is related to the corundum structure of Fig. 6, it is distorted to triclinic symmetry ($a = 4.935$, $b = 5.377$, $c = 4.903$ Å; $\alpha = 90.72$, $\beta = 119.13$, $\gamma = 63.00°$). Furthermore, from the fact that it is antiferromagnetic ($T_N = 153 \pm 5$°K), Chamberland deduced that the atomic ordering is probably not that of ilmenite, which has alternate cation basal planes of A^{2+} and B^{4+} ions. He postulated that each cation has unlike near-neighbor cations within its basal plane and is paired to a like cation along the *c* axis. This configuration requires that V^{4+}–V^{4+} bonding across a shared octahedral-site face outweigh the greater electrostatic repulsion between two 4+ ions.

Similarly, Rüdorff *et al.* (1956) were not able to prepare $CoVO_3$ at atmospheric pressure, but Chamberland (1970a) was able to react CoO and VO_2 at 65 kbar and 1000°C in platinum capsules. The structure had the same triclinic distortion of the corundum-type structure. This same distortion was also found by Syono *et al.* (1970) in one of the high-pressure forms of $CuVO_3$ and in $MnVO_3$ prepared below ~40 kbar and 1000°C (Fig. 24). However, Rea *et al.* (1972) have shown that triclinic $CuVO_3$, which has a larger pseudohexagonal *c*/*a* ratio, is a distorted ilmenite with V^{4+}–V^{4+} bonding in the basal plane.

In TiO and VO, metal–metal bonding induces the spontaneous creation of atomic vacancies, about 15% vacancies of each type occurring at the stoichiometric composition. The number of vacancies varies relatively little with annealing temperature at atmospheric pressure, but it can be substantially reduced by annealing under very high pressure (Banus, 1968). The lattice parameter of the more dense phase is larger at atmospheric pressure, indicating that creation of the vacancies has permitted stronger metal–metal bonding.

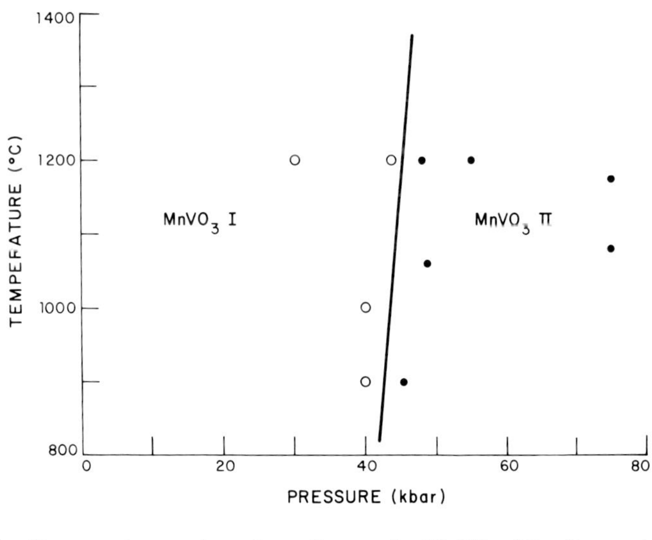

FIG. 24. Pressure–temperature phase diagram for $MnVO_3$. After Syono *et al.* (1970).

3. *Tetrahedral-site B cations*

The ions Si^{4+} and Ge^{4+} have a notoriously strong tetrahedral-site-preference energy. Since the Zn^{2+} ion also has a tetrahedral-site-preference energy, the pyroxene $ZnSiO_3$ is not stable at atmospheric pressure. It has been prepared by Hayashi *et al.* (1965) at 30 kbar and 500°C, and by Ringwood and Major (1967a) at 110 kbar and 900°C. Presumably this structure can be transformed to the ilmenite, if not to the perovskite, structure at very high pressures. Ringwood and Major found no compound formation for $ZnGeO_3$ at 35 kbar and 1100°C, but complete reaction to the ilmenite structure at 110 kbar and 900°C. It would be interesting to know whether a pyroxene structure can be stabilized at intermediate pressures and whether the ilmenite transforms to the corundum or perovskite structure at still higher pressures.

As known by those who have ever attempted to prepare single crystals of either the rare-earth orthoferrites or the rare-earth ferrogarnets from a flux, it is necessary to distinguish those conditions where the perovskite phase is stable relative to the garnet phase, and vice versa. This situation is symptomatic of the fact that where the B cation may be stable in tetrahedral coordination, the formation of $A_3B_5O_{12}$ (garnet) + A_2O_3 may be competitive with the synthesis of an ABO_3 perovskite. Since pressure favors the more dense phase (higher cation coordination), it can induce synthesis of the perovskite phase. This situation occurs in the rare-earth gallates. Keith and Roy (1954) and Geller (1957) found that at atmospheric pressure, the $LnGaO_3$ perovskite phases are formed only for Ln = La, Ce, Pr, Nd. For the other rare earths, there is no compound formation at the 1:1 composition (Schneider *et al.*, 1961). However, at high pressure (45–70 kbar and 1000°C) Marezio *et al.* (1966b, 1968) synthesized $LnGaO_3$ perovskites having Ln = Sm, Eu, Gd, Tb, Dy, Ho, Er, Tm, Yb, Lu, and Y by decomposing the respective garnets $Ln_3Ga_5O_{12}$ in the presence of NaOH, a successful flux for the growth of α-Ga_2O_3 (Remeika and Marezio, 1966) that is readily dissolved in water. Presumably the $LnGaO_3$ perovskites may also be prepared by direct reaction of Ln_2O_3 and Ga_2O_3 under the same conditions of pressure and temperature.

Although Ludekens and Welch (1952), Berezhnaya and Bukhalova (1967), and Bukhalova *et al.* (1969) have reported that $CsZnF_3$ and $CsMgF_3$ have a tetragonal-perovskite structure if prepared by solid state reactions, this structural assignment is not consistent with the observation that $CsBF_3$ compounds do not form the perovskite structure at atmospheric pressure if $r_B < 0.85$ Å (i.e., for B = Ni, Co, Fe, Mn). In a study of the pressure dependence of the structures of $CsBF_3$ compounds, Longo and Kafalas (1969b) found that no 1:1 compound is formed for B = Zn or

Mg at pressures and temperatures less than 30 kbar and 700°C. This view is supported by Babel (1965, 1967) and Steinfink and Brunton (1969), who found only the 4:3 compounds $Cs_4B_3F_{10}$ at atmospheric pressure. Above 30 kbar, $CsZnF_3$ forms in the 6H polytype structure of Fig. 4c and $CsMgF_3$ in the 9R structure of Fig. 4e, the latter transforming above 35 kbar to the 6H polytype.

H. Interlayer Compounds

$SrIrO_3$ crystallizes in the 6H polytype of Fig. 4c, but transforms to the perovskite form at 45 kbar and 1000°C. The interlayer compound Sr_2IrO_4 may be prepared at atmospheric pressure. Apparently the matching requirements of alternate layers of SrO and $SrIrO_3$ constrain the lattice parameters sufficiently to stabilize perovskite interlayers. As the proportion of SrO to $SrIrO_3$ decreases, these constraints are weakened and pressures $P < 45$ kbar are required to prepare the intermediary compounds $Sr_4Ir_3O_{10}$ and $Sr_3Ir_2O_7$, which have the structures illustrated in Fig. 9. Longo and Kafalas (1969c) have prepared these intermediates by reacting under pressure stoichiometric amounts of Sr_2IrO_4 and IrO_2 in platinum capsules for ½ hr at 1000°C. For $Sr_4Ir_3O_{10}$, synthesis required $P > 35$ kbar, whereas for $Sr_3Ir_2O_7$, $P > 10$ kbar was sufficient. At atmospheric pressure, both $Sr_4Ir_3O_{10}$ and $Sr_3Ir_2O_7$ decompose slowly at 800°C to Sr_2IrO_4 and the 6H polytype of $SrIrO_3$. The perovskite form of $SrIrO_3$ transforms slowly to the 6H polytype only above 1200°C.

We have also synthesized Ba_2RuO_4 at 65 kbar and 1200°C by the reaction of $BaRuO_3$ and BaO in platinum. The best results were obtained with a 10% excess of BaO, which was subsequently washed away. This product gave an x-ray diffraction pattern that was completely indexed on the basis of a K_2NiF_4 structure [$a = 3.99(1)$, $c = 13.43(5)$ Å].

VI. Syntheses Aided by High Pressure

Many solid state reactions are extremely slow or lead to poorly crystallized products if carried out at atmospheric pressure. Where the product is stabilized by pressure, as is the case for perovskites, high pressures may reduce reaction times to less than an hour and/or markedly increase the crystallinity of the product.

We have prepared the orthoferrites $LnFeO_3$ in ½ hr at 50 kbar and

1200°C by the reaction of Ln_2O_3 and Fe_2O_3 in platinum capsules. Normal ceramic techniques usually require several days of regrinding and firing. Shannon (1970) found it more convenient to prepare the orthorhodates $LnRhO_3$ by high-pressure techniques. In his study of the lattice parameters of this series, crystallinity of the product was very important and difficult to obtain adequately with conventional solid state reactions. Suitable crystallinity was achieved in 24 hr at 3 kbar and 1350°C in sealed platinum tubes.

In the interlayer series $(SrO)_n(SrRuO_3)$, the end members $SrRuO_3$ and Sr_2RuO_4 are both thermodynamically stable at 1 atm, in contrast to the series $(SrO)_n(SrIrO_3)$. Although this means that the two intermediary compounds $Sr_4Ru_3O_{10}$ and $Sr_3Ru_2O_7$ can be prepared at 1 atm by reacting appropriate mixtures of $SrRuO_3$ and Sr_2RuO_4, we found that to obtain single-phase materials by this process required repeated 1100°C firings and regrindings for periods up to two weeks. By contrast, 15 min at 20 kbar and 1000°C was sufficient to obtain a single-phase product of either $Sr_4Ru_3O_{10}$ or $Sr_3Ru_2O_7$ by the same reaction.

We have also synthesized $K_3Ni_2F_7$ and $Rb_3Ni_2F_7$ in 30 min by reacting mixed fluorides at 65 kbar and 800°C. Synthesis of $Rb_3Ni_2F_7$ may have required high pressure, since $RbNiF_3$ has the 6H polytype structure and requires 20 kbar to transform to perovskite. On the other hand, $KNiF_3$ is a perovskite and K_2NiF_4 is stable at 1 atm. Nevertheless, synthesis of $K_3Ni_2O_7$ at 1 atm is very difficult, if not impossible, whereas at high pressures, it is quite routine and rapid.

Synthesis of the A_xBO_3 "bronzes" is also simpler under pressure. In fact, as discussed in Section V.E, many of these compositions can only be prepared under pressure. Bierstedt *et al.* (1966) have prepared tetragonal and hexagonal (Fig. 22) A_xWO_3 bronzes (A = Ca, Sr, Ba, In and Tl) by hydrothermal techniques at 3 kbar and 600°C for 6 hr. All these bronzes are superconducting. Chamberland (1969b), using decomposition of the alkali-metal azides at high pressure (65 kbar and 750–1000°C or 1.5 kbar and 500°C) as discussed in Section V.F, has conveniently prepared various bronzes in the systems A_xWO_3 (A = Na, K), A_xMoO_3 (A = Na, K, Rb), A_xReO_3 (A = Na, K), and K_xUO_3.

Among the ilmenites, $NiMnO_3$ has been prepared in a few hours by hydrothermal reaction at 3 kbar and 700°C (Swoboda *et al.*, 1958) and more tediously by solid state reactions at low temperatures. The ilmenite $KSbO_3$ transforms to the cubic Pn3 structure of Fig. 8b if heated at 1000°C for several weeks (Spiegelberg, 1940). Treatment of the ilmenite phase for only 5 min at 800°C and 20 kbar is adequate to transform it to the disordered, cubic Im3 structure (Kafalas, 1971).

REFERENCES

Arend, H., and Kihlborg, L. (1969). *J. Amer. Ceram. Soc.* **52,** 63.

Aslanov, L. A., and Kovba, L. M. (1964). *Zh. Neorgan. Khim.* **9,** 2441 (*English Transl.: Russ. J. Inorg. Chem.* **9,** 1317).

Austin, A. E. (1969). *J. Phys. Chem. Solids* **30,** 1282.

Babel, D. (1965). *Z. Naturforsch.* **A20,** 165.

Babel, D. (1967). *Structure Bonding* **3,** 1.

Banus, M. D. (1968). *Mater. Res. Bull.* **3,** 723.

Belyaev, I. N., Lesnykh, D. S., and Lupeiko, T. G. (1969). *Zh. Neorgan. Khim.* **14,** 648 (*English Transl.: Russ. J. Inorg. Chem.* **14,** 338).

Berezhnaya, V. T., and Bukhalova, G. A. (1967). *Zh. Neorg. Khim.* **12,** 2179 (*English Transl.: Russ. J. Inorg. Chem.* **12,** 1148).

Bertaut, E. F., and Forrat, J. (1956). *J. Phys. Radium* **17,** 129.

Bertaut, E. F., and Mareschal, J. (1963). *C. R. Acad. Sci. Paris Ser. C.* **257,** 867.

Bierstedt, P. E., Bither, T. A., and Darnell, F. J. (1966). *Solid State Commun.* **4,** 25.

Bither, T. A., Gillson, J. L., and Young, H. S. (1966). *Inorg. Chem.* **5,** 1559.

Blasse, G. (1965). *J. Inorg. Nucl. Chem.* **27,** 993.

Bokov, V. A., and Myl'nikova, I. E. (1960). *Fiz. Tverd. Tela.* **2,** 2728 (*English Transl.:* (1961). *Sov. Phys.—Solid State* **2,** 2428).

Bokov, V. A., Myl'nikova, I. E., Kizhaev, S. A., Bryzhina, M. F., and Grigoryan, N. A. (1965). *Fiz. Tverd. Tela* **7,** 3695 (*English Transl.:* (1966). *Sov. Phys.—Solid State* **7,** 2993).

Bukhalova, G. A., Berezhnaya, V. T., and Okol'chishena, N. E. (1969). *Zh. Neorgan. Khim.* **14,** 1746 (*English Transl.: Russ. J. Inorg. Chem.* **14,** 917).

Chamberland, B. L. (1967). *Solid State Commun.* **5,** 663.

Chamberland, B. L. (1969a). *Inorg. Chem.* **8,** 286.

Chamberland, B. L. (1969b). *Inorg. Chem.* **8,** 1183.

Chamberland, B. L. (1970a). *J. Solid State Chem.* **1,** 138.

Chamberland, B. L. (1970b). *J. Solid State Chem.* **2,** 521.

Chamberland, B. L., and Cloud, W. H. (1969). *J. Appl. Phys.* **40,** 434.

Chamberland, B. L., Sleight, A. W., and Weiher, J. F. (1970a). *J. Solid State Chem.* **1,** 506.

Chamberland, B. L., Sleight, A. W., and Weiher, J. F. (1970b). *J. Solid State Chem.* **1,** 512.

Chapin, D. S., Kafalas, J. A., and Honig, J. M. (1965). *J. Phys. Chem.* **69,** 1402.

Clark, S. P., ed., (1966). "Handbook Phys. Constants," rev. ed., p. 153. Geological Society.

Clearfield, A. (1963). *Acta Crystallogr.* **16,** 134.

Demazeau, G., Grannec, J., Marveuf, A., Portier, J., and Hagenmuller, P. (1969). *C. R. Acad. Sci. Paris. Ser. C* **269,** 987.

Deschanvres, A., and Raveau, B. (1964). *C. R. Acad. Sci. Paris Ser. C* **259,** 3771.

DeVries, R. C., and Roth, W. L. (1968). *J. Amer. Ceram. Soc.* **51,** 72.

Dickinson, J. G., Katz, L., and Ward, R. (1959). *J. Amer. Chem. Soc.* **81,** 4109.

Dickinson, J. G., Katz, L., and Ward, R. (1961). *J. Amer. Chem. Soc.* **83,** 3026.

Donohue, P. C., Katz, L., and Ward, R. (1965). *Inorg. Chem.* **4,** 306.

Donohue, P. C., Katz, L., and Ward, R. (1966). *Inorg. Chem.* **5,** 335.

Eatough, N. L., Webb, A. W., and Hall, H. T. (1969). *Inorg. Chem.* **8,** 2069.

Gardner, R. A. (1969). Thesis, Brown Univ., Rhode Island.

Gardner, R. A., Vlasse, M., and Wold, A. (1969a). *Acta Crystallogr.* **B25,** 781.

Gardner, R. A., Vlasse, M., and Wold, A. (1969b). *Inorg. Chem.* **8**, 2784.
Geller, S. (1956). *J. Chem. Phys.* **24**, 1236.
Geller, S. (1957). *Acta Crystallogr.* **10**, 243.
Geller, S., Espinosa, G. P., and Crandall, P. B. (1970). *J. Amer. Ceram. Soc.* **53**, 352.
Gier, T. E., Pease, D. C., Sleight, A. W., and Bither, T. A. (1968). *Inorg. Chem.* **7**, 1646.
Goldschmidt, V. M. (1926). *Akad. Oslo I. Mat. -Natur.* No. 2, 7–117.
Goodenough, J. B. (1965). *Bull. Soc. Chim. Fr.* **1965** (4), 1200.
Goodenough, J. B. (1966). *J. Appl. Phys.* **37**, 1415.
Goodenough, J. B. (1967). *Czech. J. Phys.* **B17**, 304.
Goodenough, J. B. (1971). Metallic Oxides. *In* "Progress in Solid State Chemistry," (H. Reiss, ed.), Vol. V, pp. 145–399. Pergamon Press, Oxford.
Goodenough, J. B., and Longo, J. M. (1970). "Landolt-Bornstein, Tabellen, Neue Serie" Vol. III/4a, pp. 126–314. Springer-Verlag, Berlin.
Goodenough, J. B., Longo, J. M., Kafalas, J. A. (1968). *Mater. Res. Bull.* **3**, 471.
Gschneidner, K. A., and Pearson, W. B. (1968). *Mater. Res. Bull.* **3**, 951.
Gschneidner, K. A., and Valletta, R. M. (1968). *Acta. Met.* **16**, 477.
Gushee, B. E., Katz, L., and Ward, R. (1957). *J. Amer. Chem. Soc.* **79**, 5601.
Hahn, H., and Mutschke, W. (1956). *Z. Anorg. Allg. Chem.* **228**, 269.
Hardy, A. (1961). *Bull. Soc. Chim. Fr.* **1961**, 1329.
Hardy, A. (1962). *Acta Crystallogr.* **15**, 179.
Hayashi, H., Nakayama, K., Hasegawa, K., Mizukusa, M., Mizura, M., Ogiso, S., and Torii, O. (1965). *Nagoya Kogyo Gijutsu Shikensho Hokuku* **14**, 384.
Hong, H. Y-P. (1972). Private communication.
Hubbard, J. (1964). *Proc. Roy. Soc. London Ser. A* **276**, 238.
Jayaraman, A. (1965). *Phys. Rev.* **139**, 690.
Jayaraman, A., and Sherwood, R. C. (1964). *Phys. Rev.* **134**, 691.
Kabalkina, S. S., Vereschagin, L. F., and Lityagina, L. M. (1969). *Fiz. Tverd. Tela* **11**, 1040 (*English Transl. Sov. Phys—Solid State* **11**, 847).
Kafalas, J. A. (1971). *NBS Materials Research Symp., Solid State Chem., 5th* (R. S. Roth, ed.), to be published.
Kafalas, J. A., and Longo, J. M. (1968). *Mater. Res. Bull.* **3**, 501.
Kafalas, J. A., and Longo, J. M. (1970a). *Mater. Res. Bull.* **5**, 193.
Kafalas, J. A., and Longo, J. M. (1970b). Unpublished research.
Kafalas, J. A., Gatos, H. C., Lavine, M. C., and Banus, M. D. (1962). *J. Phys. Chem. Solids* **23**, 1541.
Kafalas, J. A., Longo, J. M. and Batson, D. A. (1970). Solid State Res. Rep. MIT (1970:2), Lincoln Lab. Massachusetts Institute of Technology, Lexington, Massachusetts.
Kanamaru, F., Miyamoto, H., Mimura, Y., Koizumi, M., Shimada, M., Kume, S., and Shin, S. (1970). *Mater. Res. Bull.* **5**, 257.
Katz, L., and Ward, R. (1964). *Inorg. Chem.* **3**, 205.
Keith, M. L., and Roy, R. (1954). *Amer. Mineral.* **39**, 1.
Kestigian, M., Croft, W. J., and Leipziger, F. D. (1967). *J. Chem. Eng. Data* **12**, 97.
Kleber, W., and Wilke, K.-Th. (1969). *Krist. Tech.* **4**, 165.
Klement, W., Jr., and Jayaraman, A. (1967). *In* "Progress in Solid State Chemistry" (H. Reiss, ed.), Vol. III, pp. 289–376. Pergamon Press, Oxford.
Kohn, K., Fukuda, R., and Iida, S. (1967). *J. Phys. Soc. Japan* **22**, 333.
Liebertz, J., and Rooymans, C. J. M. (1965). *Z. Phys. Chem.* (*Frankfurt am Main*) **44**, 242.

Liebertz, J., and Rooymans, C. J. M. (1967). *Solid State Commun.* **5**, 405.
Longo, J. M., and Sleight, A. W. (1968). *Inorg. Chem.* **7**, 10.
Longo, J. M., and Kafalas, J. A. (1968). *Mater. Res. Bull.* **3**, 687.
Longo, J. M., and Kafalas, J. A. (1969a). *J. Appl. Phys.* **40**, 1601.
Longo, J. M., and Kafalas, J. A. (1969b). *J. Solid State Chem.* **1**, 103. Copyright by Academic Press, New York.
Longo, J. M., and Kafalas, J. A. (1969c). 22th IUPAC Meeting, Sydney, Australia, 29 Aug. 1969.
Longo, J. M., Raccah, P. M., and Goodenough, J. B. (1969). *Mater. Res. Bull.* **4**, 191.
Longo, J. M., Kafalas, J. A., and Arnott, R. J. (1971). *J. Solid State Chem.* **3**, 174.
Ludekens, W. L. W., and Welch, A. J. E. (1952). *Acta Crystallogr.* **5**, 841.
MacChesney, J. B., Sherwood, R. C., and Potter, J. F. (1965a). *J. Chem. Phys.* **43**, 1907.
MacChesney, J. B., Potter, J. F., Sherwood, R. C., and Williams, H. J. (1965b). *J. Chem. Phys.* **43**, 3317.
MacChesney, J. B., Potter, J. F., Sherwood, R. C., and Williams, H. J. (1967). *Phys. Rev.* **164**, 779.
Marezio, M. (1969). *Trans. Amer. Crystallogr. Assoc.* **5**, 29.
Marezio, M., Remeika, J. P., and Jayaraman, A. (1966a). *J. Chem. Phys.* **45**, 1821.
Marezio, M., Remeika, J. P., and Dernier, P. D. (1966b). *Mater. Res. Bull.* **1**, 247.
Marezio, M., Remeika, J. P., and Dernier, P. D. (1968). *Inorg. Chem.* **7**, 1337.
Marezio, M., Dernier, P. D., Chenavas, J., and Jonbert, J. C. (1971). *NBS Materials Research Symp., Solid State Chem., 5th* (R. S. Roth, ed.), to be published.
Matsuo, Y. (1970). *Yogyo Kyokai Shi* **78**, 46.
Matsuo, Y., Sasaki, H., Hayakawa, S., Kanamaru, F., and Koizumi, M. (1969). *J. Amer. Ceram. Soc.* **52**, 516.
McQueen, R. G., Marsh, S. P., and Fritz, J. N. (1967). *J. Geophys. Res.* **72**, 4999.
Megaw, H. D. (1946). *Trans. Faraday Soc.* **A189**, 261.
Menyuk, N., Kafalas, J. A., Dwight, K., and Goodenough, J. B. (1969). *J. Appl. Phys.* **40**, 1324.
Michel, C., Moreau, J. M., Achenbach, G. D., Gerson, R., and James, W. J. (1969). *Solid State Commun.* **7**, 701.
Michel, C., Moreau, J. M., and James, W. J. (1971). *Acta Crystallogr.* **B27**, 501.
Moreau, J. M., Michel, C., Gerson, R. and James, W. J. (1970). *Acta Crystallogr.* **B26**, 1425.
Morgenstern-Badarau, I., Poix, P., and Michel, A. (1963). *C. R. Acad. Sci. Paris Ser. C* **256**, 692.
Morgenstern-Badarau, I., Poix, P. and Michel A. (1964). *C. R. Acad. Sci. Paris Ser. C* **258**, 3036.
Mori, S. (1965). *J. Amer. Ceram. Soc.* **48**, 165.
Morimoto, N., and Akimoto, S. (1969). *Kagaku* **39**, 131.
Negas, T., and Roth, R. S. (1969). *J. Res. Nat. Bur. Snd.* **73A**, 425.
Panyushkin, V. N., DePasquali, G., and Drickamer, H. G. (1969). *J. Chem. Phys.* **51**, 3305.
Pearson, W. B. (1967). *J. Less-Common Metals* **13**, 626.
Pierce, J. W., McKinzie, H. L., Vlasse, M., and Wold, A. (1970). *J. Solid State Chem.* **1**, 330.
Pistorius, C. W. F. T. (1966). *J. Chem. Phys.* **44**, 4532.
Popova, S. V. (1964). *Fiz. Mat. Mekh. Tr. Mosk. Konf. Molodyykh Uch 1st* p. 58, (Pub. 1968).

Post, B., and Banks, E. (1968). Brooklyn Polytech. Inst., U. S. Army Res. Office, Contract No. DA-31-124-ARO(D)-70.
Prewitt, C. T., and Shannon, R. D. (1969). *Trans. Amer. Crystallogr. Assoc.* **5,** 51.
Prewitt, C. T., and Sleight, A. W. (1969). *Science* **163,** 386.
Randall, J. J., and Ward, R. (1959). *J. Amer. Chem. Soc.* **81,** 2629.
Range, K. J., Becker, W., and Weiss, A. (1968a). *Z. Naturforsch.* **23b,** 1009.
Range, K. J., Becker, W., and Weiss, A. (1968b). *Z. Naturforsch.* **23b,** 1261.
Range, K. J., Engert, G., Engels, J., and Weiss, A. (1968c). *Z. Naturforsch.* **23b,** 1008.
Range, K. J., Engels, J., and Weiss, A. (1968d). *Z. Naturforsch.* **23b,** 1262.
Range, K. J., Lindenberg, B., Keubler, M., Leeb, R., and Weiss, A. (1969a). *Z. Naturforsch.* **24b,** 1651.
Range, K. J., Keubler, M., and Weiss, A. (1969b). *Z. Naturforsch.* **24b,** 1653.
Range, K. L., Becker, W., and Weiss, A. (1969c). *Z. Naturforsch.* **24b,** 1654.
Range, K. J., Engert, G., and Weiss, A. (1969d). *Solid State Commun.* **7,** 1749.
Range, K. J., Keubler, M., and Weiss, A. (1969e). *Z. Naturforsch.* **24b,** 1060.
Range, K. J., Engert, G., and Weiss, A. (1969f). *Z. Naturforsch.* **24b,** 1061.
Range, K. J., Poppinger, D., and Weiss, A. (1969g). *Z. Naturforsch.* **24b,** 1063.
Range, K. J., Becker, W., and Weiss, A. (1969h). *Z. Naturforsch.* **24b,** 811.
Range, K. J., Engert, G., and Weiss, A. (1969i). *Z. Naturforsch.* **24b,** 813.
Raveau, B. (1967). *Rev. Chim. Minérale* **4,** 729.
Rea, J. R., Bless, P. W., and Kostiner, E. (1972) *J. Solid State Chem.* (in press).
Reid, A. F., and Ringwood, A. E. (1969). *J. Geophys. Res.* **74,** 3238.
Reid, A. F., and Ringwood, A. E. (1970). *J. Solid State Chem.* **1,** 557.
Remeika, J. P., and Marezio, M. (1966). *Appl. Phys. Lett.* **8,** 87.
Ringwood, A. E., and Major, A. (1967a). *Nature* **215,** 1367.
Ringwood, A. E., and Major, A. (1967b). *Earth Planet. Sci. Lett.* **2,** 106.
Ringwood, A. E., and Major, A. (1968). *Phys. Earth Planet. Interiors* **1,** 164.
Ringwood, A. E., and Reid, A. F. (1968). *Earth Planet. Sci. Lett.* **5,** 67.
Ringwood, A. E., and Reid, A. F. (1969). *Earth Planet. Sci. Lett.* **5,** 245.
Ringwood, A. E., and Seabrook, M. (1962a). *Nature* **196,** 883.
Ringwood, A. E., and Seabrook, M. (1962b). *J. Geophys. Res.* **67,** 1690.
Ringwood, A. E., and Seabrook, M. (1963). *J. Geophys. Res.* **68,** 4601.
Rodi, F. (1963). Thesis: Eberhatd-Karls-Universität, Tubingen.
Rodi, F., and Babel, D. (1965). *Z. Anorg. Allg. Chem.* **366,** 17.
Rooymans, C. J. M. (1967). Doctoral Thesis, Univ. of Amsterdam, Amsterdam, The Netherlands.
Rooymans, C. J. M. (1969). *In* "Advances in High Pressure Research," Vol. 2, p. 1. Academic Press, New York.
Roth, W. L., and DeVries, R. C. (1967). *J. Appl. Phys.* **38,** 951.
Rudorff, W., Walter, G., and Becker, H. (1956). *Z. Anorg. Allg. Chem.* **285,** 287.
Schneider, S. J. (1961). *J. Res. Nat. Bur. Std.* **65A,** 429.
Schneider, S. J., Roth, R. S., and Waring, J. L. (1961). *J. Res. Nat. Bur. Std.* **65A,** 345.
Seifert, H. J., and Koknat, F. W. (1965). *Z. Anorg. Allgem. Chem.* **341,** 269.
Seifert, K. F. (1967). *Fortschr. Mineral.* **45,** 214.
Shannon, R. D. (1967). *Inorg. Chem.* **6,** 1474.
Shannon, R. D. (1970). *Acta Crystallogr. Sec., B.,* **26,** 447.
Shannon, R. D., and Prewitt, C. T. (1968). *J. Inorg. Nucl. Chem.* **30,** 1389.
Shannon, R. D., and Prewitt, C. T. (1969). *Acta Crystallogr. Sec., B.* **25,** 925.
Shannon, R. D., and Prewitt, C. T. (1970). *J. Solid State Chem.* **2,** 134.
Shannon, R. D., and Sleight, A. W. (1968). *Inorg. Chem.* **7,** 1649.

Shimada, M., Kume, S., and Koizumi, M. (1968). *J. Amer. Ceram. Soc.* **51,** 713.
Siegel, S., and Gebert, E. (1964). *Acta Crystallogr.* **17,** 790.
Sleight, A. W. (1969a). *Mater. Res. Bull.* **4,** 377.
Sleight, A. W. (1969b). *Inorg. Chem.* **8,** 1764.
Sleight, A. W., and Bither, T. A. (1969). *Inorg. Chem.* **8,** 566.
Sleight, A. W., and Gillson, J. L. (1966). *Solid State Commun.* **4,** 601.
Sleight, A. W., and Prewitt, C. T. (1970). *Mater. Res. Bull.* **5,** 207.
Sleight, A. W., Bither, T. A., and Bierstedt, P. E. (1969). *Solid State Commun.* **7,** 299.
Sokhor, M. I., Kondakov, V. G., and Fel'dgun, L. I. (1967). *Dokl. Akad. Nauk SSSR* **175,** 826 (*English Transl.:* (1968). *Sov. Phys.—Dok.* **12,** 749).
Spiegelberg, P. (1940). *Arkiv för Kemi, Min. Och Geol.* **14A**(5), 1.
Steinfink, H., and Brunton, G. (1969). *Inorg. Chem.* **8,** 1665.
Sugawara, F., Iida, S., Syono, Y., and Akimoto, S. (1965). *J. Phys. Soc. Japan* **20,** 1529.
Sugawara, F., Syono, F., and Akimoto, S. (1968a). *Mater. Res. Bull.* **3,** 529.
Sugawara, F., Iida, S., Syono, Y., and Akimoto, S. (1968b). *J. Phys. Soc. Japan* **25,** 1553.
Swoboda, T. J., Toole, R. C., and Vaughan, J. D. (1958). *J. Phys. Chem. Solids* **5,** 293.
Swoboda, T.J., Arthur, P., Cox, N. L., Ingraham, J. N., Oppegard, A. L., and Sadler, M. S. (1961). *J. Appl. Phys.* **32,** 3745.
Syono, Y. (1969). *Bussei* **10,** 366.
Syono, Y., Akimoto, S., and Kohn, K. (1969a). *J. Phys. Soc. Japan* **26,** 993.
Syono, Y., Sawamoto, H., and Akimoto, S. (1969b). *Solid State Commun.* **7,** 713.
Syono, Y., Akimoto, S., Ishikawa, Y., and Endoh, Y. (1969c). *J. Phys. Chem. Solids* **30,** 1665.
Syono, Y., Akimoto, S., and Endoh, Y. (1970). Tech. Rep. Inst. Solid State Phys. Ser. A, No. 401, Feb. 1970.
Tomashpol'skii, Yu. Ya., Zubova, E. V., Burdina, K. P., and Venevtsev, Yu. N. (1967). *Isv. Akad. Nauk SSSR Neorg. Mater.* **3,** 2132 (*English Transl.: Inorg. Mater.* (*USSR*) **3,** 1861).
Tomashpol'skii, Yu. Ya., Zubova, E. V., Burdina, K. P. and Venevtsev, Yu., N. (1968). *Kristallographiya* **13,** 987 (*English Transl.: Sov. Phys.—Crystallogr* **13,** 859.
Van Veen, J. A. R. (1971). *Mater. Res. Bull.* **6,** 1269.
Van Vucht, J. H. N., and Buschow, K. H. J. (1965). *J. Less-Common Metals* **10,** 98.
Verma, A. R., and Krishna, P. (1966). *In* "Polymorphism and Polytypism in Crystals." Wiley, New York.
Wadsley, A. D. (1964). *In* "Non-Stoichiometric Compounds" (L. Mandelcorn, ed.). Chapter 3, Academic Press, New York.
Waintal, A., and Chenavas, J. (1967a). *C. R. Acad. Sci. Paris* **264,** 168.
Waintal, A., and Chenavas, J. (1967b). *Mater. Res. Bull.* **2,** 819.
Waintal, A., Capponi, J. J., Bertaut, E. F., Contré, M. and Francois, D. (1966). *Solid State Commun.* **4,** 125.
Weiher, J. F., Chamberland, B. L., and Gillson, J. L. (1971). *J. Solid State Chem.* **3,** 529.
Wilhelmi, K. A. (1967). *Ark. Kemi* **26,** 157.
Wilhelmi, K., Johnson, O., and Lagervall, E. (1969). *Acta Chem. Scand.* **23,** 1074.
Yakel, H., Koehler, W. C., Bertaut, E. F., and Forrat, F. (1963). *Acta Crystallogr.* **16,** 957.

High-Pressure Techniques in Preparative Chemistry

C. J. M. ROOYMANS

Philips Research Laboratories, N. V. Philips' Gloeilampenfabrieken
Eindhoven, The Netherlands

I. Introduction

Application of high pressures in preparative chemistry has been known for many years; the oldest example, and one of great industrial importance, may well be the synthesis of ammonia by Haber in 1913.†

In the last decade, the use of high and very high pressures for the synthesis and crystal growth of materials, mostly inorganic, has increased markedly. This has not been confined to solid state processes only, where the synthesis of diamond has acted as a catalyst for many investigations on polymorphism in inorganic compounds; the use of liquids and gases under high-pressure conditions has also received more attention, owing to the availability of new materials and techniques for operating under these conditions, and driven by the need for specific high-grade materials which could not be produced by conventional methods. This chapter presents a survey of the state of the art, emphasizing the more recent progress, since there is already a vast bibliography on what are now classic high-pressure techniques. The main books on preparative aspects are listed in the bibliography.

Examples are given of the application of the methods, including crystal-growth processes, the synthesis of new polymorphs, and morphologies

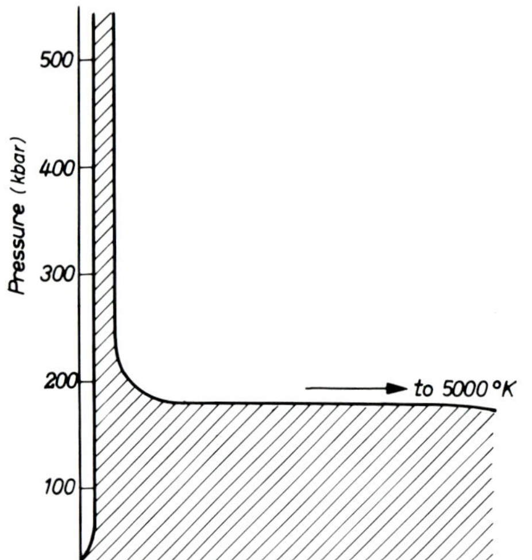

FIG. 1. The pressure–temperature range now available for chemical synthesis experiments under static conditions.

† See Kirk and Othmer (1956), Vol. I, p. 782.

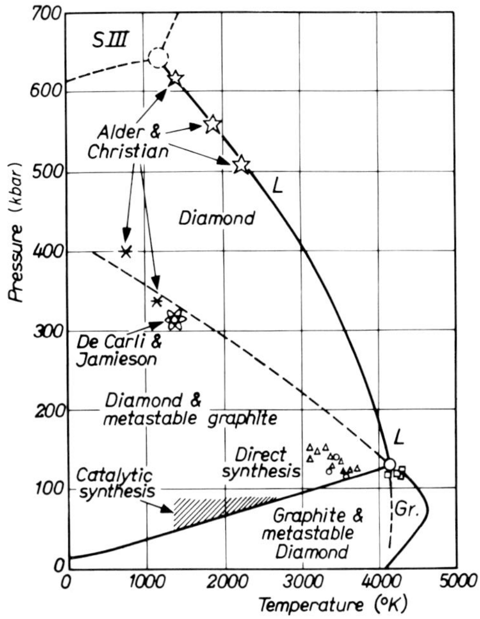

FIG. 2. The pressure–temperature diagram of the various carbon modifications. After Bundy (1963).

other than under ambient conditions. Industrial applications such as the synthesis of ammonia and polyethylene will not be discussed.

The available pressure–temperature range for static experiments is roughly outlined in Fig. 1. Pressures of up to 600 kbar† have been reached at room temperature, while at temperatures up to and above 3000°K, pressures of 2–300 kbar are possible for fairly short periods of time. Under dynamic conditions, e.g., shock waves, still higher values of temperature and pressure can be attained. Thus a combined use of static and dynamic methods can result in a survey of the equation of state of the material involved over an extensive P–T range. An example is given in Fig. 2, where the polymorphic behavior of carbon is shown for pressures up to 700 kbar and temperatures up to 4500°K. In the top left part of the figure, a so far only presumed new phase of carbon has been indicated, possibly with the white-tin structure.

In Sections II–IV attention will be given to the use of high pressures in gaseous, liquid, and solid state systems, respectively. In Section IV the possibilities for studying high-pressure phases that cannot be quenched to ambient conditions will be included. The primary technique considered is x-ray diffraction under high-pressure conditions, although other possi-

† 1 kbar (kilobar) = 1000 bars; one bar = 0.98 atm.

TABLE I

	Gaseous systems	Liquid systems	Solid state systems
Apparatus	Autoclaves	Autoclaves	Opposed-anvils, piston–cylinder, belt, girdle, multilateral devices
Pressure system	Intensifier or decomposing compounds	Intensifier or degree of fill	Direct load over area
Heating system	External or internal	External	External or internal
P–T range:			Anvils:
external	2500°C, 2.5 kbar	750°C, 5 kbar (cold-seal)	external: 700°C, 150 kbar
internal	1600°C, 4.5 kbar	500°C, 4 kbar (mod. Bridgman)	internal: 1000°C 150 kbar
			Superbelt: 3000°C, 200 kbar
Volume	5000 cm^3	Up to 150 liters (quartz growth)	Up to 500 cm^3
Pressure-transmitting medium	Inert gas (A or N_2) or participant; hydrostatic	Mostly water, hydrostatic	Pyrophyllite, talc, AgCl, quasi-hydrostatic
Examples	III–V compounds, transition-metal oxides	α-Quartz, see also Table III	Diamond, polymorphism, see Table V

bilities include resistance measurements, differential thermal analysis, and accurate density determination by a piston-displacement method. The problems of calibrating apparatus and measuring the exact pressure and temperature conditions will likewise be discussed. The classification used is certainly not unambiguous. In the supercritical region, there is no distinction between liquid and gas. In many very high-pressure reactions to be discussed in the section on solid state systems, liquid phases, such as molten metals, may well play an important role.

The main differences between the three types of systems dealt with here are outlined in Table I, the numbers given being merely rough indications of what has been attained in specific areas. The table also constitutes a concise survey of the subject of this chapter.

High-pressure technologies such as hydrostatic extrusion, high-pressure powder compaction, etc. are beyond the scope of this chapter. Interested readers are referred to the recent book, "Mechanical behaviour of materials under pressure" (Pugh, 1970), in which these aspects are dealt with in detail and with great expertise.

Shockwaves will be dealt with in Section V, while finally, some attention is given to safety aspects in Section VI.

II. Systems under High-Pressure Gas Conditions

With regard to work in this field, the long period of dormancy in which only a few devoted enthusiasts were engaged in exploratory studies has passed, and recently it has become a focus of attention. This has been mainly due to the industrial need for GaP crystals which, like many other II–VI and III–V semiconductor materials, cannot be grown directly from the melt under 1-atm conditions.

However, high-pressure techniques are essential not only in crystal growth, but also in materials synthesis. Since these two disciplines are closely intermeshed, they will not be described separately. Recent developments in apparatus and work on II–IV and III–V compounds, oxides, and other compounds will be discussed in that order.†

A. New Developments in Apparatus

As stated above, the possibilities and limitations of the autoclaves needed for these studies have been extensively described. Interested readers should consult the book by Comings (1956) or the annotated bibliography by Zeitlin (1964).

† See note added in proof on p. 132.

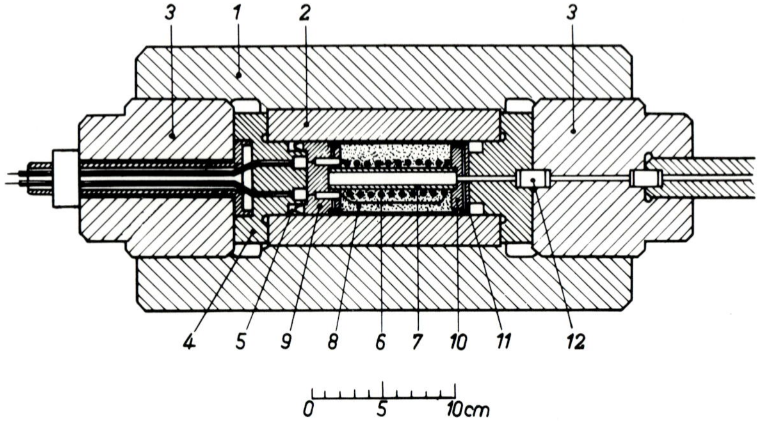

FIG. 3. Cross section of the gas autoclave designed by Wilhelmi (1968). (1) External steel cylinder, (2) internal steel cylinder, (3) screw, (4) gasket plunger, (5) U-cap of hydrofit, (6) Mo wire, (7) Morgan tube, (8) bubbled alumina, (9) brass tube with asbestos contact box, (10) furnace heat (asbestos), (11) gasket plunger cover, (12) high-pressure O-ring seal.

High-temperature autoclaves for high gas pressures have been developed by Boyer *et al.* (1966) of the Battelle Memorial Institute. Understanding of gas heat flow problems has permitted satisfactory operation of such autoclaves at pressures up to 2.5 kbar combined with temperatures up to 2500°C.

Recent developments include an internally heated autoclave for work up to 4.5 kbar and 1600°C by Wilhelmi (1968), shown in Fig. 3. This autoclave, although built mainly for preparative work, can easily be made adaptable to various *in situ* investigations of physical properties. The autoclave consists essentially of two horizontal toughened-steel cylinders, one of which is shrunk onto the other. Internal heating is by means of a molybdenum wiring, the high-pressure argon gas ensuring an excellent heat transfer. The external steel cylinder has a cooling jacket (not shown in Fig. 3).

Cold-seal pressure vessels can now operate up to 1100°C and 3 kbar. External heating is used, leading to more stringent requirements on the construction materials. The combined use of a TZM alloy (molybdenum with 0.5% titanium and 0.08% zirconium) for the inner vessel inside a Nimonic-75† rod, drilled out to give a good fit onto the vessel, as shown

† Nimonic is a trademark of Henri Wiggin and Company, Ltd., Birmingham (England).

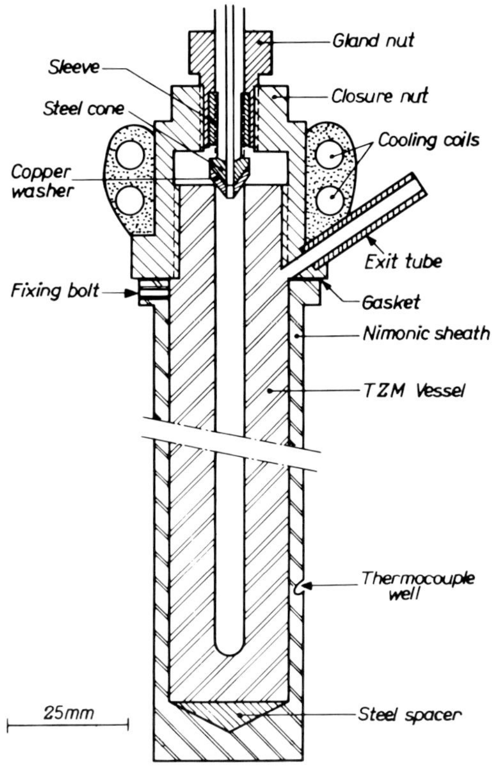

FIG. 4. Sectional view of TZM vessel, sheath, and closure. After Williams (1968).

in Fig. 4, was developed by Williams (1968). Internally heated pressure vessels, which are more complicated and expensive, are therefore no longer required for work below about 3 kbar in the temperature range up to 1100°C. Earlier cold-seal pressure vessels, used mainly for hydrothermal experiments, were limited to 1 kbar and 1200°C (Luth and Tuttle, 1963; Williams, 1966). (See also Section III.)

The necessary pressures can be generated by compressors, which is no problem, as pumping assemblies for pressures up to 20 kbar are commercially available.†

Pressure control and, in particular, temperature control present greater difficulties. As the density of the gas increases, convection currents play a crucial role in heat transport, leading to a marked shift in heat distribu-

† See, for instance, Cerveny (1963) or the pamphlets available from Kobe Steel, Ltd., Kobe, Japan, among others.

tion under high-pressure conditions as compared to atmospheric conditions. Provision has to be made for the insertion of thermocouples. In the case of an internal heating arrangement, the choice of a good insulating material is of great importance if excessive heating of the wall or autoclave heads is to be avoided. An insulating material which impedes the convection currents, e.g., bubbled alumina, must be used.

As stated previously, the combined use of high pressures and high temperatures in industrial processes such as the synthesis of polyethylene or ammonia is outside the scope of this chapter. These processes generally involve somewhat lower pressure and temperature ranges, although it is clear that industrial needs have greatly contributed to the solution of difficulties relating to construction of autoclaves and scaling-up. The state of the art in industrial chemistry has been reviewed best in the Encyclopedia of Chemical Technology (Kirk and Othmer, 1956), by Shreve, (1956), and, for the high-pressure aspects, in the book by Weale (1967).

B. II–VI and III–V Compounds

The majority of the II–VI and III–V compounds, such as ZnS (mp 1830°C), CdS (mp 1475°C), and GaP (mp 1470°C), cannot be grown from the melt under 1-atm conditions, because they will decompose into their volatile components long before the melting temperature has been attained. This makes other methods of material preparation and crystal growth necessary. We will now show how melt growth can still be accomplished. The compound materials are mostly commercially available in powder form, but even the very dense powders differ too much in specific weight from the molten ingot; the crucible space would thus be used very ineffectively. Fischer (1970) has described a method based on the preheating and melting of the elements in a small autoclave pressurized at the melting point of the compound to about 400 atm of argon. This is a method especially suitable for the II–VI compounds. Dense polycrystalline ingots of III–V compounds, notably GaP, can be made from a gallium-enriched melt at lower temperature and pressure (Bass and Oliver, 1968). A different version of that apparatus, shown in Fig. 5, was described by Fischer (1970). This latter reference also gives a survey of the state of the art for the preparation of single crystals by melt growth of II–VI and III–V compounds under pressure.

Crystal growth methods include *Bridgman* growth under high inert gas pressure using resistance or rf heating. Pressures of over 100 atm of argon or nitrogen are used for ZnS and CdS. This method does not work for III–V compounds, where there is a large difference in volatility between

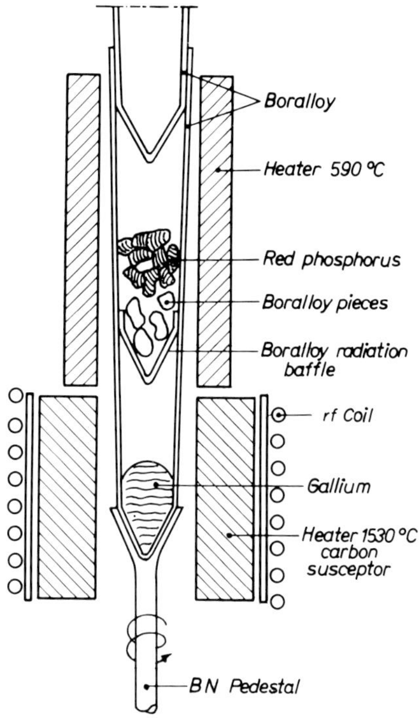

FIG. 5. GaP synthesis and crystal growth in a demountable pyrolytic BN ampoule. After Fischer (1970).

the cation and anion. At about 40 atm of P pressure in Boralloy† pyrolytic boron nitride crucibles, GaP can be grown by a vertical Bridgman method. GaN and InN, however, decompose even below 200 atm of nitrogen pressure. (This gas, because of the stability of the N_2 molecule, behaves as an inert one.)

Although it is undoubtedly the best crystal growth method, *Czochralski* pulling using an oriented seed cannot always be used directly because of the lack of visibility through the mandatory viewing windows, since some of the anion vapors are opaque. Another problem is that of force transfer to the sealed quartz tube and the danger of implosions and explosions. However, the procedure of *liquid encapsulation*, i.e., placing a liquid blanket of molten boric oxide on top of the melt and applying inert gas pressure exceeding the decomposition pressure, gives excellent new possibilities for crystal pulling under pressure through the vitreous viscous blanket. This method was first introduced by Metz *et al.* (1962) and successfully applied by Mullin *et al.* (1965) for GaAs and InAs, and by Bass

† Boralloy is a trademark of Union Carbide Corporation, N.Y.

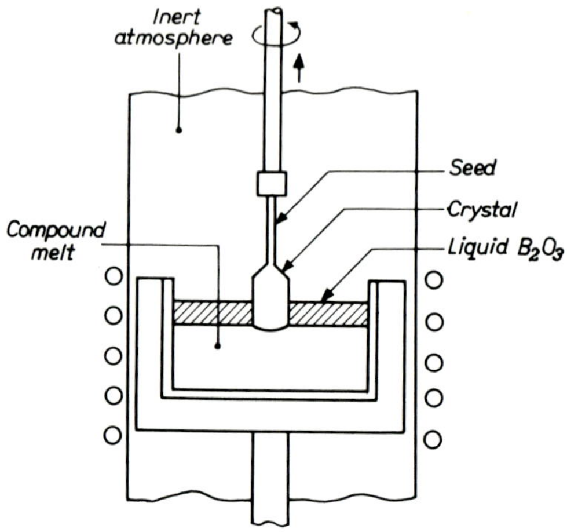

FIG. 6. Liquid encapsulation growth under pressure.

and Oliver (1968) and Mullin *et al.* (1968) for GaP. The evaporation of B_2O_3 is very slow up to 1500°C, making crystal growth runs of many hours quite feasible. The blanket is also beneficial in damping oscillations of the melt meniscus and reduces heat losses from the surface of the melt. At present, boric oxide seems to be the only acceptable encapsulant that gives rise to contamination in the ppm range only. This material might, however, not be suitable for II–VI compounds, due to the miscibility of molten B_2O_3 with II–VI melts (Meiling and Leombruno, 1968). Here, new encapsulants have to be found. A schematic arrangement suitable for liquid encapsulation growth is shown in Fig. 6. A detailed description has recently been given by Nygren *et al.* (1971). Recent NMR measurements have shown, however, that the stoichiometry of GaP crystals grown in this way may vary gradually, the top being less stoichiometric and containing more lattice defects, due to depletion of phosphorus (Peterson *et al.*, 1971).

C. Oxides

The combination of the high temperatures necessary for melting refractory oxides and high oxygen pressures is possibly one of the most dangerous encountered, and special precautions are necessary, e.g., a special explosion-protected building and remote control. Nevertheless, this combination is very useful, because many oxides, particularly those of the transition metals, lose oxygen at the elevated temperatures where compound forma-

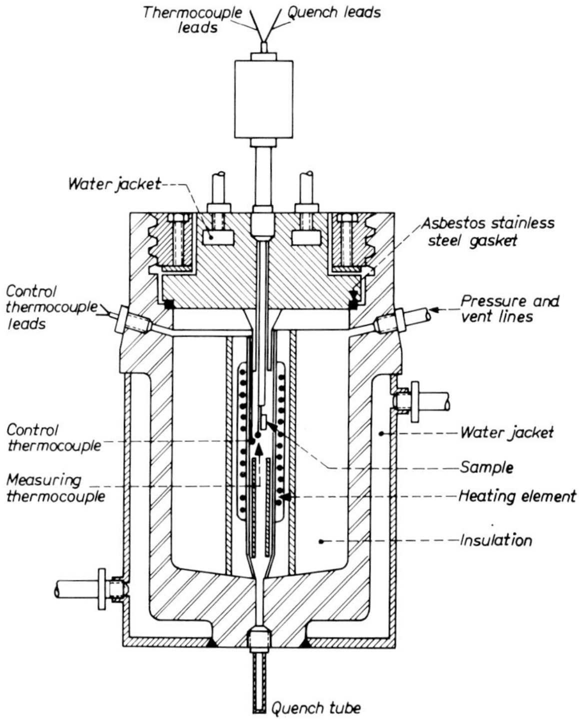

FIG. 7. Pressure vessel for high-temperature oxide studies. After Van Hook (1965).

tion, sintering, and crystal growth take place. An apparatus for phase-equilibrium studies of oxides at temperatures up to 1700°C, combined with oxygen pressures of up to 100 atm, has been constructed by Van Hook (1965). The high temperatures necessarily lead to internal heating. Figure 7 shows the Van Hook apparatus which makes use of a well-protected platinum-wound furnace, in contrast with the design of Ferretti *et al.* (1961) which uses induction heating, and is suitable for crystal-growth experiments up to 75 atm of oxygen pressure. Pressure furnaces of the Van Hook type have been used to study phase equilibria in systems such as Y_2O_3–Fe_2O_3 and BaO–Fe_2O_3, including crystal growth of $BaFe_{12}O_{19}$ (Van Hook, 1963, 1964).†

Investigations at higher pressures and lower temperatures have been carried out using the externally heated pressure vessels discussed above. Pressures up to 5 kbar and temperatures up to 750°C have been maintained for periods ranging from hours up to days. The tubing and valves tend to rust out somewhat faster than with water, but thorough drying and elim-

† See note added in proof on p. 132.

ination of all oil gives a simple, relatively safe system (Roy, 1965). These latter studies include crystal chemical and phase-equilibrium studies in the binary systems Pb–O and Cr–O (Roy, 1965) and the ternary systems Fe–Cr–O (Roy, 1965), Mg–Cr–O and Cd–Cr–O (Muller *et al.*, 1968).

Instead of using the oxygen gas itself for controlling the oxygen pressure by means of commercially available 150-atm oxygen cylinders and pressure intensifiers, the decomposition pressure of certain oxides can also be successfully applied. The decomposition of CrO_3 or MnO_2 is a convenient source, since, as described by Reed (1970), the decomposition pressures of these compounds at 700°C are 10^4–10^6 atm. This author has surveyed the role of the oxygen pressure in the control and measurement of composition in 3d metal oxides. The decomposition of PtO_2, taking place at temperatures from 400 to 500°C, has been applied in mineral studies by Huckenholz and Yoder (1970).

In experimental mineralogy, other contributions have been made which are of great importance for the preparation of inorganic compounds in which elements with variable valence are present. The fact that these contributions have been published mainly in geochemical papers has possibly prevented their widespread use.

Eugster (1957) described the principles of a method which makes it possible to have an accurately controlled oxygen atmosphere at high temperature and high total pressure. For normal atmospheric-pressure systems, the partial pressure of oxygen (P_{O_2}) is conveniently controlled by mixtures of $CO + CO_2$, as employed by Darken and Gurry (1945), or of $H_2 + CO_2$, as used by Muan (1955). Such gases cannot be employed at high total pressures. Eugster's method, which is also applicable to other gases, is based on the use of one or more solid phases, which together function as an "oxygen buffer." In principle, any solid-phase equilibrium which, at a given temperature, depends on a given fugacity of oxygen (f_{O_2}) is capable of functioning as an oxygen buffer, providing that, at that temperature, these phases are contained in the presence of water. Examples of such oxygen buffers are

$$\tfrac{3}{2}Fe + O_2 \rightleftharpoons \tfrac{1}{2}Fe_3O_4$$

$$3Fe_2SiO_4 + O_2 \rightleftharpoons 2Fe_3O_4 + 3SiO_2$$

$$4Fe_3O_4 + O_2 \rightleftharpoons 6Fe_2O_3$$

As shown in Fig. 8, derived from the work of Eugster and Wones (1962), an osmotic system can be employed in this way with platinum acting as a semipermeable membrane and gold as an outer impermeable protection. Silver tubes have also been used (Turnock, 1960). The effect of osmotic

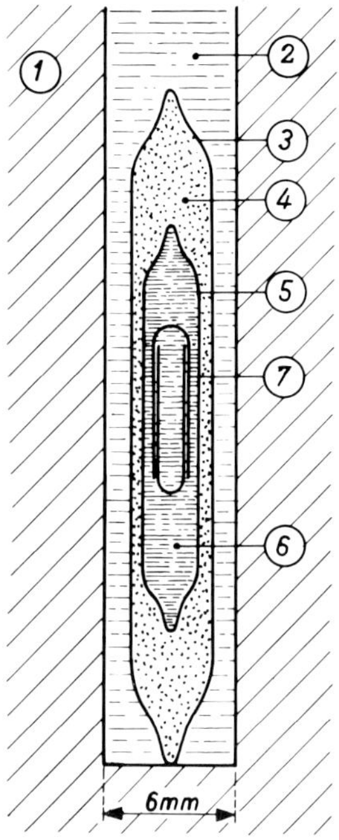

FIG. 8. Cutaway sketch of run assembly after Eugster and Wones (1962). (1) Pressure vessel; (2) pressure medium (H_2O); (3) sealed gold tube; (4) oxygen buffer in H_2O; (5) sealed platinum tube; (6) charge, present in and around (7) the open silver capsule.

diffusion tends to balance the hydrogen of the substance under study with that of the buffer until they become equal. The oxygen fugacities can be determined because of the constant dissociation of water at a given pressure and temperature. In the case of iron compounds, an open silver tube is necessary to prevent the formation of solid solutions between the iron and platinum. Among the results obtained by this technique are the synthesis and determination of the stability fields of many minerals, spinels (Turnock, 1960), Fe–Al oxides (Turnock and Eugster, 1962), and Fe–Ti oxides (Buddington and Lindsley, 1964). Carapezza (1966) suggested a method to obtain a continuously variable control of oxygen fugacity, instead of discontinuous control, namely a limitation of the Eugster method. Another method of obtaining this continuous control with high precision has been described by Shaw (1963). Unfortunately, however, this method requires a complicated modification of the standard hydrothermal techniques described in Section III. The attainable pressures and temperatures are therefore limited by the possibilities of this experimental setup. Eugster and Wones (1962) gave a table in which the relevant thermodynamic properties of various compounds are collected. The oxygen fugacity values obtainable with these buffers have been calculated as a function of temperature.

The oxygen fugacity–temperature relationships of manganese oxide and nickel oxide buffers have recently been determined by Huebner and Sato (1970).

D. Other Compounds

As stated above, the Eugster method of accurately controlling the pressure of a given gas under high temperature and high pressure is not limited to the case of oxygen and hydrogen. The buffer approach can be extended, for instance, by combining the oxygen buffer with the graphite buffer (French and Eugster, 1965). A survey of the experimental possibilities for gases of the C–O–H system was recently published (Eugster and Skippen, 1967). The control of fugacities in fluorine-bearing hydrothermal systems has been studied by Munoz (1968). This is of relevance for many rock-forming minerals, since fluorine easily substitutes for the hydroxyl group in these compounds.

A high-pressure device for gaseous hydrogen and deuterium, suitable for pressures up to 25 kbar in the temperature range from -50 to $+100°C$, has been described by Wisniewski (1970). The apparatus consists essentially of a double piston–cylinder system, in which the inner cylinder is made of an aluminum bronze, a material with a small hydrogen diffusion coefficient and not subject to destructive action by hydrogen or deuterium.

Boksha (1957) designed high-pressure chambers for crystal growth of volatile compounds such as phosphorus (in its high-pressure black modification) or selenium.† The seeded growth of selenium crystals under high argon pressures in excess of 3 kbar has been described by Harrison (1968). Other ways of growing trigonal selenium, of great interest as a nonlinear optical, piezoelectric, photoconductive, and semiconductor material, such as by hydrothermal means, are described in Section III.

III. Systems under High-Pressure Liquid Conditions

The end of the preceding section has already made clear that from an experimental point of view, there is no discontinuous transition between gaseous and liquid systems. It is also true that under the high-pressure and high-temperature conditions dealt with here, we are partly or wholly in the supercritical region where the distinction no longer has any real physical meaning.

In this section, mainly hydrothermal systems used for synthesis of ma-

† See note added in proof on p. 132.

TABLE II

SOLUBILITY (in ppm) OF SOME INORGANIC COMPOUNDS IN WATER VAPOR AT 500°C AND 1030 BAR[a]

Compound	Solubility (ppm)	Compound	Solubility (ppm)
UO_2	0.2	$CaSO_4$	20
Al_2O_3	1.8	$BaSO_4$	40
SnO_2	3.0	$PbSO_4$	110
NiO	20.0	Na_2SO_4	4300
Nb_2O_5	28.0	$CaCO_3$	120
Ta_2O_5	30	Sphalerite (zincblende, ZnS)	204
Fe_2O_3	90		
BeO	120		
SiO_2	2600		
GeO_2	8700		

[a] After Franck, 1961. Density of water under these conditions, 0.54 g cm^{-3}.

terials, crystal growth, or study of reaction kinetics are discussed. Although experiments in alkaline media are most extensively performed, interesting reactions have recently been realized in an acid environment.

As in the preceding section, more recent improvements in methods and materials will be stressed, the new user of these techniques being referred to the "classic" survey articles. The necessary apparatus, the methods, and the materials will then be mentioned, while finally, some investigations in nonaqueous systems will also be discussed. Reviews of hydrothermal chemistry have been published by Laudise and Nielsen (1961), Laudise (1962), and Nørlund Christensen (1969). Hydrothermal growth as compared to other methods of growing single crystals has been described by Ballman and Laudise (1963) and recently by Laudise (1970).

In the hydrothermal technique, water is used as a solvent for many refractory materials which are insoluble at ambient conditions. The solubility of some of these compounds at elevated pressure and temperature is given in Table II. The function of the pressure is to keep a sufficient density of the solvent.

The density is determined by the fraction of free volume of the autoclave initially filled with water at room temperature (the percent fill). The pressure–temperature relationships for pure water at a varying degree of fill are given in Fig. 9. The critical degree of filling is 32%, because the density of H_2O under the critical conditions (373°C and 221 atm) is equal to 0.32. This means that in a 32%-filled autoclave the meniscus does not

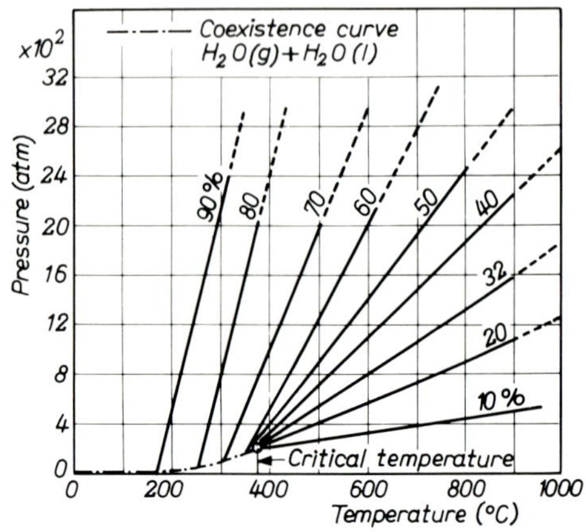

FIG. 9. Pressure–temperature relationships for water at various densities.

move during heating and disappears at 373°C. For lower degrees of filling, the autoclave "boils dry," while for the higher degrees—mostly employed in crystal growth experiments—the autoclaves fill up with liquid before the critical temperature has been reached. The most successful conditions for crystal growth are often present—as we will see later—at densities around 0.65 g cm^{-3}.

The solvent power of pure water is mostly insufficient for efficient crystallization of refractory compounds. It can be considerably enhanced by adding complexing agents to the solution. The latter compounds must be active enough to bring more of the relevant material into solution, but not so strong that the "complexate" becomes the stable solid phase.

In geochemical literature, such components, e.g., the alkali hydroxydes and ammonium chloride, are called "mineralizers." The use of hydrothermal methods for materials synthesis and crystal growth was strongly stimulated by the industrial production of α-quartz single crystals. On the basis of a study of natural growth conditions and the work of Nacken (1950), a temperature-differential method for large-scale production of quartz has been developed, which is now in use at a number of quartz-manufacturing plants all over the world. About one-third of the world's total quartz consumption of about 300,000 pounds annually is of synthetic origin.

A typical setup for hydrothermal crystal growth is given in Fig. 10. The solute to be crystallized (nutrient) is placed at the bottom of a cylindrical autoclave. A perforated metal disk or baffle, which separates the dissolving and growth zones, is inserted and suitably oriented single-

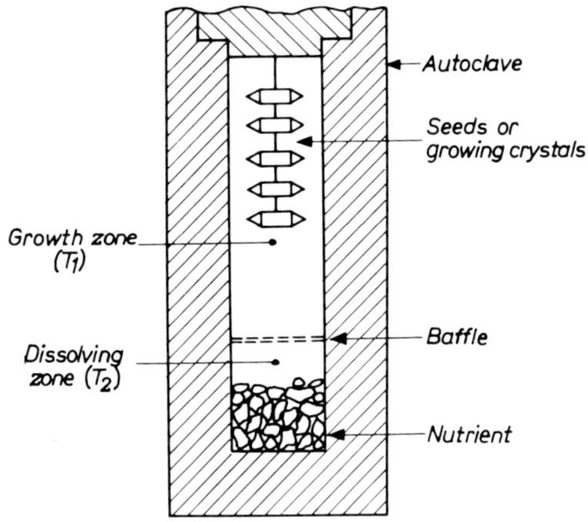

FIG. 10. Schematic setup of the temperature differential method used for hydrothermal crystal growth.

crystal seeds are placed on top. The autoclave is filled up to the desired percent fill with the appropriate solvent and placed in a furnace which heats the dissolving zone to a higher temperature than the growth zone. Because the convection currents occurring in the liquid are mainly responsible for the heat transport, the baffle effectively brings about a temperature difference between the zones.

The application of the hydrothermal method, however, is not restricted to crystal growth; it can also be very useful in reactivity studies, since the presence of the water vapor—sometimes with the mineralizer—often enables reactions to be run at much lower temperatures than those observed under "dry" conditions. In reactivity studies, isothermal systems are also used instead of the differential temperature method.

A. AUTOCLAVE DESIGNS

Most autoclaves used for materials synthesis are of the type shown schematically in Fig. 11 and known as the cold-seal, cone-closure vessel. This was first developed by Tuttle (1949). A bomb based on the same principle, but suited more especially for gaseous systems, was shown in Fig. 4. A vessel of this kind can be made from several alloys which have sufficient strength at the high temperatures to which they are exposed. Runs at 750°C and 5000 bar are quite feasible with certain nickel-based alloys.†

† For more specific information, see for instance the pamphlets by Tempres, Inc., State College, Pennsylvania.

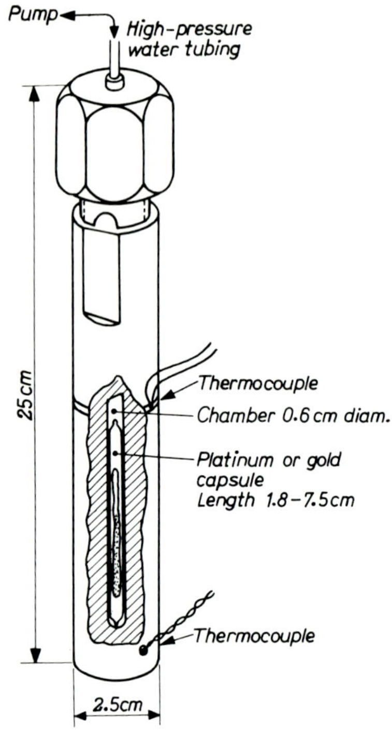

FIG. 11. Reaction vessel for hydrothermal investigations. After Laudise and Nielsen (1961), copyright by Academic Press.

The high-pressure volume is connected to a pump via a cone-seal closure which must be kept outside the furnace. A small gold or platinum capsule contains the material and the solvent. The capsules are filled with the appropriate quantity of fluid with the aid of a microburette. The permissible filling percentage is calculated from the P–V–T data for water (Fig. 9), so that the tube is filled without distension at the operating temperature and pressure. After the autoclave is sealed and pumped to the desired pressure, the capsule will collapse, but does not rupture. The pressure can be kept constant at a preset value by means of a pressure intensifier. When the operating temperature is subsequently reached, the capsule will be expanded to its original volume. The best procedure is in general to overbalance the outside pressure slightly to ensure that the capsule remains under compression. A different approach is used by Rau and Rabenau (1967a) in hydrothermal acid experiments with a quartz inner vessel under compression.

If larger quantities of a compound are to be synthesized, a useful vessel is the flat-plate closure vessel first devised by Morey and Niggli (1913).

This autoclave, shown in Fig. 12, can, if made of steel, be used for pressures up to 600 atm and temperatures near 400°C. Special alloys such as Nimonic 115 permit operation up to almost 700°C. The length can vary from 5 to 25 cm, with an internal diameter of about 2.5 cm. The liner and seal disk are made of silver or gold, depending on the material under study.

In this vessel, as well as in the others to be described, the pressure is not measured directly, but is calculated from the degree of filling and the *P–V–T* data of water.

A welded closure vessel, as described by Walker and Buehler (1950) and shown in Fig. 13, has been used quite successfully for much higher pressures up to 3000 atm. This design has the disadvantage that the liner can only be used once, because the end of the welded liner is sawn off to remove the charge at the conclusion of an experiment. This is not a serious drawback because liners are still relatively inexpensive compared with an entire vessel. A disposable liner permits a good deal of exploratory work to be done in systems where corrosion might ruin a more complex vessel. It also offers possibilities for the use of solvents other than aqueous solutions.

The modified Bridgman high-pressure design, as applied in several commercially available vessels, is probably the one most generally used for

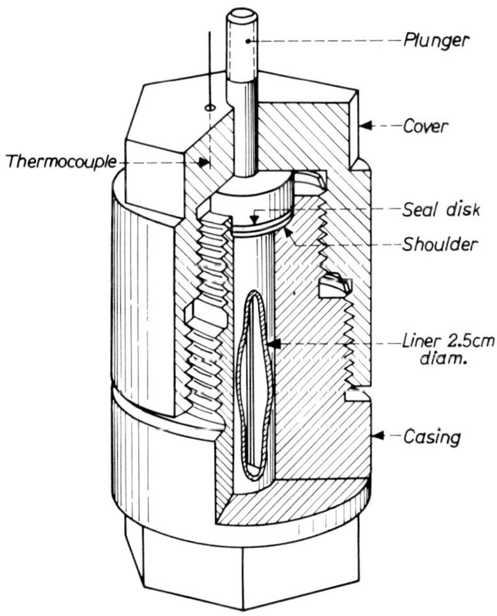

FIG. 12. Hydrothermal reaction vessel of the Morey type. After Laudise and Nielsen (1961).

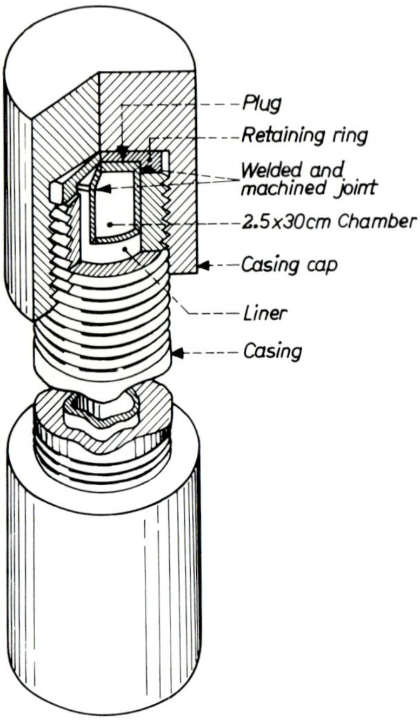

FIG. 13. Welded closure vessel. After Laudise and Nielsen (1961).

hydrothermal crystal growth. The autoclave is shown in Fig. 14. Here, the initial sealing is made by the action of the plunger against the deformable gasket. The force of the pressure acting upon the piston achieves the final sealing due to the unsupported area closure. These vessels can be used up to 4000 atm and temperatures up to, or over, 500°C, depending on the choice of the construction material.

Because of the corrosive nature of the solutions used at the pressures and temperatures of the experiment, the walls of the autoclaves will be readily attacked during the process. Lining of the vessel, generally with an inner can made of a noble metal such as silver, gold, or platinum, is necessary to extend the life of an autoclave beyond one or two experiments. The impurities introduced into the solution and into the material as a result of the corrosive action cannot, of course, be tolerated, either. In the case of growth in mildly alkaline solutions of NaOH, low-carbon steel liners are sufficient, since the relatively insoluble acmite, with the formula $Na_2O \cdot Fe_3O_4 \cdot 4SiO_2$, forms a sufficiently protective coat on the vessel walls. This is the reason for the fact, surprising at first sight, that clear, relatively pure quartz can be grown in steel autoclaves. Other solutions and combinations, however, can give rise to considerable difficulty and it is there-

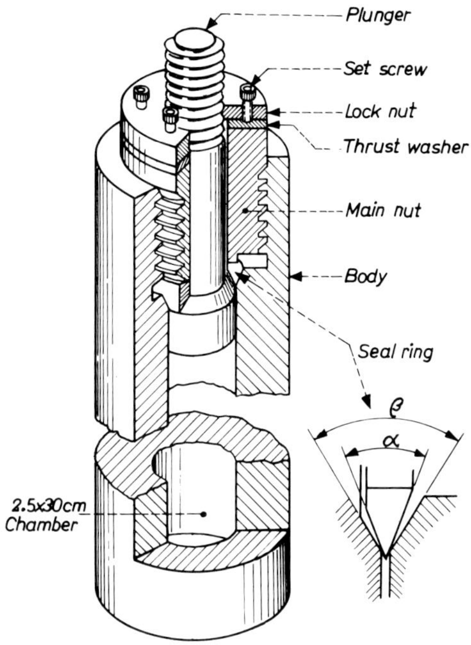

FIG. 14. Modified Bridgman vessel, Autoclave Engineers, Erie, Pa. After Laudise and Nielsen (1961).

fore not surprising that much effort has been devoted to designing alternative constructions for liners.

A disposable liner is employed by Kopp and Clark (1966) in their split autoclave. The split autoclave consists of two stainless steel halves, which are bolted together when in use. Pressures of up to 2000 atm have been applied at temperatures of about 450°C.

Liebertz (1967) designed a reliable construction for self-closing autoclaves with a silver liner. A sketch of his design is given in Fig. 15. The autoclave was tested for continuous operation for several weeks at temperatures of up to 400°C and a pressure of 1000 atm.

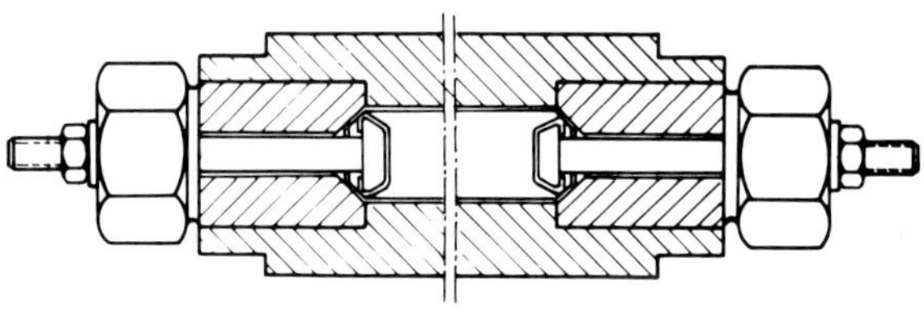

FIG. 15. Self-closing autoclave. After Liebertz (1967).

A variable-volume silver liner for hydrothermal work under accurately controlled pressure–temperature conditions was recently designed by Ashby *et al.* (1970). The liner, which fits only loosely into the autoclave, has bellow segments at each end, permitting a varying degree of filling without opening it. The liner is charged with nutrient, a seed, and the corrosive solution. The annular space between the liner and the autoclave is charged with sodium hydroxide solution to which some silica has been added. A protective layer of acmite is thus formed on the steel autoclave walls. Pressure differences developing between the internal fluid and the annular fluid during operation can be evened out by expansion or contraction of the silver bellows. The final pressure can be adjusted by the addition or withdrawal of some of the annular fluid. This liner can be removed easily from the autoclave, thus facilitating repairs, charging, and cleaning-up procedures.

A slightly different approach for the measurement of pressure generated by corrosive solutions has been elaborated by Hill *et al.* (1969). The two methods given both rely on the use of a silver diaphragm which is flexible enough to permit a pressure equilization between a corrosive solution inside the liner and pure water from a high-pressure pump line, connected to a manometer, on the outside. The design, simpler than that of Ashby *et al.* (1970), has been tested for cold-seal cone-closure vessels and Morey-type autoclaves. A schematic diagram of such a design is given in Fig. 16.

A high-pressure vessel for the determination of solubilities in aqueous systems at up to 500°C and 1000 atm with a Teflon instead of a metal gasket was designed by Gill and Marshall (1961).

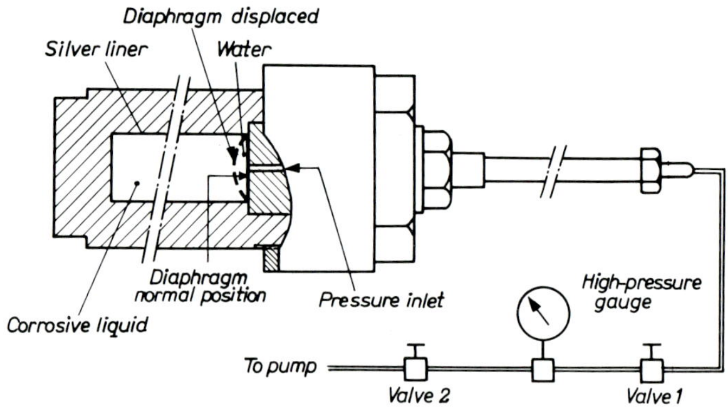

FIG. 16. Schematic diagram of simple diaphragm system for measuring saturated vapor pressure generated by corrosive solutions. After Hill *et al.* (1969).

TABLE III

Compound	Growth conditions: Mineralizer	Growth conditions: Crystallization temperature, pressure, % fill, time	Results: dimensions, shape	Ref.
Ag	4 *N* KOH	Above 400°C	Needles, l = 5 mm	Hill (1968)
Au	10 *M* HI	480 → 500°C, 65% fill	Crystals, 1–10 mm	Rabenau and Rau (1968)
Pt	10 *M* HI	480 → 500°C, 65% fill	~3 mm	Rau and Rabenau (1968)
Se	Supercritical ammonia	160–196°C, 1000 atm	Hexagonal needles, l = 8 mm	Balascio *et al.* (1967)
Te	10 *M* HI	440 → 220°C, 65% fill	~2 mm	Rau and Rabenau (1968)
CaF_2	4 *N* NH_4Cl	400–450°C, 2000–2800 atm	Octahedra, 1.5 mm	Liebertz (1965)
CuI	2 *M* HI	365 → 345°C, 7 hr	Yellow platelets 6 × 6 × 3 mm	Rau and Rabenau (1967b)
$KZnF_3$	5 *M* KOH − F_2	300°C, 4 hr	Cubes, 8 mm	Lukina and Klientova (1969)
BeO	4 *N* KOH + 0.06 *N* LiOH	500°C	Weight 1.3 g	Hill and Harker (1968)
α-Al_2O_3	1 *N* NaOH, Na_2CO_3	415°C, 2000 atm		Laudise and Ballman (1958)
α-SiO_2	1 *N* NaOH, Na_2CO_3	400°C, 1500 atm	Large crystals	Laudise and Nielsen (1961)
$CaCO_3$	2 *N* NaOH	450°C, 1400 atm, 7 days	2 mm	Liebertz (1965)
$AlPO_4$	6 *M* H_3PO_4	160°C, 7–33 atm, 60 days	Up to 80 g	Stanley (1954)
$AlAsO_4$	33 *N* H_3AsO_4	254°C, 22 days	l = 100 mm	Stanley (1956)
In_2O_3	NaOH	450°C, 8 days	3–4 mm	Litvin and Ivanova (1970)
SnO_2	15% LiOH + 15% NaOH	450°C, 75% fill	2 mm	Kuz'mina and Litvin (1963)
(Ba, Sr, Pb)SO_4	70–90% H_2SO_4	350°C, 28 days	2 mm	Liebertz (1970)
PbO	1 *N* LiOH	450 → 250°C, 14 days	10 × 5 × 0.5 mm	Rooymans and Langenhoff (1968)
$LiGaO_2$	3.5 *N* NaOH	385°C		Kolb *et al.* (1971)
TiO_2 (rutile)	10% KF	550°C, 60% fill		Kuznetsov and Panteleev (1965)
$BaTiO_3$	4 *M* NaOH	600°C, 2900 atm	Prisms; 0.4 × 0.4 × 1 mm	Nørlund Christensen (1970)

TABLE III (continued)

Compound	Growth conditions: Mineralizer	Growth conditions: Crystallization temperature, pressure, % fill, time	Results: dimensions, shape	Ref.
$PbTiO_3$	10% KF	600°C, 4–5 days	1.5 mm	Shternberg and Kuznetzov (1969)
V_2O_3	NaOH or CH_3COOH	550–700°C, 1330 atm, 10 days	1 mm	Guggenheim (see Laudise and Nielsen, 1961)
LiV_2O_4	1 *N* LiOH	500–700°C, 3000 atm, 1 day	0.75 mm	Rogers *et al.* (1967)
$PbCrO_4$	4 *N* HNO_3	$\Delta T = 20$°C, 14 days	15 × 8 × 8 mm	Rosztoczy (1965)
Fe_3O_4	NH_4Cl	515°C, 2000 atm, 2 days		Koenig (1957)
$(RE)FeO_3$	20 *M* KOH	375°C, 30 days		Kolb *et al.* (1968)
$Y_3Fe_5O_{12}$	20 *M* KOH	375°C, 200 atm, 30 days		Laudise and Kolb (1962)
$Y_3Fe_5O_{12}$	12% $FeCl_3$	550–600°C	3.3 mm	Mill″ (1962)
$NiFe_2O_4$	0.5 *N* NH_4Cl	480°C, 1300 atm		Koenig (1957)
$ZnFe_2O_4$	1.0 *N* NaOH	400°C, 80% fill	1 mm	Ballman (see Laudise and Nielsen, 1961)
$Fe_2(MoO_4)_3$	6% $FeCl_3$	450–500°C, 50% fill	1 mm	Klevtsov (1965)
$FeWO_4$	LiCl	575°C	7 × 0.5 mm	Klevtsov and Novgorodtseva (1970)
Na_2CoGeO_4	20–30 wt% NaOH	475°C	4 × 4 × 10 mm	Dem'yanets (1970)
Cu_2O	1 *N* NaOH	350°C, 4 days	1 × 1 × 2 mm	Kinoshita and Nakano (1967)
ZnO	1 *N* NaOH	400°C, 80% fill		Laudise and Ballman (1960)
ZrO_2, HfO_2	5% NH_4F	600°C, 1500 atm, 1–4 days	1–2 mm	Kuznetsov and Sidorenko (1968)
$LiNbO_3$	LiOH	500–600°C, 2000 atm, 14 days	1 × 0.2 mm	Hill and Zimmerman (1968)
$KNbO_3$, $KTaO_3$	52% KOH	600°C, 70% fill	2–3 mm	Shternberg *et al.* (1968)
$PbMoO_4$	12% NH_4Cl, 5–20% LiCl	450°C	4 × 6 mm	Dem'yanets *et al.* (1964)
CdO	NaOH	415°C, 80% fill	1 mm	Nielsen and Kolb (see Laudise and Nielsen, 1961)

$CdWO_4$	LiCl, N_4HCl	430°C, 60% fill, 6 days	7 × 4 × 2 mm	Dem'yanets and Litvin (1964)
$CaWO_4$	2 *N* NaOH	500°C, 2000 atm, 7 days	1 mm	Liebertz (1965)
Sb_2S_3	12% NH_4Cl	450°C, 5 days	1–8 mm	Popolitov (1969)
PbS	12 *M* HCl	450 → 430°C, 11 days	Cubes, 5 × 5 × 5 mm	Rau and Rabenau (1967b)
$Pb_5S_2I_6$ (and related compounds)	10 *M* HI	470–450°C, 65% fill	Needles, up to 30 mm	Rabenau and Rau (1969a)
CuS	48% HBr	450°C, 21 days	10 × 6 × 1 mm	Rau and Rabenau (1967b)
ZnS	NaOH	380°C	2 mm	Laudise and Ballman (1960)
	Na_2S	250–405°C		Laudise and Ballman (1960)
Ag_2S	$(C_2H_5)_3N \cdot HCl$ H_2S	180 → 150°C, 19 days	1–2 mm	Lewis and Fredericks (1970)
CdS	6 *M* HBr	450 → 430°C, 11 days	4 × 4 × 8 mm	Rau and Rabenau (1967b)
HgS	2 *M* HCl	330 → 320°C	Platelets, 4 × 4 × 0.4 mm	Rau and Rabenau (1967b)
HgS	NaHS	200°C, 9 days	1–2 mm	Scott and Barnes (1969)
HgS	2–8 *N* HCl	320°C, 1800 atm, 8 days	Platelets, 1–1.5 mm thick	Pajaczkowska (1970)
CuSe	48% HBr	360°C, 21 days	10 × 6 × 0.3 mm	Rau and Rabenau (1967)
ZnSe	5 *M* NaOH + 4 *M* LiOH	350°C		Kolb and Laudise (1970)
HgSe	12 *M* HCl	350°C, 10 days	10 × 10 × 0.2 mm	Pajaczkowska (1971)
SbSI	10 *M* HI	490–470°C, 2 hr	Spear-shaped, up to 30 mm	Rau and Rabenau (1967b)
SbSI	H_2S (*p*H = 6–5)	250–300°C, 400–600 atm.	13–14 cm long	Popolitov and Litvin (1969)
AuITe	10 *M* HI	450 → 150°C, 10 days	2–3 mm	Rabenau *et al.* (1970b)
$CuXTe_2$ (X = Cl, Br, I)	HCl, HBr, HI	400 → 150°C, 60% fill	Needles, up to 10 mm	Rabenau *et al.* (1970a)
$AuXTe_2$ (X = Cl, Br, I)	HCl, HBr, HI	400 → 150–100°C, 10 days	3 × 4 × 0.5 mm	Rabenau *et al.* (1970b)

A crucial problem in hydrothermal chemistry is the lack of visibility during the process. It is obvious that a study of the physicochemical properties of hydrothermal solutions, of the occurring convection currents, and of the crystal growth would be greatly enhanced if we were able to see into an operating autoclave. Turlakov *et al.* (1970) recently published a design for an experimental autoclave with transparent windows made of sapphire single crystals. Although the surfaces of the sapphiric cylinders were etched and their transparency impaired, especially in corrosive solutions at high pressures and temperatures, it proved possible to continue observation throughout experiments lasting several days.

B. Methods and Materials

As mentioned above, the great majority of chemical reactions are not studied in pure water, "mineralizers" being added. In the case of the growth of single crystals of α-quartz, the mineralizer is NaOH or Na_2CO_3, although other basic compounds, such as KOH, can also be used. In most of the other systems studied, such as CdS, $Y_3Fe_5O_{12}$, and ZnO, among others, the $(OH)^-$ ion again proved particularly effective. As can be seen from Table III, other substances, such as NH_4Cl, have also been successfully applied. The method of materials synthesis and crystal growth is practically independent of the mineralizer used; it was shown schematically in Fig. 10.

An unconventional approach has been developed by Rau and Rabenau (1967a). They have extensively investigated hydrothermal growth in concentrated hydrohalogen acids (Rau and Rabenau, 1967b). Working under these conditions has the great advantage that the experiments can be performed in closed quartz ampoules, quartz-glass being scarcely attacked by such solutions at temperatures up to 500°C and pressures up to 3000 bar. Moreover, quartz ampoules are capable of withstanding high external pressures. Thus the danger of explosions can be easily eliminated by ensuring that the external pressure on the ampoule is always larger than the inside pressure, which is generally unknown. A diagram of the apparatus is given in Fig. 17. The quartz ampoule is 50–70% filled with hydrochloric, hydrobromic, or hydroïodic acid and subsequently frozen. The other compounds are then added and the ampoule is evacuated, sealed, and placed in the autoclave. In the free volume of the autoclave, solid CO_2 is added in such quantity as to ensure a preponderant external pressure on the quartz ampoule at the required temperature. The autoclave is subsequently closed and placed in a furnace. As with the conventional method, a temperature-gradient method is employed to favor material

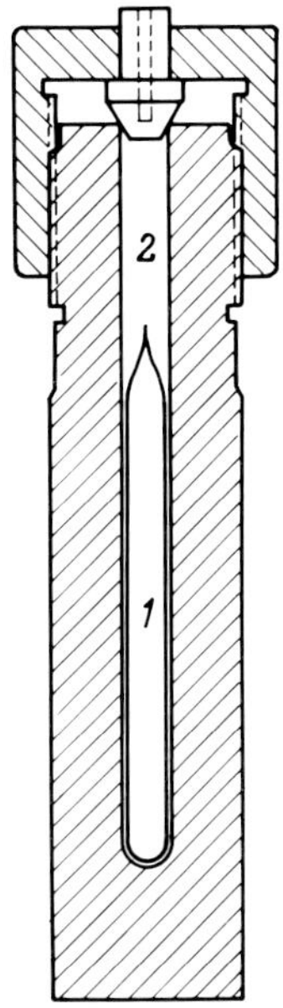

FIG. 17. Apparatus for hydrothermal crystal growth in acid environment. After Rau and Rabenau (1967a). (1) Quartz ampoule; (2) free space to be partially filled with solid CO_2. Used with permission of Pergamon Press.

transport. An autoclave made of low-alloyed steel, which can be handled conventionally, is suitable for temperatures up to 450°C and pressures up to 2000 bar. For more extreme conditions, special alloys have to be used. The method permits synthesis and growth of many compounds, some of which have not been synthesized or even known previously (Rabenau and Rau, 1969b). It is also possible to transport elements such as gold under these circumstances (Rabenau and Rau, 1968). These aspects also make the acid-hydrothermal growth method very interesting for geological and mineralogical studies. There might be a continuous transition from this hydrothermal method to the chemical transport reactions, this latter topic being beyond the scope of the present chapter, but outstandingly reviewed by Schäfer (1964).

In Table III, a survey is given of some materials synthesized and grown under hydrothermal conditions. The more mineralogically oriented studies have been omitted. The great variety of the materials and mineralizers used is quite clear from this table. A survey of some oxides and hydroxides is given in the paper by Nørlund Christensen (1969), and a survey of the rare-earth hydroxides by Mroczokowski *et al.* (1970).

Growth in a hydrothermal system, like growth in a flux, always involves the problem of the formation of inclusions and the incorporation of the solvent, here OH ions, in the lattice. Marked absorption at 3 μm, corresponding to the OH stretching frequency, has been observed in hydrothermal quartz (Dodd and Fraser, 1965) and yttrium iron garnet (Wood *et al.*, 1968). Growth in D_2O solutions shifts this absorption to the DO stretching frequency. Thus hydroxide from the growth solution is incorporated during growth. The incorporation of OH^- for O^{2-} must, however, be compensated for charge neutrality. In quartz, this can be done with substitutional Al^{3+} or Fe^{3+} ions at Si^{4+} sites. In the case of $Y_3Fe_5O_{12}$, charge compensation can be achieved by incorporation in the lattice of Ca^{2+}, a contaminant in KOH or NaOH mineralizers (Kolb *et al.*, 1967). In the case of garnet, a KOH mineralizer greatly reduces the H^+ concentration, while for quartz, the addition of LiOH, or more especially $LiNO_2$, is found to be very effective. The hydrogen concentration in quartz can be reduced in this way by more than a factor of five (Ballman *et al.*, 1966). The lithium ion directly affects the distribution constant for H^+. Neither the Li^+ concentration in the crystal nor the concentration of other possible charge-compensating ions is appreciably altered by this process. Apparently the lithium compound is adsorbed on the surface of the growing crystal. Because the H^+ concentration is proportional to the acoustic loss occurring in quartz, the addition of Li^+ is of great practical importance (Ballman *et al.*, 1966). It is not clear why hydrothermal synthesis of the tetragonal PbO crystals occurs much more readily in LiOH than in a NaOH solution (Rooymans and Langenhoff, 1968). Possibly, adsorbing phenomena are again involved. The addition of Li^+ ions is also of considerable importance for the quality of hydrothermally grown ZnO crystals. (Laudise *et al.*, 1964, Laudise and Kolb, 1969).

Solutions of Na_2S have been successfully employed for the hydrothermal growth of selenium by Kolb and Laudise (1971). Here the hydrothermal process was combined with the temperature-cycling procedure described by Scholz and Kluckow (1966) for vapor transport. In this way, it proved possible to grow spontaneously nucleated crystals up to nearly 1 cm in length.

The solid-polyphase buffer technique discussed earlier has been adapted to permit quantitative study of synthetic fluorine-bearing hydrothermal systems by Munoz (1969). The reactions studied were of a mineralogical nature, but the method could also be extended to other areas.

C. Nonaqueous Solvents

Little attention has been paid to the use of other liquids for the synthesis and growth of inorganic materials. In general, the techniques used for inorganic syntheses in liquid NH_3, SO_2, HF, or other media do not differ in principle from the techniques described above, apart from the selection of specific corrosion-free materials for the walls, heads, and liners of autoclaves. Interested readers should consult Volume I, "Principles and Techniques" in the series "The Chemistry of Non-Aqueous Solvents" (Lagowski, 1966) or "Non-Aqueous Solvent Systems" (Waddington, 1965).

Ammono–thermal experiments were performed by Liebertz *et al.* (1962). In cold-seal cone-closure vessels of the kind described previously, they tried unsuccessfully to grow AlN crystals at temperatures around 450°C and pressures up to 5000 bar. $LiNH_2$ had been added as a mineralizer. As result of these ammono–thermal experiments small crystals of $LiNH_2$ were obtained, the AlN starting material remaining unchanged.

Ammono–thermal synthesis for the preparation of amides, imides, and nitrides has been applied successfully by Juza and Jacobs (1966), Juza and Gerke (1968), and Jacobs and Juza (1969).

Balascio and others (1967) grew selenium single crystals in supercritical ammonia. A Teflon-lined hydrothermal bomb was used, and good crystals of up to 8 mm in length were obtained at 1000 bar and temperatures near 200°C. Crystals grown in this way should be compared with melt-grown crystals (Harrison, 1968) and crystals grown in hydrothermal Na_2S solutions (Kolb and Laudise, 1971).

Lewis and Fredericks (1970) have studied the crystal growth of some sulfides in supercritical H_2S in a hydrothermal autoclave of the Walker and Buehler (1950) type with a steel inner vessel. Expecially when triethylammonium chloride was added as a mineralizer, good results were obtained.

Supercritical HCl and HBr have been applied by Rabenau *et al.* (1970b, 1971) for the preparation of some gold telluride and selenide halides in the same arrangement as used earlier by these authors (Rabenau and Rau, 1969b).

IV. Solid State Systems under High-Pressure Conditions

No clear distinction is present between the systems discussed here and the gaseous and liquid systems discussed earlier. The transition is more or less continuous, gaseous and hydrothermal experiments under conditions of very high pressure having been performed in either opposed-anvil or piston–cylinder apparatus. The main distinction lies in the character of the studies, which in this case are aimed rather at the direct effect of the pressure in terms of thermodynamic stability or kinetic effects, instead of the more indirect effects of pressure, such as the prevention of decomposition, or the enhancement of the density of a solvent, as was often the

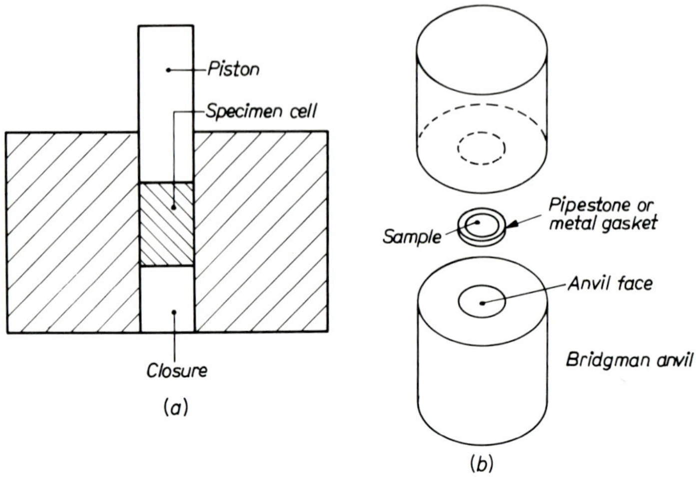

FIG. 18. (a) The piston–cylinder system and (b) the opposed-anvil design.

case in the preceding sections. What the growth of quartz was for the development of hydrothermal chemistry, the synthesis of diamonds, either catalytic or direct, has been for this particular field. Of the world's annual production of industrial diamonds, namely 50 million carats, about 40% is now of synthetic origin.

Apart from experiments directed at the crystal growth of gem-size diamonds (Strong and Chrenko, 1971), most of the effort has gone not into crystal growth, but rather into excursions into the fields of crystal chemistry, polymorphic relations and stability, completely new compounds, etc., as subsequent examples will show.

The two main principles in use for high-pressure experiments are the piston–cylinder and the opposed-anvil constructions, both shown schematically in Fig. 18. All the apparatus now in use can be derived from these two basic designs. The first presumes an internal heating arrangement, while the latter is most easily constructed with an external furnace, although very promising results have been obtained with heating arrangements between the anvils as well. The fact that this sector of high-pressure chemistry has been the subject of particularly great activity in the 1960's will make it unavoidable to restrict observations here to the main outlines and to recent progress. A fairly recent review of apparatus of these kinds has been given by Hall (1964).

We will deal in turn with opposed anvils, the piston–cylinder construction, the belt and girdle constructions, and multilateral devices. Some attention will then be devoted to the techniques of studying reversible high-pressure reactions. This will be followed by discussions of pressure and temperature calibration, pressure-transmitting media, homogeneity, and reactivity.

A. Opposed-Anvil Design

On the basis of an earlier design by Bridgman† (1935, 1952), a simple apparatus providing pressures of around 70 kbar at 500°C, 40 kbar at 700°C, and 15 kbar at 1000°C was constructed by Griggs and Kennedy (1956). The device is familiarly known as "the simple squeezer." Modifications were made by Dachille and Roy (1962), whose apparatus is shown in Fig. 19. The high pressures are supported on small areas of the anvils and the sample is thin enough to be retained entirely by friction. Much work has gone into achieving optimum anvil construction. An important point is the cone angle, which has to be of the order of 10° or less to ensure optimum distribution of the stress immediately below the anvil surface. This construction is known as the principle of massive support (Hall, 1964). A simple anvil design is limited by the ultimate compressive

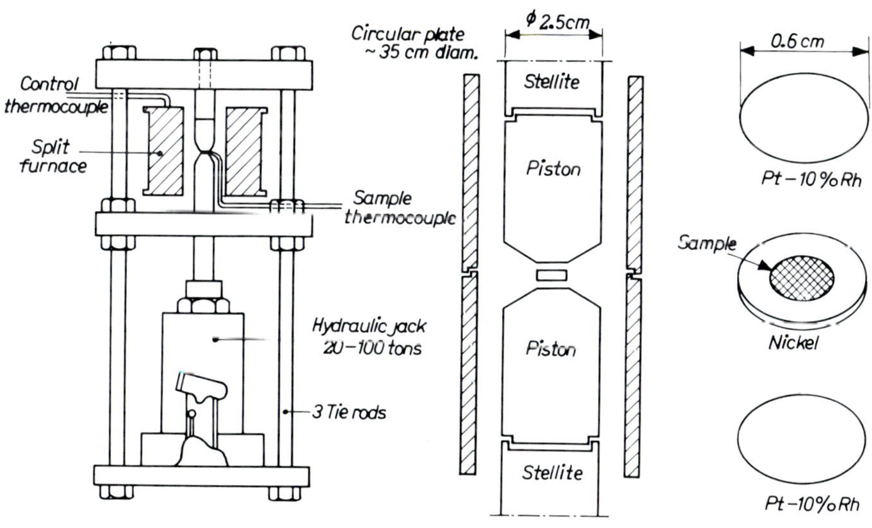

Fig. 19. Schematic of a complete opposed-anvil apparatus. After Dachille and Roy (1962).

† See Bridgman (1964) for a collection of experimental papers.

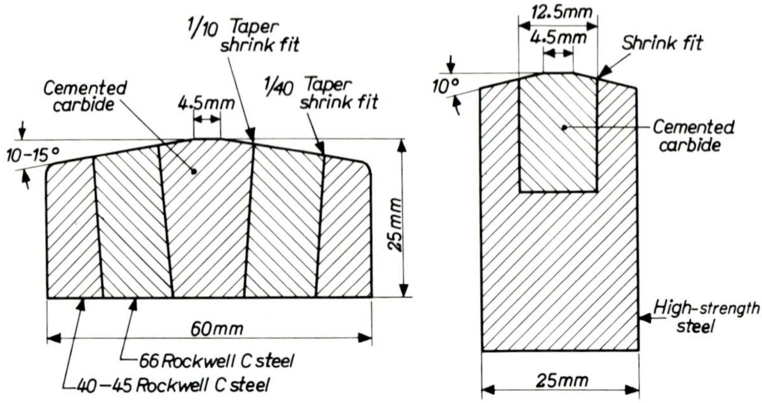

FIG. 20. Supported anvil constructions. After Dachille and Roy (1962).

strength of the material, about 60 kbar, a limit which has been very closely approached in practice. Supported anvils in which lateral support for the central part is derived from interference and a slight taper expand the range considerably. Two possible constructions are shown in Fig. 20. A survey of the various materials and designs is given in Fig. 21. Cemented tungsten carbide pistons and inserts—with a low cobalt percentage— are used most. Diamond anvils have also been used for x-ray diffraction under high pressures (see Section IV.E). The nickel, soft iron, or pyrophyllite

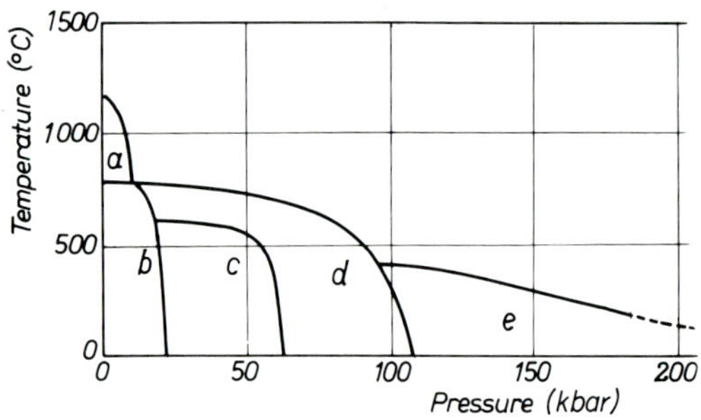

FIG. 21. Diagram illustrating useful P–T regions of various anvil materials and designs. The overlap of some areas is not shown. The regions are labeled for anvils of the following types: (a) sintered polycrystalline alumina or mullite; (b) René 41 steel; (c) tool steels of type HS-66 or Speed Star; (d) carbide inserts in René 41; (e) carbide inserts in the tool steels of type HS-66 or TK. Area (e) also includes, at room temperature, diamond anvils and the graded-support carbide anvils (Drickamer type) reported to approach the 500-kbar range. After Dachille and Roy (1962).

ring around the sample is partially squeezed out during application of the pressure. The high friction between the ring and pistons prevents complete forcing-out. The sample is then contained between two thin disks of platinum or another inert metal (thickness about 0.05 mm) and the deformed ring. In the case of reactive material, the sample can also be wrapped completely in platinum foil. Sealed tubes make it possible to extend hydrothermal techniques into the 50-kbar and higher pressure range (Dachille and Roy, 1961). Powdered samples are most easily handled if prepressed in a specially constructed mold (Dachille and Roy, 1962). The specimens processed between Bridgman anvils are rather small, the thickness being usually around 0.2 mm, with a wafer diameter of 2–6 mm. A detailed discussion of the geometry of the sample and ring and its consequences for the attainable pressure and hydrostaticity will be found in Myers *et al.* (1963).

As shown in Fig. 19, the common design makes use of an external heating system, which limits the pressure–temperature area. A sample assembly incorporating a resistance furnace in the form of a thin metal or graphite strip makes it possible to effect syntheses at much higher P–T conditions. Such a device has been reported also by Dachille and Roy (1961). Successful operation and recovery of sufficient sample to permit x-ray and optical examination has been realized by, among others, Ringwood and Major (1966). Runs at 150–200 kbar simultaneously with temperatures of approximately 1000°C are now quite feasible. Sufficient accuracy is still a problem. Flash-heating could yield even higher temperatures, bringing the diamond synthesis well within reach.

Vereshchagin *et al.* (1969a) recently published the operational details of a three-stage, high-pressure apparatus in which the first stage consists of normal Bridgman anvils and the other two serve to create pressure in a quasiplastic solid medium such as pyrophyllite, used to back up the conical surfaces of the anvils. This is reported to create a substantially larger working volume, namely up to 250 cm^3 at 160 kbar.

Pressure and temperature calibration are discussed in Section IV. F, and pressure homogeneity in IV. G.

B. Piston–Cylinder Design

Pressure in these designs is generated by a piston of circular cross section moving into a blind hole, transmitting pressure quasihydrostatically to the sample via solid media of low shear strength. The calculated pressure, known from simple geometric considerations, will differ from the true pressure owing to pressure- and temperature-dependent friction. The

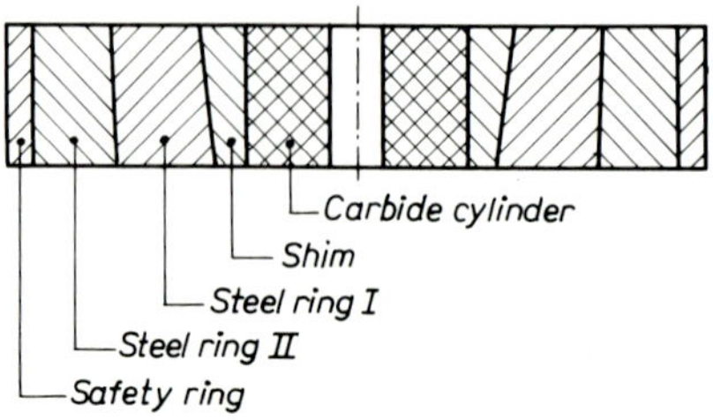

FIG. 22. Compound cylinder. After Hall (1958).

pressure attainable for the simple setup of Fig. 18 with presently available high-strength materials is about 20 kbar. The double-ended device constructed by Coes (1962) was, from a historical point of view, the first one with a range considerably in excess of 20 kbar. Coes used an alumina cylinder force-fitted into a steel ring. Compound cylinders utilizing tungsten carbide expand the range very substantially. Figure 22 shows such a cylinder consisting of an inner tungsten carbide ring and two shrink-fitted steel rings. A tapered shim is introduced to facilitate assembling. Pressures of up to 45 kbar are perfectly possible in repetitive application. Pressures up to 55 kbar can in fact be generated, but the lifetime of the cylinder is then reduced to a small number of cycles. A clamping force is then desirable to prevent the cylinder from splitting in a plane perpendicular to

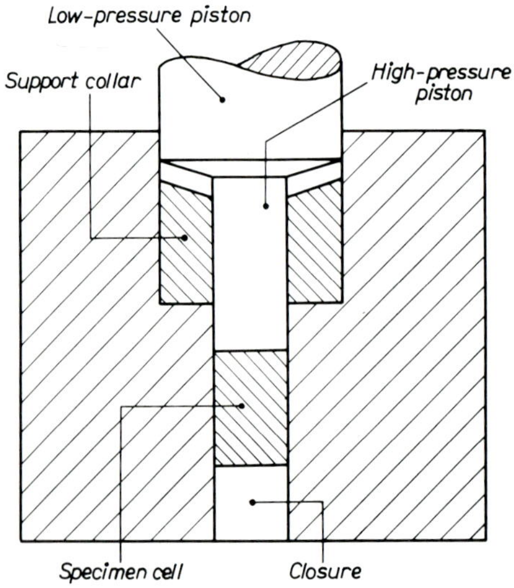

FIG. 23. Supported-piston system.

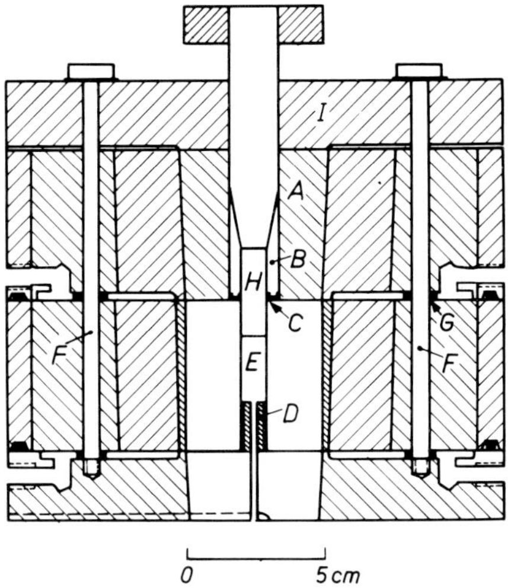

FIG. 24. Double-stage piston system. After Boyd (1962). (*A*) Steel supporting stage core; (*B*) volume filled with KBr cell wrapped in indium foil; (*C*) sealing ring, *R*i 50; (*D*) steel power lead, R_c 60, with 0.1-mm-thick pyrophyllite insulating sleeve; (*E*) volume filled by sample and furnace assembly; (*F*) steel bolts (six), R_c 30, insulated with tape; (*G*) neoprene washers; (*H*) cemented carbide piston, 1–3% Co; (*I*) end plate, 4340, R_c 50. All carbide parts are stippled and, with the exception of (*H*), contain 6% Co.

the centerline of the apparatus (Hall, 1958). Haygarth and Kennedy (1967) have investigated the crushing strength of cemented tungsten carbide–cobalt pistons. Under normal conditions, the limit is about 50 kbar. The life and attainable pressure can be enhanced up to 65 kbar however, by taking measures to ensure that the unconfined length of the piston is as small as possible.

Still higher pressures can be achieved by the "double-staging" technique, in which a smaller piston is immersed in a suitable medium under confining pressure produced by a larger piston. A setup for a supported or stepped-piston concept is given in Fig. 23, and a more realistic drawing showing the complexity of the device as realized by Boyd (1962) in Fig. 24. Pressures of up to 70 kbar are then possible. The heating is always built in so that no direct limits are set for the attainable temperatures and certainly not in the case of runs of short duration, where the heat sink is no problem.

A simple piston–cylinder apparatus for 2000°C, but only 20 kbar, has been designed by Robinson *et al.* (1964).

A double, symmetrically opposed, supported-piston construction in which diamond synthesis has been achieved was reported by Giardini *et al.* (1960). Pressures up to 100 kbar and temperatures up to 3000°C were claimed for this construction. The use of a tapered-gasket support which enabled the pressures to be reached at a lower press load than in the Giardini design was reported by Bradley *et al.* (1965).

C. The "Belt" and "Girdle" Devices

The "belt" (Hall, 1960) is probably the most successful high-pressure, high-temperature apparatus known. It was the apparatus in which the diamond synthesis was achieved in the U. S. for the first time.† Scaled-up versions of this device are still responsible for the main share of synthetic diamond. An exploded view of the "belt," which also explains the nickname, is given in Fig. 25. In a way, it is a hybridization of the anvil and the piston–cylinder designs, the principle of massive support now being applied to the anvil as well as to the piston, with an internal heating arangement. Two conical pistons, both made of cemented tungsten carbide, are pushed into opposite ends of a specially shaped chamber. The chamber and pistons receive lateral support from stressed binding rings. Axial

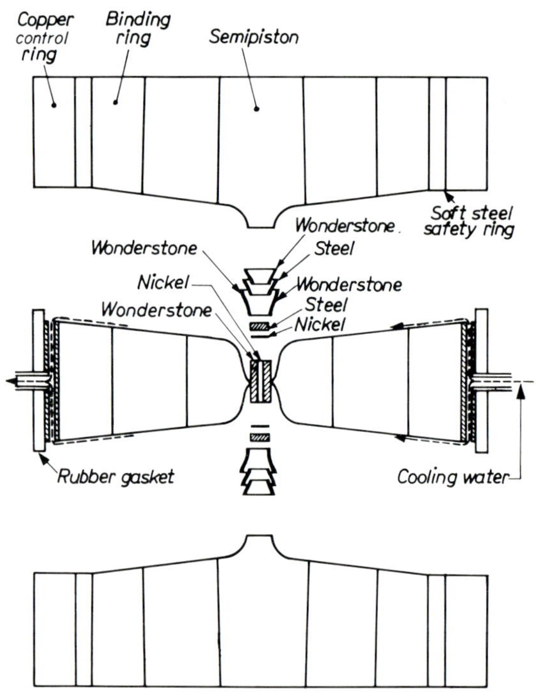

Fig. 25. The "belt," a high-temperature, high-pressure apparatus; "exploded" assembly. After Hall (1960).

† For a description of the apparatus used simultaneously in Sweden for the diamond synthesis, see Von Platen (1962).

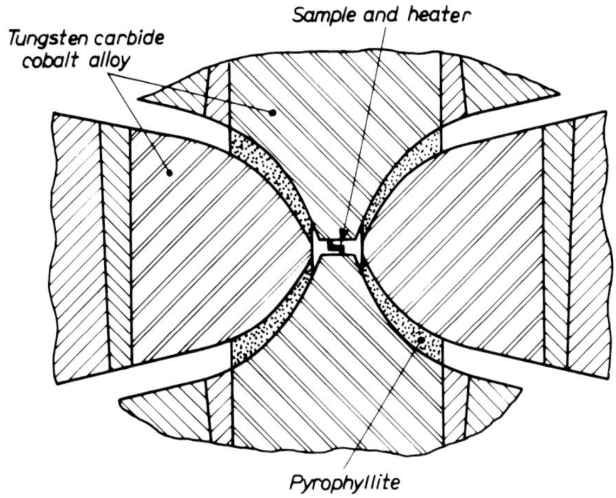

FIG. 26. The "superbelt," a high-compression belt apparatus. After Bundy (1963).

motion of the pistons is accomplished, while still maintaining a pressure seal, by the use of a specially shaped sandwich gasket made of pyrophyllite† and a steel cone. This sandwich gasket increases the stroke and hence the sample size. In addition to its use in the gasket, the pyrophyllite functions as a pressure-transmitting medium with excellent thermal and electrical insulating properties. The equipment, described by Daniels and Jones (1961), resembles the belt design, but the large relative compression of the gasket is here due to an extrudable consolidation ring of Teflon or polyethylene. Other differences are in the decompression caps and the ram support. This device is generally known as the "girdle." Pressures above 100 kbar simultaneously with temperatures above 3000°C were claimed.

Bundy (1963) developed a "superbelt" capable of pressures up to 200 kbar combined with transient temperatures up to about 5000°K using an electrical flashheating technique. In this apparatus, of which the center part is shown schematically in Fig. 26, a direct transformation of graphite to diamond could be effected.

It was Hall's opinion (1958) that anvil and piston–cylinder designs could not be scaled up sufficiently to make high-pressure syntheses economically feasible. He was the first to start the development of multilateral devices which would not be subject to the restrictions inherent in basically two-dimensional devices and would also ensure better hydrostaticity in the pressurized sample.

† Pyrophyllite, a mineral of volcanic origin, has the formula $Al_2(OH)_2Si_4O_{10}$; it is related to talc, which has the composition $Mg_3(OH)_2Si_4O_{10}$.

D. Multilateral Devices

In 1958, Hall described a tetrahedral-anvil apparatus, developed after the belt device (however, the description of the anvil device was published before that of the belt device). In this apparatus four massive support anvils are located at the ends of four rams, enclosing a solid piece of the gasket material in which the sample cavity and the heating arrangement are found. The original machine was rated at 200 tons per cylinder, but commercially available machines have since been built with rams of up to 2000 tons capacity. A detailed view of the inner tetrahedron, in most cases made of pyrophyllite again, with a possible heating element and sample holder is given in Fig. 27. Instead of four separate rams, it is also possible to fix three anvils in a wedge-type bottom-containing ring, and apply external force to only one anvil on top. A compact multianvil apparatus of this kind, suitable for placing in a normal press, has been designed by Lloyd *et al.* (1959). A diagram of such a device is given in Fig. 28. A specific difficulty is the friction between the sliding anvils and the conical ring. Very thin Teflon sheets have been found to be satisfactory lubricants. Pressure and temperature limits are the same as for the original Hall design. The three lower anvils, which can now no longer move independently, are somewhat easier to guide. Hall (1962), however, has designed a simple anvil-guide device for the original tetrahedral press, which automatically allows for a uniform advancement of all anvils, thus preventing asymmetric gasket formation and unbalanced loads on various

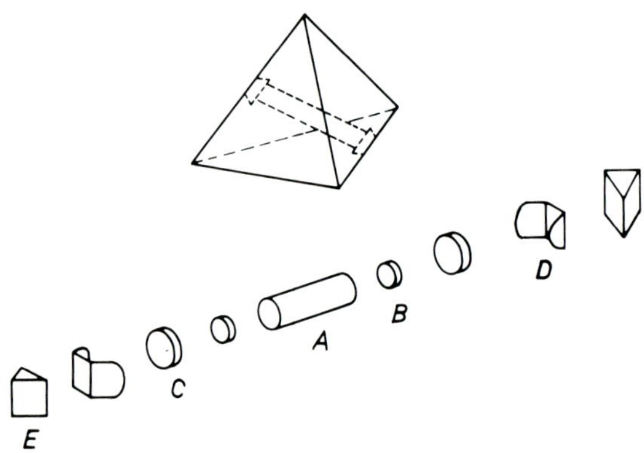

Fig. 27. Edge-to-edge heater geometry for pyrophyllite tetrahedron in Hall's (1958) device. (*A*) Specimen heater; (*B*) end closures; (*C*) anti-extrusion disks; (*D*) electrical contact tabs; (*E*) end wedges.

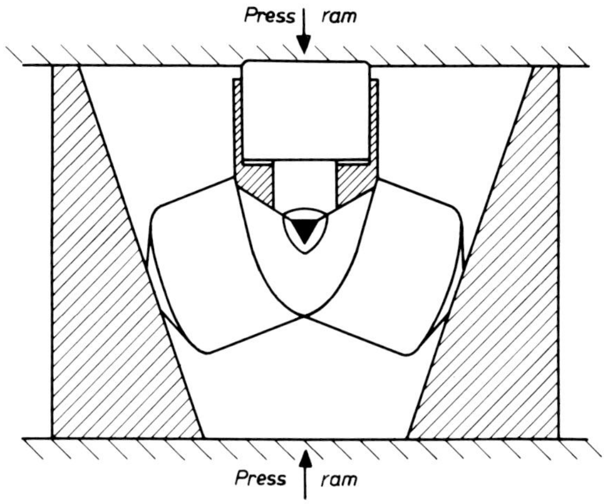

FIG. 28. Compact tetrahedral anvil design. After Lloyd *et al.* (1959).

parts of the press. The design and performance of the tetrahedral anvil apparatus have been described in detail by Lees (1966).

Specific discussions of the efficiency of this device as related to anvil and tetrahedron size are given by Banus and Nye (1964) and Lees (1965). Multianvil devices, however, are not restricted to devices of the tetrahedral type. A lot of work has been done on cube units, in which six rams act on the faces of an inner pyrophyllite cube. Here again, the dimensions can vary, the cube edges ranging from 0.5 cm up to 5 cm on an edge, the latter with six 2000-ton rams. Hexagonal presses have also been made along lines similar to the Lloyd *et al.* (1959) design, i.e., two wedge-type containing rings, each containing three anvils. The first press of this type was developed by the NBS group (Houck and Hutton, 1963). In the compact cubic anvil apparatus designed by Osugi *et al.* (1964), an external ram force applied in only one direction is converted into six equal components of force. The six anvils are forced to advance uniformly to the center of the pyrophyllite cube via tetragonal bipyramidal guide blocks and the trapezoid end-blocks of the four horizontally located anvils. Here, too, pressures up to 100 kbar can be generated. A vertical cross section of this apparatus, which is now commercially available, is shown in Fig. 29.† A novel design, in which the pressure is exerted not axially but radially, was constructed by Witteman and Werkman (1963). The high pressure is

† Multianvil devices are being advertised, by, among others, Bliss Co. formerly Barogenics, Grand Rapids, Michigan, and Kobe Steel, Ltd., Kobe, Japan.

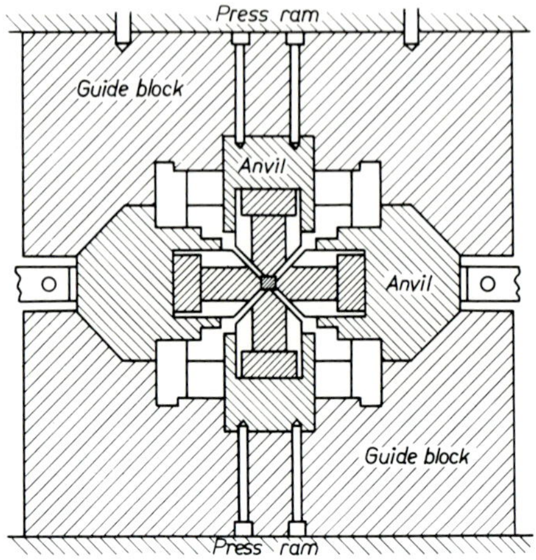

FIG. 29. Vertical cross section of compact cubic anvil apparatus. After Osugi *et al.* (1964).

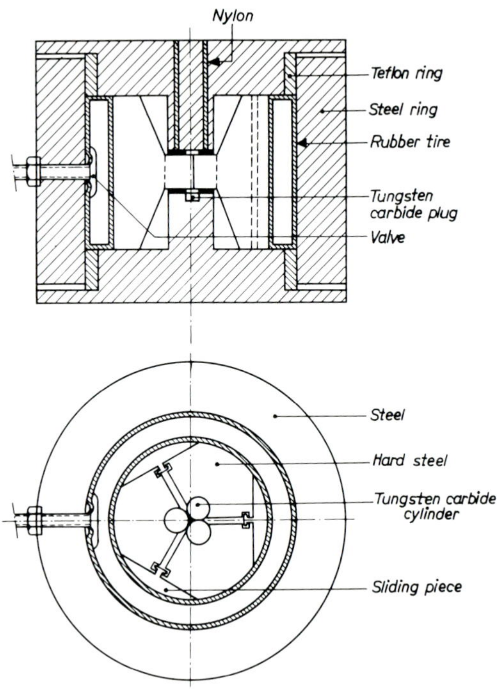

FIG. 30. Cylindrical high-pressure apparatus. After Witteman and Werkman (1963).

generated by a tire. The rather elongated shape of the high-pressure volume, formed as three coaxial cylindrical bars of tungsten carbide are brought together, may be a disadvantage. Pressures up to 80 kbar have been attained. Figure 30 shows the device.

E. Specific Apparatus

Apart from the general-purpose high-pressure apparatus described so far, a great number of devices has been designed and constructed for specific studies on the properties of materials as a function of pressure or as a function of the interatomic distances of the substance under investigation. These studies have been concerned with optical and magnetic properties, conductivity and superconductivity, ESR, NMR, and Mössbauer techniques, and neutron and x-ray diffraction, among other subjects. To mention all the advances made in this field would take us far beyond the scope of the present chapter. Interested readers should consult such journals as the *Review of Scientific Instruments*, *Journal of Physics E* (*Scientific Instruments*), *Pribory i Tekhnika Eksperimenta* (*Instruments and Experimental Techniques*) and *High Temperature—High Pressures*, in which the great majority of these articles have been published. We will, however, make an exception with regard to techniques which are of direct relevance for the preparation of new phases by giving clues to the occurrence of phase transformations or chemical reactions. Many of these reactions are reversible in nature, i.e., release of pressure leads directly to the phase that is stable at ambient conditions. Insight into the structure and properties of the high-pressure phase can therefore only be obtained from *in situ* experiments.

1. *X-ray diffraction*

The most efficient method in this respect is x-ray diffraction under pressure, since it gives direct information on the structure and interatomic distances (compressibility) of the high-pressure phase. It is not surprising, therefore, that at least 50 articles have been published on this topic alone, statring with the classic article by Jacobs (1938). The main difficulty in this technique is that while the x-ray tube and the camera are in 1-atm surroundings, the x rays have to penetrate the pressure barrier without

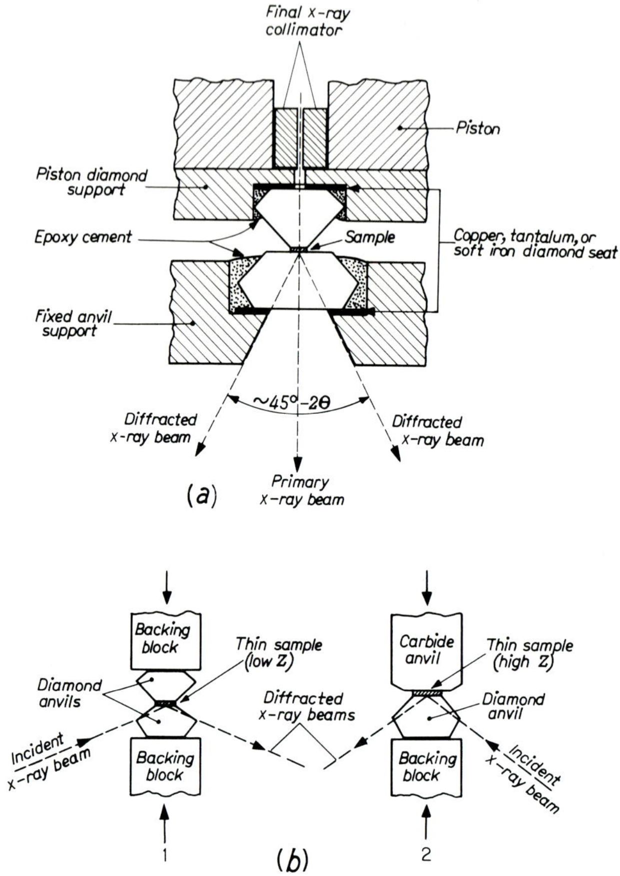

FIG. 31. (a) Schematic diagram of diamond Bridgman-anvil x-ray units with x-ray path coincident with pressure axis. (b) Schematic diagram of diamond Bridgman anvils for use with a diffractometer. The axis of rotation for the goniometer is perpendicular to the plane of the diagram and in the center of the sample face toward the x-ray beam. With a sample of high-Z elements, only one diamond anvil is required. After Banus (1969).

losing too much of their intensity by absorption. Moreover, the conditions at the sample have to be reproducible and suitable for pressure and temperature calibration. The apparatus can be roughly divided into the following categories:

(a) High-pressure vessels, made of x-ray-transparent but high-strength materials, mainly beryllium (Davis and Walawender, 1968), boron carbide (Meyer *et al.*, 1967), or diamond (Kasper *et al.*, 1960; Bradley *et al.*, 1964), containing the sample surrounded by gas, a liquid, or a viscous medium such as zinc stearate.

(b) Constructions consisting of opposed anvils between which the sample is compressed, the x rays being allowed to enter either directly via the anvils or at right angles to them. In the first case, the anvils have to be transparent (diamonds); this design has been used by, among others, the National Bureau of Standards in some outstanding studies; see, for instance, Weir *et al.* (1969), Bassett *et al.* (1967) and McCartney (1970). Anvils do not have to be transparent in the other case, where gasketing with suitable materials is the main problem (Jamieson and Lawson, 1962a; Perez-Albuerne *et al.*, 1964; McWhan and Bond, 1964; Saito and Ozaki, 1969). Schematic sketches of the two possibilities are given in Fig. 31.

(c) Multianvil designs. A specific version of the tetrahedral design already discussed has been constructed by Barnett and Hall (1964), a belt-type device by Freud and Sclar (1969), and, finally, a hexahedral anvil device by Contré (1969).

Readers interested in these techniques are also referred to the survey articles by Jamieson and Lawson (1962b), McWhan (1969), Weir *et al.* (1969b), and, especially, Banus (1969). The technique of high-pressure x-ray diffraction is certainly not very easy, great skill and experience in both high-pressure and x-ray diffraction techniques being prerequisites.

2. *Other methods*

Another helpful tool in detecting the occurrence of phase transitions is measurement of the *change in volume* and/or *heat capacity* (*DTA*) during compression, as first applied by Bridgman and later refined by Kennedy and La Mori (1962), Pistorius (1965), and Genshaft *et al.* (1967).

One of the most often applied and simplest methods, also first extensively used by Bridgman,† is based on the *change in resistivity* of a compound in the event of a phase transition. The pressure calibration of most high-pressure apparatus relies on resistivity changes (see below). The necessary set up is simple and can be easily arranged for opposed-anvil, piston–cylinder, or other designs. Recent descriptions of apparatus designs usable at varying temperatures can be found in papers by King (1969), Slutskii (1969), and Leman (1968). In the last of these devices, electrical resistance measurements can be made on two samples in separate circuits simultaneously. The center assembly in this high-pressure resistance cell, resembling that designed by Drickamer and Balchan (1962), is sketched in Fig. 32.

† See Bridgman (1964).

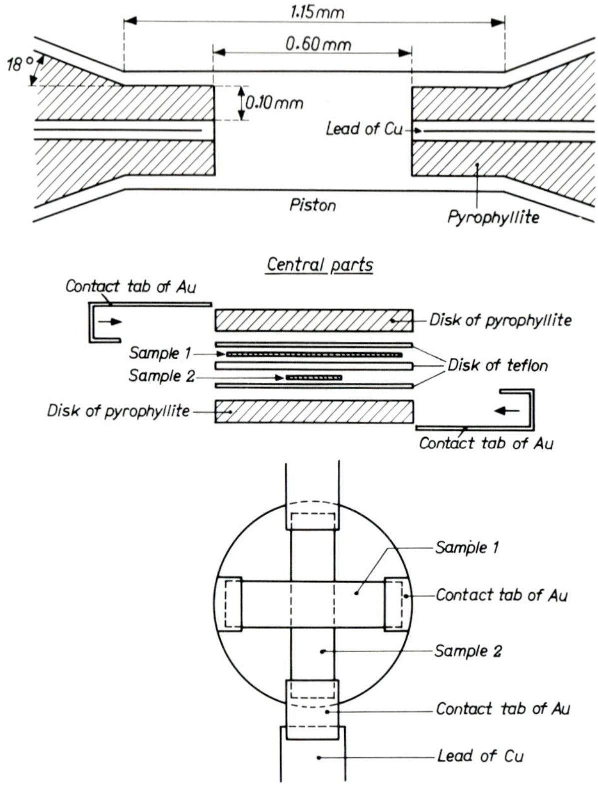

FIG. 32. Center assembly of the high-pressure resistance cell. After Leman (1968).

The measurement of volume and resistivity changes by an ac inductive technique has been successfully performed by Giardini *et al.* (1964a,b, 1965). The sensitivity to volume changes was shown to be at least 0.1%.

Reference may be made to the recent review papers by Drickamer *et al.* (1969) and Holzapfel (1970) for the Mössbauer technique, and to a paper by Bloch and Pavlovic (1969) for magnetic measurements.

F. PRESSURE AND TEMPERATURE CALIBRATION†

1. *Pressure calibration*

Although it may not always be of great importance to know exactly the pressure at which a certain chemical reaction has taken place, it seems

† See note added in proof on p. 132.

relevant to discuss here concisely the measuring devices available. The three chief means of measuring the pressure are: (1) dead-weight gauges; (2) elastic-deformation gauges; (3) changes in the electrical resistance of some materials.

Precision measurements require a free-piston (dead-weight) gauge as a primary standard. In general, this method is limited to the lower end of the high-pressure range. The maximum value measured with a freely rotating piston is the transition in cesium around 42 kbar, by Kennedy and La Mori (1962). Above this pressure, even a carbide piston suffers distortion and binds in the surrounding cylinder. The transition occurring in bismuth around 25 kbar has been measured by Heydemann (1967) in a specially designed piston gauge. At higher pressures, namely toward 100 kbar, pressures may be estimated by correcting for friction, but this procedure is subject to considerable inaccuracy, as shown, for instance, by Pistorius *et al.* (1967) and Tamayama and Eyring (1967). A specific device for the measurement of pressures up to 100 kbar using a free piston was reported by Vereshchagin *et al.* (1966).

Elastic-deformation gauges comprise Bourdon gauges and strain gauges. Bourdon-tube pressure gauges all work on essentially the same principle, i.e., the tube, whether in the form of a *C*, a spiral, or a helix, has an elliptical or oval cross section. When pressure is applied to the tube, its cross section will distort, tending to become circular and thus causing a displacement of the end. Bourdon gauges, which constitute a secondary method because they have first to be calibrated against a free-piston gauge, provide the most generally useful method of measuring fluid pressures. Bourdon gauges are available for pressures up to 7 kbar.

Strain gauges, while simple and convenient, are not well adapted to the measurement of pressures too high for thin-walled vessels since they are sensitive to hydrostatic pressures and therefore confined to surfaces subjected to constant pressure, usually exposed.

The sensitivity of wire-resistance gauges to applied pressure is exploited in the manganin pressure gauge (Adams *et al.*, 1937; Newhall, 1962). The operation of this gauge depends upon the variation of the resistance of a coil of wire with pressure. Manganin wire is preferred for this purpose, because it has a pressure coefficient of resistance of 2.4×10^{-6} parts bar^{-1}, coupled with a fairly low temperature coefficient of resistance of the order of 10^{-5} parts °C^{-1}. The resistance of a manganin coil varies linearly with pressure up to 8 kbar, a deviation from linearity of about 1% being present at 25 kbar. Zeto and Vanfleet (1969) have extended the use of a manganin coil under hydrostatic pressure up to 60 kbar. For most high-pressure apparatus, with solid media systems, neither direct calculation of the

TABLE IV

HIGH-PRESSURE CALIBRATION POINTS AT ROOM TEMPERATURE

	Point (kbar)	Reference
Mercury (freezing at O°C)	7.570 ± .001	(Yasunami, 1967)
Bi I–II	25.5 ± .06	(Heydemann, 1967)
Tl II–III	36.7 ± 0.3	(Kennedy and LaMori, 1962)
Cs II–III	41.7 ± 0.5	(Kennedy and La Mori, 1962)
Ba II–III	55 ± 2	*a*
Bi III–V	77 ± 3	*a*
Sn	100 ± 6	*a*
Ba III–IV	123–125	(Vereshchagin *et al.*, 1970)
Pb	135–138	(Vereshchagin *et al.*, 1970)
	110–160 (up only)	*a*

[a] Value accepted at the Symposium on accurate characterization of the high-pressure environment, National Bureau of Standards, Gaithersburg, Maryland, October, 1968; Lloyd (1971).

pressure nor a manganin-coil resistance method can be employed, due to the fact that these solid state media are only quasihydrostatic. A calibration based on fixed points is then necessary. This type of calibration depends on knowing the pressure at which sharp phase transformations occur, their occurrence being marked by abrupt changes in volume and/or electrical resistance of the calibrant. A great deal of work has been done to find reliable values for these discontinuities; the new, commonly accepted standards are collected in Table IV. The accuracy diminishes with increasing pressure.

A recent review of measuring devices and their calibration in ultrahigh-pressure apparatus has been published by de Malherbe and Firth (1970).

One problem, however, is that the assembly and the conditions of the real experiment are not the same as during calibration. This is especially true for experiments in which we are concerned with the combination of high pressures and high temperatures. The losses due to friction and the mechanical properties of the pressure-transmitting medium, e.g., pyrophyllite, will certainly be both pressure-dependent and temperature-dependent. This means that a room-temperature pressure calibration as discussed above is only of limited value. One way to overcome this objection is to use the resistance method as a calibrant, but now for the fusion temperature as a function of pressure for elements such as zinc, germanium, silicon, etc., as proposed by, among others, Lees and Williamson, 1965.

Hardly any reliable values for the fusion temperatures as a function of pressure are known so far.

An apparatus for measuring pressures of about 100 kbar at temperatures up to 2000°C, and relying on the change in frictional torque acting on a measuring piston, has been described by Vereshchagin *et al.* (1969b).

2. *Temperature calibration*

Normal thermocouple readings have to be corrected for the pressure dependence of the electromotive force. Hanneman and Strong (1965) have published values for the relative temperature corrections for Pt/Pt–10 Rh, Pt/Pt–13 Rh, chromel–alumel, and iron–constantan thermocouples as a function of temperature and pressure. The magnitude of these corrections for the range up to 1300°C and 50 kbar can exceed 40°C in some cases. For chromel–alumel thermocouples, the ΔT is maximal around 10°C. Wentorf (1969) tried to measure the temperature at high pressures by thermal noise, as first suggested by Garrison and Lawson (1949). Although this method is simple in principle, its success depends largely on careful attention to experimental detail and it is, moreover, not very convenient. The method does, however, give an independent check on the effect of pressure on the emf of thermocouples. The results of Wentorf (1969) were in satisfactory agreement with earlier experiments by Hanneman and Strong (1965), Freud and LaMori (1971), and Bell *et al.* (1971), among others. A detailed study of the behavior of thermocouples in the single-stage piston–cylinder apparatus, together with many experimental details, has been published by Mao and Bell (1970). A simple device for constant-temperature control was constructed by Wilhelmi and Burger (1969).

G. Pressure Homogeneity and Reactivity

The degree of homogeneity of the pressure over the whole area of the sample is another problem to consider. In liquid and gaseous systems, the pressure will be really hydrostatic; the solid media, however, will only lead to a quasihydrostatic pressure transfer. Many studies have been concerned with the choice of a particular material for the container and for this medium, depending on the pressure and temperature conditions of the experiment and the material to be studied.

1. *Containing materials*

Room-temperature experiments up to about 60 kbar can be successfully performed in a real hydrostatic environment by using a 1–1 mixture of *n*-pentane with either isopentane or isoamyl alcohol in a Teflon or steel capsule (Jayaraman *et al.*, 1967; Barnett and Bosco, 1967).† Carbonization makes use of these organic liquids virtually impossible above, say, 200°C. A material commonly used at high temperatures is the mineral pyrophyllite. This material is not homogeneous from origin, and the resulting time-dependent flow pattern can be rather complex (see, for instance, Lees, 1966). Many papers have been published on the advantages and disadvantages of either predrying, preheating, or prepressing the pyrophyllite before use in the high-pressure apparatus (see, for instance, Houck and Hutton, 1963; Hall, 1964; Chao and Ruoff, 1970). General information on this remarkable material—it is also known as "wonderstone"—was given by Nell *et al.* (1937), and Deer *et al.* (1962). In addition to talc, which is somewhat similar in its composition and properties to pyrophyllite, other hydrostatic media (each with its own problems), are silver chloride, indium, boron nitride, etc.

The lack of hydrostaticity can be due to the pressure–transmitting medium or to the apparatus used (see next subsection), but Wayne *et al.* (1970) have shown that nonisotropic strains in the sample can also be caused by the use of porous samples. Care must be exercised in interpreting "hydrostatic" measurements on porous samples.

2. *Apparatus-dependent factors*

In the description of the opposed-anvil apparatus, it has already been noted that the configuration of the assembly has a great influence on the attainable pressure and on the pressure distribution over the sample area. Myers *et al.* (1963a, b) have described the relevant factors in detail. Their data include detailed mappings of the pressure distribution for different wafer geometries. The optical studies by Lippincott and Duecker (1964), who used materials with pressure-sensitive absorption bands in a diamond-anvil apparatus, and the work of Schock and Katz (1967), who studied the pressure gradient in such a cell with the known phase transitions in some halides (see Fig. 33), also made it clear that the pressure gradients can be considerable. This may lead to a pressure multiplication effect at the center of about 2.5 times the value calculated from the simple pressure = load ÷ area relationship (Myers *et al.*, 1963b).

Tamayama and Eyring (1967) have studied the pressure losses in a piston–cylinder device due to interfacial friction between cylinder and piston, between cylinder and the solid, pressure-transmitting medium, and due to internal friction in that medium. The experiments showed that the last factor was largest.

A detailed study of deformations occurring in tetrahedron samples has

† See note added in proof on p. 132.

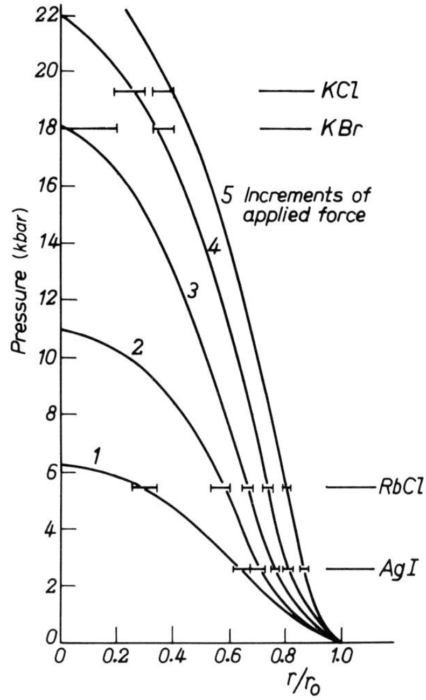

FIG. 33. Form of pressure gradient in diamond cell; r/r_0 is the normalized radius from anvil center to edge. Successive curves represent equal increments of applied force. Experimental error for each transition point is shown. After Schock and Katz (1967), used with permission of Pergamon Press.

been made by Adler and Margolin (1963) and Ashcroft and Lees (1963); see also Lees (1965b, 1966). The latter authors inserted fine wire meshes of soft bronze into split tetrahedra and studied the nodes in the mesh after a high-pressure experiment. The "hydrostatic" inner part of the tetrahedron to be used for the sample occupies only 7% of the volume of the original tetrahedron. Another study on pressure distribution in a tetrahedral-anvil device using sleeved calibration specimens at various positions within the cell was made by Deaton and Graf (1963).

The pressure homogeneity in the hexahedral anvil device has been studied by Samara *et al.* using the same technique (1964). The resulting pressure distribution is shown in Fig. 34. The inductive-coil technique has also been used by this group for analysis of nonhomogeneous material environment in these types of devices (Giardini, 1967).

H. Reactivity and Materials Synthesis

The possibilities for a chemical compoumd either reacting to form new compounds or transforming to other crystal structures are obviously not the same as for the 1-atm situation. This is due not only to thermodynamic

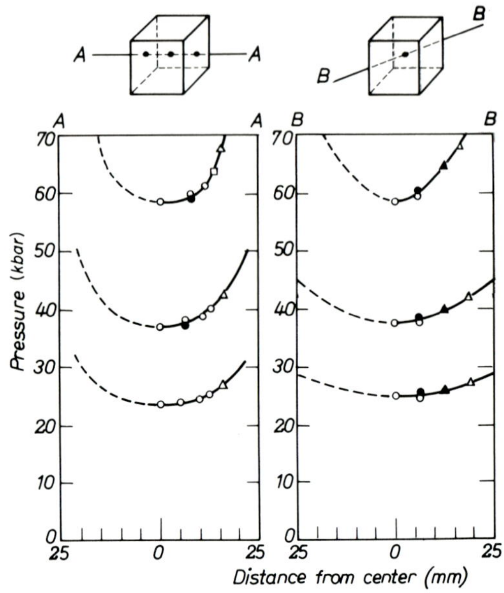

FIG. 34. Pressure distribution at different locations within standard pyrophyllite container with preformed gaskets in hexahedral-anvil high-pressure apparatus. After Samara *et al.* (1964).

factors, a chemical reaction being pushed in a direction with an overall higher density, but also to kinetic factors: closer contact between the grains, enhancement of the reactivity by pressure gradients, etc. The relevant factors will not be reviewed here. Interested readers can be referred to the Proceedings of the Internal Symposia on the Reactivity of Solids (Amsterdam, 1960; Munich, 1964; and Schenectady, 1968), and, more especially, to the papers by Dachille and Roy (1961) and Rooymans (1965) and the survey article by Rooymans (1968a). At the last of the above conferences (1968), the chemical processes in high-pressure systems formed one of the main topics.

To give here a survey of the various materials made under conditions of high and very high pressures is an even more impossible task, since hundreds of compounds have been investigated and the occurrence of a phase transformation is a rule rather than an exception. Some typical examples of different types of elements and compounds have been collected in Table V, but interested readers may be referred to a number of fairly recent review articles: Rooymans (1968b, 1969, 1970), Seifert (1968), and Kleber and Wilke (1969).

TABLE V

HIGH-PRESSURE BEHAVIOR OF SOME ELEMENTS AND INORGANIC COMPOUNDS

Compound or element	Normal structure	Necessary pressure and temperature conditions		High-pressure structure	Remarks
		P (kbar)	*T* (°C)		
C	Graphite	60	1500	Diamond	Catalytic
	Graphite	130	3000	Diamond	Direct transition
Si	Diamond	120	20	White tin	Only stable at high *P*
				Bcc	After unloading
CdS	Wurtzite	30	20	"NaCl"	Only stable at high *P*
KCl	"NaCl"	20	20	"CsCl"	Only stable at high *P*
$NaAlO_2$	"Ordered" wurtzite	40	400	"Ordered" rock salt	Quenchable
$CdTiO_3$	Ilmenite	25	700	Perovskite	Quenchable
Li_2MoO_4	Phenacite	10	400	Spinel	Quenchable
$(Mg_{0.8}Fe_{0.2})_2SiO_4$	Olivine	170	800	Spinel	Quenchable
$FeCr_2S_4$	Spinel	30	500	"NiAs"	Quenchable
SiO_2	Quartz	20	500	Coesite	Quenchable
	Quartz	125	1200	Stishovite (rutile)	Quenchable

The general rule is that reactions which occur at high pressure under room-temperature conditions will be partly or wholly reversible in character; in other words, pressure unloading leads directly either to the original substance or to a metastable intermediate formed during decompression. In cases where both high pressures and high temperatures are needed, the high-pressure products can be retained by dropping the temperature before unloading. In the latter case, study of the new phase(s) can be performed by conventional techniques, while in the former, sophisticated measurements *in situ* will be necessary. Some of these devices have been described here.

V. Shock Waves

All the experiments and procedures mentioned so far are concerned with static pressure. It is, however, perfectly feasible by the use of directed detonations to subject solids to very high-pressure forces due to shock waves. With shock-wave techniques, pressures can be attained which are one to two orders of magnitude larger than those attainable by conventional static methods. Their duration is of the order of a few microseconds. In the framework of this review, we will restrict our observations to the principles of the method, and refer readers to existing reviews on the subject, such as the article by Batsanov in this book.

The method used to generate high-intensity shock waves in the laboratory generally involves high explosives. The chemical energy is released as a detonation wave. This wave initiates a chemical reaction which supplies the necessary energy to maintain the shock in a stable state. Typical velocities for such detonation waves are around 8 mm μsec,$^{-1}$ generating shock pressures on a solid in the range 200–600 kbar, depending on the impedance of the solid. A plane-wave generator is usually circular in cross section and composed of fast- and slow-detonation explosives. The energy can be transferred directly, but higher pressures can be obtained by placing a thin metal plate at the end of the explosive and accelerating it up to about 5 mm μsec^{-1} before it strikes a target plate of the material under investigation, where it then induces shock pressures of up to 2 Mbar. For still higher pressures, convergent detonation waves may be used. A schematic diagram for a plane-wave shot assembly is shown in Fig. 35. Many solid state reactions effected by static high-pressure means have also been achieved by shock-wave techniques. The first application which comes to mind is the synthesis of diamonds (De Carli and Jamieson, 1961). In fact, nature had already done this much earlier. There are clear indi-

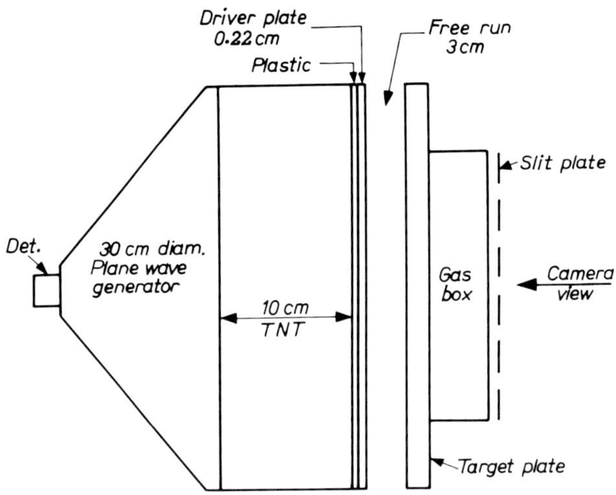

FIG. 35. Profile of a typical shot assembly. A range of pressures can be obtained by varying the dimensions and materials; for example, the system in the figure using steel or nickel driver plates gives a pressure of 1.0 Mbar in a brass target, which can be increased to 1.5 Mbar by exchanging the 10-cm TNT charge for a 15-cm Cyclotol charge. After McQueen (1964).

cations that the diamonds present in metallic or oxidic objects of meteoritic origin were formed there by the shock waves originating from the collision with the earth. This is especially true for the Canyon Diablo meteorite, the starting point for the diamond-synthesis experiments by Moissan and the General Electric team. Other shock-induced phases are the ϵ-phase of iron and the high-pressure phases of bismuth, iodine, and Zn_2SiO_4, among many others.

Nevertheless, it is not recommended as an attractive laboratory preparative method, unless all other methods fail, its advantage being the extreme pressures which can be attained. In addition to the chapter in this book by Batsanov, interested readers are referred to the papers by Hamann (1962), Deal (1962), McQueen (1964), Bradley and Munro (1965), and Skidmore (1965), in which the laboratory techniques are dealt with in detail.

VI. Safety

In addition to the hazards present in high-temperature activities, we are concerned here with the extra risks due to the high-pressure aspects, i.e., to the extra energy which is stored in the pressurized sample, the $P\Delta V$ in thermodynamic terms. It is not the pressure in itself which is dangerous, but the product. It will be obvious from this that the greater hazards occur in high-pressure work on gases. At a few kilobars, gases represent the storage of dangerously large amounts of energy. Precautions must be taken against the blowing-out of valve spindles, observation windows, and tube connections, which may be ejected like bullets or shells from a gun barrel. Especially for high-pressure work on gaseous systems, it is highly advisable to have a separate room—or better, a specially constructed building—with remote control and TV observation.

Possible failure of Bourdon gauges may likewise be mentioned, not only for these systems, but also for hydrothermal chemistry, where, however, the total stored energy is, as a general rule, considerably smaller. From a safety point of view, we are mainly concerned here with possible rupture of the vessel—which must obviously never be made of brittle materials—and escaping steam. Where fragmentation is felt to be a danger, pits dug in the ground or at least armor-plate barricades should be designed.

Since the volume of the sample and the medium directly surrounding it is quite small in solid state systems, the energy stored in it by compression will remain small even if the sample is under a pressure of 100 kbar. In fact, more energy will be stored in the compressed oil and in elastic stresses in components of the apparatus than in the sample itself. The likeliest hazard in solid state systems—such as the anvil and piston–cylinder devices—lies in possible failure of the brittle, cemented tungsten carbide parts, leading to small sharp flakes or splinters flying from the apparatus for distances of meters. Thick perspex or armor plating around the apparatus is necessary.

ACKNOWLEDGMENTS

The author is indebted to Dr. W. Albers, Mr. J. Goorissen, Dr. J. Liebertz, Prof. Dr. A. Rabenau, and Mr. F. Westendorp for helpful discussions.

The kind assistance of Mr. M. van Hout in preparing Table III is gratefully acknowledged.

Thanks are also due to Mr. P. Vissers for his care in the preparation of the figures.

BIBLIOGRAPHY

1. Bridgman, P. W. (1952). "The Physics of High Pressure." Bell, London.
2. Comings, E. W. (1956). "High Pressure Technology." McGraw-Hill, New York.
3. Hamann, S. D. (1957). "Physico-Chemical Effects of Pressure." Butterworth, London.
4. Bundy, F. P., Hibbard, W. R., Strong, H. M. (eds.) (1961). "Progress in Very High Pressure Research." Wiley, New York.

5. Wentorf, R. H. (ed.) (1962). "Modern Very High Pressure Techniques." Butterworth, London.
6. Giardini, A. A., and Lloyd, E. C. (eds.) (1962). "High Pressure Measurement." Butterworth, London.
7. Gschneidner, R. A., Hepworth, M. T. and Parlee, N. A. D. (eds.) (1964). "Metallurgy at High Pressures and High Temperatures." Gordon & Breach, New York.
8. Bridgman, P. W. (1964). "Collected Experimental Papers," Vols. 1–7. Harvard Univ. Press, Cambridge, Massachusetts.
9. Zeitlin, A. (1964). Annotated Bibliography on High Pressure Technology. Publ. jointly with the Amer. Soc. Mech. Eng., New York.
10. Bradley, R. S., and Munro, D. C. (1965). "High Pressure Chemistry." Pergamon Press, Oxford.
11. Jonassen, H. B., and Weinberger, A. (eds.) (1965). "Technique of Inorganic Chemistry," Vol. 6. Wiley (Interscience), New York, esp. the article by S. E. Babb, Jr. on Techniques of High-Pressure Experimentation.
12. Bradley, C. C. (1966). "High Pressure Methods in Solid State Research." Butterworth, London.
13. Weale, K. E. (1967). "Chemical Reactions at High Pressures." Spon, London.
14. Bradley, R. S. (ed.). *Advan. High Pressure Res,* **1** (1966); **2** (1969); **3** (1969).
15. Pugh,H. Ll. D. (ed.) (1970). "Mechanical Behaviour of Materials Under Pressure." Elsevier, Amsterdam.
16. High Pressure Bibliography, 1900–1968, Vols. I and II (1971). PB-191 174, PB-191 175 of the U.S. Dep. of Commerce. Nat. Tech. Information Service, Springfield, Virginia.
17. Lloyd, E. C. (ed.) (1971). "Accurate Characterization of the High-Pressure Environment." Symp. Nat. Bur. Std. Gaithersburg, Maryland, October 1968. NBS spec. publ. 326, U.S. Govt. Print. Off., Washington D.C.
18. Ulmer, G. C. (ed.) (1971). "Research Techniques fór High Pressure and High Temperature." Springer-Verlag, Berlin, Heidelberg, New York.

REFERENCES

Numbers in square brackets at the end of some references refer to entries in the preceeding Bibliography.

Adams, L. H., Goranson, R. W., and Gibson, R. E. (1937). *Rev. Sci. Instrum.* **8,** 230.
Adler, Ph., and Margolin, H. (1963). *In* "High Pressure Measurement" (A. A. Giardini and E. C. Lloyd, eds.), p. 95. Butterworth, London. [6]
Ashby, C. T., Berry, J. W., and Deutschman, A. J. (1970). *Amer. Mineral.* **55,** 1800.
Ashcroft, K., and Lees, J. (1963). *Nature* **198** (4884), 957.
Balascio, J. F., White, W. B., and Roy, R. (1967). *Mater. Res. Bull.* **2,** 913.
Ballman, A., and Laudise, R. A. (1963). *In* "The Art and Science of Growing Crystals" (J. J. Gilman, ed.), p. 231. Wiley, New York.
Ballman, A. A., Laudise, R. A., and Rudd, D. W. (1966). *Appl. Phys. Lett.* **8,** 53.
Banus, M. D. (1969). *High Temp. High Pressures* **1,** 483.
Banus, M. D., and Nye, S. D. (1964). *Rev. Sci. Instrum.* **35,** 1319.
Barnett, J. D., and Bosco, C. D. (1967). *Rev. Sci. Instrum.* **38,** 957.
Barnett, J. D., and Hall, H. T. (1964). *Rev. Sci. Instrum.* **35,** 175.
Bass, S. J., and Oliver, P. E. (1968). *J. Crystal Growth* **3,** 286.
Bassett, W. A., Takahashi, T., and Stock, P. W. (1967). *Rev. Sci. Instrum.* **38,** 37.

Bell, P. M., England, J. L., and Boyd, F. R. (1971). *In* "Accurate Characterization of the High-Pressure Environment" (E. C. Lloyd, ed.), Symp. Nat. Bur. Std. Gaithersburg, Maryland, October 1968, p. 63. NBS spec. publ. 326, U.S. Govt. Printing Off., Washington D.C. [17]

Bloch, D., and Pavlovic, A. S. (1969). *Advan. High Pressure Res.* **3,** 41. [14]

Boksha, S. S. (1957). *Sov. Phys. Crystallogr.* **2,** 195.

Boksha, S. S. (1972). *J. Crystal Growth* **12,** 113.

Boyd, F. R. (1962). *In* "Modern Very High Pressure Techniques" (R. H. Wentorf, ed.), p. 151. Butterworth, London. [5]

Boyer, C. B., Orcutt, F. D., and Conaway, R. M. (1966). *Chem. Eng. Progr.* **62,** 99.

Bradley, R. S., and Munro, D. C. (1965). "High Pressure Chemistry," Chapter 7, p. 141. Pergamon, Oxford. [10]

Bradley, R. S., Grace, J. D., and Munro, D. C. (1964). *Z. Krist.* **120,** 349.

Bradley, R. S., Munro, D. C., and Whitfield, M. (1965). *J. Sci. Instrum.* **42,** 714.

Bridgman, P. W. (1935). *Phys. Rev.* **48,** 825.

Bridgman, P. W. (1952). *Amer. Acad. Arts. Sci. Proc.* **81,** 165.

Bridgman, P. W. (1964). "Collected Experimental Papers," Vols. 1–7. Harvard Univ. Press, Cambridge, Massachusetts.

Buddington, A. F., and Lindsley, D. H. (1964) *J. Petrol.* **5,** 310.

Bundy, F. P. (1963). *J. Chem. Phys.* **38,** 631.

Carapezza, M. (1966). *Atti Acad. Lincei. Rend. Classe Sci. Fis. Mat. Nat.* **40,** 646 (*English Transl.:* (1969). *Geochem. Int. R. USSR* **6,** 819).

Cerveny, W. J. (1963). *Ind. Eng. Chem.* **55,** 40.

Chao, C. C. and Ruoff, A. L. (1970). *Rev. Sci. Instrum.* **41,** 720.

Coes, L. L. Jr. (1962). *In* "Modern Very High Pressure Techniques" (R. H. Wentorf, ed.), p. 137. Butterworth, London. [5]

Comings, E. (1956). "High Pressure Technology." McGraw-Hill, New York. [2]

Contré, M. (1969). *High Temp. High Pressures* **1,** 339.

Dachille, F., and Roy, R. (1961). *Reactiv. Solids Proc. Int. Symp. 4th, Amsterdam, 1960* (J. H. de Boer *et al.*, eds.), p. 502. Elsevier, Amsterdam.

Dachille, F., and Roy, R. (1962). *In* "Modern Very High Pressure Techniques" (R. H. Wentorf, ed.), p. 163. Butterworth, London. [5]

Damen, J. *et al.* (1972). *J. Crystal Growth* **12.**

Daniels, W. B., and Jones, M. T. (1961). *Rev. Sci. Instrum.* **32,** 885.

Darken, L. S., and Gurry, R. W. (1945). *J. Amer. Chem. Soc.* **67,** 1398.

Davis, B. L., and Walawender, M. J. (1968). *J. Phys. Chem. Solids* **29,** 1177.

Deal, W. E. Jr. (1962). *In* "Modern Very High Pressure Techniques" (R. H. Wentorf, ed.), p. 200. Butterworth, London. [5]

Deaton, B. C., and Graf. R. B. (1963). *Rev. Sci. Instrum.* **34,** 45.

De Carli, P. S., and Jamieson, J. C. (1961). *Science* **133,** 1821.

Deer, W. A., Howie, R. A., and Zussman, J. (1962). *In* "Rock Forming Minerals" W. A. Deer *et al.*, eds.), Vol. 3, p. 115. Longmans, London.

de Malherbe, M. C., and Firth, B. W., (1970). *Metals Mater.* **4,** 1.

Dem'yanets, L. N. (1970). *Sov. Phys. Crystallogr.* **15,** 339.

Dem'yanets, L. N., and Litvin, B. N. (1964). *Sov. Phys. Crystallogr.* **9,** 497.

Dem'yanets, L. N., Garashina, L. S., Litvin, B. N. (1964). *Sov. Phys. Crystallogr.* **8,** 639.

Dodd, D. M., and Fraser, D. B. (1965). *J. Phys. Chem. Solids* **26,** 673.

Drickamer, H. G., and Balchan, A. S. (1962). *In* "Modern Very High Pressure Techniques" (R. H. Wentorf, ed.), p. 25. Butterworth, London. [5]

Drickamer, H. G., Fung, S. C., and Lewis, G. K. (1969). *Advan. High Pressure Res.* **3,** 1. [13]

Eugster, H. P. (1957). *J. Chem. Phys.* **26,** 1760.
Eugster, H. P., and Skippen, G. B. (1967). *In* "Researches in Geochemistry" (P. H. Abelson, ed.), Vol. II. Wiley, New York.
Eugster, H. P., and Wones, D. R. (1962). *J. Petrology* **3,** 82.
Ferretti, A., Wickham, D. G., and Wold, A. (1961). *Rev. Sci. Instrum.* **32,** 566.
Fischer, A. G. (1970). *J. Electrochem. Soc.* **117,** 41C.
Franck, E. U. (1961). *Angew. Chem.* **73,** 309.
French, B. M., and Eugster, H. P. (1965). *J. Geophys. Res.* **70,** 1529.
Freud, P. J., and LaMori, P. N. (1971). *In* "Accurate Characterization of the High-Pressure Environment" (E. C. Lloyd, ed.), Symp. Nat. Bur. Std. Gaithersburg, Maryland, October 1968, p. 67. NBS spec. publ. 326, U.S. Govt. Print Off. Washington D.C. [17]
Freud, P. J., and Sclar, C. B. (1969). *Rev. Sci. Instrum.* **40,** 434.
Garrison, J. B., and Lawson, A. W. (1949). *Rev. Sci. Instrum.* **20,** 785.
Genschaft, Yu. S., Livshits, L. D., and Ryabinin. Yu. N. (1967). *Sov. Phys. Techn. Phys.* **12,** 126.
Giardini, A. A. (1967). *Trans. A. S. M. E., J. Basic Eng.* **89,** 554.
Giardini, A. A., Tydings, J. E., and Levin, S. B. (1960). *Amer. Miner.* **45,** 217.
Giardini, A. A., Poindexter, E. H., and Samara, G. A. (1964a). *Rev. Sci. Instrum.* **35,** 713.
Giardini, A. A., Poindexter, E. H., and Samara, G. A. (1964b). *Trans. A.S.M.E., J. Basic Eng.* 736.
Giardini, A. A., Poindexter, E. H., Samara, G. A., and Barnett, J. D. (1965). *Rev. Sci. Instrum.* **36,** 1742.
Gill, J. S., and Marshall, W. L. (1961). *Rev. Sci. Instrum.* **32,** 1060.
Griggs, D. T., and Kennedy, G. C. (1956). *Amer. J. Sci.* **254,** 722.
Hall, H. T. (1958). *Rev. Sci. Instrum.* **29,** 267.
Hall, H. T. (1960). *Rev. Sci. Instrum.* **31,** 125.
Hall, H. T. (1962). *Rev. Sci. Instrum.* **33,** 1278.
Hall, H. T. (1964). *In* "Metallurgy at High Pressures and High Temperatures" (K. Gschneidner *et al.*, eds.), p. 133. Gordon & Breach, New York. [7]
Hamann, S. D. (1962). *Rev. Pure Appl. Chem.* **10,** 139.
Hanneman, R. E., and Strong, H. M. (1965). *J. Appl. Phys.* **36,** 523.
Harrison, J. D. (1968) *J. Appl. Phys.* **39,** 3672.
Haygarth, J. C., and Kennedy, G. C. (1967). *Rev. Sci. Instrum.* **38,** 1590.
Heydemann, P. L. (1967). *J. Appl. Phys.* **38,** 2640.
Hill, V. G. (1968). *J. Electrochem. Soc.* **115,** 720.
Hill, V. G., and Harker, R. I. (1968). *J. Electrochem. Soc.* **115,** 294.
Hill, V. G., and Zimmerman, K. G. (1968). *J. Electrochem. Soc.* **115,** 978.
Hill, V. G., Harding, W. R., and Zimmerman, K. G. (1969). *Rev. Sci. Instrum.* **40,** 1352.
Holzapfel, W. B. (1970). *High Temp. High Pressures* **2,** 241.
Houck, J. C., and Hutton, U. O. (1963). *In* "High Pressure Measurement" (A. A. Giardini and E. C. Lloyd, eds.), p. 221. Butterworth, London. [6]
Huckenholz, H. G., and Yoder, H. S. (1970). *Carnegie Inst. Washington Yearb.* **69,** 182.
Huebner, J. S., and Sato, M. (1970) *Amer. Mineral.* **55,** 934.
Jacobs, H., and Juza, R. (1969). *Z. Anorg. Allg. Chem.* **370,** 248.
Jacobs, R. B. (1938). *Phys. Rev.* **54,** 325.
Jamieson, J. C., and Lawson, A. W. (1962a). *J. Appl. Phys.* **33,** 776.
Jamieson, J. C., and Lawson, A. W. (1962b). *In* "Modern Very High Pressure Techniques" (R. H. Wentorf, ed.), p. 70. Butterworth, London. [5]

Jayaraman, A., Hutson, A. R., McFee, J. H., Coriell, A. S., and Maines, R. G. (1967). *Rev. Sci. Instrum.* **38,** 44.

Juza, R., and Gerke, H. (1968). *Z. Anorg. Allg. Chem.* **363,** 245.

Juza, R., and Jacobs, H. (1966). *Angew. Chem.* **78,** 208.

Kasper, J. S., Hilliard, J. E., Cahn, J. W., and Phillips, V. A. (1960). WADC Tech. Rep. 59-747 G. E. Res. Lab., Schenectady, N. Y.

Kennedy, G. C., and La Mori, P. N. (1962). *J. Geophys. Res.* **67,** 851.

King, E. (1969). *J. Sci. Instrum.* **2,** 59.

Kinoshita, A., and Nakano, T. (1967) *Jap. J. Appl. Phys.* **6,** 656.

Kleber, W., and Wilke, K. Th. (1969) *Kristall. Techn.* **4,** 165.

Kirk, R. E., and Othmer, D. F. (eds.) (1956). "Encyclopedia of Chemical Technology." Wiley (Interscience), New York. (With 1957 and 1960 supplements.)

Klevtsov, P. V. (1965). *Sov. Phys. Crystallogr.* **10,** 370.

Klevtsov, P. V., Novgorodtseva, N. A., and Kharchenko, L. Yu. (1970). *Sov. Phys. Crystallogr.* **15,** 532.

Koenig, J. (1957). Final Rep. NUAFRC-TR-57-190, Contract No. AF-19 (604)-1419, AFCRC 1955–1957, Cambridge, Massachusetts.

Kolb, E. D., and Laudise, R. A. (1970). *J. Crystal Growth* **7,** 199.

Kolb, E. D., and Laudise, R. A. (1971). *J. Crystal Growth* **8,** 191.

Kolb, E. D., Wood, D. L., Spencer, E. G., and Laudise, R. A. (1967). *J. Appl. Phys.* **38,** 1027.

Kolb, E. D., Wood, D. L., and Laudise, R. A. (1968). *J. Appl. Phys.* **39,** 1362.

Kolb, E. D., Caporaso, A. J., and Laudise, R. A. (1971). *J. Crystal Growth* **8,** 354.

Kopp, O. C., and Clark, G. W. (1966). *Rev. Sci. Instrum.* **37,** 372.

Kuz'mina, J. P., and Litvin, B. N. (1963). *Sov. Phys. Crystallogr.* **8,** 379.

Kuznetsov, V. A., and Panteleev, V. V. (1965). *Sov. Phys. Crystallogr.* **10,** 369.

Kuznetsov, V. A., and Sidorenko, O. V. (1968). *Sov. Phys. Crystallogr.* **13,** 651.

Lagowski, J. J. (ed.) (1966). "The Chemistry of Nonaqueous Solvents," Vol. I, Principles and Techniques. Academic Press, New York.

Laudise, R. A. (1962). *Prog. Inorg. Chem.* **3,** 1.

Laudise, R. A. (1970). "The Growth of Single Crystals." Prentice-Hall, Englewood Cliffs, New Jersey.

Laudise, R. A., and Ballman, A. A. (1958). *J. Amer. Chem. Soc.* **80,** 2655.

Laudise, R. A., and Ballman, A. A. (1960). *J. Phys. Chem.* **64,** 688.

Laudise, R. A., and Kolb, E. D. (1962). *J. Am. Ceram. Soc.* **45,** 51

Laudise, R. A., and Kolb, E. D. (1969). *Endeavour* **28** (105), 114.

Laudise, R. A., and Nielsen, J. W. (1961). *Solid State Phys.* **12,** 149.

Laudise, R. A., Kolb, E. D., and Caporaso, A. J. (1964). *J. Amer. Ceram. Soc.* **47,** 9.

Lees, J. (1965a). *Rev. Sci. Instrum.* **36,** 564.

Lees, J. (1965b). *Nature* **208** (5005), 31.

Lees, J. (1966). *Advan. High Pressure Res.* **1,** 2. [14]

Lees, J., and Williamson, B. H. J. (1965). *Nature* **208** (5007), 278.

Leman, G. (1968). F. O. A. 4 Rapport C 4361-23. Res. Inst. of Nat. Defense, Stockholm, Sweden.

Lewis, L. C., and Fredericks, W. J. (1970). *J. Crystal Growth* **7,** 120.

Liebertz, J., Eckerlin, P., Rabenau, A., and Tillack, J. (1962). Unpublished results.

Liebertz, J. (1965). *Chem. Ing. Techn.* **37,** 830.

Liebertz, J. (1967). *Chem. Ing. Techn.* **39,** 1294.

Liebertz, J. (1970). *Kristall. Techn.* **5,** K23.

Lippincott, E. R., and Duecker, H. C. (1964). *Science* **144,** 1119.

Litvin, B. N., and Ivanova, L. I. (1970). *Sov. Phys. Crystallogr.* **14,** 988.

Lloyd, E. C. (ed.) (1971). "Accurate Characterization of the High-Pressure Environment." Symposium Nat. Bur. Stds., Gaithersburg, Maryland, 1968. NBS spec. publ. 326, U.S. Govt. Print Off. Washington D.C. [17]

Lloyd, E. C., Hutton, U. O., and Johnson, D. P. (1959). *J. Res. Nat. Bur. Std.* **63 C,** 59.

Lukina, M. M., and Klientova, G. P. (1969). *Sov. Phys. Crystallogr.* **14,** 314.

Luth, W. C., and Tuttle, O. F. (1963). *Amer. Mineral.* **48,** 1401.

Mao, H. K., and Bell, P. M. (1970) *Carnegie Inst. Washington Yearb.* **69,** 207.

McCartney, J. H. (1970). *EHPRG Conf., Leeds.*

McQueen, R. G. (1964). *In* "Metallurgy at High Pressures and High Temperatures" (K. Gschneidner *et al.*, eds.) p. 44. Gordon & Breach, New York. [7]

McWhan, D. B. (1969). *Trans. A.C.A.* **5,** 39.

McWhan, D. B., and Bond W. L. (1964). *Rev. Sci. Instrum.* **35,** 626.

Meiling, G. S., and Leombruno, R. (1968). *J. Crystal Growth* **3, 4,** 300.

Metz, E. P. A., Miller, R. C., and Mazelsky, R. (1962). *J. Appl. Phys.* **33,** 2016.

Meyer, H. O. A., Bell, P. M., and England, J. L. (1967). *Carnegie Inst. Washington Yearb.* **66,** 541.

Mill", B. V. (1962). *Sov. Phys. Crystallogr.* **7,** 521.

Morey, G. W., and Niggli, P. (1913) *J. Amer. Chem. Soc.* **35,** 1086.

Mroczkowski, S., Eckert, J., Meissner, H., and Doran, J. C. (1970). *J. Crystal Growth* **7,** 333.

Muan, A. (1955). *J. Metals* **7,** 1.

Muller, O., Roy, R., and White, W. B. (1968). *J. Amer. Ceram. Soc.* **51,** 693.

Mullin, J. B., Straughan, B. W., and Brickell, W. S. (1965). *J. Phys. Chem. Solids* **26,** 782.

Mullin, J. B., Heritage, R. J., Holliday, C. H., and Straughan, B. W. (1968). *J. Crystal Growth* **3, 4,** 281.

Munoz, J. L. (1968). *Carnegie Inst. Washington Yearb.* **67,** 170.

Munoz, J. L. (1969). *Amer. Mineral.* **54,** 943.

Myers, M. B., Dachille, F., and Roy, R. (1963a). *In* "High Pressure Measurement" (A. A. Giardini and E. C. Lloyd, eds.), p. 17. Butterworth, London. [6]

Myers, M. B., Dachille, F., and Roy, R. (1963b). *Rev. Sci. Instrum.* **34,** 401.

Nacken, R. (1950). *Chem. Z.* **74,** 745.

Nell, L. T., Jacobs, H., Allan, J. T., and Bozzoli, G. R. (1937). "Wonderstone," Union of S. Africa, Dept. of Mines, Bull. no. 8, Public Geol. Survey Div.

Newhall, D. H. (1962). *Instrum. Control Systems* **35,** 103.

Nørlund Christensen, A. (1969). *Rev. Chim. Minér.* **6,** 1187.

Nørlund Christensen, A. (1970). *Acta Chem. Scand.* **24,** 2447.

Nygren, S. F., Ringel, C. M., and Verleur, H. W. (1971). *J. Electrochem. Soc.* **118,** 306.

Osugi, J., Shimizu, K., Inoue, K., and Yasunami, K. (1964). *Rev. Phys. Chem. Japan* **34,** 1.

Pajaczkowska, A. (1970). *J. Crystal Growth* **7,** 93.

Pajaczkowska, A. (1971). *J. Crystal Growth* **8,** 137.

Perez-Albuerne, E. A., Forsgren, K. F., and Drickamer, H. G. (1964). *Rev. Sci. Instrum.* **35,** 29.

Peterson, G. E., Carnevale, E., and Verleur, H. W. (1971). *Mater. Res. Bull.* **6,** 51.

Pistorius, C. W. F. T. (1965). *J. Phys. Chem. Solids* **26,** 1003.

Pistorius, C. W. F. T., Rapoport, E., and Clarck, J. B. (1967). *Rev. Sci. Instrum.* **38,** 1741.

Popolitov, V. I. (1969) *Sov. Phys. Crystallogr.* **14,** 461.

Popolitov, V. I., and Litvin, B. N. (1969). *Sov. Phys. Crystallogr.* **13,** 483.

Pugh, H. Ll. (ed.) (1970). "Mechanical Behaviour of Materials under Pressure." Elsevier, Amsterdam.

Rabenau, A., and Rau, H. (1968). *Naturwiss.* **55,** 336.

Rabenau, A., and Rau, H. (1969a). *Z. Anorg. Allg. Chem.* **369,** 295.

Rabenau, A., and Rau, H. (1969b). *Philips Tech. Rev.* **30,** 89.

Rabenau, A., Rau, H., and Rosenstein, G. (1970a). *Z. Anorg. Allg. Chem.* **374,** 43

Rabenau, A., Rau, H., and Rosenstein, G. (1970b). *J. Less-Common Metals* **21,** 395.

Rabenau, A., Rau, H., and Rosenstein, G. (1971). *Monatsh. Chem.* **102,** 1425.

Rau, H., and Rabenau, A. (1967a). *Mater. Res. Bull.* **2,** 609.

Rau, H., and Rabenau, A. (1967b). *Solid State Commun.* **5,** 331.

Rau, H., and Rabenau, A. (1968). *J. Crystal Growth* **3, 4,** 417.

Reed, T. B. (1970). *In* "The Chemistry of Extended Defects in Non-Metallic Solids" (L. Eyring and M. O'Keeffe, eds.), p. 21. North-Holland Publ., Amsterdam.

Ringwood, A. E., and Major, A. (1966). *Earth Planet. Sci. Lett.* **1,** 241.

Robinson, L. B., Vahldiek, F. W., and Lynch, C. T. (1964). *In* "Metallurgy at High Pressures and High Temperatures" (K. Gschneidner *et al.*, eds.), p. 180. Gordon & Breach, New York. [7]

Rogers, D. B., Gillson, J. L., and Gier, T. E. (1967). *Solid State Commun.* **5,** 263.

Rosztoczy, F. E. (1965). *Amer. Mineral.* **50,** 755.

Rooymans, C. J. M. (1965). *Reactiv. Solids Proc. Int. Symp. 5th, Munich, 1964* (G. M. Schwab, ed.), p. 100. Elsevier, Amsterdam.

Rooymans, C. J. M. (1968a). *Reactiv. Solids Proc. Int. Symp. 6th, Schenectady, New York, 1968* (J. W. Mitchell *et al.*, eds.), p. 754. Wiley (Interscience), New York.

Rooymans, C. J. M. (1968b). *Philips Res. Rep. Suppl. no. 5.*

Rooymans, C. J. M. (1969). *Advan. High Pressure Res.* **2,** 1. [14]

Rooymans, C. J. M. (1970). *In* "Les Propriétés Physiques des Solides sous Pressions," p. 393. C.N.R.S., Paris.

Rooymans, C. J. M., and Langenhoff, W. F. Th. (1968). *J. Crystal Growth* **3, 4,** 411.

Roy, R. (1965). *Bull. Soc. Chim. Fr.* **1965,** 1065.

Roy, R. and Tuttle, O. F. (1956). *Phys. Chem. Earth* **1,** 138.

Saito, S., and Ozaki, Y. (1969). *Jap. J. Appl. Phys.* **8,** 797.

Samara, G. A., Henius, A., and Giardini, A. A. (1964). *Trans. A.S.M.E., J. Basic Eng.* **86,** 729.

Schäfer, H. (1964). "Chemical Transport Reactions." Academic Press, New York.

Schock, R. N., and Katz, S. (1967). *J. Phys. Chem. Solids,* **28,** 1985.

Scholz, H., and Kluckow, R. (1966). *Crystal Growth, Proc. Intl. Conf., Boston, 1966.* (H. S. Peiser, ed.), p. 475. Pergamon Press, New York.

Scott, S. D., and Barnes, H. L. (1969). *Mater. Res. Bull.* **4,** 897.

Seifert, K. F. (1968). *Fortschr. Miner.* **45,** 214.

Shaw, H. R. (1963). *Science* **139,** 1220.

Shternberg, A. A., and Kuznetsov, V. A. (1969). *Sov. Phys. Crystallogr.* **13,** 647.

Shternberg, A. A., Lapsker, Ya. E., and Kuznetsov, V. A. (1968). *Sov. Phys. Crystallogr.* **12,** 838.

Shreve, R. N. (1956). "The Chemical Process Industries," 2nd ed. McGraw-Hill, New York.

Skidmore, I. C. (1965). *Appl. Mater. Res.* **4,** 131.

Slutskii, A. B. (1969). *Instrum. Exp. Tech.* **6,** 1582.

Stanley, J. M. (1954). *Ind. Eng. Chem.* **46,** 1684.

Stanley, J. M. (1956). *Amer. Mineral.* **41,** 947.

Strong, H. M., and Chrenko, R. M. (1971). *J. Phys. Chem.* **75,** 1838.
Tamayama, M., and Eyring, H. (1967). *Rev. Sci. Instrum.* **38,** 1009.
Turlakov, V. N., Safronov, G. M., Kleschev, G. V., and Simonov, A. V. (1970). *Inorgan. Mater.* **6,** 1216.
Turnock, A. C. (1960). *Carnegie Inst. of Washington Yearb.* **59,** 98.
Turnock, A. C., and Eugster, H. P. (1962). *J. Petrology* **3,** 533.
Tuttle, O. F. (1949). *Bull. Geol. Soc. Amer.* **60,** 1727.
Ulmer, G. C. (ed.) (1971). "Research Techniques for High Pressure and High Temperature." Springer Verlag, Berlin, Heidelberg, New York. [18]
Van Hook, H. J. (1963). *J. Amer. Ceram. Soc.* **46,** 248.
Van Hook, H. J. (1964). *J. Amer. Ceram. Soc.* **47,** 579.
Van Hook, H. J. (1965). *Rev. Sci. Instrum.* **36,** 1119.
Vereshchagin, L. F., Zubova, E. V., Buimova, I. P., and Burdina, K. P. (1966). *Sov. Phys. Dokl.* **11,** 585.
Vereshchagin, L. F., Sermerchan, A. A., Kuzin, N. N., and Sadkov, Yu. A. (1969a). *Sov. Phys. Dokl.* **13,** 1165.
Vereshchagin, C. F., Zubova, E. V., Konyaev, Yu. S., and Dovbnya, A. A. (1969b). *Instrum. Exp. Tech.* **6,** 1584.
Vereshchagin, L. F., Semerchan, A. A., Kuzin, N. N., and Sadkov, Yu. A. (1970). *Sov. Phys. Dokl.* **15,** 292.
Von Platen, B. (1962). *In* "Modern Very High Pressure Techniques" (R. H. Wentorf, ed.), p. 118. Butterworth, London. [5]
Waddington, T. C. (ed.) (1965). "Non-Aqueous Solvent Systems." Academic Press, New York.
Walker, A. C., and Buehler, E. (1950). *Ind. Eng. Chem.* **42,** 1369.
Wayne, R. C., Samara, G. A., and Lefever, R. A. (1970). *J. Appl. Phys.* **41,** 633.
Weale, K. E. (1967). "Chemical Reactions at High Pressures." Spon, London.
Wentorf, R. H. (1969). General Electric R. and D. Center Rep. 60-C-106, March 1969.
Weir, C. E., Piermarini, G. J., and Block, S. (1969a). *Rev. Sci. Instrum.* **40,** 1133.
Weir, C. E., Piermarini, G. J., and Block, S. (1969b). *Trans. A.C.A.* **5,** 106.
Wilhelmi, K. A. (1968). *J. Sci. Instrum.* **2,** 1240. Copyright by The Inst. of Physics.
Wilhelmi, K. A., and Burger, W. (1969). *Acta Chem. Scand.* **23,** 414.
Williams, D. W. (1966). *Mineral. Mag.* **35,** 1003.
Williams, D. W. (1968). *Amer. Mineral.* **53,** 1765.
Wisniewski, R. (1970). *Rev. Sci. Instrum.* **41,** 464.
Witteman, W. J., and Werkman, Th. (1963). *Philips Res. Rep.* **18,** 447.
Wood, D. L., Kolb, E. D., and Remeika, J. P. (1968). *J. Appl. Phys.* **39,** 1139.
Yasunami, K. (1967). *Rev. Phys. Chem. Japan* **37,** 1.
Zeto, R. J., and Van Fleet, H. B. (1969). *J. Appl. Phys.* **40,** 2227.
Zeto, R. J. *et al.* (1972). Rev. Sci. Instr. **43,** 132.
Zeitlin, A. (1964). "Annotated Bibliography on High-Pressure Technology." Publ. jointly with the Amer. Soc. Mech. Eng., New York. [9]

Notes Added in Proof

P. 75. Readers interested in these matters are also referred to Ulmer (1971).

P. 81. Recently J. Damen *et al.* (1972) have had published "A furnace for crystal growth at high gas pressures," which deals with crystal growing experiments in oxidizing atmospheres up to 150 atm.

P. 84. The recent article by S. S. Boksha (1972) on "High gas pressure crystal growth of SbSI, $NiCO_3$, and black phosphorus" deserves to be mentioned.

P. 114. Readers interested in matters of calibration and homogeneity should certainly consult the book "Accurate Characterization of the High Pressure Environment," edited by E. C. Lloyd (1971), in which many papers on these topics have been published.

P. 118. Recently a pressure cell with ten electrical leads for liquid hydrostatic pressures to 60 kbar has been described by R. J. Zeto *et al.* (1972).

Syntheses under Shock-Wave Pressures

S. S. BATSANOV†

Institute of Thermophysics, Academy of Science
Novosibrisk, U.S.S.R.

I. Introduction

The intensive implementation of physical methods in chemistry has for a long time been of an analytical character, i.e., they have been used to study the composition and structure of substances, but not to change them. The application of physical methods in synthetic chemistry has now become possible, primarily due to the advances in high-pressure techniques.

The creation and implementation of high static pressures, which have made it possible to carry out many interesting physical and chemical transformations, are the traditional direction of the development of these techniques. However, the use of explosives as sources of high pressures has made it possible to increase the ranges of pressure and temperature by one or more orders of magnitude, thus extending the potential possibilities of these methods.

† Present address: State Committee for Standards, Moscow, U.S.S.R.

II. Thermodynamics of Shock Compression

Shock waves are formed when bodies traveling at high relative velocities collide, or as a result of the detonation of explosives. The magnitude of the pressure developed depends on the geometry of the experiment, the density of the substance under investigation, and the speed with which the shock wave is distributed in it. The pulse pressure can be varied from 10 thousand to 10 million bar by changing these parameters. The temperatures developed under these pressures can also be changed within wide limits—from several degrees to 10,000 deg.

The principal characteristic of shock compression is the extremely short duration of its action. Due to the great speeds of shock wave transmission, the time of action of high pulse pressures and temperatures varies (depending on the experimental parameters and the properties of the substance) from a few tenths of a microsecond to 10 μsec. After the relief of pressure, the body cools, not to the initial temperature, but, depending on the substance and the value of the pressure, to a temperature a few degrees to thousands of degrees higher than the initial one. Subsequent cooling occurs according to the usual laws.

Thus, if the temperature and pressure in static installations are determined mainly by the design, in dynamic experiments the same parameters depend basically on the properties of the substance which is being compressed. Therefore, though the compression of a body under dynamic conditions allows for much greater variety in the thermodynamic parameters than does compression in a static state, the determination of these parameters requires knowledge of the mechanical properties (shock adiabatics) of the mixture.

Work was carried out in our laboratory on the theoretical calculations of the adiabatics of inert bodies (Doronin and Stupnikov, 1970). This offers opportunities for the determination of pressures and temperatures along with the kinetic study of the processes in reacting systems.

The problem of obtaining the shock adiabatic is one of finding the dependence of pressure P upon the mass speed U of the substance. Let x and $1 - x$ be the weight concentrations of the first and second components of a mixture, respectively. Then, from the laws of conservation of mass and momentum for the mixture and for each of its components under the same pressure, the following relation between the mass speeds of the mixture $U_{1,2}$ and of the components U_1 and U_2 can be obtained:

$$U_{1,2}^2 = xU_1^2 + (1 - x)U_2^2 \quad (1)$$

This expression has the physical meaning of the law of conservation of

TABLE I

MARBLE–PARAFFIN SYSTEM

x_{marble}	P (kbar)	$U_{1,2}$ (km sec^{-1})	
		Experimental	Calculated
77.3	40	0.59	0.59
74.8	51	0.73	0.72
72.3	71	0.93	0.92
77.4	81	0.97	0.96
78.0	97	1.09	1.10
78.0	105	1.13	1.15
74.9	153	1.59	1.59

energy, i.e., the energy of a unit mass of a mixture under shock compression is equal to the sum of the energies of unit masses of its components compressed by shock waves to the same pressure. An expression which is analogous to (1) can also be obtained for a mixture containing three or more components.

In Table I, the results of the calculation, using formula (1), of the shock adiabatic of the mixture of marble and paraffin are compared with the results of experiments (Dremin and Karpukhin, 1960).

A porous material, such as a powder, can be looked upon as a mixture of a single-phase substance with air. Therefore expression (1) becomes

$$U^2_{\text{porous}} = U^2_{\text{single phase}} + \tfrac{5}{6}(m-1)(P/\rho_0) \tag{2}$$

where m is the porosity of the substance, i.e., the relation of the volume of powder to the total volume of the body, and ρ_0 is the density of the single-phase substance. Table II gives the results of the calculations using Eq. (2) for some powderlike metals and compounds. Doronin and Stupnikov (1970) applied the idea of the additive calculation of shock adiabatics to compounds by considering them as condensed mixtures.

Bearing in mind the law of conservation of energy for a shock wave, we may write for identical pressures:

$$U^2_{1,2} = p\,\Delta V_{1,2} \qquad \text{and} \qquad U_c^2 = p\,\Delta V_c \tag{3}$$

where $\Delta V_{1,2}$ is the change in volume of a mixture for a change from its initial state to a given pressure, and ΔV_c is the change for a compound of identical composition.

TABLE II

POWDER–AIR SYSTEM

Powder	Porosity	P (mbar)	U_1 (km sec^{-1})	
			Experimental	Calculated
Cu	1.57	2.626	5.01	4.99
	1.57	7.01	8.60	8.60
	2.00	2.204	5.40	5.41
	2.00	5.95	9.27	9.26
	3.01	1.582	6.02	5.95
	4.00	1.260	6.42	6.29
	4.00	3.54	10.94	10.78
Al	1.43	1.391	6.08	6.09
	2.08	1.003	6.56	6.69
	2.98	0.702	6.98	7.01
W	1.03	1.98	1.74	1.76
	1.03	5.01	3.26	3.32
	1.76	0.315	1.12	1.15
	1.76	1.32	2.39	2.42
	2.96	0.91	2.93	2.99
	3.99	0.185	1.67	1.67
	4.30	2.05	5.63	5.68
LiF	1.55	1.239	6.19	6.15
	2.08	0.138	2.40	2.31
	2.08	0.925	6.59	6.57
	3.00	0.655	7.03	6.95
	4.68	0.430	7.44	7.33

Since the difference in shock compression energies for the mixture and the compound should depend on pressure, one may assume to a first approximation that this dependence is linear:

$$U_{1,2}^2 - U_c^2 = kpV_c^0 \tag{4}$$

where V_c^0 is the initial volume of the compound.

After determining the value of the coefficient k, we have

$$U_{1,2}^2 - U_c^2 = \tfrac{5}{6}p(V_{1,2}^0 - V_c^0) \tag{5}$$

This formula can be used for shock adiabatic calculations and therefore for calculating pressures using known (or additively calculated) adiabatics of inert mixtures of the same composition. Physically, Eq. (5) means that under high pressures, shock adiabatics of mixtures and compounds

TABLE III

ADDITIVE CALCULATION OF MASS VELOCITIES FOR COMPOUNDS

	LiI		NaI		KI		RbI	
P	U_{exp}	U_{cal}	U_{exp}	U_{cal}	U_{exp}	U_{cal}	U_{exp}	U_{cal}
50	0.39	0.40	0.46	0.44	0.61	0.50	0.62	0.50
100	0.72	0.75	0.79	0.82	1.00	0.97	0.99	0.96
150	—	—	1.06	1.10	1.32	1.31	1.28	1.28
200	1.27	1.26	1.32	1.35	1.60	1.59	1.54	1.55
250	—	—	1.55	1.57	1.85	1.85	1.76	1.78
300	1.71	1.65	1.74	1.76	2.07	2.08	1.97	1.99
400	2.11	1.98	—	—	—	—	—	—

run in parallel in P–V coordinates and are $\frac{1}{6}(V_{1,2}^0 - V_c^0)$ apart, i.e., they differ by the difference between the initial volumes of the compound and mixture.

Table III gives examples of calculations, using formula (5), of mass velocities of alkali metal iodides, which are compared with experimental values (Kalinin and Zharkov, 1968).

The good agreement between experimental and theoretical data in Tables II and III enables us to determine pulse pressures in compounds and in binary mixtures with 2–5% precision using the additively calculated values of mass velocities U.

This approach represents the first attempt to work out "chemical" methods of pulse pressure calculation. Further experimental and theoretical development of this subject is being carried out now in our laboratory. We have no doubts about the suitability of this approach, since any compound consists of several elements whose shock adiabatics are either known or can be obtained in the near future. The experimental determination of the adiabatics of the millions of known compounds presents an unrealizable task.

III. Transmission of Shock Waves in Single Crystals

The passage of a shock wave through a crystal is accompanied by a very fast (approximately several km sec^{-1}) translation of the compressed state of the substance, the depth of the shock wavefront being about an order of magnitude greater than the lattice spacing (Kormer, 1968). On the wave-

TABLE IV

CHANGE OF DISLOCATION DENSITY AS A RESULT OF PULSE PRESSURE

Crystal	P (kbar)	ρ_{initial} (cm^{-2})	$\rho_{\text{compression}}$ (cm^{-2})
LiF	85	10^5	10^{11}
NaCl	60	10^5	10^{10}
KBr	~60	10^5	10^{10}
CsI	~70	10^5	10^9
MgO	80	10^4–10^5	10^{12}
CaF_2	60	10^5	10^7

front boundary, due to the high gradient of P and T, a sort of "grinding" of the crystal takes place, and as a consequence, numerous lattice defects occur: dislocations, vacancies, etc.

Experimental determinations of coherent scattering fields in crystals subjected to the effects of explosions (Klein, 1966) showed that the size of the smallest single-crystal region which is not subject to further fractioning in the shock wavefront is of the order of 100 Å and therefore the number of atoms in it should be about 10^3–10^4. Assuming that dislocations take place on the boundaries of these regions, we find the following: in a 1 cm^3 volume, the number of dislocations should be about 10^{18}–10^{19}, the dislocation density being 10^{12}–10^{13} cm^{-2}. After removal of the pulse load, some of the dislocations will be cured by later heat release and this density may decrease by one or two orders of magnitude. Thus the residual density of dislocations in shock-compressed single crystals should not be more than 10^{11}–10^{12} cm^{-2}.

If, as a result of the passage of shock waves through a single crystal, greater fractioning takes place, the crystallite size may be of the order of 10 Å, i.e., it may contain 10^2–10^3 atoms. In this case, the crystal may transform into a disordered glassy state (see De Carli and Jamieson, 1959).

Experimental data on the density of dislocations (ρ) in inorganic crystals exposed to shock pressure are given in Table IV (Batsanov, 1970).

It is of interest that those crystals that have no phase transformation under the pressures used preserve their previous dislocation structure, on the background of which new dislocations appear. In the case of KBr, after impulse compression, only the new dislocation structure is observed. This is because a phase transformation of this crystal leads to the disappearance of the previous dislocation structure.

The physical cause for the increase in dislocation density of crystals under shock compression is the fractioning and disorientation of crystallites (Milton and De Carli, 1963; Altshuler *et al.*, 1967; Batsanov *et al.*, 1968b, 1969c, in press). The development of dislocations and the disorientation of crystallites result in an increase in the free surface of the crystal and therefore in a change in its molecular and physical properties. It is natural to expect, for example, that in the case of shock-compressed crystals, endothermic effects will be weaker (less energetic), while exothermic effects will be stronger than those of the initial substance. Other possible indicators of stored energy are the temperature and time required for recrystallization of the compressed sample (Bergmann and Barrington, 1966).

To conclude, we note that the changes in dislocation density in single crystals after the passage of shock waves vary for different crystallographic directions. In the case of alkali-metal halides, we have shown (Yaroslavtsev *et al.*, in press) that the growth of dislocation density is a minimum for the ⟨110⟩ direction and a maximum for ⟨111⟩. The change in microhardness (which, after the passage of shock waves through a crystal, increases by 1.3–2 times) and in other physical properties also depends on orientation.

In addition to the changes in physical properties as a result of shock compression of crystals (of alkali-metal halides in particular), the chemical behavior of the substance changes slightly and, in particular, alkalinity appears, caused by a charge transfer from anions to cations. The concentration of such chemical defects is small (a fraction of a percent) and generally speaking depends on the intensity of the shock waves. A more detailed description of defect formation will be given later.

It is of importance to note that all the properties of crystals that change due to shock compression return to their initial state after annealing, the rate of return depending on the temperature. Restoration of the properties also depends on the crystal orientation (Yaroslavstev *et al.*, in press; Klein *et al.*, 1963).

IV. Transmission of Shock Waves in Polycrystalline Materials

The physical picture of the transmission of shock waves in powders is much more complicated than in the case of solid crystals. This problem is of the greatest interest for chemists since polycrystalline materials are used most often in chemical research. In general, we can say that in a qualitative sense, the shock compression of powders practically always gives the same result (if no chemical decomposition takes place): shock waves force the air out and compress the powder more or less tightly.

The density of the sample obtained depends on the values of pressure and temperature as well as on the plastic properties of the material. It is also to be expected that there will be a different texturing of the powder and a different orientation of crystal grains depending upon the configuration of the shock waves.

Due to strong gradients of P and T at the grain–air boundaries, fractioning of particles may occur (as has already been mentioned). The size of these particles in the case of shock-compressed SiC and B_4C can reach 0.1–1.0 μm (Bergmann and Barrington, 1966). Qualitatively the same result was obtained in the compression of MgO and Al_2O_3 (Heckel and Youngblood, 1968). In both cases, crystallographic study of shock-compressed material showed that the coherent scattering regions are less than the real volume of the particles, indicating that fractioning of particles takes place more effectively in larger grains. Knowing the surface of polycrystalline sample it is easy to determine the value of stored energy, which proves to be 0.51 cal g^{-1} for MgO and 0.37 cal g^{-1} for Al_2O_3. A calorimetric study showed good agreement with the crystallographic determination of stored energy.

The fractioning of particles and development of dislocations, vacancies, and other defects should in principle reduce the density of the substance. However, the high final temperature (in cases of sufficiently stable substances) may bring about annealing of defects, and as a result, the shock-compressed powder will improve its microstructure; its density will approach the single-crystal value. Such improvement of structure was found in aluminum oxide (Adadurov *et al.*, 1965), nitrates and sulfates of calcium, strontium and barium (Batsanov *et al.*, 1966b) and fluorides of alkaline-earth metals (Adadurov *et al.*, 1967). In cases of thermally unstable compounds, the high final temperature results in the development of defects and even in the decomposition of the substance (see Batsanov *et al.*, 1966b; Dremin and Breusov, 1968).

From the chemical point of view, the development of defects in shock-compressed powderlike materials is always accompanied by the lowering of the valence of some of the elements in the compound which may result eventually in complete dissociation of the atoms. Therefore we can say that shock waves possess reductive properties.

The lowering of valence in samples quenched after shock was observed for the first time by Ryabinin (1956). Magnetic susceptibility changes were observed in ZnS, corresponding to the transition of a considerable part of Zn(II) into Zn(I), as a result of 15 consecutive blasts (Nomura and Todisawa, 1965). An analogous change of valence was observed in europium salts, where we found a transition of Eu(III) to Eu(II) (Lapshin and Batsanov, 1968; Batsanov *et al.*, 1969b), and in shock-compressed TiO_2 where we found development of defects containing Ti(III). This, by the way, caused a sharp increase in catalytic activity of titanium dioxide powder treated by shock waves (Batsanov *et al.*, 1967a).

In the case of more complex compounds the lowering of atomic valence is accompanied by destruction of polyatomic anions, formation of free radicals, their recombination, and other processes. Detailed study showed that the mechanism of shock decomposition was not always equivalent to ordinary thermal disintegration (Batsanov *et al.*, 1966a; Boldyrev *et al.*, 1967; Lapshin *et al.*, 1970).

V. Inorganic Syntheses in Shock Waves

Having reviewed briefly the physical features of the passage of shock waves through a substance, we now proceed to describe the results of applying pulse techniques to problems of inorganic synthesis.

Chemical syntheses of inorganic compounds in shock waves was first carried out by Japanese scientists. Kimura (1963) made an attempt to form zinc ferrite through shock compression of a mixture of the corresponding oxides.† In 1963, Horiguchi and Nomura obtained titanium carbide by the dynamic compression of titanium and carbon powders. Two years later, the authors reported on their attempts at the synthesis of WC and Al_4C_3 (Horiguchi and Nomura, 1963, 1965) by shock compression. Finally, in 1966, Horiguchi synthesized zinc orthosilicate by the dynamic compression of zinc and silicon dioxide.

It was mentioned in these studies that, in explosions of small charges, association reactions take place to a small extent or not at all. However, after additional heating of the compressed mixture, the latter reacts faster or at a lower temperature than the initial mixture of the same components. Apparently, this effect is due to fine fractioning and surface development, and to the mixing of particles to create favorable conditions for physical and chemical processes to take place.

Our experiments on the synthesis of chromium chalcogenides (Batsanov and Zolotova, 1968) may serve as indirect evidence of the last statement. The reaction of chromium and sulfur required the use of 150 g of hexogen in standard cylindrical ampoules (the experiment is described in Batsanov *et al.*, 1971a). The Cr + Se and Cr + Te reactions required 300 and 500 g

† Later, this reaction was examined by (Adadurov *et al.*, 1969b), who used Mössbauer spectroscopy to study the products obtained.

charges, respectively. This fact correlates well with the temperatures required for the initiation of association reactions (and therefore with activation energies) of chromium with chalcogens under ordinary conditions: sulfur reacts at 380°, selenium at 400°, and tellurium at 500°C. If chromium–selenium or chromium–tellurium mixtures are blasted with smaller charges of explosives, the x-ray powder patterns of compressed mixtures show only the strongest lines of CrSe and CrTe; the reflections of the initial components have practically the same intensity as before shock compression. In addition, a calorimetric study of "undercompressed" samples showed that endothermic effects (the melting of selenium and tellurium) are weak or not present at all, while exothermic effects of chemical reactions are found with the same or even greater intensity as in the control mixture.

The cause of this phenomenon lies in the fractioning of crystallites as a result of the passage of shock waves. Stored energy will naturally facilitate processes of further breakdown of crystal structure and in addition increase the energy released in processes of particle aggregation.

An analogous dependence of the completeness of the chemical reaction on the size of the explosive charge was observed in the case of tin (Batsanov *et al.*, 1969d), where the synthesis of SnS took place with 100 g of hexogen charge, SnSe with 150 g, and SnTe with 500 g. Chemical reaction temperatures under normal conditions for these substances are 250°, 270°, and 420°C, respectively.

Measuring final temperatures in the Sn + S system in accordance with the scheme of Batsanov *et al.* (1969a) showed that the shock-compressed mixture at 0.1 sec after the explosion had $T = 1100°C$, whereas shock-compressed SnS (which had been previously synthesized by usual methods) had a final temperature of only 130°C. Such a sharp increase of temperature is attributed to the exothermic character of the association reaction of tin with sulfur.

We have studied the shock interaction of tin with chalcogens using other experimental methods. The possibility of a reaction with chalcogens in the form of thin films on tin plates was studied by the use of shock waves transmitted at right angles to the film (Batsanov *et al.*, in press). It was found that the compound SnS (or SnSe or SnTe) was formed by the interdiffusion of atoms of sulfur (or selenium or tellurium) and tin. An increase of the impulse at the expense of the duration of the explosion load led to the growth of the diffusion area: for example, with an impulse of 3.2 kg sec^{-1}, the layer of sulfide extends to a distance of 1 μm from the surface, while with an impulse of 7.8 kg sec^{-1}, it goes to a depth of 3 μm.

Explosion methods can be used not only to obtain diffusion covering, e.g., covering metals with chalcogenides, but also as a preliminary preparation for subsequent thermoprocessing. It was discovered that the 50-hr annealing of a sample compressed by a 3.2 kg sec^{-1} impulse gave a diffusion area of 9 μm, while for an uncompressed sample the diffusion area after this same annealing was only 4 μm. Thus, shock compression increases the speed of thermodiffusion during subsequent annealing.

As was noted before, chemical combination initiated by shock compression can be accompanied by a sharp heat discharge, which increases still more the final temperature corresponding to a given pressure and the given equation of state of the substance. As a result, the heat remaining after relief of the pressure can seriously affect the structure and even the nature of the products of chemical synthesis. For instance, in the shock synthesis of $Ln_2O_3 + 3H_2O$, the formation of trihydroxide $Ln(OH)_3$ was observed for the cerium group of rare-earth metals only. Starting with Eu, only monohydroxides are obtained (Batsanov *et al.*, 1967b). Since in the lanthanide series, the thermal stability of the hydroxide decreases with the ionic radius of the metal, the decrease in the amount of complexed water is actually the result of the thermodynamic instability of trihydroxides of heavy lanthanides at sufficiently high temperatures.

In our opinion, molecular combination under shock compression is of particular interest from the point of view of synthetic chemistry. The examples given of the production of ferrites and silicates, along with the synthesis of double fluorides (Dremin and Breusov, 1968) and garnets (Adadurov, 1969a) carried out in the Moscow Chemical Physics Institute, give sufficient grounds for optimism. Shock-wave synthesis used to combine those molecules that cannot otherwise be made to react without decomposition, seems to be a promising synthetic tool.

In our laboratory the shock method was used to synthesize, in standard ampoules using small charges of hexogens (50–150 g), both halide and mixed halide complexes of platinium in accordance with the scheme $2KX + PtY_2$, where X and Y are halogens. A series of such compounds was synthesized for the first time. The details of the experiments will be covered elsewhere.

The synthesis of mixed dihalides of tin was also interesting. The usual chemical methods, such as the action of interhalogen XY on a metal by fusion or co-crystallization of SnX_2 and SnY_2, were used to synthesize the following compounds: $SnCl_{0.5}Br_{1.5}$, SnClI, and SnBrI (Batsanov *et al.*, 1971a). However, varying the experimental conditions did not make it possible to obtain the chlorobromide with the composition SnClBr. The treatment of $SnCl_2$ and $SnBr_2$ with shock compression made it possible to

TABLE V

UNIT CELL PARAMETERS OF SnXY COMPOUNDS

Compound	a (Å)	b (Å)	c (Å)
SnClCl	7.79	9.20	4.43
SnClBr	8.27	10.01	4.77
SnClI	8.45	10.07	4.87

obtain a compound of this formula, as well as to study it and determine its structure. In Table V, the cell parameters of mixed chlorohalides of tin which crystallize in the structural form of $PbCl_2$ are given.

As is seen in the table, subsequent substitution of bromine and iodine for an atom of chlorine leads to a systematic increase in the parameters of the orthorhombic cell, which testifies to the correctness of the structural determinations.

Comparison of the results of the chemical and shock syntheses of chlorobromides of bivalent tin indicates a difference in the compositions of the compounds obtained, which is indirect evidence of chemical reaction proceeding in the compressed state of the system, and not after the pressure has been relieved as a result of residual energy. However, this cannot be regarded as definite evidence, since there is no real proof at all that chemical reactions take place within the front of a shock wave. We hope to deal with this problem from the point of view of the "additive shock adiabatics" which have been considered in Section I.

An unusual result of shock synthesis was obtained in the system $CuBr_2 + Cu$, where, as a result of shock compression, a bromide of monovalent copper with reduced interatomic distances was formed. The cell parameter of shock-synthesized CuBr is 5.643 Å, whereas the standard value is 5.690 Å. In addition, the density and dielectric permeability appeared to be greater than those of the standard substance, i.e., 5.26 and 5.06 g cm^{-3} and 17.8 and 10.0, respectively. The thermal analysis showed that the transition from the sphalerite phase to the wurtzite phase took place at a lower temperature, 375°C for a shock-compressed sample, than for a standard sample, where such a transition took place at 395°C. After $\frac{1}{2}$ hr of heating to a temperature of 400°C (in the course of thermal analysis), compressed CuBr resumed in part its original properties: ϵ returned to 13.1, and the cell parameters became 5.664 Å. After repeated heating, all physical properties conformed fully to those of the standard substance. This peculiar change of the cell size without alternation of its symmetry (this phenomenon can be formally called "isomorphous transition") was observed not only with copper bromide, but also with BN, where there was a change of the cell parameters c from 6.66 to 6.56 Å (Batsanov *et al.*, 1970), and with calcium fluoride and cadmium fluoride, where the changes of the cell parameters were, respectively, from 5.54620 to 5.4592 and from 5.3839 to 5.3819 Å. In all of the cases, the physical parameters and physical properties of "compressed" substances resumed their initial values after more or less prolonged annealing.

The shock method can be used to synthesize solid solutions as well as simple compounds. The system RbCl–CsCl is of interest with regard to the application of shock-wave techniques. In this system the Rb and Cs atoms are crystallographically different and have coordination numbers six and eight, respectively. On being shock-compressed, this mixture, irrespective of the ratio of components, forms a solid solution of the composition 2RbCl:CsCl (excess quantities of RbCl or CsCl do not interact), in which the cesium atom reduces its coordination number (Batsanov *et al.*, 1969a). Thus, the lattice becomes a sort of crystalline trap for CsCl, which undergoes a phase transition. Such a method of fixing phase transitions should be of interest.

We shall not dwell on the results of research on and evidence for new phase transitions because of the lack of space, but we note that shock compression has proved to be a very fruitful method of carrying out atomic and electronic transformations.

The material reviewed in this chapter shows that a new field is opening in science—the physics and chemistry of pulse pressures—and is just at the initial stage of its development, where there seem to be more questions than answers. However, it should be clear that chemists now have available a new technique for studying the physiochemical properties of materials.

REFERENCES

Adadurov, G. A., Breusov, O. N., Dremin, A. N., and Lazarev, A. I. (1965). *Fiz. Gor. Vzryva* **4**, 130.

Adadurov, G. A., Aliyev, Z. G., Borodko, J. G., Breusov, O. N., Dremin, A. N., Muranevich, A. Kh., and Khrameyeva, N. P. (1967). *Izv. Akad. Nauk. SSSR, Neorg. Mater.* **3**, 932.

Adadurov, G. A., Bavine, T. V., Breusov, O. N., Dremin, A. N., Klopova, E. N., and Tatsyi. (1969a). "O Chim. Protsessach Pry Udarnom Szjatii." Deponent, Viniti.

Adadurov, G. A., Novikov, G. V., Ovanesyan, N. S., Truhtanov, V. A., and Shehtman, V. M. (1969b). *Fiz. Tverd. Tela.* **11**, 1988.

Altshuler, L. V., Brazhnik, M. I., German, V. I., and Mirkin, L. I. (1967). *Fiz. Tverd. Tela.* **9**, 3063.

Batsanov, S. S. (1970). *Izv. Akad. Nauk. SSSR, Neorg. Mater.* **6**, 697.

Batsanov, S. S., and Deribas, A. A. (1965). *Fiz. Gor. Vzryva No.* **1**, 103.
Batsanov, S. S., and Zolotova, E. S. (1968). *Dokl. Akad. Nauk. SSSR* **180**, 93.
Batsanov, S. S., Derbeneva, S. S., *et al.* (1966a). *Fiz. Gor. Vzryva* **3**, 110.
Batsanov, S. S., Deribas, A. A., and Kutolin, S. A. (1966b). *Izv. Akad. Nauk SSSR, Neorg. Mater.* **2**, 87.
Batsanov, S. S., Boreskov, G. K., *et al.* (1967a). *Kinet. Catal.* **8**, 1348.
Batsanov, S. S., Deribas, A. A., and Kustova, G. N. (1967b). *J. Neorg. Chim.* **12**, 2283.
Batsanov, S. S., Doronin, G. S., Koshevoi, V. P., and Stupnikov, V. P. (1968a). *Fiz. Gor. Vzryva* **4**, 108.
Batsanov, S. S., Zhdan, P. A., and Kolomiychouk, V. N. (1968b). *Fiz. Gor. Vzryva* **4**, 274.
Batsanov, S. S., Doronin, E. M., and Ovsyannikova, I. A. (1969a). *Phys. Gor. Vzryva* **5**, 283.
Batsanov, S. S., Malyshev, E. M., Kobets, L. I., and Ivanov, V. P. (1969c). *Fiz. Gor. Vzryva* **5**, 444.
Batsanov, S. S., Lapshin, A. I., and Ovsyannikova, I. A. (1969b). *J. Strukt. Chem.* **10**, 268.
Batsanov, S. S., Shestakova, N. A., Stupnikov, V. P., Litvak, G. S., and Nigmatullina, V. M. (1969d). *Dokl. Akad. Nauk. SSSR* **185**, 330.
Batsanov, S. S., Moroz, E. M., and Kuzutin, V. P. (1970). *J. Strukt. Chim.* **11**, 156.
Batsanov, S. S., Lyachova, V. F., and Moroz, E. M. (1971a). *J. Neorg. Chim.* **16**, 231.
Batsanov, S. S., Zhdan, P. A., Kolomiychouk, V. N., Lapshin, A. L., and Nigmatullina, V. M. (1971b). *Phys. Chem. Methods Treatment Mater.* **4**, 139.
Batsanov, S. S., Ovsyannikova, I. A., and Shestakova, N. A. (to be published).
Bergmann, O., and Barrington, J. (1966). *J. Amer. Ceram. Soc.* **49**, 502.
Boldyrev, V. V., Zarko, E. E., and Deribas, A. A. (1967). *High Energy Chem.* **1**, 177.
De Carli, P., and Jamieson, J. (1959). *J. Chem. Phys.* **31**, 1675.
Doronin, G. S., Stupnikov, V. P. (1970). *Izv. SO(Siberia) Akad. Nauk SSSR* **3**, 102.
Dremin, A. N. and Breusov, O. N. (1968). *Usp. Khim.* **37**, 898.
Dremin, A. N. and Karpukhin, A. I. (1960). *J. Appl. Mech. Tech. Phys.* **3**.
Heckel, R., and Youngblood, J. (1968). *J. Amer. Ceram. Soc.* **51**, 398.
Horiguchi, Y. (1966). *J. Amer. Ceram. Soc.* **49**, 519.
Horiguchi, Y., and Nomura, Y. (1963). *Bull. Chem. Soc. Japan* **36**, 486.
Horiguchi, Y., and Nomura, Y. (1965). *Chem. Ind.*, 1791.
Horiguchi, Y., and Nomura, Y. (1966). *J. Less-Common Metals* **11**, 378.
Kalinin, V. A., and Zharkov, V. N. (1968). "Equation of State of Solid Bodies under High Pressures and Temperatures," Nauka.
Kimura, Y. (1963). *J. Appl. Phys. Japan* **2**, 312.
Klein, M., and Rudman, P. (1966). *Phil. Mag.* **14**, 1199.
Klein, M., Rough, F., and Simons, C. (1963). *J. Amer. Ceram. Soc.* **46**, 356.
Kormer, S. B. (1968). *Usp. Fiz. Nauk.* **94**, 641.
Lapshin, A. I., and Batsanov, S. S. (1968). *J. Appl. Spectrosc.* **8**, 1033.
Lapshin, A. I., Borodaevsky, and Batsanov, S. S. (1970). *Chim. Vys. Energyi* **4**, 154.
Milton, D., and De Carli, P. (1963). *Science* **140**, 670.
Nomura, Y., and Tobisawa, S. (1965). *Appl. Phys. Lett.* **7**, 126.
Ryabinin, J. (1956). *J. Tech. Phys.* **26**, 2661.
Yaroslavtsev, V. A., Zhdan, P. A., Kolomiychouk, and Batsanov, S. S. (to be published).

Modern Methods of Growing Single Crystals of High-Melting-Point Oxides

A. M. ANTHONY

Centre National de la Recherche Scientifique
Orleans, France

R. COLLONGUES

Université de Paris
Paris, France

Introduction

The first real scientific application of artificially produced single crystals dates back to the first World War, when quartz and tourmaline crystals were used in ultrasonic detection apparatus. It was only after the second

World War that the industrial use of single crystals enjoyed a significant development, notably with the advent of semiconductors. This industrial development coincided with a desire on the part of researchers to obtain a better understanding of the properties of the solid state, namely an understanding of the physical phenomena (magnetism, energy transfer, etc.) from the point of view of the physicists, and a knowledge of the materials from the point of view of metallurgists and chemists.

For a long time, whatever knowledge existed of the properties of solids was based on experiments with sintered materials or with polycrystalline materials. In such cases, all the properties measured are greatly affected by grain boundaries and porosities, so that very large discrepancies can be found in the results obtained by different experimenters. Indeed, it would not be an exaggeration to say that measurements made on sintered or polycrystalline materials more often indicate the properties of boundaries and the grain porosity of the material than its intrinsic properties. Only by studying single crystals is it possible to acquire an understanding of the intrinsic properties of the solid.

There are many different methods of growing single crystals. Every year, several international congresses are devoted to describing new growth methods, and entire journals are devoted to the publication of the results obtained. In fact, the very number of the methods used can lead to two conclusions:

1. For any given material, there is no unique crystal-growing method. Depending on the mechanical, physical, or chemical characteristics desired in the material, one method may be preferable to another.
2. It is almost never possible to adapt a given method to a new problem without some changes. Each material poses its own crystal-growth problems, and preparing a single crystal of some new material is invariably a long and tedious process.

It is therefore most important for the future development of solid state physics and chemistry to have at one's disposal an extensive range of crystal-growing methods from which the experimenter can choose the one best suited to his material.

For any given material, the transition from the initial crystallization state to the final crystallization state (the single crystal) can take place in two ways:

1. By progressive evolution and migration of the grain boundaries.
2. By total destruction of the initial state of crystallization (change of state) and recrystallization in the desired form (single crystal).

Methods of the first type have been applied until now to metals, with

TABLE I

A. Growth from the pure melt (temperature gradient, heat transfer)

moving crucible	moving crystal	moving interface	moving material
Bridgman–Stöckbarger	Czochralski	Kyropoulos	Verneuil

Advantages	*Disadvantages*
Rapid growth rates	Quality may be poor
Simple apparatus	Large temperature gradients
Large crystals obtainable	Crystals may require annealing
	Constitutional supercooling for heavily doped crystals

B. Growth from solution (concentration gradient, material transfer)

liquid water (<100°C)	liquid water (>100°C)	fluid water (>T_c)	molten oxide or salt	vapor carrier
water-soluble crystals	hydrothermal growth		fluxed-melt growth	vapor-phase growth

Advantages	*Disadvantages*
Almost isothermal growth environment	Possible contamination by container or transport medium
Slow growth, good quality	Growth rate slow for production
Crystals unstrained	

some very fine crystals being obtained by critical cold-hardening or secondary recrystallization. It is worth noting that some successful attempts to grow crystals of nonmetallic materials have been made in the last few years.

Methods of the second type are by far the most important for nonmetallic materials. The changes of state which occur are solid–liquid–solid or solid–vapor–solid.

In the case of high-melting-point oxides, the main difficulty obviously involves the change-of-state temperature. Fusion or sublimation requires very-high-temperature-generating facilities. The problem essentially is then to adapt the crystal-growth method selected to produce the crystal characteristics required at the high temperature needed.

When the temperature involved is very high, the difficulties encountered are such that many experimenters prefer to resort to an auxiliary constituent (flux) to lower the working temperature.

It is possible to distinguish between two major families of methods (Table I) (White, 1964a):

1. High-temperature growth methods with or without a change of state. These methods involve only the material itself.
2. Low-temperature methods involving an auxiliary substance.

Having made this distinction, it will be possible later to compare the characteristics of the crystals obtained by the two methods.

I. High-Temperature Crystal Growth

A. Review of High-Temperature Crystal-Growing Methods

These methods of growing single crystals can be classified in several ways. As in any other classification, a certain measure of arbitrariness is introduced. It is proposed here to adopt a classification based on the nature of the phases involved:

Single-crystal growth from the vapor phase.
Single-crystal growth from the liquid phase.
Single-crystal growth in the solid state.

1. *Single-crystal growth from the vapor phase*

The principle of the method is extremely simple. It consists in subliming the substance and then condensing it on a collector.

This method has been applied primarily to low-melting-point, high-vapor-pressure metals (Zn, Cd, Hg). Sulfides, such as CdS and ZnS, also lend themselves well to this type of preparation. Among the highly refractory single crystals, mention may be made of the silicon carbide crystals obtained by this method. As a rule, the crystals obtained are small (although 25-mm-long samarium single crystals have been prepared recently). The method is particularly useful for preparing trichites.

Even at very high temperatures, refractory oxides usually have very low vapor pressures, which explains why this method has found only limited application for growing single crystals of high-melting-point oxides. However, an example of growth by this method is the case of single crystals of WO_3.

2. *Single-crystal growth from the liquid phase*

This group of methods is by far the most important for growing single crystals of high-melting-point oxides without the use of an auxiliary substance. The usual method consists in melting the oxide and recrystallizing it under certain conditions leading to the formation of a single crystal.

The oxides considered here are single or mixed superrefractory oxides with melting temperatures usually above 2000°C (Table II).

Regardless of the method, the crystal-growing system consists of four main components: the starting material, the liquid, the crystal, and the solid–liquid interface. The spatial growth of the crystal can take place only by a process of displacing one of the four components. It is proposed to adopt the classification suggested by White (1964) and Collongues (1968) and based on the nature of the moving component in the system.

a. Displacement of the crystal–liquid interface. This method is known variously as the Bridgman, Stöckbarger, or Kyropoulos method, the respective name indicating the modifications these authors have introduced.

TABLE II

Oxide	*T* (°C)	Oxide	*T* (°C)	Oxide	*T* (°C)
BeO	2550	Al_2O_3	2050	CeO_2	2600
CaO	2575	La_2O_3	2300	HfO_2	2780
MgO	2800	Sm_2O_3	2350	ThO_2	3210
NiO	2000	Y_2O_3	2410	UO_2	2870
SrO	2410	Yb_2O_3	2350	ZrO_2	2680

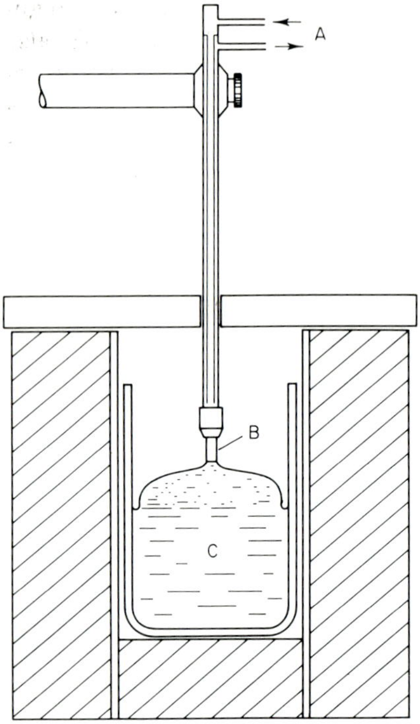

FIG. 1. Kyropoulos apparatus (White, 1964a), A, Water; B, seed crystal; C, molten salt.

The melt is placed in a crucible at a temperature slightly higher than its melting temperature (Fig. 1). A strongly cooled seed crystal is brought to the surface of the liquid and cooling takes place through it. Crystallization begins at the surface of the seed crystal and proceeds within the melt. The size of the crystal will be determined primarily by the furnace temperature control. Certain changes in the relative positions of the seed crystal and the crucible are sometimes necessary if crystallization takes place with a marked change in volume. Until now, this technique has been used mainly for producing large crystals of alkali halides. In such cases, the rate of growth is of the order of 1 cm hr^{-1}. The crystals are of good optical quality, but are crystallographically imperfect. In the specific case of single crystals obtained from high-melting-point oxides, two basic difficulties are encountered:

1. The first and most important difficulty is that above 2300°C it is nearly impossible to find crucibles capable of containing a melted oxide.
2. The other difficulty is maintaining a solid seed crystal—usually a poor heat conductor—at the surface of the melt.

When this method is used for superrefractory oxides, it is in a somewhat modified form: The material is melted in the "self-crucible" technique, and the walls of this "self-crucible" play the part of the seed crystal. The growth of stabilized zirconia single crystals by this method will be described later.

b. Displacement of the crystal. This is the so-called pulling or Czochralski method. It was first used for growing single-crystal metallic films. As in the method just described in *a*, the melt is maintained at a temperature slightly above the melting point, and heat is extracted by means of a seed crystal. But in this case, the crystal, in the course of its growth, is drawn out of the melt at a rate such that the crystal–liquid interface stays in a fixed position (Fig. 2). In order to avoid stresses and high dislocation densities, the system is carefully controlled so as to maintain a flat interface, making it necessary to compensate for losses due to radiation from the crystal faces by an auxiliary heat input. The Czochralski method is probably currently the most important method for growing single crystals from a melt. It is the major technique for growing silicon and germanium crystals used for semiconductors.

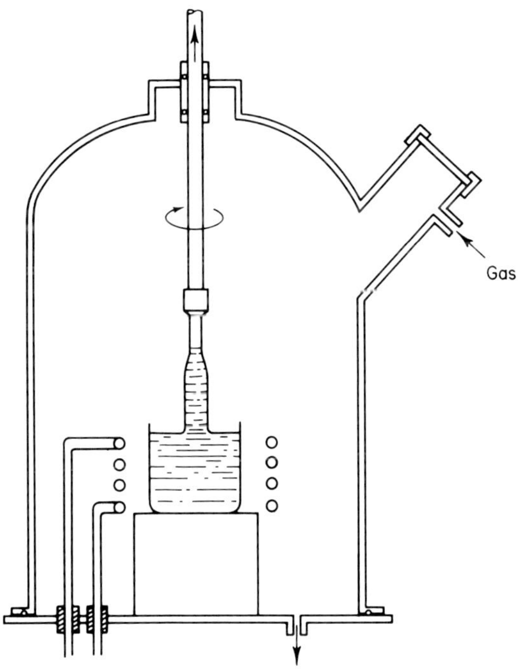

FIG. 2. Czochralski apparatus (White, 1964a).

Here again, the basic problem in the case of high-melting-point oxides is the incompatibility of the melt with the crucible used. Thus crystals of $CaWO_4$ have been produced in platinum-alloy crucibles (melting point around 1500°C), while crystals of alumina-based substances (rubies, garnets, lanthanide aluminates—melting point about 2000°C) have been grown in iridium crucibles. In the case of materials with a melting point above 2300°C, there is no known crucible compatible with the melt, so that the operation must be carried out by utilizing the "self-crucible" technique.

Several examples will be described later, including the production of laser materials in iridium crucibles, and the production of stabilized zirconia crystals by the self-crucible technique.

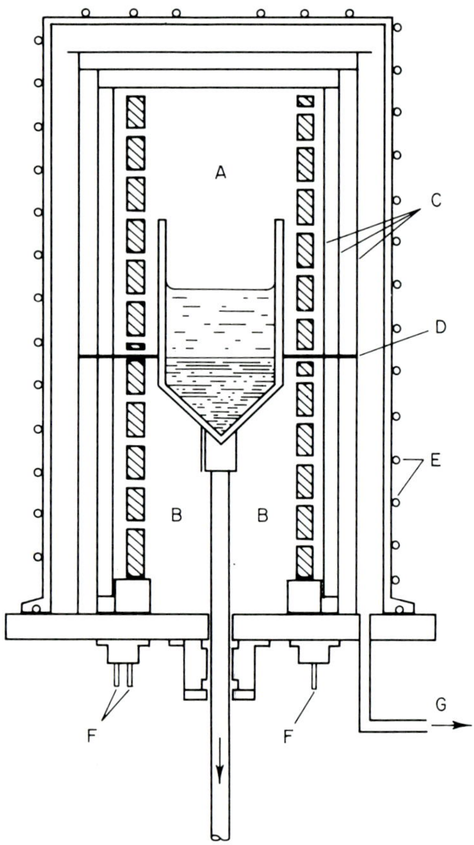

FIG. 3. Bridgman–Stöckbarger apparatus (White, 1964a). A, Hot zone; B, cool zone; C, radiation shields; D, baffle; E, water cooling; F, power leads; G, vacuum.

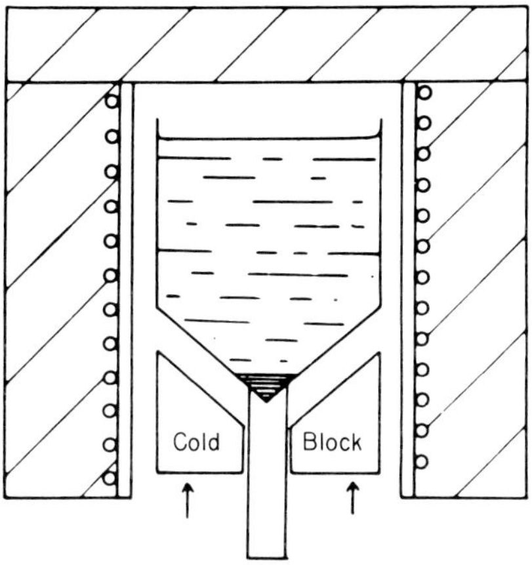

FIG. 4. Stöber–Strong apparatus (White, 1964a).

c. Displacement of the liquid. This is known as the Bridgman method and was developed first by Stöckbarger.

The method consists in lowering the crucible containing the melt through a space providing a steep temperature gradient. Nucleation occurs naturally at various points in the coldest region, i.e., the bottom of the crucible. To limit the number of nuclei produced, the crucible is cut to a fine point (Fig. 3), the cone apex angle being an important factor in the crystal quality. It should be observed that no seed crystal is used in this method. A temperature gradient is provided, as a rule, by an auxiliary furnace placed beneath the main furnace. In a variation on the Bridgman method (the Stöber–Strong method), the bottom of the crucible is surrounded by a block of cooled copper (Fig. 4).

These methods have been used chiefly for growing single crystals of various alkali or alkaline-earth halides, ferrites of the spinel type, and calcium tungstate.

As for high-melting-point oxides, the crucible problem exists in this method also.

d. Other liquid-displacement methods (melting-zone and floating-zone). These methods, well-known as means of purification, have also been used for the growth of single crystals. In the horizontal melting-zone method, the material is contained in a boat, and a single crystal seed is placed at one end. The melting zone is brought into contact with this seed crystal and then moved along the boat (Fig. 5). The quality of the crystal is controlled mainly by the rate of displacement.

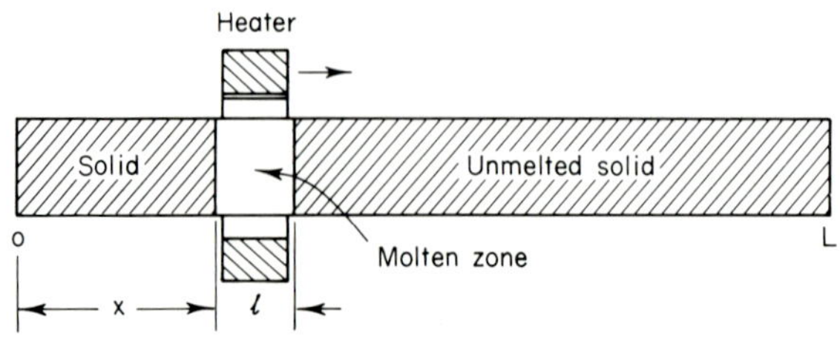

FIG. 5. Melting-zone technique.

In the case of high-melting-point oxides, the main problem arises from the possibility of contamination by the boat. In such cases, preference is given to the vertical melting-zone, or the floating-zone, method (Fig. 6). The starting material is prepared in the form of a bar secured at either end. The process of crystal growth by displacement of the zone takes place as before. Naturally, stability of the zone depends on the material and on its physical properties, especially its surface tension at the melting temperature.

This unusually attractive method also introduces a purification of the material by the process of zone melting. A few examples of application to crystal growth from alumina, ruby, and Ga_2O_3 (gallium oxide) will be described.

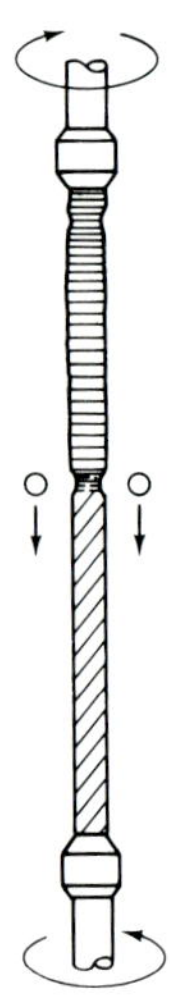

FIG. 6. Floating-zone technique.

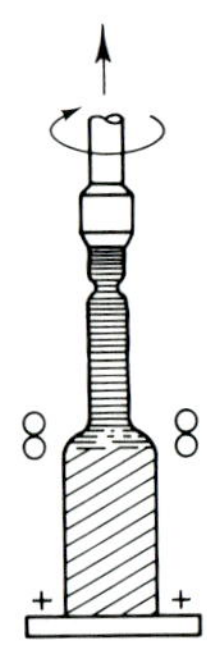

FIG. 7. Modified floating-zone technique.

There exists a variation of the floating-zone method which is, in fact, a combination of the floating-zone method and the pulling method. It consists in using a large-diameter bar as the starting material. The upper end of the bar is melted and a crystal is drawn from this melted drop. The drawing rate and the rate of displacement of the initial bar are inversely proportional to the square of the diameters (Fig. 7).

e. Displacement of the starting material. This is the Verneuil method, which was the first method used to produce artificial single crystals from high-melting-point oxides in 1904 and was used to prepare ruby and

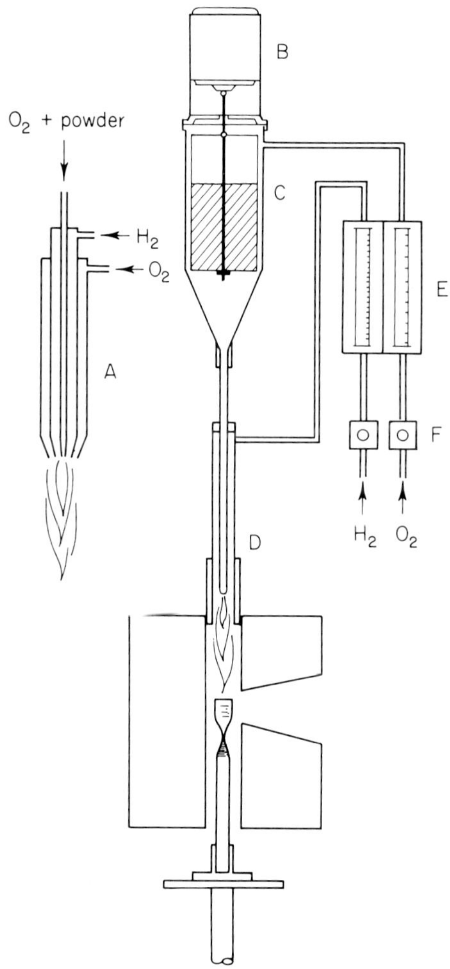

FIG. 8. Verneuil apparatus (White, 1964a). A, Tricone burner; B, vibrator; C, feed hopper; D, standard burner; E, flow gauges; F, gas controls and regulators.

sapphire crystals. Verneuil's apparatus has remained virtually unchanged since that time. The principle of the method is to cause the starting material, in a fine powder form, to pass through the flame of an oxyhydrogen torch (Fig. 8). After melting has taken place in the flame, the droplets feed the melted upper part of the growing crystal. Crystallization may begin either on a piece of previously melted material, or preferably on a seed crystal. The automatic lowering of the crystal is controlled so as to maintain the interface at a fixed distance from the flame. Recent modifications to Verneuil's apparatus have involved the following:

1. A number of features assuring greater physical and chemical homogeneity and improved crystal quality: feed method, better control of the gaseous outputs, rotation of the support, possible annealing to eliminate thermal gradients.
2. Revised system geometries: support axis perpendicular to torch axis to obtain disks or cylinders, possibility of providing a time-varying feed-powder composition to obtain concentric layers of different composition.
3. New heating methods to enable single crystals to be obtained from oxides with a melting temperature of over 2200°C, which would be impossible with an oxyhydrogen torch (plasma furnace) or a flameless device, for the purpose of obtaining growth in a controlled atmosphere (image furnace or electric arc furnace).

Two recent applications will be described: growing single crystals of oxides of transition elements by transfer through an electric arc, and of CaO (melting temperature 2600°C) using the plasma torch as the means of heating.

3. *Growth in the solid state*

If a polycrystalline solid is heated for a sufficient length of time at a temperature below its melting temperature, an increase in the average size of the crystals will be observed. It is possible to distinguish among three different processes, depending on the circumstances of such growth:

1. Primary recrystallization: germination and growth of new crystals during the process of heating a plastically deformed material.
2. Growth of the crystals: an increase in average size without germination of new crystals. This usually follows the primary recrystallization. The driving force is the reduction in the area of the grain boundaries and hence in free energy of the solid. In theory, the steady state is reached when the whole of the solid has been transformed into a single crystal. In prac-

tice, the size of the crystals obtained depends on the number of active nuclei capable of growing at the expense of the polycrystalline matrix. Following primary recrystallization with a very pronounced texture, the number of such nuclei is small, and large crystals can be obtained.

3. Intermittent growth: the development of very large crystals in a matrix having two constituent elements. This occurs when growth of the grains is inhibited by the second constituent, which may be a second-phase inclusion, a porosity, or any break in continuity able to prevent the migration of the boundaries. Intermittent growth occurs when this break in continuity vanishes (through dissolution in the matrix, evaporation, etc.), thereby freeing the grain boundaries.

These three processes have been fully identified in metals making the growth of some very beautiful crystals possible. Although these processes have been less thoroughly investigated in the case of nonmetallic solids, they nonetheless exist. Primary recrystallization has been observed with alkali halides or oxides that can be deformed plastically under certain conditions. Enlargement of the grains is also known, though the conditions existing at the grain boundaries are not always clearly specified. Finally in the case of ceramics and refractory oxides in particular, intermittent growth is the most effective way of obtaining single crystals.

Consider the interface separating a polycrystalline aggregate from a single crystal (Fig. 9). It is well known that the rate of migration of a grain boundary increases with increasing temperature and is inversely proportional to its radius of curvature. It will therefore be necessary for the radii of curvature at the single-crystal–polycrystal interface to be small, i.e., for the crystals of the polycrystalline aggregate to be small. For a given rate of growth, it will then be possible to reduce the temperature, thus making it possible to decrease the vapor or dissociation pressure and to reduce the number of flaws of thermal origin. To ensure that growth of the single crystal continues under satisfactory conditions, the growth of the crystals in the polycrystalline aggregate must be limited throughout the operation. This is accomplished first, by reducing the mean quadratic

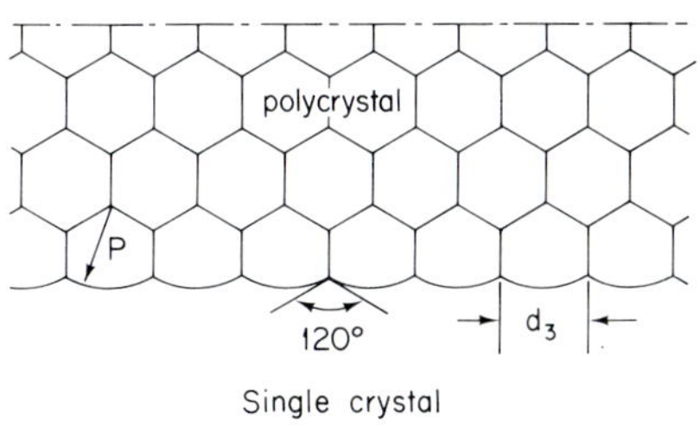

FIG. 9. Schematic illustration of the growth front of a single crystal in a polycrystalline aggregate (Lacour and Paulus, 1968). Used with permission of North-Holland Publishing Co., Amsterdam.

error between the aggregate's crystals and second, by maintaining the aggregate at a temperature below that of the growth front, i.e., by establishing a thermal gradient.

These growth methods in the solid state are particularly useful in the case of materials with noncongruent melting characteristics, or in which one element has a high vapor pressure, or which can be prepared only by sintering.

The case of barium ferrite single crystals will be discussed.

B. Examples of the Growth at High Temperature of Single Crystals of High-Melting-Point Oxides

Here we give a few examples to illustrate the methods described previously: growth from the vapor phase, the liquid phase, and the solid state.

1. *Growth from the vapor phase*

We shall describe the growth of single crystals of tungsten oxide (WO_3) by sublimation and condensation (sublimation temperature, 1350°C).

Tungsten oxide (WO_3) is interesting because of its ferroelectric and semiconductor properties at certain temperatures. It exhibits numerous

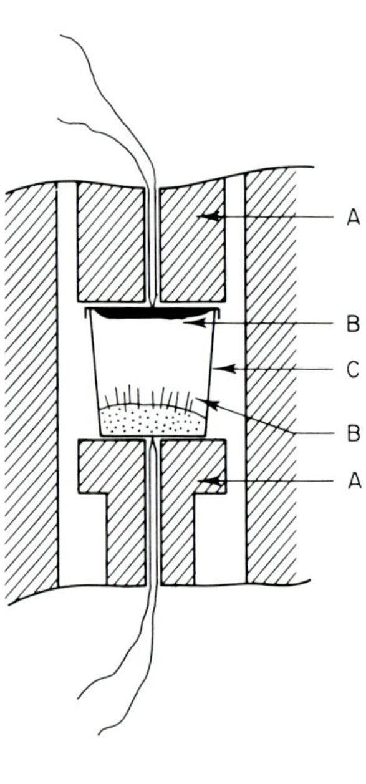

FIG. 10. Growth of WO_3 crystals (Le Bihan and Vacherand, 1969). A, Alumina; B, WO_2; C, platinum crucible.

FIG. 11. Single-domain and polydomain WO_3 single crystals (Le Bihan and Vacherand, 1969).

changes of phase, which makes it even more interesting when these are related to the changes in its properties. This study requires the growth of single crystals which are obtained by sublimation in air or in oxygen in a platinum crucible. The crucible is slightly conical in shape, with a diameter of 2–3 cm and a height of 3 cm. It is two-thirds filled with a charge of WO_3 powder. A platinum/platinum–rhodium thermocouple allows the temperatures at the top and bottom of the crucible to be measured. The assembly is placed in a vertical furnace so that the bottom of the crucible is at 1370°C and the top is at 1340°C. Crystals develop on the crucible lid and immediately above the charge (Fig. 10). These are lamellar crystals which, depending on conditions, reach thicknesses varying from a few microns to $\frac{1}{2}$ mm. The platelets usually have sides 1 cm long, but can attain a length of 2–3 cm. Some have a multidomain structure, others a single-domain structure (Fig. 11).

Chemical analysis shows that their formula corresponds to $WO_{3\mp0.002}$. It is also possible to prepare distinctly nonstoichiometric crystals in an argon atmosphere. Indeed, their green color appears to be caused by a loss of oxygen. Single crystals of molybdenum oxide (MoO_3) have also been grown by this method (Fig. 12).

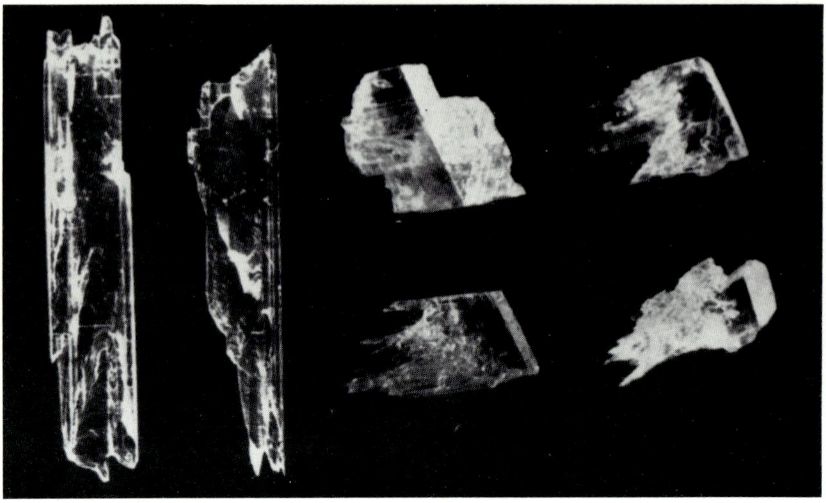

FIG. 12. Single crystals of MoO_3 from the upper and lower parts of the sublimate (Wanklyn, 1968). Used with permission of North-Holland Publishing Co., Amsterdam.

2. *Growth from the liquid phase*

The previously described study of the methods of growing single crystals of oxides from melts has revealed a major difficulty due to the high melting temperatures of the oxides in question, namely the difficulty of finding a crucible capable of containing the melted oxide without reacting with it.

Crucibles made of noble metals (platinum, platinum alloys, iridium) are generally used. These are suitable when the melting or evaporating temperature of the oxide lies below the melting temperature of the metal, but when the temperature required is above 2300°C, it becomes essential to resort to methods not involving the use of a crucible. A few examples of these two categories of methods will now be given.

a. Crystal growth in crucibles. Among the numerous classes of materials crystallized by methods involving melting in a crucible, the laser materials are chosen for discussion, since they are currently the subject of exhaustive research.

Before proceeding further, it may be worth while to review the principal properties which materials capable of laser emission must possess. First, these materials must fluoresce with an intense emission and a narrow bandwidth under the anticipated experimental conditions; second, they must have very good optical qualities. Some of the materials which have produced a laser effect are given in Table III.

The choice of materials capable of operating continuously with a low input power is limited by the following considerations:

1. The matrix must be a single crystal of the most perfect optical quality and be free from flaws, inclusions, or impurities which can act as light-diffusing centers. It must also be hard and chemically stable, to enable it to be suitably polished.
2. Florescent ions, primary rare-earth and transition elements possessing several energy levels, including an upper level broad enough to permit pumping, must occupy clearly defined positions in the matrix structure. It is important to avoid electrostatic effects due to differences in charge between the substituted ions and the matrix ions, as is the case with calcium tungstate doped with trivalent rare-earth elements. In such cases, the charges must be compensated with sodium, for example.

Under such conditions, the spectrum of the fluorescent ions must have narrow bands consistent with the crystal's symmetry. As a rule, the

TABLE III

Crystal	Fluorescent ion	Crystal	Fluorescent ion
Al_2O_3	Cr^{3+}	$SrWO_4$	Nd^{3+}
CaF_2	U^{3+}	$CaMoO_4$	Nd^{3+}
	Nd^{3+}		
	Ho^{3+}	$SrMoO_4$	Nd^{3+}
	Sm^{2+}		
	Dy^{2+}	$PbMoO_4$	Nd^{3+}
	Tm^{2+}		
SrF_2	U^{3+}	$La_{0.5}No_{0.5}MoO_4$	Nd^{3+}
	Nd^{3+}		
	Tm^{3+}	$Y_3Al_5O_{12}$	Nd^{3+}
	Sm^{2+}		Ho^{3+}
BaF_2	U^{3+}	$Y_3Ga_5O_{12}$	Nd^{3+}
	Nd^{3+}		
LaF_3	Nd^{3+}	$Gd_3Ga_5O_{12}$	Nd^{3+}
$CaWO_4$	Nd^{3+}	$CaNb_2O_6$	Nd^{3+}
	Pr^{3+}		
	Tm^{3+}	$Ca_5(PO_4)_3F$	Nd^{3+}
	Ho^{3+}		
	Er^{3+}		

greater the symmetry of the crystal, the greater the degeneration of the electron states and the fewer the transitions observed in the fluorescence.

3. An optimal environment exists for each ion capable of fluorescing. The Cr^{3+} ion, for instance, will fluoresce only in matrices in which it substitutes for Al^{3+}, Ga^{3+}, and Mg^{2+} and not in others.

Of the rare-earth ions, those with an atomic number close to that of Gd^{3+} will fluoresce in substances of the hydrate type, a larger number will fluoresce in fluorides, and all of them in oxides. In oxides, the intensity of emission of a rare-earth ion tends to increase with the bonding energy, especially when the ion's emission is not masked by interactions with the matrix.

4. The concentration of the fluorescent ion is also important. In the case of ruby, where transitions take place toward the fundamental state, an excess of Cr^{3+} can make it difficult to obtain population inversions for the two levels considered, and the spectrum reveals additional fine lines.

In the case of rare-earth elements in which the effects of the crystal field are less important, the opposite effect often occurs because emission becomes simpler and more concentrated.

5. Finally, the processes involving energy transfer phenomena between the matrix and the rare-earth ions, or between the transition ions and rare-earth ions, are often of interest. This has been demonstrated for Sm^{3+} in calcium tungstate, or for the transfer between Cr^{3+} and Nd^{3+} in yttrium aluminum garnet.

In practice, laser matrices capable of receiving dopant ions can be classified into these main groups: corundum and ruby; garnets; alkaline earth fluorides; tungstates, calcium molybdates and niobates, strontium, and lead; and new matrices like the apatites.

(1) Ruby and garnet (Margotin, 1969). Ruby ($Al_2O_3 + Cr_2O_3$ solid solution) has the rhombohedral structure of corundum (Al_2O_3) and its melting temperature is 2050°C. Yttrium aluminum garnet $Y_3Al_5O_{12}$ (YAG) is cubic and its melting temperature is 1970°C.

High-purity alumina is prepared by calcinating ammonium alum, $Al_2(SO_4)_3(NH_4)_2SO_4 \cdot 24H_2O$, which undergoes several fractionated crystallizations. By cocrystallization with the corresponding chromium alum, good uniformity of distribution of the chromium can be ensured. As a rule, the products used attain purities of 99.999%.

In most cases, the operation is carried out under controlled atmosphere so as to avoid the dissolution of nitrogen or oxygen in the melt and the formation of bubbles in the crystal obtained (Fig. 13). The means of heating

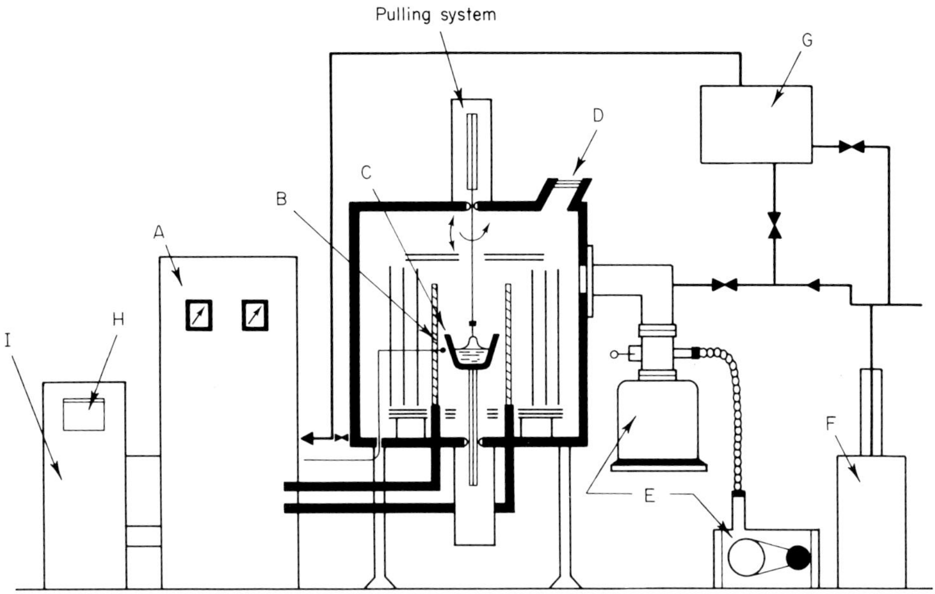

FIG. 13. Pulling furnace for producing single crystals (Margotin, 1969). A, Power supply; B, thermocouple; C, crucible; D, window; E, vacuum system; F, gas purifier; G, gas analyzer; H, recorder; I, regulation.

can be either a metal resistor or an induction coil. Even at temperatures close to 2200°C, the useful life of the crucible will permit several operations.

The crucible is usually one-half or two-thirds full. Temperature regulation is realized either by optical pyrometry or by means of a tungsten/tungsten–rhenium thermocouple, and can be controlled within 0.5°C.

The drawing rate may be less than 1 mm hr^{-1} and may vary up to 10 or 15 mm hr^{-1}, but is commonly about 5 mm hr^{-1}. Rotation speeds are between 1 and 40 rpm. A number of other parameters also significantly affect the quality of the crystal. An example is the temperature gradient in the crucible: The temperature measured at the top of the crucible can be kept 50°C below the temperature measured on the melt. The crucible may or may not be held stationary during the drawing process, provided the temperature gradient is maintained. Of course, all these parameters depend on the volume of the crystal, which governs the concentration of the dopant element. It is commonplace to note variations of the order of 10–20% in the average content from one end of the crystal to the other.

It is routinely possible to draw ruby or garnet crystals with a diameter of 30 mm and lengths of 150 mm or 80 mm, respectively. In rather more special operations, it has been found possible to draw ruby crystals 200 mm long and 16 mm in diameter with outstanding optical properties. Ruby boules are usually drawn 60° from the crystal axis, while garnet boules are drawn along the [111] axis.

(*2*) *Fluorite* (*Guggenheim, 1963: Petit Le Du, 1970*). The melting point of fluorite is 1402°C. Doped with neodymium, it constitutes a well-known laser generator. For this application the crystal matrix should be of great purity; more specifically, its oxygen content should be as low as possible, and it should be capable of accepting the effective quantity of neodymium injected in the form of NdOF or Nd_2O_3 without deterioration of the lattice. The purity of the material is evaluated by spectrographic and radiochemical analyses.

The starting material used is a synthetic powder in which the following impurities were revealed by spectrographic analysis: $Mg > 10^{-4}$, $Si > 10^{-6}$, $Fe \approx 10^{-6}$, $Al \approx 10^{-6}$, $Sr \approx 10^{-6}$.

This fluorite powder also contains a considerable quantity of adsorbed water, whose presence, like that of the oxygen, is a major obstacle to growing a good quality crystal. The water is eliminated by heating the powder, before melting, to 400°C for 6 hr while argon purified by means of molecular sieves is passed over it. About 2½% of the water is retained. After being dehydrated in this way, the powder is melted in a graphite boat, then purified by zone-melting to get rid of the principal impurities by evaporation and segregation. Traces of lime are eliminated by adding PbF_2, according to the reaction

$$CaO + PbF_2 \rightarrow CaF_2 + PbO$$

Because the PbO and the PbF_2 are volatile at the fluorite's melting temperature, the residual lead content is of the order of 10^{-6} This zone-melting usually takes place with a slow rate of transfer (<25 mm hr^{-1}). Three or four zone passages will usually suffice to obtain crystals of excellent limpidity, hence free from lime, and with a notably reduced content of other impurities. For instance, in the case of magnesium, spectrographic analysis at three points along a 15-cm-long bar revealed an increasing concentration from end to end of 10^{-6}, $>10^{-6}$, and 10^{-5}, respectively.

The starting material prepared in this way is crystallized by Bridgman's method. A graphite crucible (inner diameter 12 mm, height 120 mm) is arranged vertically inside a tubular silica-glass vessel, under purified argon, and heated by a two- or three-coil induction heater and a 7-kW, 3-MHz high-frequency generator.

It should be noted that graphite is not an ideal crucible material, because of its reactivity with calcium fluoride in the presence of the residual impurities it contains and the neodymium which is added. A process of degradation and blackening may be observed toward the stem of the crystal, probably due to the formation of such compounds as CaC_2 (black), NdC_2 (yellow), Al_4C_3 (yellow-green), Fe_3C (grey), and SrC_2 (black). As a result, a 90-mm crystal can be used only over approximately half its length.

The fluorescence effect at 1.0885 μm is obtained by doping with neodymium in the form of NdOF or, preferably, of Nd_2O_3, the quality of this doping being controlled on each crystal by means of the fluorescence spectrum. The coefficient of segregation of neodymium in the fluorite is of the order of 0.95. The dopant is mixed with crushed fluorite and placed at different points along the initial bar. The search for the optimum dopant value led us to explore a wide range of concentrations [from 0.05 to 5% (Nd/Ca atoms) for NdOF and from 0.05 to 0.6% for Nd_2O_3] and finally to adopt 0.2% in both cases. Mother-alloy doping was attempted with no substantial improvement.

Acceptable growth rates lie between 5 and 25 mm hr^{-1}. One of the important factors determining the rate of crystallization is the appearance of internal stresses which make the fluorite particularly brittle; the slowest rates are therefore better. However, by using a crucible of high thermal inertia and adopting very slow cooling, thus creating a sort of *in situ* annealing, it is possible to attenuate these stresses and their effects to a great extent.

The useful portion of the single crystals obtained in this way is 40 mm long and 12 mm in diameter. Rods extracted from these crystals have exhibited a laser effect with a pulse excitation threshold of 50 J at ordinary temperatures.

(*3*) *Calcium niobate* (*Ballman et al., 1963: Petit Le Du, 1970*). Calcium niobate $Ca(NbO_3)_2$ crystallizes in the orthorhombic system. Its melting point is 1560°C. Crystals doped with neodymium can be used as lasers, emitting continuously at ordinary temperatures, under better conditions than calcium tungstate.

The starting material is prepared by melting together, in an iridium crucible in air, niobium oxide Nb_2O_5 and calcium carbonate $CaCO_2$, according to the reaction

$$Nb_2O_5 + CaCO_3 \rightarrow Ca(NbO_3)_2 + CO_2$$

Purification is then effected by successive drawings at the rate of 25 mm hr^{-1}, with argon containing 5–10% oxygen passing through. The presence of oxygen is indispensable to avoid the creation of oxygen vacancies in the lattice, which would be indicated by the appearance of a blue-black color. The ideal condition would be manipulation in a pure oxygen atmosphere, but the deterioration of iridium, with the attendant contamination of the melt, would then take place too rapidly. The limiting value for the oxygen content was established at 10% for an acceptable iridium oxidation rate, on the basis of the clarity of the ingots obtained.

The technique of purification by successive drawings is preferred to the zone-melt technique, despite the greater rapidity and effectiveness of the latter. This is due to the difficulty of finding a crucible material which will not deteriorate too much and will have no excessive reactivity with the melted calcium niobate under the operating conditions dictated by the physicochemical properties of the latter. So far, only iridium has given relatively satisfactory results in the case of the drawing technique, though the niobate tends to adhere strongly to the crucible walls after crystallization, which raises a complex problem of emptying and cleaning. This problem was solved by developing a method that includes a mechanical discharging followed by a dissolution in sodium tungstate (Na_2WO_4), itself soluble in water. The usable regions of the purified ingots are selected by optical examination in an intense beam of light in order to reveal and eliminate flaws or precipitates detrimental to the quality of the final product.

Undoped single crystals weighing about 50 g and with various orientations are commonly obtained by the Czochralski drawing technique, the drawing and rotation rates being 11 mm hr^{-1} and 120 rpm, respectively. The equipment used consists of a twin-coil inductor fed by a 22-kW, 500-kHz high-frequency generator. The cylindrical iridium crucible is 40 mm in outer diameter, 40 mm high, and has a wall thickness of 2 mm. It is surrounded and surmounted by various alumina jackets and screens which provide both the heat insulation and an initial annealing of the crystal formed. The complete assembly is placed in a vacuum-tight vessel consisting of a silica-glass tube with water-cooled double walls. The growth of the crystal can be observed through a tinted window which is also water-cooled and which is protected against internal condensation by a retractable metal baffle. The drawing system consists of a mechanical device which allows the carrier of the seed crystal to translate and rotate at variable speeds. The vessel is connected to a secondary oil diffusion vacuum pump fitted with a liquid nitrogen trap. Following discharge, flushing with the appropriate gas is carried out, namely with a 90% argon/10% oxygen mixture in the case of calcium niobate.

There is no major difficulty in obtaining undoped single crystals under these conditions, although this would appear not to be the case when neodymium has to be injected into the matrix. The initial doping tests were made with double sodium niobate and with $NaNd(NbO_3)_4$, the presence of sodium being justified in order to establish the charge compensations on the assumption that one Na^+ and one Nd^{3+} can substitute for two Ca^{2+}. The same problem had been solved in similar fashion for calcium tungstate by doping with the double salt $NaNd(WO_4)_2$. In the course of 1% doping attempts in the melt (Nd/Ca atoms), one may note that, during the growth of the crystal, perturbations appear which seem to be due to a rapid change in the composition of the crucible's contents, with the probable appearance of several phases. Furthermore, while a radiochemical analysis of the crystals obtained shows that the neodymium has been inserted into the lattice with a segregation coefficient very close to unity, it also shows that sodium is present only to one-tenth of this extent.

After annealing for 15 hr at 1400°C while flushing with pure oxygen, a doped single crystal can be cut out and shaped in the usual way. It exhibits a laser effect with a pulse threshold of 0.28 J and functions in the continuous mode at ordinary temperatures. This result, obtained with a material containing one tenth as much sodium as neodymium, suggests that full compensation of the charges in the crystal lattice is not a paramount requirement for growing single crystals of a quality adequate for producing a laser effect under satisfactory conditions.

Other crystals of similar compounds obtained by the Czochralski drawing technique are $CaWO_3$ (Deutschbein *et al.*, 1966) and $Ba_2NaNb_5O_{15}$ (Zupp *et al.*, 1969).

b. Crystal growth without a crucible. As stated, the existence of a crucible is in most cases the main limiting factor in the production of a large number of refractory oxides. In the case of materials with melting temperatures above 2300°C, it is necessary to resort to methods which dispense with a crucible, which use a cooled crucible, or which involve a "self-crucible." As a matter of fact, there is little to distinguish between the last two of these methods, since the material remains solid on the cooled wall of the crucible and the melt is isolated from the latter by a layer of the same solid material. In effect, therefore, we are operating in a "self-crucible."

The advantages which these methods have over the previous ones are that there is no temperature limitation and that any contamination of the material by the crucible is eliminated. Very often, however, they have the drawback of making regulation difficult and hence of producing crystals of poorer crystallographic quality.

Three examples of crystal-growing operations without a crucible will now be described: the first concerns materials having a melting temperature lying between 1500 and 2100°C (e.g., gallium oxide and alumina); the second concerns materials with melting temperatures lying between 2000 and 2500°C (e.g., stabilized zirconia); and the third relates to materials with a melting temperature above 2500°C (e.g., lime).

(*1*) *Gallium oxide* (*mp* = *1780°C*) *and alumina* (*mp* = *2050°C*). The method chosen here is the floating-zone method. This method has been used for several years for purification and crystal growth of high-melting-point metallic materials, such as tungsten. In this case, the heating method used is either induction or electron beam bombardment. Heating by direct induction is difficult with alumina, because its resistivity is still high at the melting point, so a susceptor is necessary (Petit Le Du and Aubree, 1967). Electron beam bombardment in a vacuum causes dissociation of the oxides. It is therefore necessary to use a heating method which permits melting of the material in an oxidizing atmosphere and does not depend on the electrical characteristics of the material. Heating by concentrated radiation will allow crystal growth under such conditions (Kooy and Courvenberg, 1962; Revcolevschi and Collongues, 1968; Saurat *et al.*, 1970). The technique used here is the image-furnace technique. In the experiment described, the image furnace is of the biellip-

FIG. 14. Floating-zone imaging furnace apparatus (Revcolevschi and Collongues, *C. R. Acad. Sci.* Paris **266,** 1767, 1968.

tic type (Fig. 14). The mirrors are 390 mm in diameter, and the focal lengths are 143 and 850 mm. The basic problem to be resolved is the stability of the hot source. In most image furnaces used for operations other than crystal growth, the light source is a graphite-electrode arc. Changes in the position of the crater and the lack of homogeneity of the electrodes preclude the possibility of maintaining a constant (to within ±10°C) temperature on the material to be processed. In addition, the limited useful life of the electrodes makes long-duration experiments impossible. The advent of short-arc xenon lamps capable of continuous operation for several dozen hours and ensuring a sample temperature constant to within ±5°C has made it feasible to use the image furnace for high-temperature crystal-growing operations (Fig. 15). The mechanical translating device permits vertical transfer of a melted zone at rates of 2–50 mm hr^{-1}. In most cases, the translation, and hence the growth rate, is of the order of 10 mm hr^{-1}. The bars are rotated in opposite directions at 60 rpm. Temperature variations measured in the melt are under 8°C.

The starting materials (alumina of 99.99% purity, gallium oxide of

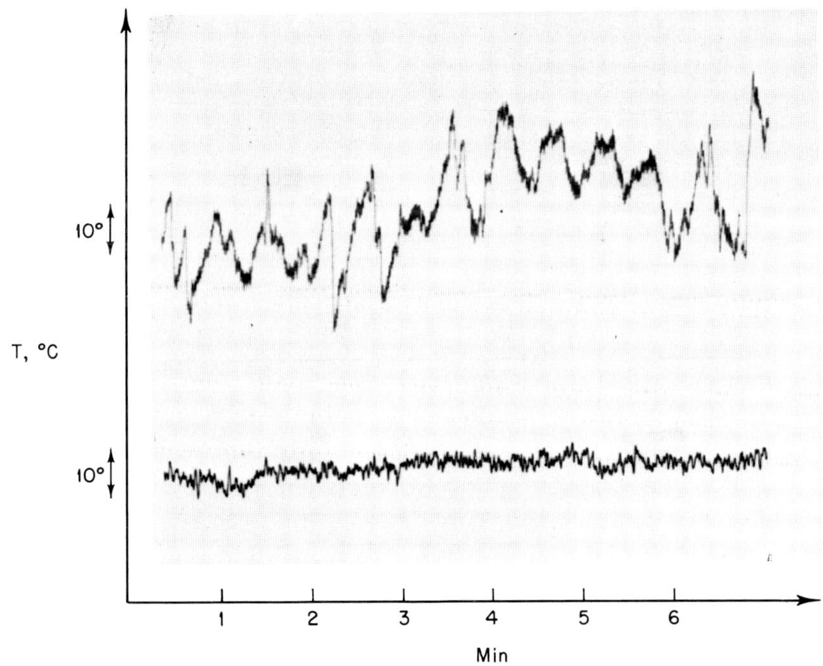

FIG. 15. Temperature fluctuations near 2000°C in the hot zone of the imaging furnace. Top: graphite arc source; bottom: xenon arc lamp source.

FIG. 16. Solid–liquid interface during floating-zone growth of Ga_2O_3 (Saurat *et al.*, *C. R. Acad. Sci. Paris* **271**, 58, 1970).

99.999% purity) are sintered to form bars or rods 2–8 mm in diameter. These rods are placed in the securing jaws of the translating device if they are long enough. If not (the great advantage of the method is the possibility of obtaining crystal growth on small quantities of substance), the upper rod is suspended from a platinum wire. The two rods are initially located on either side of the hot zone to melt their tips, after which they are moved closer together in order to form the melting zone (Fig. 16).

By comparison with the other methods of growing crystals at high temperatures, the method described here offers five main advantages:

1. It is possible to grow crystals on any material irrespective of its electrical characteristics.
2. Contamination is totally absent.
3. It is possible to carry out crystal growth in any atmosphere inside an ordinary silica-glass tube. For example, it is possible to prepare rutile crystals of different stoichiometries.
4. It is possible to grow crystals from amounts of substance markedly smaller than those used in other methods. The efficiency of the operation is

very high, nearly all the starting material being converted into a single crystal, especially if one takes the precaution of initiating the crystallization on a seed crystal. If the operation is performed without a seed crystal, one can observe a polycrystalline zone at the start of growth from which a crystal of preferred orientation develops. This zone will not appear if crystallization begins on a seed crystal. It has been found possible in this way to prepare alumina single crystals doped with ^{17}O starting with quantities of substance not exceeding 100 mm^3.

5. It is possible to obtain single crystals of pure oxides from material of commercial-grade purity, since the operation introduces a purification process by the mechanism of zone melting. Thus, alumina single crystals of very great purity have been produced from commercial-grade alumina (purity 99.5%) after several passages through the floating zone (Fig. 17).

More complex systems can readily be devised (Field and Wagner, 1968). For instance, heat can be applied on both sides of the sample (Fig. 18).

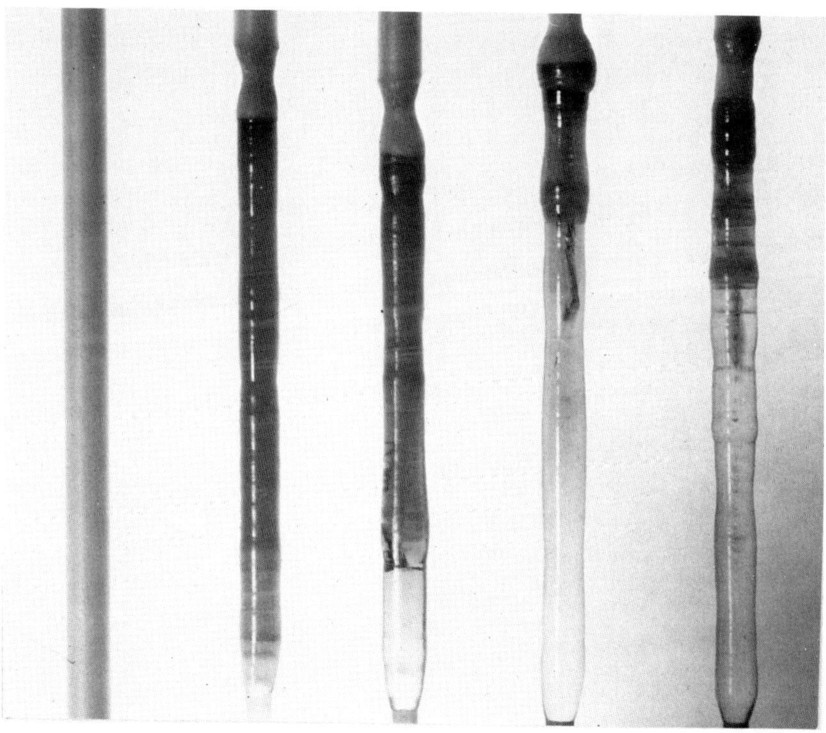

FIG. 17. Purification of alumina during crystal growth by zone floating (Dhalenne *et al.*, *C. R. Acad. Sci. Paris* **272,** 538 1971).

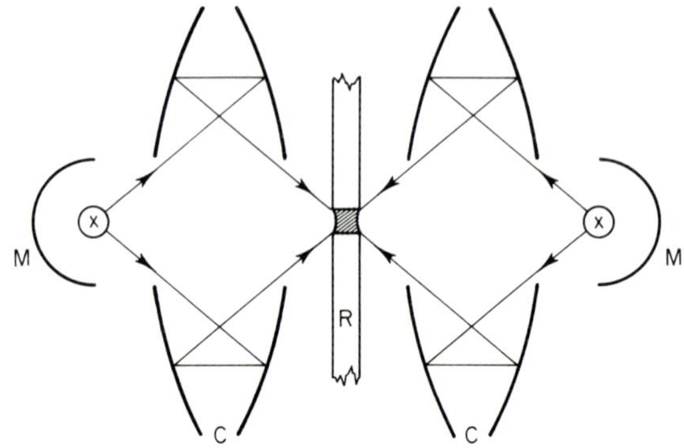

FIG. 18. Optical system of dual compound furnace (Field and Wagner, 1968). C, Compound mirror assemblies; X, xenon lamps; R, rod being float-zoned; M, backup mirrors. Used with permission of North-Holland Publishing Co., Amsterdam.

For greater effectiveness, two spherical mirrors M may be positioned behind the sources, thus increasing the flux at the common focal point by 25%. Naturally, the two systems must be rigorously aligned. A furnace used for many crystal-growing experiments consists of two 5-kW xenon lamps and mirrors having diameters of 90 cm and focal lengths of 35 cm. The maximum flux is 450 cal cm^{-2} sec^{-1} and the mean flux is 190 cal cm^{-2} sec^{-1}. The sections of the materials processed are 16 mm in the case of a material melting at 1500°C, and 8 mm for one melting at 2000°C. At 1500°C, temperature fluctuations are of the order of ±4°C. Ruby crystals several centimeters long and 6 mm in diameter have been obtained. Dislocation densities are of the order of 1.6×10^4 cm^{-2}.

For crystal growth by the floating-zone method without a crucible, hence without contamination, any energy concentrating method can be used: For example, electron beam bombardment. Mention may be made of experiments with a hollow cathode gun for growing single crystals of mixed aluminum and yttrium oxides, $YAlO_3$ and $Y_3Al_5O_{12}$ (Fig. 19) (Class, 1968).

The cathode is a hollow sphere truncated at its top and bottom to provide access to the focal point. The material to be processed consists of two rods, the lower one being the crystal support and the upper one feeding the melt zone. These rods rotate in opposite directions and are movable along the vertical axis. The melt zone is limited vertically by alumina cylinders, which can be heated to about 2000°C in the region of the melt zone in order to anneal the crystal. These cylinders are also used to protect the sample from cathode sputter.

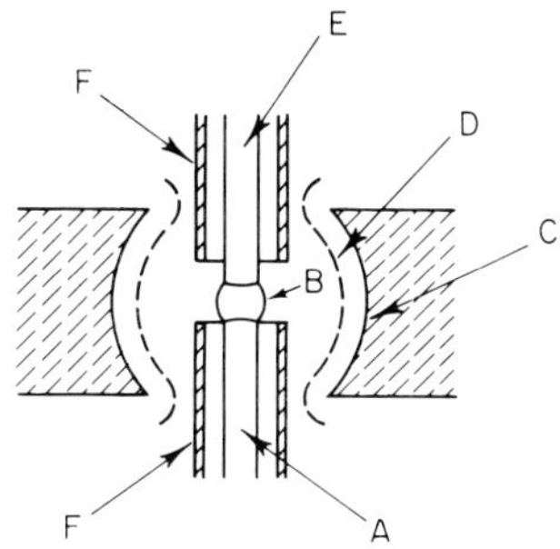

FIG. 19. Schematic illustration of hollow cathode gun (Class, 1968). A, Crystal; B, molten zone; C, cathode; D, edge of negative glow plasma; E, feed rod; F, sputtering shields. Used with permission of North-Holland Publishing Co., Amsterdam.

The Al_2O_3 and Y_2O_3 powders are first carefully dried by heating in air at 500°C for 24 hr, then pressed and sintered at 1350°C for 12 hr to obtain a density corresponding to 70% of the theoretical density. The growth is accomplished under reduced oxygen pressure (0.2 mm). These experiments demonstrate the difficulty of using this method for producing single crystals from stoichiometric mixed oxides, for the following reasons:

1. The equimolecular compound $YAlO_3$ does not crystallize immediately upon solidification. It appears in addition to the garnet $Y_3Al_5O_{12}$. Only during growth does $YAlO_3$ develop alone. This compound thus exhibits incongruent melting.
2. The garnet $Y_3Al_5O_{12}$ cannot be prepared in the form of single crystals from the stoichiometric mixture; a $YAlO_3 + Y_3Al_5O_{12}$ mixture is invariably obtained. It is therefore necessary to use a mixture containing a surplus of alumina. It seems doubtful that the alumina loss occurs by evaporation. It is more likely that the maximum melting temperature does not exactly correspond to the composition $Y_2O_3/Al_2O_3 = 3/5$.

This difficulty did not appear in the case of the crystals prepared by the Czochralski drawing technique, due to the very large volume of the melt, which limited rapid variations in composition.

(*2*) *Stabilized zirconia.* For this material, the melting temperature of which is close to 2400°C, and which has relatively low resistivity above 2000°C (about 1 ohm-cm), the method used is the Kyropoulos or Czochralski self-crucible method (Michel *et al.*, 1968).

For growth to occur, it is first necessary to have a stable melt. The heating method used is direct induction at high frequency through the oxide. At 2000°C, the resistivity of the lime-stabilized zirconia (5 mole% < CaO < 20 mole %) is of the order of 1 ohm cm^{-1}. The frequency must be chosen to achieve a reasonable penetration depth without undue dissipation of the energy in too large a volume. To obtain penetration depths

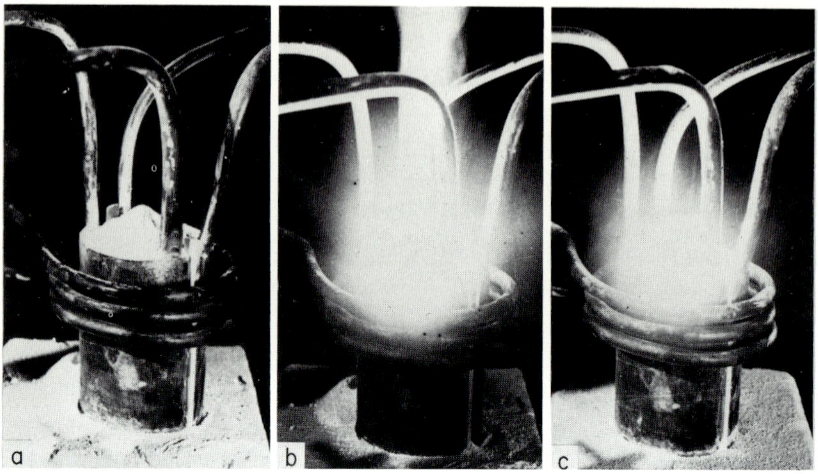

FIG. 20. Self-crucible melting. (a) Load in the cooled crucible, (b) starting by combustion of the metal, (c) melting sustained. (Michel *et al.*, *C. R. Acad. Sci. Paris* **266,** 1602, 1968).

of the order of 1 cm, it is necessary to use frequencies of the order of 1 MHz. The frequency finally employed is 8 MHz. The material to be treated is placed in a cooled copper crucible. The crucible consists of two water-cooled half-shells electrically insulated from each other by a layer of air and placed inside the inductor element.

The starting materials are pure ZrO_2 and pure calcium oxide. Both are used in powder form. The crucible is at first two-thirds filled with zirconium oxide. To allow the induced currents to flow through the material, preheating is necessary (Fig. 20). In order to avoid any impurity coming in contact with the melt, the preheating is done as follows. Zirconium chips are mixed with the zirconia powder. These chips can easily be heated by induction. When the proper temperature is reached, they suddenly oxidize, and the exothermic combustion of the metal is sufficient to raise the powder temperature, cause it to melt, and allow the induced currents to flow. A stable bath of melted zirconium oxide is then obtained. A film of the solid material remains in contact with the cooled crucible, isolating the melt in the crucible and protecting it from contamination. Indeed, the existence of this solid film is enough to contain the melt inside the two shells and prevent it from pouring through the slit separating the two half-cylinders. The operation is therefore carried out by the self-crucible technique. The lime is then gradually introduced into the melt. During this operation, a change in the electrical characteristics of the charge may be observed, making it necessary to adjust the generator's electrical parameters. This method makes it easy to achieve the desired composition for the ZrO_2–CaO solid solution.

Single crystals can be obtained by two methods:

1. By slow cooling. The power dissipated through the charge is gradually lowered according to a predetermined program. Crystallization is initiated at the bottom of the crucible and then travels upward. Depending on the cooling conditions, it is possible to obtain crystal growths at rates varying from 1 mm hr^{-1} to 1 cm hr^{-1}. Alternatively, the crucible can be slowly lowered inside the inductor. The crystals obtained are perfectly transparent and exhibit an excellent state of crystallization. Their size may reach 10 mm.
2. By drawing from the melt (Fig. 21). The drawing system includes two translation/rotation motors permitting a uniform growth of the drawn crystal. The lower end of the drawbar is electrically insulated and is cooled by circulating water. The possible drawing rates range from 10 to 20 mm hr^{-1}.

Thus, by using a unique apparatus based on a heating mode adapted to the material involved, it is possible to achieve crystal growth by shifting

FIG. 21. Apparatus for drawing single crystals from self-crucible melts (Michel *et al.*, *C. R. Acad. Sci. Paris* **266,** 1602, 1968).

the crystal–liquid interface, by moving the liquid along a temperature gradient, or by drawing.

(3) *Calcium oxide: crystal growth at temperatures above 2500°C.* As an example, the production of single crystals of CaO (mp = 2600°C) by the Verneuil method as adapted to the plasma furnace will be considered (Galtier and Collongues, 1967). The principle on which the high-frequency current plasma furnace is based is well known. A 12-kW generator operating at a frequency of 8 MHz is used (Fig. 22).

The starting material is calcium carbonate, which was preferred to the oxide because of the latter's sensitivity to atmospheric humidity, which

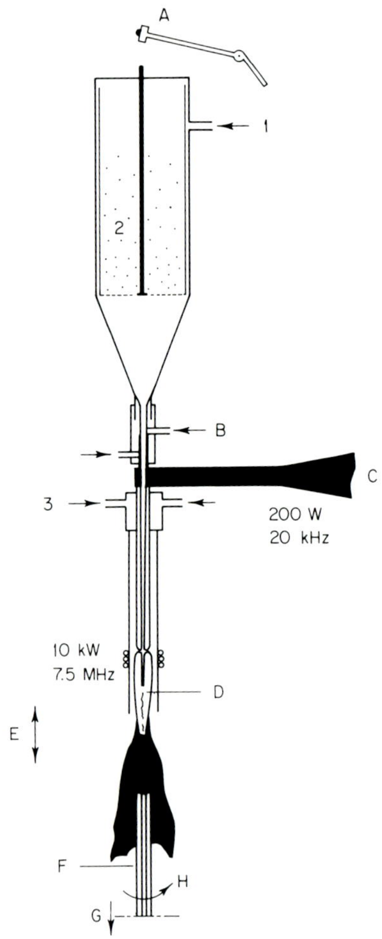

FIG. 22. Verneuil apparatus with plasma furnace (Galtier). A, Fifteen percussions per minute, 5 mg each percussion; B, water cooling; C, ultrasonic vibrator; D, plasma; E, optical control; F, support; G, translation 1–100 mm hr^{-1}; H, rotation 0.2–2 rps; (1) argon, 5 liters min^{-1}, 3 bar; (2) $CaCO_3$, 1–10 μm; (3) oxygen, 35 liters min^{-1}, 0.35 bar.

makes it very difficult to distribute the powder uniformly. The carbonate is 99.99% pure with a 10-μm grading.

The basic problem in the Verneuil method is the powder distribution. Numerous tests with continuous-type feed mechanisms were performed, but the best results are obtained with a discontinuous-type tapping device using sifters. The powder fed at each tap is controlled to within 0.1 mg.

The powder injector consists of three cylindrical stainless steel tubes terminating in a copper nozzle. The outer jackets are used for circulating the cooling water and the central duct for injecting the powders into the plasma. The outer diameter is equal to the apparent diameter of the plasma, i.e., 16 mm. The injector nozzle is placed level with the first coil of the inductor. A fraction of the powder passing through the plasma is carried along by the fluid and deposits on the injector nozzle. In the course of an operation of long duration, the accumulated powder disrupts the plasma stability. To avoid the formation of this deposit, the injector is subjected to ultrasonic vibration at a frequency of 20 kHz.

To allow for the high-frequency power losses within the plasma (about 10 kW), the discharge chamber is a transparent silica-glass tube with inner and outer diameters of 26 mm and 30 mm, respectively. The chamber is cooled only by the convection currents of the plasma gas. The essential requirement for success of the operation is rigorous coaxiality of the injector and the chamber to within less than 0.1 mm. Provided these conditions are met, all the injected powder passes through the plasma and the tube remains completely unaffected, even after several hundred hours of operation.

All the experiments were carried out using oxygen as the plasma-generating gas and argon as the powder-vectoring gas. The gas pressure is adjusted by membrane-type pressure reducers, the output being monitored by rotameters.

The crystal support is a copper tube closed at its upper end, cooled by circulating water, and kept in place by a centering device. This support is rotated at a speed varying between 0.2 and 2 rps, and is translated vertically at a rate varying between 1 and 10 mm hr^{-1}. Displacement of the crystal in relation to the plasma can be established within 0.1 mm by means of a lens which projects an image of the plasma and the crystal on a screen.

Determining the optimum growth rate has been the object of numerous tests. This depends on several parameters: the power used, the gas flow rate, the amount of powder injected, the accuracy of the centering, and the distance of the crystal from the center of the plasma. For the conditions stipulated previously, the tolerable maximum rate is 2.5 mm hr^{-1}.

It must be remembered that in this method, more than in any other, the temperature gradients are extraordinarily high. Attempts to adapt an auxiliary furnace proved a failure because such a furnace would have to ensure controlled cooling of the crystal from temperatures above 2000°C. The best results have been obtained with auto-heat-insulation of the crystal: for this purpose the injector is modified to obtain a slightly divergent jet of powder. At the end of the operation, the crystal is encased in a polycrystalline envelope, the whole then being cooled in the plasma flame. The envelope is then eliminated by abrasion.

The single crystals obtained are cylinders 8 mm in diameter, and the heights are limited only by the duration of the operation. Depending on the conditions of growth and cooling, they may exhibit a substructure or be virtually perfect.

Other single crystals have been produced by the Verneuil method adapted to the plasma furnace, an example being single crystals of rutile TiO_2 (Chase and van Ruyven, 1969). The 24-kW generator was operated at a frequency of 4 MHz. The inductor used was a flat four-turn disk. The oxide TiO_x obtained had a variable composition ($1.96 < x < 2$). Its color was dark blue. After oxidation in air at 1300°C, a colorless, transparent crystal was obtained.

Another method in which the starting material is transferred is the arc transfer process (Fig. 23) (Drabble, 1968). This very new method has been used in producing oxides of transition elements. The arc is produced between two cylindrical electrodes having a common vertical axis. One of the electrodes is consumable. The other is the support for the growing crystal. The position of the upper electrode is adjusted according to the crystal growth rate. The direction of transfer for a given material is determined by the polarity of the discharge, and growth can be produced on the upper or lower electrode. The crystal diameter is more uniform in the former case, but the section is limited. In the case of oxides of transition metals, transfer occurs from the cathode to the anode. In contrast, cerium and lanthanum oxides are transferred from the anode to the cathode. The voltage is approximately 10 V $\pm$ 0.1 V, and the current is a few amperes, controlled to within 0.1%.

The consumable electrode consists of the sintered oxide. However, in many cases (iron and nickel oxides), the metal itself can be used as an electrode, in which case the operation takes place in an oxidizing atmosphere.

Since most of the energy is dissipated by radiation, the current to be used varies as the square of the diameter. The efficiency of the transfer is shown in Fig. 24. It can be observed that for small arc lengths, 80% of the ma-

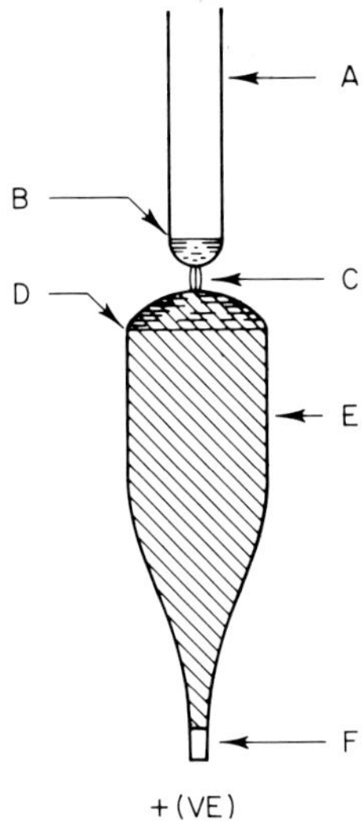

FIG. 23. Arc transfer process (Drabble, 1968). A, Feed electrode (taper consumed); B, cathode melt; C, arc; D, anode melt; E, growing crystal; F, seed crystal. Used with permission of North-Holland Publishing Co., Amsterdam.

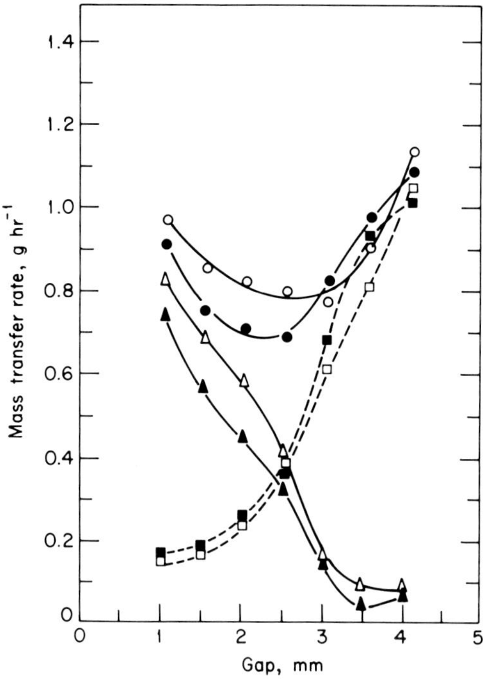

FIG. 24. Variation of mass transfer rates with gap for iron electrodes in air: Open points: cathode at the top, solid points: cathode at the bottom. Current: 6.0 A. ○, Rate of mass leaving the cathode; △, rate of mass transferred from cathode to anode; □, rate of mass loss from the arc. (Drabble, 1968.) Used with permission of North-Holland Publishing Co., Amsterdam.

terial leaving the cathode is recovered on the anode. However, the efficiency diminishes very quickly as the distance between the electrodes increases. For a fixed distance, transfer increases with increasing current intensity. The relationship is roughly linear. Taking account of the fact that the square of the diameter is proportional to the intensity, one could say that the linear growth rate is roughly independent of the intensity. Customary linear growth rates are of the order of 2–5 mm hr^{-1}. Some of the crystals prepared by this method are listed in Table IV.

The structural qualities of the crystals are good, but the main problem is the stoichiometry. Crystals of NiO produced in air contain 1–2% of surplus oxygen, while iron oxide (Fe_3O_4) crystals contain a little Fe_2O_3. To achieve stoichiometry, it is preferable to operate with reduced pressure

TABLE IV

Oxide	Electrodes	Atmosphere
CeO_2	Sintered 1600°C	Air or air–argon
CoO	Sintered 1200°C; reduced area in H_2 at 600°C	Air
NiO	Sintered 1400°C; reduced area	Air
La_2O_3	Sintered 1500°C	Air
TiO_2	Sintered 1400°C; reduced area in H_2 at 600°C	Argon
FeO	Iron	Air
$NiFe_2O_4$	Sintered NiO + FeO	Air
NiO + ϵ-Li_2O	Sintered NiO + Li_2O_3	Air
Al_2O_3	Al_2O_3 powder in Al_2O_3 tube, Al core	Transfer, but growth inhibited by resistivity of Al_2O_3
ZrO_2	Extruded; Zr core	Same as above
$CaWO_4$	Extruded; W core	Transfer, but unstable arc; poor quality of electrodes

(10^{-3}–10^{-2} Torr) with argon–oxygen mixtures. For volatile systems, on the contrary, it is possible to operate under pressures of several bar.

The usual limitation for oxides is the electrical conductivity, which must be adequate for both electrodes. The arc can easily be started with a conductor which burns afterward: In the case of nickel oxide, for example, a surface film of nickel can readily be obtained by heating beforehand in a reducing flame. Once the arc is established, the temperature rise in the electrode is sufficient to maintain adequate conductivity. In other cases, as with alumina, resistivity is too high for it to be possible to maintain the arc long enough. Transfer is nevertheless possible. Certain difficulties can be alleviated by providing auxiliary heating to raise the temperature of the electrodes and provide controlled cooling for the crystal.

3. *Growth in the solid state: single crystals of barium ferrite* ($BaFe_{12}O_{19}$) (*Lacour and Paulus, 1968*)

Barium ferrite $BaFe_{12}O_{19}$ (BaO, $6Fe_2O_3$) has a hexagonal structure consisting of structural elements superposed along the sixfold axis. Growth rates along the direction perpendicular to the sixfold axis are about ten times faster than along this axis. Growth in the polycrystalline aggregate can be achieved by several different techniques.

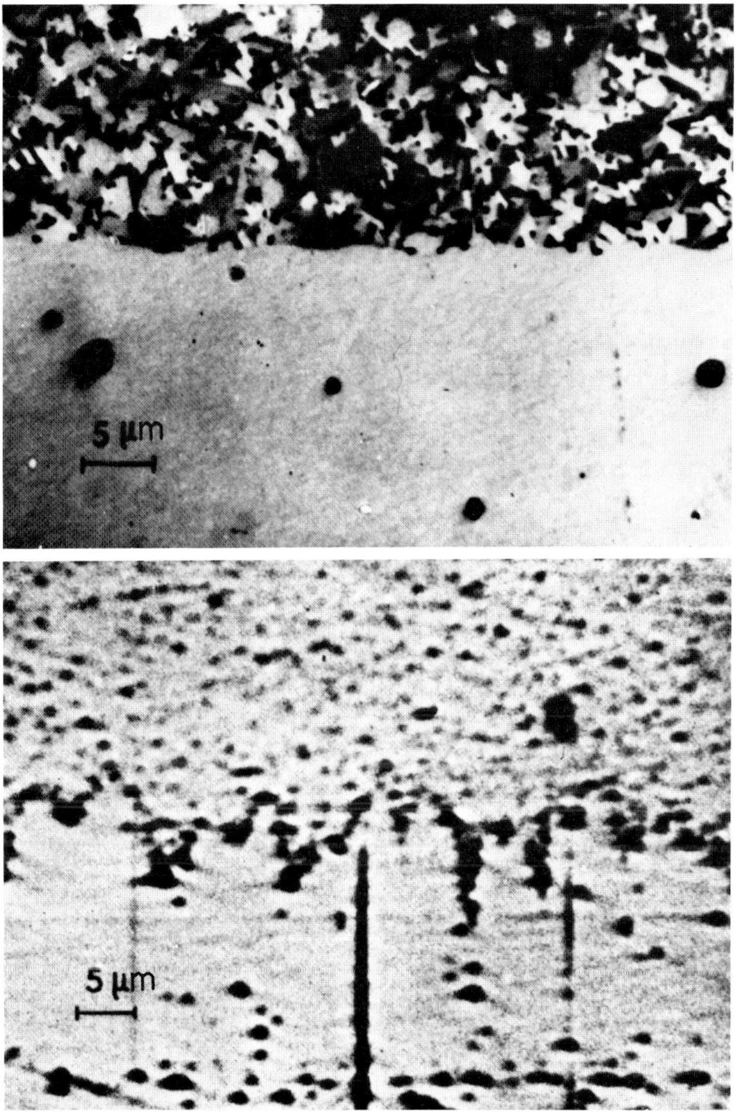

FIG. 25. Growth interfaces with and without a temperature gradient for single crystals of barium ferrite. Growth perpendicular to the sixfold axis (Lacour and Paulus, 1968). Used with permission of North-Holland Publishing Co., Amsterdam.

a. Growth without a thermal gradient. In accordance with the conditions previously indicated, the mean quadratic difference in the diameters of the crystal aggregate must be reduced to a minimum by carefully controlled protracted grinding (10 days).

Growth is carried out at 1250°C in air. A seed crystal is held against the aggregate throughout the growth by a pressure of 20 bar. Growth rates are very slow, about 0.5 μm hr^{-1}.

If growth takes place perpendicular to the sixfold axis, the growth interface will be irregular because of the presence of residual porosities, but will be straight if there is propagation along this axis.

b. Growth in a thermal gradient without a seed crystal. The sample is placed in a tubular furnace having a thermal gradient, with the seed crystal in the hot zone. The temperature gradient is close to 150°C cm^{-1}. The sample is a ferrite cylinder presintered at 1125°C in air. The rate of transfer of the sample for a given temperature must be less than or equal to the rate of migration of the interface. The growth rate is approximately 20 mm hr^{-1} at the chosen temperature (1410°C).

Figure 25 compares the uniform interface obtained with a temperature gradient and the irregular interface obtained without a temperature gradient, for growth perpendicular to the sixfold axis.

c. Growth in a thermal gradient with a seed crystal. In order to assist development of one single crystal, a well-oriented crystal is selected after a sufficient amount of growth has occurred. The direction of rapid growth of this crystal is made parallel to the axis of a rod, and the crystal is welded to the latter, while at the same time, a break (a void or screen) is provided in front of the other crystals. The welding is done at 1190°C for 5 min at a pressure of 10 bars. The selected crystal then develops in the polycrystalline matrix and it is possible to obtain crystals 5 mm in diameter and 20 mm long.

At the start of the processing, a rough recrystallization may be observed, after which a few crystals, whose direction of rapid growth is close to the rod axis, develop at the expense of their neighbors.

II. Low-Temperature Crystal Growth (in Solution)†

High-temperature growth methods cannot always be used. Apart from the high cost of the equipment required for high-melting-point refractory

† See Gilman (1963), White (1965), Laurent (1969), and Frank *et al.* (1968).

oxides, the latter sometimes decompose before melting, as is the case with garnets, or else they undergo a series of crystalline transformations with the risk of twinning or fracturing when cooled. Crystals can also be obtained by means of low-temperature methods in which growth takes place by precipitation from a solution. This process consists in dissolving the substance from the solid state and then inducing it to crystallize by cooling it or by evaporating the solution. The aim must be to establish a concentration gradient with diffusion toward the nucleus crystal. A knowledge of the solvent–solute phase diagrams is important for defining the experimental conditions of growth. Different techniques are used, depending on the solvent. If the solvent is nonaqueous, flux growth occurs. When water is used as the solvent, only high-pressure techniques known as hydrothermal growth can be used for high-melting-point oxides, since the latter are not soluble in water at atmospheric pressure.

A. Flux Growth

This technique consists in defining the liquidus curves of the solvent–solute phase diagram and in understanding the kinetics of the crystal growth.

1. *Flux growth criteria*

a. Phase diagrams (*Fig. 26*). Phase diagrams are determined by various methods which are now well known (Levin *et al.*, 1964). The most accurate results are obtained by differential thermal analysis. In principle, it is desirable to be well acquainted with the phase diagrams. Obviously, the substance B from which we wish to obtain single crystals must not produce a solid solution with the other constituents of the system in question, or react with the solvent. Assume that the substance B melts congruently. The role of A, the solvent, will be to lower the melting point of solute B, if necessary down to the eutectic temperature. As a rule, the aim is to operate at the lowest possible temperature. A consequence of this is that the eutectic temperature T_E must be low, and therefore the solvent must have a low melting point. The experiment will often be limited by an upper temperature fixed by technical or physical considerations, such as the existence of an allotropic transformation. Let T_P be this temperature; in this case, the mixture will have to have the richest composition in solute compatible with this limit. The greater the amount of solute in the mixture, the smaller the influence of diffusion and the greater the rate of growth of the crystal.

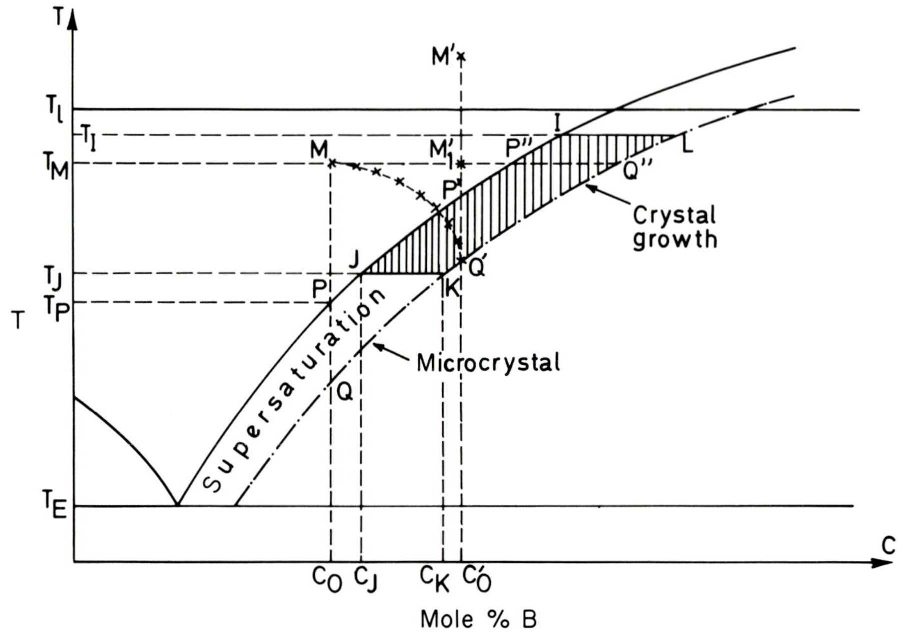

FIG. 26. Phase relations in binary systems, illustrating the difference between the two methods of flux growth, isothermal flux evaporation and slow cooling.

If substance B melts noncongruently, it will be necessary to operate with compositions richer in A than in the case just discussed. Experiments will then show that crystal growth is slower. This is due to greater diffusion. Hence, when a substance exhibits noncongruent melting, another substance C is often added in order to lower the temperature of the liquidus zone.

It should be noted that many experimenters have not made a systematic study of the phase diagrams in the course of tests for obtaining single crystals. The diagrams studied and the single crystals obtained are listed in Table V.

On the basis of a portion of the liquidus determined for a solute–solvent mixture, as shown diagrammatically in Fig. 26, there are three ways of achieving crystallization: cooling, evaporation, and using a temperature gradient.

Consider a mixture of composition C_0, usually 10–15 mole %, of B; this is brought to a temperature T_M, about 100–200°C above the liquidus temperature T_P, in order to obtain a thorough solution. Let M be the location of the solution on the phase diagram. The solution can then be cooled on the curve MPQ. At P, a solid–liquid equilibrium line is reached. In fact, to achieve solidification, the point P must be exceeded and a metastable supersaturated zone must be reached which is bounded by a line QQ′Q″, where spontaneous crystallization is initiated. Although this is the zone in which the crystals appear, all of it cannot be used for crystal growth, but only the portion IJKL. In this portion, solution parameters such as diffusion, viscosity, and rate of growth as compared to the rate of nucleation are optimal for limiting the number of new nuclei and causing growth of those that have already appeared. Thus if the solution is cooled from the point M at concentration C_0, only microcrystals will precipitate between the points P and Q. Choice of the initial concentration is important and should be situated between C_J and C_K (Fig. 26). Only by cooling along a path such as M_1′P′Q′ is it possible to reach the crystallization zone. In order to obtain complete dissolution at the concentration C_0 (about 20–30%), it is necessary to operate from M_1′, i.e., at a temperature which is sometimes well above $T_{p'}$. To reach the zone P′Q′, it is possible to evaporate the solution M and to pass in succession from M to M_1′ and from M_1′ to P′Q′, or else to follow the path MQ′ by simultaneously lowering the temperature and evaporating. Alternatively, it is possible to operate at a fixed temperature and to concentrate the solution only by evaporation along MM_1′P″Q″. The crystals appear between P″ and Q″. The utility of the method consists in operating at a fixed temperature; in other words, without varying parameters like the diffusion constants or the growth

TABLE V

Phase diagram	Crystal obtained	Ref.
$KF–BaTiO_3$	$BaTiO_3$	Karan (1954) Karan and Skinner (1953)
$B_2O_3–BaO–Y_3Fe_5O_{12}$	$Y_3Fe_5O_{12}$ (YIG)	Linares (1962, 1964)
$PbF_2–ZrO_2$	ZrO_2	Anthony and Vutien (1965)
$Li_2O–MoO_3–BeO$	BeO	Austerman (1963)
$K_2B_4O_7–TiO_2$	TiO_2	Naumova and Anikin (1966)
$PbO–PbF_2–Al_2O_3$	Al_2O_3	Wilke (1965)
$WO_3–M_2O(MO)–Al_2O_3$ (M = Li, Na, K, Sr, Ba)	Al_2O_3	Voronkova *et al.* (1967)
$PbMoO_4–MgO–Fe_2O_3$	$MgFe_2O_4$	Viting and Khomyakov (1965)
$PbF_2–YF_3–Fe_2O_3–Y_2O_3$	$Y_3Fe_5O_{12}$	Sato and Hukudo (1963)
$PbO–Fe_2O_3–Y_2O_3$	$Y_3Fe_5O_{12}$	Nielsen and Dearborn (1960)
$PbO–PbF_2–Y_2O_3–Al_2O_3$	$Y_3Al_5O_1$ (YAG)	Bakradze *et al.* (1968)
$PbF_2–PbO–Ga_2O_3–Sm_2O_3$	$Sm_3GA_5O_{12}$	Schieber (1964a,b)
$Na_2SO_4–CaWO_4$	$CaWO_4$	Cobb and Wallis (1967)

rate. The essential thing here is to choose correctly the constant temperature at which evaporation takes place in order to reach the zone IJKL. If the temperature is too high, the solvent evaporates too quickly and the crystal has no time to form. If the temperature is too low, only microcrystals are obtained. In the case in Fig. 26, the temperature must lie between T_I and T_J.

Yet another method consists in maintaining a temperature gradient along the vessel in which crystallization is taking place, such that the top or bottom of the crucible is at a temperature and concentration corresponding to the crystallization zone. This is the method most widely used in pressure-crystallization or hydrothermal synthesis.

It is interesting to note that while a few liquidus curves have been systematically determined, the range of the supersaturation zone and definition of the crystal growth zone have not yet been systematically investigated, these zones being somewhat imprecisely defined in general. Thus, for alumina in lead fluoride, the portion of the liquidus line up to 30% alumina has been determined, and the zone of crystal growth would appear to lie between 1000 and 1100°C for a concentration of 25 mole % of alumina (Vutien and Anthony, 1970).

Once the crystallization zone is identified, the size and quality of the crystals obtained will depend on the crystal growth kinetics.

b. Crystallization kinetics. The formation of a crystal in a liquid takes place in two stages, the first of which is nucleation. This consists in the production of a nucleus around which growth will occur. The growth process is the second stage. Study of the kinetics of crystallization is a matter of theory rather than experiment, and it is beyond the scope of this section to go into the various theories in detail. It will suffice to recall here the most generally accepted conclusions.

The existence of the supersaturation zone is related to the onset of an energy barrier which must be overcome in order for a stable seed to form and which is in effect the surface energy of the small particles formed. The size of the seeds produced is a matter for statistical calculations, by which it can be shown that, from the energy standpoint, only those seeds that exceed a certain critical radius can grow. Growth is related to certain kinetic factors, such as diffusion of the solute toward the crystal–solution interface, sorption of the solute at the surface, formation of the crystal lattice, and the reverse process of dissolution. Depending on the relative magnitudes of these factors, the crystal either will grow rapidly or will not grow. The transfer of heat and matter plays an important part in the growth process.

The heat transfer involves conduction through the liquid and solid phases, convection through the liquid, and elimination of the latent heat of crystallization. In the case of fluxes, elimination would appear to take place by dissipation within the solution rather than by conduction through the solid. In a supersaturated solution, the solution temperature is lower than that of the interface, while the solute concentration is higher in the solution than at the interface. A consequence of this is that single crystals will grow faster in a flux which favors a large supersaturated zone, since the larger this zone is, the greater will be the temperature variation corresponding to elimination of the latent heat of crystallization.

The transfer of matter is also of some importance since the solute concentration is often not very high. The rate of matter transfer through the solution depends on the difference in solubility between the growth source and interface, on the diffusion coefficient, and on the extent of the flux mass of flow. It should be noted, however, that the choice of solvent affects the solubility more than the diffusion. As for the diffusion rates through liquid fluxes, they are of the order of 10^{-4}–10^{-5} cm^2 sec^{-1}.

Finally, the probability of adsorption and formation of the crystallized phase from a solute diffusing toward the interface depends on factors which are still incompletely understood, among which are certain crystal properties such as dislocations and surface discontinuities which facilitate the attachment of the solute. The low-index planes, and those of maximum density, usually grow more slowly than high-index planes.

The problems of interface kinetics are complex, and it is difficult to know beforehand either the growth rate or the morphology of a crystal. Mention may be made of the work done by Cobb and Wallis (1967) on the growth of refractory oxide single crystals in melted fluxes. Cobb and Wallis consider that the cooling rate, the temperature gradient, the speed of rotation, and the dimensions of the crucible are parameters of the system, whereas solubility, diffusion, viscosity, surface energy, and crystallization and dissolution rates are independent variables. These factors were determined experimentally when synthesizing sapphire or ruby crystals, the flux used being lanthanum fluoride.

There are two important aspects of the role of impurities in crystal growth: uniformity of distribution of the impurities and solvent inclusions. Segregation of the impurities is due to the difference in the impurity concentrations in the crystal and in the solution. As a rule, solubility is higher in the liquid than in the crystal. This causes the impurities to be rejected from the growth interface and to be diffused beyond it. It is difficult to obtain crystals free from solvent ions; the latter substitute on normal sites, or they appear in the form of inclusions if the growth is disturbed by a temperature fluctuation. To reduce the presence of impurities, growth must be slow, temperature must be regulated, and a solvent must be selected that cannot readily penetrate into the crystal matrix.

The crystal structure is also a determining factor; thus a compact structure free from vacancies does not tend to capture impurities. Conversely, when crystals have several types of sites with ions of different valence, such as spinels, many impurities may get in. For instance, the addition of bivalent ions will result in large crystals of yttrium iron garnet ($Y_3Fe_5O_{12}$) (YIG). In the case of anisotropic crystals, impurities are absorbed preferentially; thus, doping corundum with magnesium produces a yellow color indicating selective absorption on the (104) planes, whereas manganese is absorbed only on the (012) and (113) planes and produces a pink color (Wallace and White, 1967). Furthermore, if small quantities of La_2O_3 are added to corundum in a PbF_2–Bi_2O_3 bath, it is possible to grow equidimensional crystals (stirring of the bath is recommended). A study of the variation in the rate of growth of YIG along the 110 direction as a function of the speed of rotation showed that there was a critical speed beyond which the growth rate remained constant. For a cooling rate of 5°C hr^{-1}, the critical speed is 40 rpm (Laudise *et al.*, 1962).

c. *Choice of the solvent.* There are numerous criteria for choosing a solvent, some essential and others desirable. The solvent must possess the following attributes: fairly high dissolving capacity, chemical inertness with respect to the solute, and a low melting point far removed from the boiling point.

It is furthermore desirable for the solvent to possess the following characteristics: low vapor pressure at the growth temperature; low toxicity; high solubility in water or aqueous reagents; a common ion with the solute or ions of a size large enough to prevent their insertion into the crystal lattice.

In order to determine the degree of solubility of an oxide in a solvent, it is possible to plot the liquidus curves of the solute–solvent phase diagrams. If the solvent is a mixture, the liquidus line will relate to a ternary system and be fairly difficult to plot. If systems such as PbO–Fe_2O_3–Y_2O_3 (Nielsen and Dearborn, 1958) and BeO–Li_2O–MoO_4 (Austerman, 1963) have been established, simpler measurements will sometimes suffice in order to obtain some information about the dissolution properties of a constituent in a solvent. It is possible, at a given temperature, to perform crystallizations in small crucibles containing solvents with known solute–solvent contents, and then to analyze the crystals obtained after cooling.

Increase in the size of the crystals indicates that the liquidus temperature has been reached. In the case of a volatile solvent, a sealed crucible must be used so as not to modify the solute–solvent ratio during the experiment.

It may be worth mentioning the way in which the solubility of calcium carbide in various melted fluxes was determined (Barber, 1961). Here, the crucible features a separating element consisting of a platinum gauze on which the solute is placed. The melted solvent is maintained above this gauze by gas pressure at the bottom of the crucible. When equilibrium is reached, the gas pressure is reduced, after which the composition of the solution collected at the bottom of the crucible can be determined by chemical analysis. The major difficulty here lies in the design of such apparatus.

Another method consists in making compact samples with varying proportions of solvent and solute and placing them in a furnace equipped with a window through which the sample is observed with a microscope. The temperature of melting is also measured.

Whenever a choice has to be made among various fluxes, preference should be given to the one that dissolves the greatest quantity of solute. A study of the solubility of BeO in various fluxes at 1000°C, for instance, gave the following results for 100 moles of solvent: 0.35 mole in KVO_3, 0.90 mole in PbO, and 22 moles in $K_2W_2O_7$. The latter was the one chosen.

There must be chemical inertness between the solvent and the solute. This can be checked either by calculation or by analysis. With the former method, it is sufficient to obtain the sign of the free-enthalpy variation ΔG in the assumed solvent–solute reaction under the experimental conditions. Thus it has been shown (Anthony and Vutien, 1965) that while it is possible to use the flux $PbO–PbF_2$ to get alumina single crystals, this flux cannot be used to prepare zirconia because of the formation of lead zirconate at 1000°C ($\Delta G < 0$). On the other hand, since no reaction is likely to occur with lead fluoride, this flux was adopted with success. However, this calculation is approximate because it assumes that the chemical potentials of the constituents in solution are known. One is usually led to make assumptions, and this calculation is significant only on condition that the free-enthalpy value found is very highly positive (no reaction) or very strongly negative (possible reaction).

Similarly, it is possible to obtain information on the possible reactions experimentally. The mixture is brought to the crystallization temperature in a small crucible, after which the reaction by-products are analyzed by either chemical or physical methods (x-ray-fluorescence, spectrography, radioactivation, etc.).

The solvent may not react with the solute, but may introduce itself by substitution or insertion into the crystalline lattice of the solute. There are two ways of limiting these solid solutions: Either the solvent should have a common ion with the dissolved substance (YVO_4 in V_2O_5, or $CaWO_4$ in Na_2WO_4), or the solvent should contain ions large in relation to the solute ions. This is the case with lead and bismuth (Pb^{4+}, r = 1.26 Å; Bi^{3+}, r = 1.20 Å), very often used in the form of PbF_2 or Bi_2O_3. These ions have a very large radius compared with that of other metal ions (e.g., Al^{3+}, 0.57 Å or Zr^{4+}, 0.82 Å).

It should be noted that lead salts are among the best solvents used. No explanation for this has yet been found, but it would appear that the following two factors play a part: the covalent nature of the linkage modes and the polarizability of the ion Pb^{2+}.

In the particular case where it is necessary to obtain doped single crystals, some of the solvent ions may become inserted into the crystal lattice to compensate for the lack of charge, or surplus charge resulting from the presence of foreign ions. For instance, Fe^{3+} ions may replace the Ti^{4+} ions in $BaTiO_3$ single crystals prepared in a potassium fluoride flux. The solvent F^- ions compensate for the lack of a positive charge and occupy the positions of the O^{2-} ions. The presence in such single crystals of Nb^{5+} ions will cause the inclusion of K^+ ions at the sites previously occupied by the Ba^{2+} ions.

These insertions may modify some of the physical properties of the crystals. Thus it has been shown (Vutien and Anthony, 1970) that in praseodymium aluminate the lead content is 0.5%, whereas up to 5% of bismuth enters the lattice. In the case of the latter, however, the charges are not unbalanced and the color of the crystals is not affected by this insertion.

As already mentioned, impurities affect both the growth rate and the morphology of crystals. Contamination can come either from the crucible or from the solvent. The impurities penetrate either as large inclusions or into the crystal lattice in positions of substitution or insertion. It is essential to use only the purest possible starting chemicals since even an impurity content of 10 ppm can affect the quality of the crystals.

Few data are available on the viscosity of melted salts at around 1000°C. In the cases of non-glass-forming ionic substances, an order of magnitude for the viscosity is 0.1 P at 1000°C (water = 0.01 P).

The vapor pressure of the solvent must not be too high at the temperature of the experiment; in other words, the melting and boiling points must be very different. However, when the solvent does not have this characteristic, it is occasionally possible to circumvent the problem. An example of this is given in the case of solvents like M_2O–MoO_3 or M_2O–WO_3, where M is an alkali metal. The flux can be considered to be an M_2MoO + MoO_3 mixture. Austerman (1963) showed that the active solvent of such a system is MoO_3, since the dissolving capacity depends on how much more MoO_3 there is than M_2MoO_4. But MoO_3 and WO_3 oxides are highly volatile at high temperatures. The addition of ions of alkali metals reduces the vapor pressure, with no other apparent effect. Such mixtures can be used as a flux. The relationship between solubility and the surplus of MoO_3 would appear to show that an association between MoO_3 and M_2O exists even in the melted state.

It is also necessary to be able to separate without difficulty the flux from the single crystals obtained. The method usually consists in dissolving the matrix with water or another suitable reagent. A mechanical method consists in placing the crucible and its gangue in a high-frequency field. The rapid heating of the crucible causes the gangue to melt in contact with the crucible and to be readily detachable (Makram *et al.*, 1966). Another technique is to pour out the surplus flux quickly at the end of crystallization, when the crystals will usually remain fixed to the sides and bottom of the crucible, or, alternatively, they can be sorted inside the eliminated flux (Sasaki, 1964).

TABLE VI

Flux	Melting point (°C)	Boiling point (°C)	Solvent
NaCl	801	1465	Water
KCl	775	1407	Water
KF	860	1500	Water
PbF_2	840	1290	Nitric or acetic acid
$PbCl_2$	500	954	Hot water
B_2O_3	450	—	Bases
Bi_2O_3	817	—	Acids
PbO	888	1470	Nitric or acetic acid
V_2O_5	690	—	Acids
$Na_2B_4O_7$	741	—	Water (slightly)
BaO–B_2O_3	870	—	Acids
PbO–B_2O_3	493	—	Acids
Li_2O–WO_3	695	—	Water
Li_2O–MoO_3	532	—	Water
Na_2O–MoO_3	499	—	Water
Na_2O–WO_3	620	—	Water
K_2O–MoO_3	480	—	Water
K_2O–WO_3	550	—	Water

Table VI lists the principal solvents used, together with their melting points, some boiling points, and the reagents used for their solution.

2. *Choice of the method*

The solute–solvent mixture is placed in a crucible which is then raised to a temperature very much higher than the saturation temperature of the solution. The mixture is kept at this temperature for several hours in order to eliminate all nuclei, i.e., until complete solution. Two conditions are then required to obtain single crystals of sufficient size: the introduction of a seed crystal after the complete solution of the material, and a slow rate of concentration of the solution either through slow cooling, evaporation, or use of a suitable temperature gradient.

a. Slow cooling. The principle of this method is simple: An unsaturated solution is cooled slowly and becomes saturated, with crystallization beginning between 20 and 40°C below the saturation temperature. The cooling rate $d\theta/dt$ is chosen as a function of the solubility coefficient $ds/d\theta$. As a rule $d\theta/dt$ varies between 1 and 5°C hr^{-1}. The concentration of the solution C_0 is important, as already indicated. If, starting from C_0, the cooling path MPQ (Fig. 26) is followed, only microcrystals will be obtained; if, starting at C_0', $M_1'Q'$ is followed, a zone of supersaturation is crossed where the nucleation rate and growth rate parameters allow crystals to appear.

Ideally, the supersaturation conditions should remain the same throughout the growth of the single crystal. Hence the cooling rate should increase with the volume of the crystal, but as a rule, it remains constant.

The volume of the crucible is also an important factor: The size of the crystals obtained for a given cooling rate and temperature cycle increases with the volume of the molten substance. This is related to the volume/area ratio which, as it increases, reduces the role of the crucible walls useful for nucleation and produces more stable temperature conditions. Using a crucible 20 cm in diameter and 25 cm high, 150 g crystals of YAG and 250 g crystals of YIG have been obtained in a PbF_2–PbO–B_2O_3 flux using a cooling cycle of 0.5°C hr^{-1} between 1300 and 950°C (Grodkiewicz *et al.*, 1967). To allow for this effect, the crucible must be filled to the maximum, which can be done by premelting an initial charge and then filling the crucible once more.

For numerous solvents, it is necessary to seal the crucible to avoid evaporation. Studies have shown the utility of working under gas pressure in order to reduce possible dissociations and occlusions. Thus a preparation of yttrium iron garnet in PbO or a mixture of PbO–PbF_2 gives crystals of better quality if produced under oxygen pressure (Makram, 1969). The presence of bivalent iron acts as an impurity and modifies the magnetic properties of the garnet; it is contained either in Fe_2O_3 or is the consequence of the dissociation of the ferric oxide. By calculation, at the melting point of Fe_2O_3 (1565°C), the oxygen equilibrium pressure is 10 bar. Under the conditions of preparation, Fe_2O_3 is in the liquid state, so that a pressure of 15 bar should prevent the formation of FeO. By operating at about 1300°C and cooling to 900°C, at the rate of 1 to 3°C hr^{-1}, one obtains crystals with an FeO content of less than 0.05% by weight, whereas without pressure, the content reaches 1–2%. The absence of Fe^{2+} in the crystals is desired in order to obtain a very narrow ferrimagnetic resonance line (0.28 Oe), and furthermore, the crystals show fewer cracks and inclusions. Similarly, maintaining the PbF_2/PbO ratio constant is a factor favoring a uniform distribution of gallium in $Y_3Fe_{5-x}Ga_xO_{12}$. The technique of growth by cooling is the hardest to control because a variation in the temperature affects many of the solution parameters (growth and nucleation rates, diffusion, viscosity, density).

It is also possible to prepare thin films by the cooling method (Linares, 1968). The technique is derived from the tilting boat method developed for GaAs. The YIG crystals, a suitable solvent (lead or barium borate, sodium tungstate), and a substrate are placed inside a platinum boat which can rotate around a perpendicular axis through its middle, and are then heated to 1000–1100°C until the solvent is saturated. The boat is then tilted so that the solvent reaches the substrate, after which a slow cooling causes the film to grow on the substrate. Crackfree films 0.5–100-μm thick have been obtained on gallium and gadolinium garnet substrates. Growth rates range from 25 to 75 Å sec^{-1}, and the width of the ferrimagnetic resonance line is under 1 Oe.

b. Evaporation. In this method, one operates at fixed temperature and the solution is concentrated by evaporation. When the solvent is sufficiently volatile, the path $MM_1'P''Q''$ (Fig. 26) is followed. It is often preferable to carry out a controlled evaporation in order to limit the growth rates (the crucible being covered with a lid with a small hole in it). The arrangement is shown in Fig. 27. This technique is used especially when it is desired to dope crystals to a specific content or when the solvent–solute system exhibits a narrow range of crystallization, an example being magnesia in lead fluoride. Magnesium fluoride will form below 1150°C.

Examples of this method include beryllium oxide with Li_2MoO_4 and MoO_3 flux (Austerman, 1963), various oxides such as TiO_2, ZrO_2, HfO_2, CeO_2, and ThO_2 in $Na_2B_4O_7$ (Harari *et al.*, 1967), and lanthanum aluminates (Vutien and Anthony, 1969). The evaporation can be done in conjunction

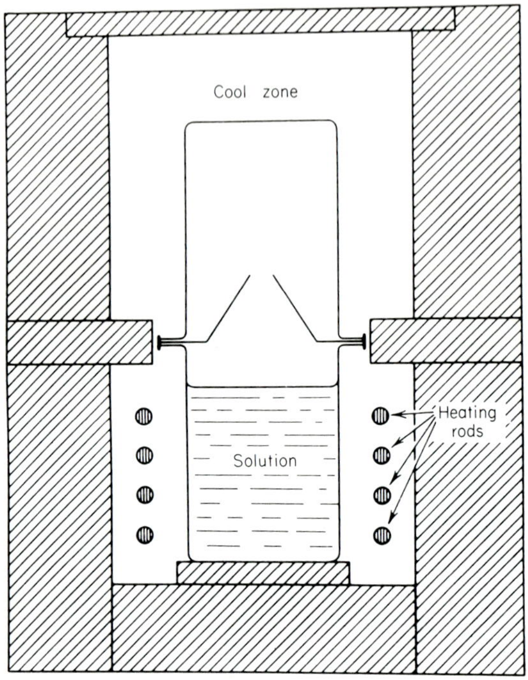

FIG. 27. Apparatus for growth by evaporation (White, 1965).

with slight cooling. The curve MQ′ (Fig. 26) is followed. This method also makes it possible to obtain small crystals from the solution–solute vapor phase, and small crystals of Al_2O_3, $MgAl_2O_4$, ZnO, and MgO have been obtained in this way (Butcher and White, 1965).

c. Temperature gradient. The growth of single crystals can be assisted by the existence of a temperature gradient which generates convection currents. This method, used during hydrothermal synthesis, has been extended to flux synthesis. Technically speaking, therefore, it can be divided into three parts:

1. The lower zone, which contains the product from which a single crystal is to be obtained and which is the hottest zone. The solution is therefore more concentrated in this zone and tends to rise by convection. Circulation is enhanced if the density of the solute is lower than that of the solvent.
2. The central zone, which is the solute transport zone. This transport process can be controlled by means of a baffle-type wall or by a suitably shaped crucible.

3. The upper zone, which is the cold zone into which a single seed crystal is placed. Heat is dissipated by radiation, and consequently the single crystal grows as the solution becomes more saturated.

Factors such as viscosity, diffusion, specific heat and thermal conductivity of the solution play a part, but unfortunately, the data available on the subject are incomplete, and the exact mechanisms involved are in most cases incompletely understood.

The principal advantages of this method are that crystal growth occurs under isothermal conditions, and constant solution concentration (particularly important in the case of doped crystals). Figure 28 illustrates this technique. A variation of this technique consists in establishing a sharp temperature gradient in the region of the crystal to be grown. This permits crystallization at a clearly defined point.

The traveling solvent technique, analogous to the melted-zone technique, consists in confining the solution narrowly between the polycrystalline

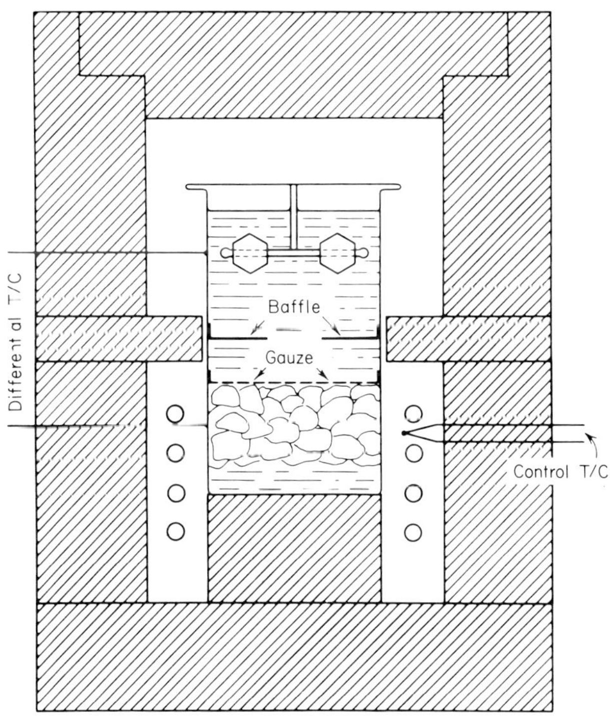

FIG. 28. Apparatus for growth utilizing temperature gradient (White, 1965).

substance and the crystal, so that by utilizing a temperature gradient, dissolution, diffusion of the solute, and crystallization on the crystal occur. $CaMoO_4$ in $LiSO_4$ was obtained in this way. Brissot and Belin (1971) obtained quite large calcite crystals.

The Czochralski technique, adapted to the case of a solution, is possible provided the flux has low volatility and a lower density than the crystal. Yttrium iron garnet (10 × 40 mm) in a bath of $BaO–B_2O_3$ was obtained in this way by drawing at the rate of 1 mm day^{-1}. A satisfactory setup has been described for preparing $NiFe_2O_4$ (Smith and Elwell, 1968).

Electrolytic growth techniques form part of the growth-in-solution processes and consist in obtaining crystals from the decomposition of homogeneous electrolytes. Crystals of UO_2, sodium molybdate bronzes, and spinels containing vanadium have been obtained in this way.

3. *Apparatus*

a. The crucible. Many of the fluxes used will dissolve metallic oxides including those considered to be refractory. Hence most of the metals protected by oxide films are not suitable. Platinum is used in most cases. It is not necessary to operate in chemically pure (99.99%) platinum crucibles, and it is usually preferred to use platinum alloyed with 0.5% rhodium. The incorporation of rhodium increases the mechanical and thermal strengths. Pure iridium or alloys containing iridium are not recommended, since they can be attacked by certain fluxes.

The shape of the crucible will depend on that of the furnace. However, other factors may also dictate the choice, such as the need to obtain a specific temperature variation during the growth of a single crystal by the temperature gradient method. For the slow cooling process, the crucible capacity will vary between, say, 50 and 200 cm^3. Although crucibles with a capacity of 1–2 liters have to be used to obtain very large single crystals, economic considerations prevent this technique from becoming widespread.

To avoid the sublimation of certain fluxes during the experiment, the crucible is fitted with a lid. Better sealing can be obtained by using a flanged lid closely fitting the rim of the crucible. When the crucible must be perfectly sealed, it is usual to employ a flanged crucible which can be covered with a platinum disk of the same diameter. The flange and the lid are then arc-welded or welded with an oxyhydrogen torch. The filler metal used is palladium, which has a melting point lower than that of platinum. In such cases, the lid must be pierced in order to avoid overpressures when heating the crucible. The hole is then plugged with a small platinum stopper once the maximum temperature has been reached.

The inner surface of the crucible must be polished, since any imperfection could become the site of nucleation, and it is essential for the number of nucleation sites to be as low as possible in order to obtain large single crystals.

Special precautions must be taken when using lead salts as the flux. These substances are readily reduced to metallic lead, which forms a low-melting-point alloy with platinum. For this reason, the crucible must be dehydrogenated beforehand.

b. The furnace and its regulating system. The temperatures used for flux growths seldom exceed 1500°C. Furnaces with electrical heating elements having a useful life of about 1000 hr are available for use up to this temperature. Up to 1300°C, Kanthal elements would appear to be satisfactory, but for higher temperatures, silicon carbide or Superkanthal elements are employed. In order to obtain rigorous temperature control, the use of two elements is recommended. The principal element permits the working temperature to be reached, while the other element permits the temperature to be regulated with care; for instance, it may be located in a portion of the furnace where it will cause a predetermined temperature gradient to be established along the crucible.

The furnace must possess great thermal inertia, since thermal fluctuations should be as small as possible. It should be filled to the maximum with refractory bricks, and the whole arrangement must be cooled by circulating water at constant temperature.

While furnaces of fairly conventional design have produced satisfactory and usable crystals, the ever more exacting requirements concerning the quality of the crystals have compelled operators to design growth furnaces with greater care on the basis of results obtained with brief tests.

The use of gas furnaces is to be discouraged; first, because it is impossible to define and keep constant the power dissipated at all points on the furnace, and second, because the combustion gases diffuse through the furnace and could contaminate the solution.

Temperature control is of great importance in the growth of a crystal. Specific arrangements for temperature-control systems are well known and it is easy to obtain a temperature constant to within ±0.2°C up to 1000°C. The problem of slow cooling is more complex, for not only must the temperature be decreased at a slow rate, but it must also be controlled. Thus, although ordinary apparatus will permit in some cases a cooling rate of 1°C hr^{-1}, it happens that the temperature variations are of this order, too. For this reason, more complex systems are employed, which differ according to the furnace used. As a rule, the furnace is connected to an autotransformer which is supplied with stabilized current and driven by a motor governed by a programmed regulating system. This arrangement makes it possible to obtain a temperature stable to within $\pm 0.01°C$ for a cooling rate of $1°C\ hr^{-1}$.

4. *Examples of flux growth*

The growth of crystals in melted salts is a widely employed method which can be used for preparing such varied crystals as sulfides, selenides, tellurides, halides, and above all, oxides (single, double, or multiple oxides). A few growth trials will often be sufficient to obtain the required crystal, and for this reason the authors were no longer concerned with increasing the quality or size of the crystals. Without doubt, research has been most extensive with regard to garnets, since both size and quality are of major importance in their particular applications.

a. Growth by the cooling process: garnets. Consider for example the preparation of $Y_3Al_5O_{12}$ (YAG) crystals, with an incongruent melting point of 1970°C, and of $Y_3Fe_5O_{12}$ (YIG) crystals, with an incongruent melting point of 1555°C. Workers at the Bell Telephone Company have succeeded in obtaining crystals of up to 150 g (YAG) and 250 g (YIG) (Van Uitert *et al.*, 1965 and Grodkiewicz *et al.*, 1967).

The garnets are prepared from a flux composed of 44 wt% PbO, 53 wt% PbF_2, and 3 wt% B_2O_3. The method used is that of slow cooling at the rate of $0.5°C\ hr^{-1}$ between 1300 and 950°C. Below 950°C, the crystals tend to redissolve. However, it is useful to separate the crystals from the flux at a high temperature and to cool them slowly to avoid thermal shock. This was the aim in designing the crucible and furnace (Fig. 29), which make it possible to drain the flux at high temperature, rotate the crucible, and establish a negative temperature gradient to induce growth on the bottom of the crucible. The platinum crucible has an inside diameter of 20 cm, a height of 25 cm, and a thickness of 0.02 cm. It is mounted on a sectioned, hollow pedestal that is locked together by a replaceable liner. The pedestal parts, the muffle, and the top plug are all made of cast alumina. The crucible has a 0.20-cm inside diameter and a 0.10-cm-long tube in the bottom. When the crucible is properly positioned, it is concentric with a platinum–10% iridium drain tube. The crucible and pedestal are mounted on a thrust bearing so that they can be rotated to enhance mixing of the melt components. During the run, the pedestal liner accommodates a porous brick plug. Oxygen is blown up through the liner to cool the drain tube and the bottom of the crucible and also to sweep lead vapors from the furnace. When the melt has cooled to 950°C, the air hose mounting and the porous brick plug are removed. A 30-cm-deep tray which has a 10-cm-high pipe mounted vertically in its center is then inserted

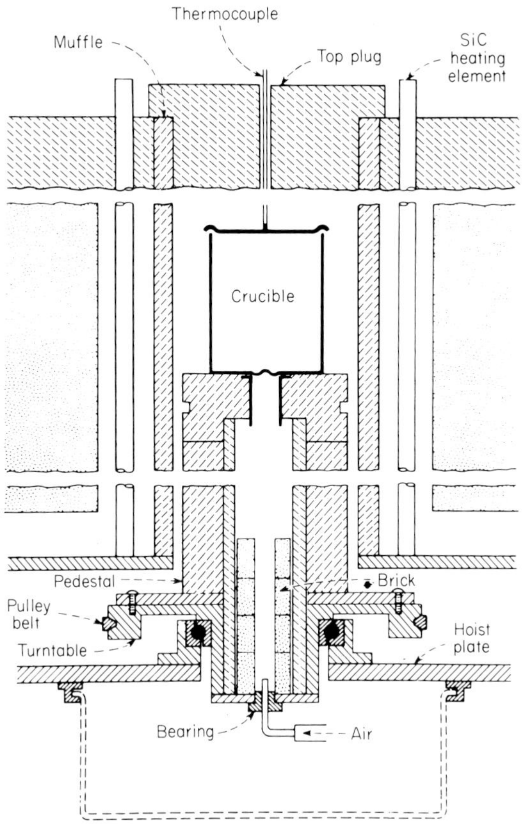

FIG. 29. Arrangement of furnace, pedestal, and crucible for crystal growth (Grodkiewicz *et al.*, 1967, used with permission of Pergamon Press).

below the hoist platform. A polished steel mirror is laid alongside the vertical pipe to allow viewing up through the hollow pedestal to guide the spear. Once the spear is in contact with the crucible, a slight tap on its end punctures the platinum insert. The crucible takes about 30 min to drain.

By using a mixture of 1694 g Y_2O_3, 2397 g Fe_2O_3, 4 g CaO, 6021 g PbO, 4926 g PbF_2, and 279 g B_2O_3, 2800 g of crystals were obtained, five crystals attaining a weight of 200 g each. For neodymium YAG with 720 g of Y_2O_3, 1220 g of Al_2O_3, 253 g of Nd_2O_3, 3556 g of PbO, 4346 g of PbF_2 and 279 g of B_2O_3, two large crystals and one small one, of total weight 248 g, were found at the bottom of the crucible. Tolksdorf (1968) has proposed a technique fairly similar to that described here which makes it possible to extract the crystals at the end of the operation and to resupply the melt and reduce the solvent losses.

A substantial improvement in the quality of the crystals is obtained by operating under oxygen pressure (Makram, 1969). A study of the effect of divalent ions on crystal growth was undertaken. Addition of BeO, MgO, MnO, ZnO, CdO, or CaO to the melt consistently resulted in the growth of large garnets. Comparable effects are obtained by adding SiO_2. The quality of the crystals varied with the divalent ion present. Small ions such as Be^{2+} and Mg^{2+} caused extensive cracking. Good crystals were obtained using the other divalent metal oxides mentioned or SiO_2. However, of these, all but the Ca^{2+} additions impaired the infrared transmission properties of the YIG crystals. The best results are obtained with CaO additions ranging from 0.1 to 1.0 wt % of the crystal components. Small SiO_2 additions best improved the optical quality of YIG grown from a lead borate flux. We have found that when PbF_2 is a flux component, equally high transparency is obtained using CaO additions. A CaO addition of 0–5 wt% is optimum. About one tenth of the Ca^{2+} enters the crystal, compensating for the presence of Si^{4+} or F^-.

b. Growth by the evaporation process: magnesia. Growth by the flux-evaporation, temperature-gradient process was applied successfully to grow small magnesia crystals (T_F = 2900°C, 5–6 mm thick) (Webster and White, 1969). This is a common crystal size for the flux growth process. In the case of magnesia, it is difficult to use the cooling technique in the presence of a PbF_2 flux because magnesium fluoride forms below 1150°C and the boiling point of PbF_2 (T_E = 1290°C) limits the working temperature.

Control of the evaporation rate can be achieved either by choice of growth temperature or by restricting the effluent by an orifice in the crucible lid. In the latter case, it is important to obtain a positive seal between the lid and crucible in order to eliminate unintentional and irreproducible losses of vapor. In practice, the latter process is preferred since it permits growth to be studied at various temperatures without consequent alteration of the growth rate. The effluent vapor was normally allowed to escape into the furnace, producing a glassy lead silicate layer on the furnace bricks.

Crystals grown by spontaneous nucleation were obtained up to 1 cm^3 in size by evaporation at 1230°C for periods of two to three weeks. The products often appeared cloudy unless grown very slowly, keeping the linear growth rate at approximately 0.3 mm day^{-1} by decreasing the orifice size.

c. Growth by the temperature gradient process: magnesia. Growth by the temperature gradient technique made it possible to seed crystals while minimizing the danger of the seeds dissolving before growth had commenced. The apparatus employed for this process is shown in Fig. 30. Source material in the form of large, crystalline lumps of fused magnesia was placed in the bottom of a 500-ml crucible and held in position by a platinum gauze spot-welded to the crucible walls. A mixture of magnesia powder and lead fluoride was then melted into the crucible almost to the brim. The $MgO–PbF_2$ ratio in the charge was chosen so that the solution was just unsaturated at the temperature in the growth zone. Magnesia seed crystals, selected from large pieces of the fused material, were either placed on the surface of the fused charge, so that they floated freely, or were held in position by platinum wires.

The cooled crucible was sealed by welding on a lid in the form of a platinum disk, and the crucible was placed in a furnace constructed to give the required temperature gradient between the top and bottom. For this purpose, a baffle in the furnace interior helped to maintain the higher base temperature. The crucible was left in position at constant-temperature conditions for periods of up to three weeks.

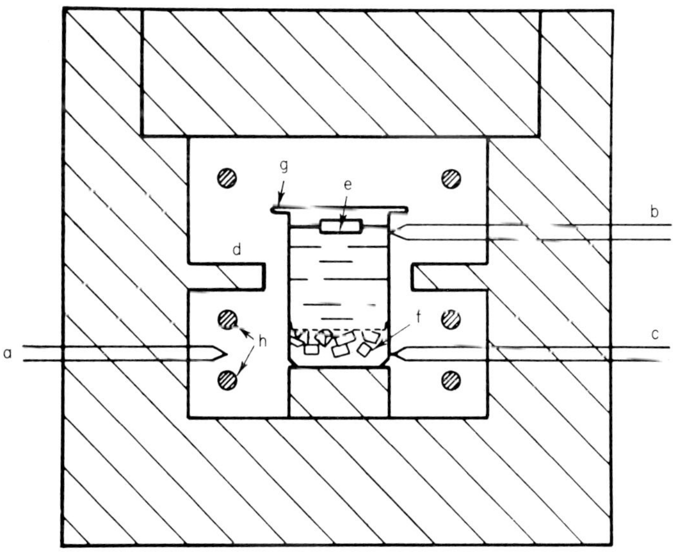

FIG. 30. Furnace and crucible assembly used for flux melt growth (Webster and White, 1969). (a, b, c) Control and monitoring thermocouples; (d) external baffle; (e) seed crystal; (f) source material; (g) sealed platinum crucible; (h) heating elements. Used with permission of North-Holland Publishing Co., Amsterdam.

The temperature gradient between the top and bottom of the crucible was monitored by means of Pt/Pt–13% Rh thermocouples or, better still, Platinel 2 thermocouples, welded to the outer wall 1.5 cm from the top and bottom edges.

The following temperature conditions were found favorable for growth on the seed crystals:

	Seeds floating	Seeds immersed
Crucible top	1210°C	1200°C
Crucible base	1225°C	1235°C

The rate of growth was approximately 2 mm week^{-1}, and layers 5–6 mm in thickness were obtained on seed crystals cleaved to expose clean (100) faces. With larger temperature gradients, spontaneous nucleation occurred at the top of the crucible. Growth features and the morphology of the crystal products will be described.

5. *Results*

Substances prepared by flux techniques are extremely varied and numerous so that it is impossible here to list them completely [see White (1965) and Laurent (1969)]. Instead, the techniques for preparing simple oxides will be summarized, indicating, at the same time, the families of complex oxides to which the in-flux growth technique was successfully applied (Table VII).

Among the more complex oxides, the following families were prepared:

Perovskite structure: $BaTiO_3$, $NaNbO_3$, $PbTiO_3$, $LnAlO_3$, $LnCrO_3$ (Ln = lanthanide element); *Flux:* PbO, PbF_2, KF, $PbCl_2$, ...

Scheelite structure: MWO_4, $MMoO_4$ (M = Mg, Ca, Sr, Ba, Zn, Cd); *Flux:* Na_2WO_4, NaCl, Na_2MoO_4, ...

Spinel structure: $MgAl_2O_4$, $MgFe_2O_4$, MFe_2O_4 (M = Ni, Co, Zn, Mg, Cd).

Garnet: RFe_5O_{12} (R = Y, Ln); RAl_5O_{12} (R = Y, Ln); *Flux:* PbO, PbF_2, ...

βAl_2O_3 *Structure:* MO, $6Fe_2O_3$ (M = Pb, Ba, Sr), $M_2'O$, $12X_2O_3$ (M′ = Na, K; X = Al, Ga); *Flux:* PbO, PbF_2, BaO–B_2O_3, ...

Silicates: Zn_2SiO_4, Al_2SiO_5, $BeAl_2Si_6O_{16}$, Mg_2SiO_4, . . . ; *Flux:* PbO, Li_2O, MoO_3, ...

Other substances: $CaCO_3$, Al_2BeO_4, YVO_4, $Bi_4Ti_3O_{12}$, $ThTi_2O_6$, $Cd_2Nb_2O_7$, $M(BO_2)_3$ (M = Sm, Gd, Tb, Dy, or Er), $LnAl_3B_4O_{12}$, ...

TABLE VII

Substance	Melting point (°C)	Flux	Remarks	Ref.
Al_2O_3	2050	PbF_2	There is ample literature[a] concerning the production of crystals of pure Al_2O_3, rubies or sapphires, crystal growth usually being attained by isothermal evaporation	Laudise *et al.* (1962) Timofeeva (1965) White and Butcher (1965)
		$PbO-PbF_2$	The crystals obtained by the flux technique are of better quality than those obtained by the Verneuil method	Chase and Osmer (1967)
		$Bi_2O_3-PbF_2$	Adding very small quantities of La_2O_3 to the flux improves the quality of the crystals	Chase and Osmer (1967)
		$PbO-B_2O_3$ or $Na_2O-B_2O_3$	Slow cooling	Nelson and Remeika (1964) Barks and Roy (1967)
		Na_2O-WO_3 or $SrO-WO_3$	Spontaneous crystallization does not occur in a sodium tungstate flux	Voronkiva *et al.* (1967, 1968)
		$BiF_3-B_2O_3$	Slow cooling of the flux to 1300°C	Grodkiewicz and Nitti (1966)
		Cryolite	A cryolite flux permits sapphire crystals at the formation of sapphire crystals at temperatures between 1050 and 950°C	Arlett (1967)
BeO	2550	$Li_2MoO_4-MoO_3$	The solubility of BeO in the flux increases with the concentration of MoO_3	Austerman (1963)
		$Li_2MoO_4-MoO_3$	Adding small quantities of $LiPO_3$ improves the quality of the crystals, while adding $LiBO_2$ slows their growth rate	Austerman (1964)
		$Na_2O-B_2O_3$	The aspect of the crystals varies with the concentration of B_2O_3	Linares (1967)

TABLE VII (continued)

Substance	Melting point (°C)	Flux	Remarks	Ref.
CaO	2575	NaCl	Heating to 950°C in a platinum crucible	Setoguchi and Sakamoto (1966)
CeO_2	2600	PbF_2 or BiF_3	Small quantities of B_2O_3 are added to the flux	Grodkiewicz and Nitti (1966)
		$Li_2O–WO_3$ or $Li_2O–MoO_3$, $Na_2O–B_2O_3$ or $BaO–B_2O_3$	Slow cooling; The morphology of the crystals depends on the flux: cubic or octahedral crystals with edges a few millimeters long are obtained	Linares (1967)
CoO	1800	NaCl	Obtained under the same conditions as CuO and NiO, the size of the crystals usually varies with NiO, CoO, CuO	Pakter (1968)
Cr_2O_3	2265	$Li_2O–B_2O_3$	Slow cooling	Barks and Roy (1967)
Ga_2O_3	1725	PbF_2 or $PbO–PbF_2$	Isothermal evaporation of the flux	Timofeeva (1965) Chase and Osmer (1967)
		$PbO–B_2O_3$	Slow cooling	Nelson and Remeika (1964)
		$Na_2O–WO_3$ or $SrO–WO_3$	Single crystals are obtained either by slow cooling and spontaneous crystallization, or by the temperature gradient technique	Voronkova *et al.* (1968)
MgO	2800	PbF_2	The size of the crystals is greater than that obtained with borate	Webster and White (1969) Nielsen and Dearborn (1958, 1960)
NiO	2000	NaCl, PbF_2	Slow cooling; the crystals are of small size and of octahedral aspect	Pakter (1968) Hill and Wanklyn (1968)
ThO_2	3210	PbF_2	PbO, Bi_2O_3, or B_2O_3 is added to the flux; perfect crystals free from inclusions can be obtained	Chase and Osmer (1967) Grodkiewicz and Nitti (1966)

TABLE VII (continued)

Substance	Melting point (°C)	Flux	Remarks	Ref.
ThO_2	3210	$Li_2Mo_2O_7$ or $Na_2B_4O_7$	Slow cooling	Harari *et al.* (1967) Linares (1967)
TiO_2	1830	PbF_2	The flux can be pure lead fluoride; small quantities of B_2O_3 or alkali borates can be added to the PbF_2; the flux evaporates slowly	Grodkiewicz and Nitti (1966) Berkes and White Harari *et al.* (1967)
		NaCl	The crystals are usually smaller in size than NiO or Fe_2O_3 crystals obtained under identical conditions	Pakter (1968)
UO_2	2870	$Na_2B_4O_7$	Isothermal heating to 1100°C in hydrogen; the crystals are cubic or octahedral, and red in color	
Y_2O_3	2410	PbF_2	Platelets 5 mm long are obtained	Hart and White (1968)
ZnO	1975	PbF_2	Slow cooling between 1150°C and 800°C; at these temperatures, however, the flux evaporates rapidly	Nielsen and Dearborn (1958, 1960)
		$PbO–PbF_2$	The temperature gradient is created by a jet of cold air blowing in through the top of the crucible	Chase and Osmer (1967)
ZrO_2	2680	PbF_2	The procedure here is to evaporate the flux; slow cooling can also be used	Anthony and Vutien (1965)
HfO_2	2780	$Na_2B_4O_7$	The cooling rate is 2°C hr^{-1} between 1050 and 690°C for ZrO_2, 4°C hr^{-1} between 1250 and 1000°C for HfO_2	Harari *et al.* (1967)

[a] The references concerning this substance are too numerous to be listed in full here.

Thus a large number of oxide base crystals can be prepared by the flux method. The size of the crystals is usually under 10 mm and the adoption of slow rates (over several days) ensures that crystals free from thermal stress are obtained. The equipment required is inexpensive compared with that required for crystal growth by the hydrothermal or high-temperature techniques. The flux growth method is a valuable technique which can be readily developed in any laboratory and used for growing a wide variety of crystals.

B. Crystallization and Synthesis by the Hydrothermal Method†

Crystallization of high-melting-point refractory oxides is impossible in water at atmospheric pressure. Hydrothermal synthesis makes use of the fact that certain solids that are insoluble in water at atmospheric pressure are soluble at high pressure, i.e., at around 500°C and 1500 bar. Under such conditions, growth takes place at relatively low temperatures, which reduces the internal stresses associated with steep temperature gradients and makes it possible to prepare phases which are unstable at the melting point. Finally, compared to the flux method, it has the advantage of offering rapid growth rates (6 mm day^{-1}).

The chief drawback is the cost of this technique, due to the need for apparatus which can withstand high pressures and corrosion. Application of the hydrothermal growth method requires information about the thermodynamic equilibrium of the solvent–solute system and about the growth kinetics; in addition, a suitable autoclave must be designed.

Hydrothermal reactions can be divided into two main groups: recrystallization reactions and hydrothermal syntheses. In the latter group, solid phases, solutions, and gases can be made to react. The by-products of the synthesis are recovered in the form of crystalline phases. The growth technique consists in establishing a temperature gradient along the autoclave such that the temperature at the bottom assists dissolution of the material whereas solubility is low at the top. Under the effect of the convection currents, material is transferred from the regions in the apparatus where solubility conditions are best toward the regions where they are less favorable. This transfer is often assisted by substances called mineralizers.

1. *Growth and hydrothermal synthesis criteria*

a. Physical properties of water. The physical properties of water can be described satisfactorily by the temperature and density parameters.

† See Ballman and Laudise (1963), White (1964), Roy and White (1968), Christensen (1969).

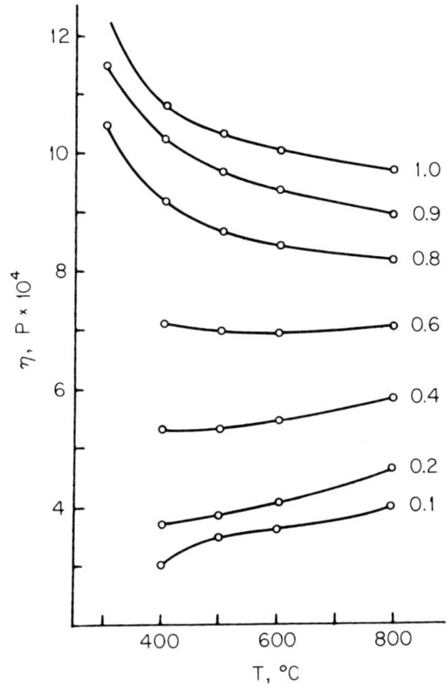

FIG. 31. Viscosity of aqueous solutions (Christensen, 1969). Values next to the curves are density (in gm^{-3}) $\times 10^4$.

Figure 31 shows the relationship existing between the viscosity of water and the two parameters in question. It will be observed that, irrespective of the density of the solution, the viscosity of water remains low in comparison with that of the molten salts.

Figure 32 gives the P–T diagram of water at constant volume. It will be remembered that the critical conditions are reached in a closed vessel only when the latter is 32% full of water and the temperature rises. If the vessel is over 32% full, the water will fill it completely before the critical temperature is reached, whereas if it is less than 32% full, it will be completely filled with water vapor before the critical temperature is reached. The use of a mineralizer to facilitate transfer of the material modifies these curves somewhat. A mineralizer is any compound added to the aqueous solution that speeds up crystallization. The mineralizer usually operates to increase the solubility of the solute through the formation of species which ordinarily cannot be present in the water.

Thus in aqueous solutions, even at 400°C and 2000 bar, the solubility of α quartz is too low to permit growth within a reasonable space of time. Alkali additions like NaOH, Na_2CO_3, KOH, and K_2CO_3 are highly effective in such cases. A minimum concentration exists, but if it increases, the growth rate is not greatly increased and new phases can be produced. For quartz, for example, NaOH to 0.25 *M* is a correct concentration. With concentrations of about 4 *M* NaOH or 2 *M* KOH, sodium and potassium silicates form along with the α quartz.

The mineralizers can be either bases like NaOH, Na_2CO_3, or KOH, or strong acids like NH_4Cl, HCl, HBr, or HI. Thus the solubility of KTN ($KTa_{0.65}Nb_{0.35}O_3$) is very low in water, and a great many mineralizers have been investigated between 350 and 500°C at 650 bar (Marshall and Laudise, 1967). In an acid medium (HCl, HF, HBO_2), KTN decomposes and yields oxides or oxychlorides. In alkali media (hydroxides or carbonates), when the cation M^+ is Li, Na, Rb, or Cs, one obtains

$$K(Ta, Nb)O_3 + M^+ \rightarrow M(Ta, Nb)O_3 + K^+$$

$Ba(OH)_2$ reacts in similar fashion, while the other alkaline-earth hydroxides are regarded as too insoluble for practical purposes. A weak base such as NH_4OH has no effect, while organic bases often decompose under hydrothermal conditions. Potash at 650°C and 1200 bar was finally chosen; the solubility of KTN is increased by 40% with a 15 *M* solution of KOH.

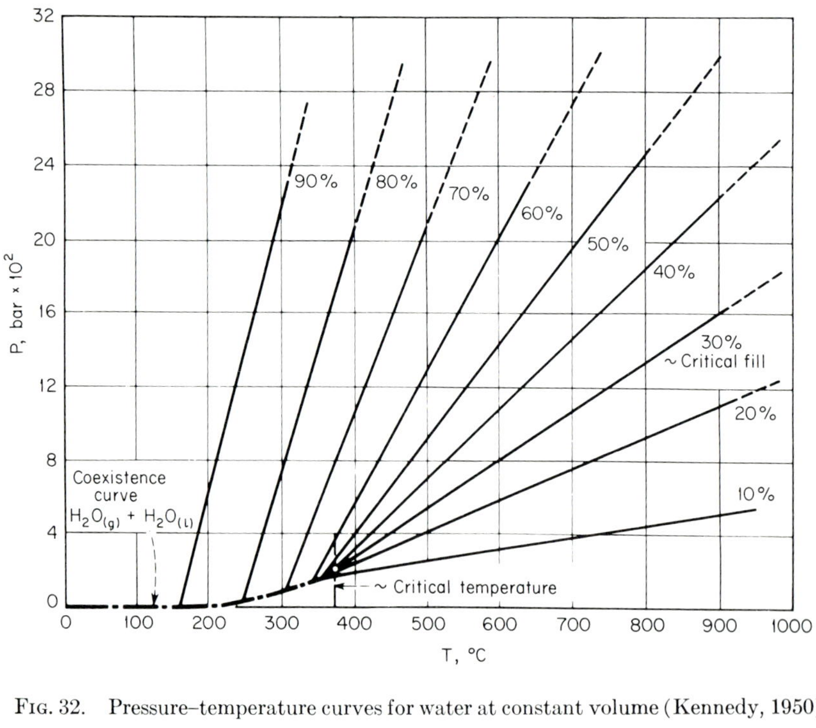

FIG. 32. Pressure–temperature curves for water at constant volume (Kennedy, 1950).

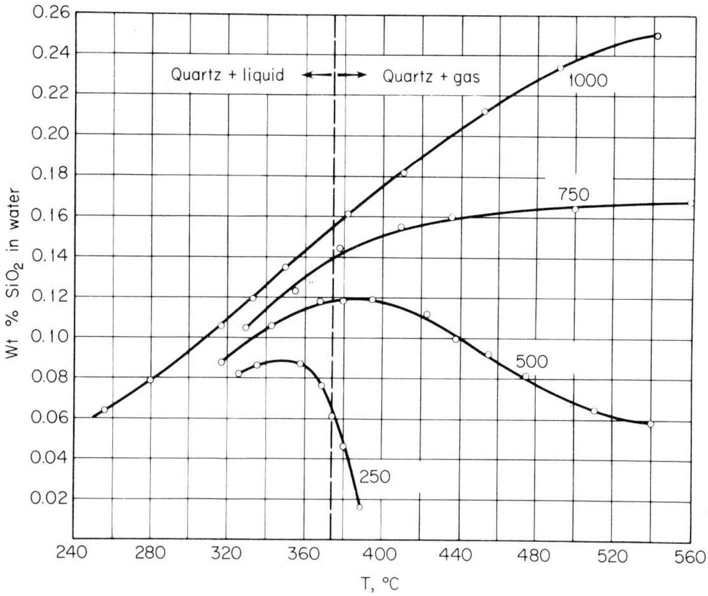

FIG. 33. Solubility of quartz versus temperature (Kennedy, 1950). Values on the curves give pressure in bars.

b. The water–solute system. The water–solute or water–solute–mineralizer phase diagrams should be examined. Certain water–solute–mineralizer diagrams are already known. If no data are available, the weight-loss method can provide useful information. Samples of known weights are placed in the autoclave and maintained at constant temperature. By performing a series of weight-loss measurements at different temperatures, pressures, and degrees of filling, it is possible to determine the conditions which will favor dissolution of the material.

After the solubility curve conditions are reached, crystallization proceeds in a manner similar to that described for the flux method (Fig. 26). Because of its piezoelectric applications, alpha quartz, which is stable only below 573°C, has been the subject of numerous solubility determinations. Figure 33 shows the Kennedy (1950) curve plotted for SiO_2–H_2O. The most desirable regions are those where the temperature coefficient exhibits a marked gradient, which permits heavy supersaturation. The temperature coefficient can be either positive or negative, permitting growth in each case.

c. Kinetics. Most data concerning growth rates have been accumulated for α quartz. The most important parameters affecting growth are the crystallization temperature, the thermal gradient along the autoclave, the degree of filling, the mineralizers, and the orientation and size of the nucleus.

To a first approximation, the logarithm of the growth rate is proportional to the reciprocal of the crystallization temperature. For a 30°C gradient and an 85% fill, the growth rate of alpha quartz is 0.5 mm day^{-1} at 350°C, becoming 2 mm day^{-1} at 400°C. The growth rate varies almost linearly with the temperature gradient. At 350°C with an 85% fill, the rate of 0.5 mm day^{-1} for a 20°C gradient becomes 1.5 mm day^{-1} for a 60°C gradient. One may observe an increase in the rate with greater opening of the baffle which separates the autoclave into two almost isothermal regions: one for the melt and the other for growth.

The rate of growth varies linearly with the degree of filling up to 80%, but in the case of alpha quartz, increases very rapidly above this figure. At 385°C, with fills of 66% and 74%, the rates are 0.5 mm day^{-1} and 1 mm day^{-1}, respectively, and exceed 2 mm day^{-1} above 80%.

The influence of the mineralizers is primarily to increase solubility and to affect the degree of supersaturation. Thus, for alumina at 405°C and 1600 bar, the solubility is 0.2 *M* in an 0.5 *M* NaOH solution and 0.25 *M* in a 1.0 *M* sodium carbonate solution. However, supersaturation is much greater in the carbonate than in NaOH, thereby increasing the rate of growth.

The anisotropic character of a material will affect the growth rate. For alpha quartz, it has been found that

$$\text{plane } 0001 > \text{plane } 01\bar{1}1 > \text{plane } 1\bar{1}01 > \text{plane } 10\bar{1}0$$

A relationship also exists between the area of the nucleus and the area of the powder to be dissolved: The area of the latter must be more than five times greater than that of the nucleus.

Finally, phenomena like transfer, diffusion, adsorption, and chemisorption play important, but as yet not well understood roles.

2. *Apparatus*

The apparatus presents two basic problems: (1) design of the autoclave, which must withstand high pressures and corrosion from the aqueous medium at the crystallization temperature, and (2) temperature control and measurement.

a. Autoclave. A wide variety of commercially available autoclaves and closure systems exists. The ideal autoclave must have a closure system which is robust enough to withstand the high pressures and simple enough to make it easy to manipulate. The construction material must have high

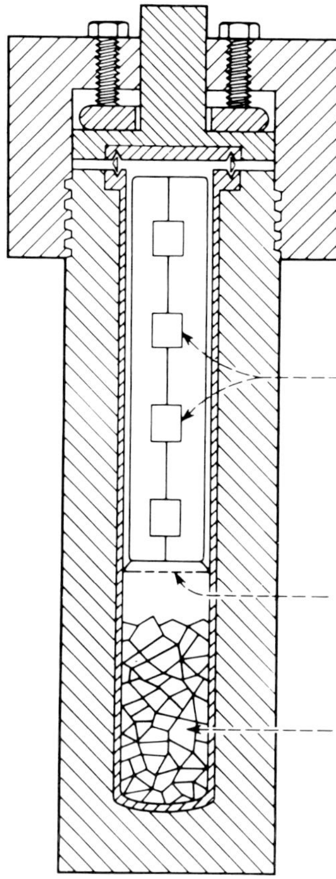

FIG. 34. Schematic of hydrothermal system (Gilman, 1963).

mechanical strength and great resistance to corrosion by acid or base solutions. The different parts are shown in Fig. 34; the actual autoclave design must be closely related to the pressures to be used. For temperatures of around 300°C and pressures of 10 bar, a thick Pyrex or quartz tube is quite adequate. For higher pressures, Ballman and Laudise (1963) review the equipment in use.

For pressures up to 500 bar and temperatures near 400°C, steel vessels with a flat plate closure are useful. The seal is made by the pressure of a plunger against a metal disk or gasket that butts against the shoulder of the vessel body. Such a vessel is quite easily assembled, and the plate gasket requires a minimum of mechanical precision.

Ordinary low-carbon steel may be used for vessels when corrosion is not expected. If corrosion is a problem, silver or platinum liners can be placed

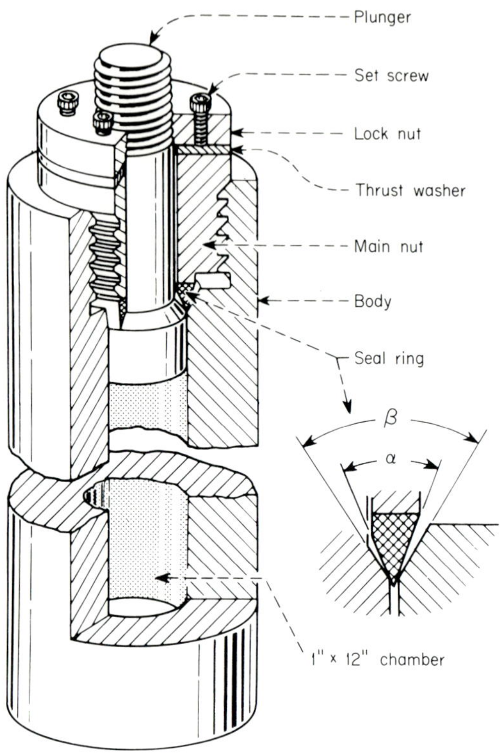

FIG. 35. Bridgman vessel (Gilman, 1963).

in the vessel and used with sealing rings of the same material. For higher temperatures, special alloys such as Inconel X and Udimet permit operation near 700°C.

A welded closure as described by Walker and Buehler (1950) has been used quite successfully for temperatures up to 450°C and a pressure of 3000 bar. The vessel ordinarily consists of an inner liner welded at both ends, and an outer tube of heavier construction.

At the conclusion of an experiment, one end of the welded liner is sawed off to remove the charge. The liner is then driven out of the outer casing and discarded, and a new liner is inserted. Thus for a series of experiments rather extensive machine shop services are needed. However, welded liners are still relatively inexpensive compared with an entire vessel. The disposable liner permits a good deal of exploratory work to be done in systems where corrosion might ruin a more complex vessel.

The closure developed by Bridgman (1914) (as modified in several commercially available vessels) is perhaps the most generally useful design for hydrothermal growth systems. The initial seal that closes the vessel is made by mechanically tightening the plunger against the deformable gasket by the set screws in the head. The unsupported area closure makes its seal when the force of the generated pressure acts upon the piston, moving it upward against the deformable gasket. Figure 35 illustrates one of several commercially available versions. Such vessels can be used quite successfully in the 500°C, 3700-bar range, the construction material being chosen to meet the particular experimental conditions.

In principle, the cone closure (Fig. 36) is quite similar to the flat plate closure that is used at medium pressures. It is usually reserved for vessels of small internal diameter (~5 mm), and the sealing surface is at a much lower temperature than the crystallization zone. The seal is made against a conical seat by a cone-shaped insert that is forced down by a screw cap. The force on the cone seat depends on the torque applied to the screw cap.

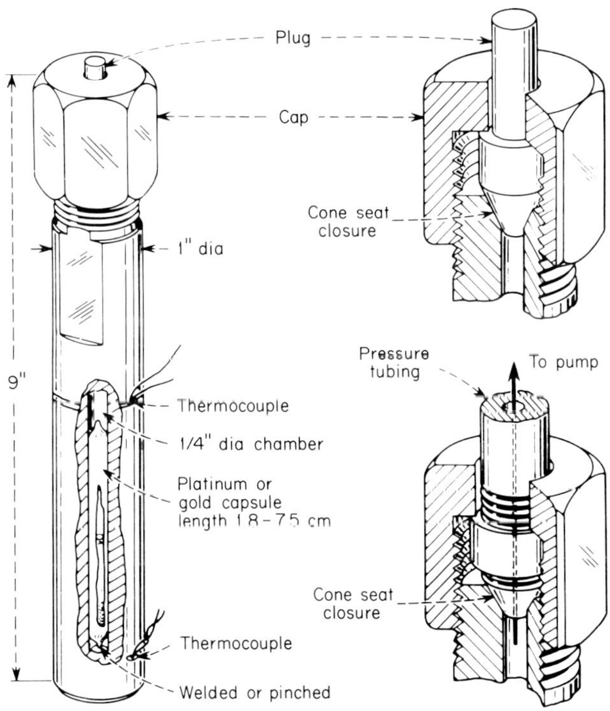

FIG. 36. Reaction vessel with a cold cone seat closure (Roy and Tuttle, 1956, used with permission of Pergamon Press).

This limits the closure (at high pressures) to vessels of approximately 7 mm internal diameter or less.

Because the cone seat can be in a relatively cool zone, it permits crystallization at high temperatures and pressures with little creep. Pressures up to 2400 bar can be held quite easily at 750°C, and higher temperatures at somewhat lower pressures.

Although vessels of this type are most often used for the study of phase equilibria, crystal growth experiments have been successfully undertaken in them. It should be noted that Boksha (1968) describes a chamber capable of operating at pressures up to 30,000 bar.

An inert system was designed in each of the pressure systems discussed, except the low-pressure glass system; noble-metal liners or plating may be used to prevent or inhibit corrosion. Small platinum, gold, or silver tubes welded at both ends contain the charge and have been used in both the medium- and high-pressure vessels. If both the inner noble-metal tube and the outer vessel are filled to the same percentage of their free volume (per cent fill, or degree of fill) with the aqueous solvent, the noble-metal tube will, in principle, be under no pressure. This "pressure-balancing" technique permits the use of thin-walled tubing.

Although flat plate closures with inert liners have been used quite successfully in the medium-pressure range, attempts to design a completely inert system in the high-pressure range have not been completely successful. Further improvements are needed before the high-pressure range can be fully exploited.

b. Temperature and pressure control and measurement. Furnaces for hydrothermal vessels depend on the degree of temperature control required and the temperature gradients to be used for crystallization. Hot plates heated with nichrome strip heaters are often adequate up to about 400°C. For higher temperatures or controlled temperature differentials, ceramic tubes wound with nichrome or Kanthal wire are efficient. The spacing of the winding may be varied to produce a variety of temperature gradients.

For rough temperature control, a variable transformer, such as a Variac, may be used to regulate the furnace. More sensitive control can be obtained with an on–off type of controller, and further improvement can be obtained by the use of saturable core reactors.

Temperature measurements are usually made by strapping chromel–alumel thermocouples to the outside of the vessel. For internal measurements, sheathed thermocouples of low heat capacity have been used. Gradients through the walls of the vessel are surprisingly small (of the order of a few degrees). The pressure generated in the vessel can be measured by a Bourdon gauge or by a strain-gauge cell. In most laboratory experiments, however, the pressure is estimated from the degree of filling of the free volume in the vessel and the temperature of operation.

Kennedy's (1950) *PVT* data for pure water are shown in Fig. 32. Measurements made in the Na_2O–SiO_2–H_2O system (used to grow quartz) indicate a pressure reduction of about 15% of the pure water system. The extent to which pressure is reduced at any given temperature as compared to that with a pure water system will depend on the mineralizer concentration and the solubility of the solute. Greater pressure reduction occurs for higher solute and mineralizer concentrations. In such a case, one needs some *PVT* data on these systems.

The obvious safety hazards associated with hydrothermal crystallization and high-pressure equipment can be minimized by reasonable precautions. The entire furnace assembly may be placed in a pit below the ground, or a thick-walled barricade can be constructed to deflect possible autoclave fragmentation. Rupture disks should be incorporated in the vessel. These are available in a wide range of bursting pressures. Even for low-pressure, low-temperature experiments, minimum precautions should at least include shielding of personnel from live steam or hot solution in case a vessel fails.

Roughly speaking, a hydrothermal operation proceeds as follows. The material to be dissolved is placed at the bottom of the autoclave (dissolving zone) and the nucleus is suspended at the top level with the lid (crystallization zone). A baffle may be used to separate the two zones. After charging, the solvent is placed in the autoclave to the required degree of fill, after which the whole system is closed and placed in a furnace at the necessary temperature. When thermal equilibrium is reached, the material dissolves and saturates the solution. The top of the autoclave is cooler, resulting in a supersaturation of the solution in the region of the nucleus and permitting growth. The temperature gradient along the autoclave causes convection currents to supply saturated solution to the level of the nucleus. The following conditions have been used for alpha-quartz crystals of up to 800 g:

Dissolving temperature	400°C
Crystallization temperature	360°C
Temperature differential	40°C
Degree of fill	80%
Solution	1.0 *M* NaOH
Baffle aperture	5%
Pressure	1700 bar

We now take a closer look at the hydrothermal growth of corundum and ruby.

3. *Growth of corundum and ruby*

Marais (1969) has given a detailed description of the results of his investigation into the hydrothermal growth of Al_2O_3.

a. Investigation of the Al_2O_3–H_2O system. Ruby consists of a crystalline variety of alumina: corundum or αAl_2O_3, doped with more or less red Cr^{3+} ions. Corundum crystallizes in the rhombohedral system.

Corundum is a result of the gradual dehydration of gibbsite [$Al(OH)_3$] under hydrothermal conditions and constitutes the stable phase at temperatures above 450°C. The stability domains of the other varieties were determined by Ervin and Osborn in 1951 (Fig. 37). It should be noted that corundum exists regardless of the pressure at temperatures up to 400°C. The data obtained by Laudise and Ballman (1958) in investigating the Al_2O_3–H_2O–($NaOH$ or Na_2CO_3) systems and Marais's tests in connection with the Al_2O_3–H_3O–(K_2CO_3 or Rb_2CO_3) systems have confirmed the

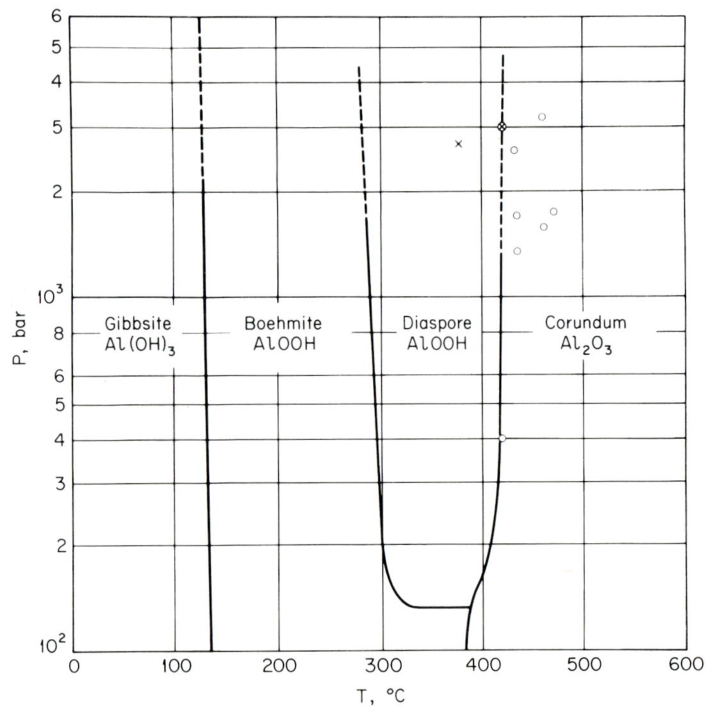

FIG. 37. Systems of Al_2O_3–H_2O (Gilman, 1963). (——) Ervin and Osborne (1951); (– – –) extrapolated; with Na_2CO_3; ○ corundum formed; × diaspore formed;

TABLE VIII[a]

T (°C)	Solubility in g liter^{-1} of K_2CO_3 solution			
	2 M	3 M	4 M	6 M
450	58	76	92	128
470	61.7	83	106	136.5
Difference	3.7	7	14	8.5

[a] Degree of fill, 85%; time, 48 hr.

temperature obtained by Ervin and Osborn. On the other hand, it was found impossible to obtain alpha alumina with the Al_2O_3–H_2O–$Na_2B_4O_7$ system.

Although the range in which the formation of corundum occurs is known, it is also necessary to ascertain the solubility of corundum as a function of the parameters defining the hydrothermal system, i.e., the nature and concentration of the solution, the temperature, and the pressure. Values have been established by Yamaguchi (1962) and by Barns *et al.* (1963) in the Bell Telephone laboratories for the solubility in alkali and alkaline-earth hydroxides, in alkali carbonates, and in certain alkali halides. An investigation of the solubility of corundum in potassium carbonate as a function of concentration and temperature, by measuring the loss of weight suffered by a fragment of αAl_2O_3, has resulted in the observations reported in Table VIII.

These results lead to the following remarks:

1. Since an increase in the concentration of the solution is accompanied by a reduction in pressure, this factor obviously has very little effect on the solubility.
2. The difference in solubility between 450°C and 470°C reaches a maximum for a 4 M solution. Experiment has indeed confirmed that the highest growth rates are obtained with this concentration.
3. The solubility of the other crystalline varieties of alumina were not investigated since these varieties are converted *in situ* into αAl_2O_3 after a few hours.
4. An examination of the values given in the literature and those obtained from our own tests reveals no zone of retrograde solubility—at least not in alkali carbonate solutions.

b. Apparatus. Four autoclaves were constructed with the parameters given in the accompanying table.

Capacity (ml)	P_s (bar)	T_s (°C)
45	2000	500
150	2000	500
150	4000	550
500	4000	550

Sealing is provided by a modified Bridgman point (with an unsupported area closure) positioned by a double block with a differential screw thread. This system facilitates opening and closing operations, but requires the use of large wrenches.

The first two autoclaves are made of Z 20 CN 17.2 grade stainless steel, the last two of Z 20 CDV 13 grade refractory steel. They are fitted with an inner lining of pure silver to avoid contamination of the crystals by the alloy metals.

The autoclaves rest on a heating plate of appropriate capacity (2–4 kW). A precise temperature gradient across the bottom and top is obtained by providing an internal screen pierced with holes over about 10% of the section. The temperature of the crystallization zone is adjusted as required by means of a low-power heating coil. The entire system is heat-insulated by refractory bricks.

The temperature is regulated to within ± 1°C by a six-way blind regulator controlled by Fe–Cr thermocouples with their junctions inserted into small-diameter blind holes drilled into the wall of the autoclave. The system is monitored by a six-way logger. The temperatures inside the cavity are measured with a special device enabling two thermocouples to be inserted into the autoclave.

The autoclave consists of a pressure-proof metal body, the top of which receives the closure carrying the two thermocouples ("thermocoax" elements brazed to a conically threaded sheath). Fitted to the bottom is a tube for connecting the device to the autoclave. Sealing between the main body, the thermocouple-bearing closure, and the interconnecting tube is ensured by flanged closures.

The pressure measurements are made with an electrical pressure gauge based on the change in resistance of a manganin coil. The resistivity of manganin alloy (Cu, 80–84%; Mn, 4–5%; Ni, 1–2%, the remainder being iron) varies linearly with the pressure up to 13,000 bar. Variation in resistance is 2.48×10^{-6} ohm ohm^{-1} bar^{-1}. The instrument consists of a pressure-gauge block containing the measuring resistor and the comparison resistor which is connected to the autoclave via a pressure transmitter. The resistors form the branches of a Wheatstone bridge, the imbalance of which is measured. The instrument is graduated directly in pressure readings. Accuracy is to within ± 100 bar up to 12,000 bar.

c. Influence of various factors on growth. There are many factors governing hydrothermal growth and, except for the initial orientation of the nuclei, it is impossible to make them vary independently. The attempt to obtain large crystals has led to an investigation of the conditions permitting rapid growth rates.

(1) Run. After cleaning, the mother substance (small fragments of Verneuil corundum) is placed in the autoclave. The amount of mother substance is determined in relation to the volume of the solution and the degree of solubility. After the drilled screen is placed in position, the quantity of solution needed to obtain the required degree of fill is poured in. The crystal nuclei are in the form of platelets of a given orientation. They are placed in the autoclave suspended from a silver support. The autoclave is then closed and brought to the necessary temperature; at the end of the run, it is cooled and opened. The nuclei, the thickness and weight of which have been determined beforehand, are freed of the layer of boehmite which invariably covers them.

The nuclei are then subjected to an x-ray examination by the Lambot method, to a dislocations count, and to a chromium determination (in the case of a ruby growth), as the case may be.

In the following, the orientation of the nuclei is designated by the three Miller indices related to the Niggli–Mauguin morphological rhombohedra, $a = 3.5$ Å, $\alpha = 85° 46'$, $c/a = 1.3652$.

(2) Influence of the nature of the solution. The runs were performed under the following conditions: mother substance, corundum; concentra-

TABLE IX

Solution	Growth rate (μm day^{-1} face^{-1})		
	[100]	$[\bar{1}1\bar{1}]$	$[10\bar{1}]$
Na_2CO_3	141	78	6.1
$NaHCO_3$	0	0	—
$Na_2B_4O_7$	0	0	—
Li_2CO_3	0	0	0
K_2CO_3	17	21.6	3.1
Rb_2CO_3	35.3	67.5	1.5

tion, 2 *M*; degree of fill, 76.5%; dissolution temperature, 455°C, growth temperature, 425°C (temperatures measured in the autoclave wall).

With the exception of lithium carbonate, all the alkali carbonates give appreciable growth rates (Table IX). Even when the trial conditions (temperature and pressure) were modified, no result was obtained with $Na_2B_4O_7$.

(*3*) *Influence of the concentration of the solution.* The experimental conditions were: mother-substance, corundum; solution, Rb_2CO_3; degree of fill, 85%; dissolution temperature, 475°C; crystallization temperature, 425°C, (temperatures measured in the autoclave wall); orientation, (100).

Results are given in Table X. The growth rate reaches a maximum for $C = 4\ M$. Over this interval, the pressure has little effect on the growth rate.

TABLE X

Concentration (M)	Pressure (bar)	Growth rate (μm day^{-1} face^{-1})
2	2130	132
3	1790	160.7
4	1330	161
6	550	132

(*4*) *Influence of the temperature gradient* (*Fig. 38*). The experimental conditions were: mother substance, corundum; solution, 4 *M* Rb_2CO_3 or K_2CO_3; degree of fill, 85%.

It will be noted (Table XI) that the temperature gradient greatly influences the crystallization rate, since rates in excess of 1 μm day^{-1} face^{-1} were obtained. The crystals produced with a steep temperature gradient

TABLE XI

		Growth rate (μm day^{-1} face^{-1})			
$T_D - T_C$	ΔT	(100)	(122)	(112)	Solution
450–425	25	14.3	92.9	89.3	Rb_2CO_3
475–425	50	121	386	150	Rb_2CO_3
500–425	75	336	557	600	Rb_2CO_3
500–450	50	260	1100	760	K_2CO_3

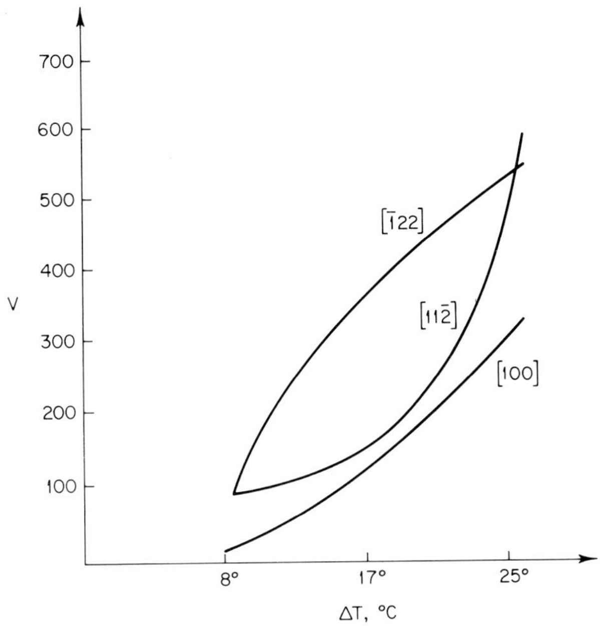

FIG. 38. Growth rate V (in μm day^{-1} face^{-1}) of Al_2O_3 in 4 *M* Rb_2CO_3; degree of fill, 85%; $T = 442$°C (Marais, 1969).

proved very turbid and unusable, and the conditions under which they were obtained cannot be used to grow a good quality ruby.

(5) *Influence of the crystallization temperature.* The experimental conditions were: mother substance, corundum; solution, 4 *M* Rb_2CO_3; degree of fill, 85%; temperature differential, 17°C.

The results are shown in Table XII.

TABLE XII

T_C in autoclave (°C)	Growth rate (μm day^{-1} face^{-1}) (100)	(122)	(112)
416	32	39.3	10.7
442	57	93	21
468	190	262	79.5
494	321	514	250

(6) *Influence of the nature and orientation of the seed crystal.* In the case of nuclei cut from Verneuil boules invariably having marked disorientations, the crystal quality has never been improved; moreover, the growth zones are always crackled.

Best results were obtained with nuclei cut from ruby bars produced by the Czochralski drawing method. An oriented ($\bar{1}22$) platelet cut from a "Linde" ruby with the following characteristics was used: orientation, 60°; axis variation along crystal axis, $\pm 1'$; dislocations, 10^4–10^5 cm^{-2}.

The hydrothermal conditions were as follows: mother substance, ruby doped with 500 ppm of Cr; solution, 4 *M* K_2CO_3; degree of fill, 85% pressure, 1330 bar; dissolution temperature, 480°C; crystallization temperature, 450°C; temperature differential, 15°C; duration, 673 hr.

The results were: initial thickness, 1.35 mm; final thickness, 24.5 mm; growth rate, 400 μm day^{-1}; initial weight, 5.532 g; final weight, 68.090 g.

The crystal obtained is very clear and has well-developed natural faces (211), (311), (111) (Fig. 39). Maximum disorientation of the lattice planes, as measured by the Lambot–Vassamillet method, was under 8′ of arc. The dislocation density measured on (111) was 10^5 cm^{-2}.

(7) *Doping with chromium.* To obtain a ruby which will meet the requirements for stimulated emission, first, the distribution of the $Cr^{3\pm}$ ions

FIG. 39. Ruby crystals grown hydrothermally (Marais, 1969).

must be uniform and, second, contamination of the crystal by fluorescence-inhibiting ions like Fe^{3+} must be avoided. This latter condition can be satisfied by having crystallization take place in capsules made of a noble metal. Initial tests were carried out in silver capsules with screwed covers of smaller volume than that of the autoclave cavity, and using the same degree of fill outside as inside the capsule. The growth rates remained very low because of the reduced useful volume of the autoclave and undoubtedly also because the presence of a nonnegligible mass of highly heat-conducting metal made it difficult to obtain an adequate temperature gradient inside the capsule. Subsequently, silver capsules were used with an outer diameter very close to the inner diameter of the autoclave (clearance 0.5 mm), the closure of which consisted of a double-spigoted silver-foil lid. These vessels give complete satisfaction with respect to crystal growth and purity.

For the initial tests, sodium bichromate was injected into the solution. This results in a nonuniform distribution of the chromium ions. When using a solution containing 0.25 g liter^{-1} of chromium, for instance, one observes an accumulation of chromium on the nucleus–crystal interface, followed by a pale yellow zone containing 1200 ppm of chromium, and finally a pink zone in which the chromium content is 3500 ppm. Using ruby as the mother substance, the chromium distribution is uniform except at the nucleus–crystal interface.

4. *Results*

In recent years, there has been revived interest in the use of recrystallization and hydrothermal syntheses, and there have been reports of numerous single or binary oxides, a few complex oxides like tourmaline and apatite, and halides. The tabulation of crystals produced by hydrothermal growth cannot be as complete as for the solution crystals. Table XIII gives a summary of results obtained with single oxides, and indicates the families of complex oxides to which hydrothermal growth has been applied.

The number of double or complex oxides prepared by hydrothermal growth or synthesis is very high. Table XIV lists the most important high-melting-point oxides.

Finally, in some cases, one can combine high temperatures and high pressures to obtain metastable phases under different conditions. At pressures from 40,000 to 120,000 bar, at around 1500°C, in the presence of water, a new variety of TiO_2 with an α-PbO_2 type of structure is obtained (Kestigian *et al.*, 1964). Also noteworthy is the preparation of the

TABLE XIII

Compound	T (°C)	Pressure (bar)	Mineralizer	Remarks	Ref.
Al_2O_3	400–700	3500	Na_2CO_3, KCO_3, Pb_2CO_3	Cr doping difficult	Laudise and Ballman (1958)
			$NaHCO_3$, $KHCO_3$, NaOH	Numerous aqueous inclusions	Kashkurov and others (1968)
BeO	500	—	4 N KOH + 0.06 N NaOH	—	Hill and Harker (1968); Newkirk and Smith (1965)
γGa_2O_3	150	—	—	—	Pohl (1968)
GeO_2	435	550	—	—	Christensen (1967)
	450	1000–4100	—	—	Harvill and Roy (1969)
SnO_2	400	2200	2–4 N NH_4Cl	—	Liebertz (1965)
	700	4100	2 M KOH	ΔT = 100°C, 9.6 × 10^{-4} cm hr^{-1} for the (001) plane	Harvill and Roy (1966)
TiO_2	450–550	600	KF, NaF, or NH_4F	Size <2 mm	Kuznetsov (1968)
	600–700	4100	9.15 M H_2SO_4	Dislocations 10^4 cm^{-2}	Harvill and Roy (1967)
ZrO_2	600	600	KF, NaF, or NH_4F (10%)	Size 1 mm	Kuznetsov (1968)
HfO_2	600	600	10% NH_4F	Size 1 mm	Kuznetsov (1968)
ZnO	200–500	2500	NaOH, KOH, N_2OH, NH_4Cl	Size 8–9 mm by spontaneous growth; several cm on a seed crystal	Kuzmina and Antanova (1964)

TABLE XIV

$Y_3Al_5O_{12}$	Puttbach *et al.* (1967)
$LiGaO_2$	Marshall and Laudise (1967b)
$MTiO_3$ (M = Ba, Pb)	Christensen and Rasmussen (1963)
$MZrO_3$ (M = Sr, Ba, Pb)	Kuznetsov (1968)
$LiNbO_3$	Hill and Zimmerman (1968)
$KNbO_3$ and $KTaO_3$	Shternberg *et al.* (1968)
K(Nb, Ta)O_3	Marshall and Laudise (1967a)
LiV_2O_4	Rogers *et al.* (1967)
$MMoO_4$ (M = Mg, Ca, Sr, Ba, Zn, Cd, Pb, Mn, Fe, Co)	Sleight and Chamberland (1968) Litvin *et al.* (1966)
$NiMoO_4$	Klevtsov (1965)
$Fe_2(MoO_4)_3$	Marshall (1967)
$In_2(MoO_4)_4$	Klevtsov (1968)
$CaWO_4$	Liebertz (1965)
$SrWO_4$	Dem'Yanets *et al.* (1966)
$CdWO_4$	Dem'Yanets and Litvin (1968)
$Ln_2W_2O_9$ (Ln = lanthanide)	Klevtsov *et al.* (1967)
$MLn(WO_4)_2$ (Ln = lanthanide; M = Li, Na, K)	Kharchenko and Klevtsov (1968)
$LnWO_4Cl$ (Ln = lanthanide)	Kharchenko *et al.* (1967)

metastable phase of αGa_2O_3 in NaOH at 1000°C and 44,000 bar (Remeika and Marezio, 1966). Similarly, 1.5-mm crystals of CrO_2 have been grown between 900° and 1300°C at pressures of 60,000–65,000 bar. However, these are still no more than laboratory curiosities, and the surface of this high-pressure, high-temperature realm has barely been scratched yet.

To summarize, it can be concluded that growth or synthesis by hydrothermal methods is valuable for the following reasons. There is little contamination by the solvent, dislocations are less than 10^4 cm^{-2}, and metastable phases can be prepared. In any event, the cost of the apparatus is matched only by the quality of the crystals obtained. The major drawbacks still to be overcome are undeniably the high installation cost, the slow growth rates, and the difficulty of achieving uniform doping.

III. Comparison between Crystals Obtained with the Two Methods

An important problem in crystal growth is how to characterize crystal quality. This can be done in either of two ways: by making a systematic search for a given type of defect, or by testing the property for which the

TABLE XV

1. Structure; space group, position of atoms
2. Purity
   - (a) Nature of impurities and concentration
   - (b) Location of impurities (substitutional, interstitial, foreign phases)
   - (c) Valence of impurities
   - (d) Homogeneity of impurity distribution
3. Nature of principal phase
   - (a) Stoichiometry and homogeneity of stoichiometry
   - (b) Valence of components of principal phase
4. Perfection
   - (a) Low-angle grain boundaries
   - (b) Dislocations
   - (c) Vacancies
5. Nature of surface

crystal was specifically grown. The first of these methods is of a general nature, and Table XV summarizes the properties which characterize crystal perfection.

Destructive (D) and nondestructive (N) methods exist for studying these crystal properties, but because of the technical difficulties involved in growing crystals, one usually prefers the nondestructive methods. The methods most commonly used are listed in Table XVI.

TABLE XVI

Method	Destructive or nondestructive	Information given
Chemical analysis Spectrographic analysis	D	Composition
X-ray analysis	N	Structure (some information on composition)
X-ray fluorescence spectroscopy	N	Composition
Electron diffraction	N	Structure, surface detail
Electron microscopy	N	Surface detail
Electron beam x-ray spectroscopy	N	Composition
Optical spectroscopy IR → UV	N	Structure and composition
Electron spin resonance	N	Purity (structure)
Optical examination	N	Imperfections, surface detail
Etching, decorating	N/D	Perfection
Measurement of specialized physical properties (electrical or magnetic)	N	Perfection, purity

Regardless of the crystal-growth method used, the crystals obtained will have a number of defects of various kinds. We shall now examine in succession the principal defects observed in crystals grown by the two different methods (at high and low temperatures).

A. Defects Characteristic of Crystals Grown at High Temperature†

There are several ways of classifying these defects, some of which are based on the nature of the defects themselves (physical, chemical, etc.), and others on the effect they have on the crystal properties.

In order to underline the importance of the effect which the growth technique has on the quality of the crystals, the defects will be classified according to the conditions under which they are formed during the successive phases of the crystal-growing operation. We will consider here only crystals which are obtained by the melting and crystallization methods.

1. *Defects introduced during melting*

The melting operation in itself has no effect on the physical properties of the crystals. On the other hand, the conditions under which it is carried out can profoundly affect their chemical characteristics. Essentially, two effects may be noted: (1) a change in the crystal's impurity content with respect to the starting material; (2) a nonstoichiometry of the crystal.

a. Change in the impurity content. Depending on the method used, the crystal may be either contaminated or purified with respect to the starting material. As a rule, contamination is due to parts of the apparatus coming in contact with the melt (the supports, and above all, the crucibles). In fact, at very high temperatures, only operations carried out without a crucible or by the self-crucible technique can result in crystals of high purity. Operations carried out in a crucible, even if the latter is made of a noble metal (platinum or iridium), will often result in a crystal containing fine, dispersed metal particles which act as diffusion centers and are highly detrimental for optical applications.

In most cases, the impurity which is introduced is troublesome, not so much in itself, but because of the alterations, such as stresses, which it induces in the matrix.

On the other hand, certain high-temperature crystal-growing operations

† See Cockayne (1968).

TABLE XVII

IMPURITY CONCENTRATIONS (× 10^{-6}) ALONG AN ALUMINA CRYSTAL

Impurity determined	Position along a 60-mm bar				
	Head	Head + 20 mm	Head + 50 mm	Head + 55 mm	End
Chromium	3	2	2.4	3.3	4.7
Cobalt	0.006	0.014	0.1	0.3	0.3
Europium	0.005	0.27	—	—	1.7
Iron	2	14.3	240	620	640
Gallium	6.5	16	—	—	107
Hafnium	0.017	0.20	1.25	2.8	2.6
Lanthanum	0.005	0.1	—	—	0.9
Manganese	0.5	1.3	—	—	26
Scandium	0.01	0.03	1.2	3.1	3.7
Zinc	0.05	0.05	2	4.4	4.8
Zirconium	0.4	0.32	16	25	33

have been found to result in a reduced impurity content. In most cases, this purification is due to evaporation of the impurities, which are more volatile than the basic material. This can be troublesome in the case of doped materials: When growing ruby crystals, for instance, a substantial decrease in the chromium content is observed during the melting process. In other cases, the purification process is inherent in the method itself, examples being the zone-melting or zone-floating techniques. Analysis of an alumina crystal obtained in this way reveals substantial purification over the entire length of the crystal (Dhalenne *et al.*, 1970) (Table XVII).

b. Nonstoichiometry. This occurs fairly often in the case of oxide crystals grown at very high temperatures and is usually revealed by an oxygen vacancy, in which case it may have one of two effects: either (1) the crystalline structure of the material remains the same, the small proportion of oxygen vacancies showing a disorderly distribution (e.g., $TiO_2 \rightarrow TiO_{2-x}$), or (2) a new phase (suboxide) appears, being the result either of too great a shortage of oxygen or of an orderly pattern of the vacancies (e.g., $TiO_2 \rightarrow Ti_nO_{2n-1}$).

If the alteration is not too great, the stoichiometric composition can be attained by heating the crystal in oxygen at moderate temperature. Thus crystals of titanium oxide (TiO_2) are yellow if produced in pure oxygen, and black if produced under low oxygen pressure. In either case, they can be decolored by heating in pure oxygen at about 1000°C.

2. *Defects introduced during solidification*

These flaws generally stem from changes in the shape or orientation of the solidification interface. The changes can be due to thermal, mechanical, or chemical factors.

a. Thermal factors. These are gradual or rapid temperature variations at the solid–liquid interface. The gradual changes are due to the increased heat transfer between the crystal and its environment as the crystal grows; for example, in the case of a crucible-type operation, the radiation from the walls of the crucible absorbed by the crystal increases as the latter's volume grows. The most important alteration resulting from this process is a variation in the crystal's diameter: it decreases as the crystal tends to heat up with increasing absorption of the energy radiated by its environ-

FIG. 40. Growth striations in a (Cockayne, 1968) Cr-doped $MgAl_2O_4$ crystal revealed using ^{51}Cr as a radioactive tracer (× 2.6). Upper part, 50 rpm; middle, 100 rpm; lower, 150 rpm. Used with permission of North-Holland Publishing Co., Amsterdam.

ment. The safest way of avoiding such variations is to operate under conditions such that the crystal's presence has little effect on the thermal state of the environment, as by drawing a small crystal from a very large melt of high thermal inertia.

Rapid temperature variations are due either to inadequate regulation or to unstable heat convection. They are particularly marked in the case of crystals grown from doped materials (Cockayne and Gates, 1967), since the resulting variations in the rate of growth modify the distribution of the dopant. This causes striations in the crystal, consisting of zones which are alternately richer and poorer in dopant. The curvature of the striations depends on the rate of rotation of the crystal during growth (Fig. 40).

b. Mechanical factors. In most techniques for growing crystals from melts, rotation is imparted to the growing crystal. This motion modifies the shape of the crystallization interface. A very common defect in crystals drawn from melts or grown by the zone-floating technique (Fig. 41) is faceting. This consists of the development of flat surfaces at the interface. The facets occur when the cooling isotherm becomes parallel to a single-index crystallographic plane. Their formation is attributed to nucleation difficulties on such planes. The facet planes most usually observed are the $(10\bar{1}1)$ and (0001) planes in rubies, the (111) plane in spinel, and the (211) plane in garnets (Bardsley and Cockayne, 1967; Charvat *et al.*, 1967; Basterfield *et al.*, 1968).

In order to avoid the formation of these facets, the shape of the cooling isotherm must be modified. The best way is to try to obtain a completely flat interface. With drawing techniques, this can be done by increasing the rate of rotation of the crystal, often to as much as 150 rpm (Cockayne *et al.*, 1968). The interference-fringe patterns obtained for different rates of rotation when growing doped $Y_3Al_5O_{12}$ garnets clearly show greater crystal perfection for high rotation rates (Fig. 42). Another convenient way of obtaining a substantially flat interface is to increase the ratio of the melt diameter to the crystal diameter.

Facets are seldom a nuisance in themselves. However, the impurities will very often be found to segregate in the plane of the facets: In the case of ruby, the chromium content of the facets is only 2% higher than in the overall crystal, but in the case of neodymium-doped garnet, the concentration of neodymium on the facets is 20% higher than it is in the matrix. With chromium-doped spinel, there is increased chromium concentration on the facets. These variations in composition result in local variations in the crystal parameters, and produce strains in the crystals.

The use of high rotation speeds is not without its drawbacks, since crystals grown in this way exhibit a spiral radial strain pattern. This would

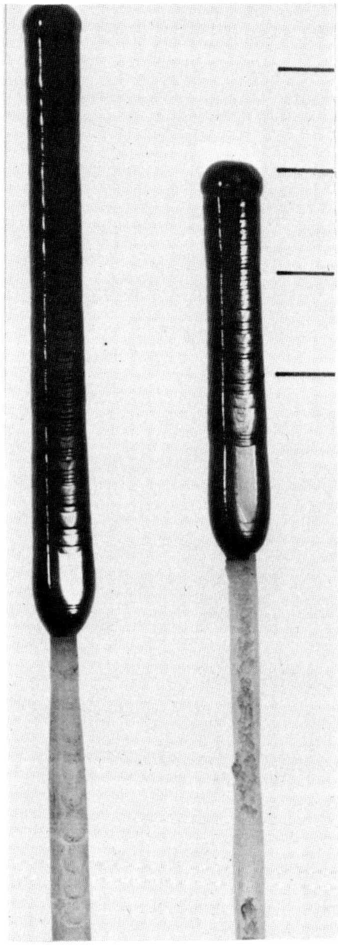

FIG. 41. Faceting in a ruby crystal (Revcolevschi, 1970).

appear to be due to the hydrodynamic effects of the rotation on the liquid. Here again, the phenomenon occurs only at higher rotation speeds, provided that the ratio of melt diameter to crystal diameter is high enough. It occurs chiefly in heavily doped crystals and is troublesome only because it modifies the distribution of the impurities or the dopants. It can lead to domain structures in ferroelectric materials (Parfitt and Robertson, 1967).

c. Chemical factors. These result from the coefficient of liquid–solid division for different impurities. Liquid droplets rich in impurities may

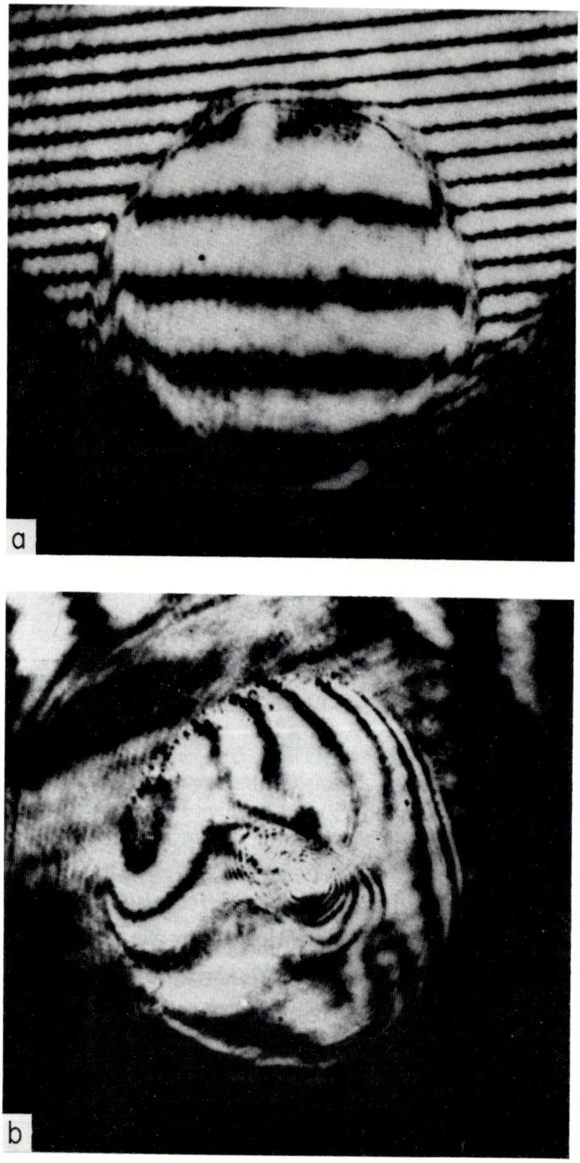

FIG. 42. Interference fringes viewed through $Y_3Al_5O_{12}$ single crystal doped with 0.6 atom % Nd grown at (a) 150, (b) 50 rpm (Cockayne, 1968). Used with permission of North-Holland Publishing Co., Amsterdam.

become trapped at the interface during crystallization, and in extreme cases, may join to form a continuous pattern and impart a cellular structure to the crystal. This phenomenon would seem to be particularly liable to occur at high rotation and growth rates. In the case of chromium- or molybdenum-doped garnets therefore, growth rates have been reduced to as low as 0.5 mm $hr.^{-1}$ Dissolved gases like argon or nitrogen can also produce this phenomenon.

Furthermore, it has already been shown that the presence of impurities is a factor in creating defects of all types.

3. *Defects introduced during cooling*

These are primarily due to strains and tensions of all sorts (structural, thermal, chemical), and the resulting defects profoundly affect the properties of the crystals.

a. Structural strains. Obviously, it is impossible to use melts to grow crystals from oxides exhibiting allotropic transformations, such as zirconium oxide. However, in a few cases where the transformation has little effect on the structure, this growth method is possible, examples being $LaAlO_3$ (Fay and Brandle, 1967) and $Ba_2NaNb_5O_{15}$. More often than not however, transformation induces heavy strains and, very often, twinning (Fig. 43). It is possible to eliminate such twinning by carrying out the cooling under mechanical stress.

b. Thermal stresses. A relatively steep thermal gradient is the principal characteristic of a high-temperature crystal-growing technique. It produces tensile or compressive stresses depending on the relative temperatures of the inner and outer portions of the crystal. The magnitude of these stresses depends on the shape of the isotherms, and the most important defects occur when these stresses exceed the yield stress.

If the material happens to possess a range of plastic deformation, these stresses produce dislocations (Figs. 44 and 45) that can recombine into polygonization walls. On the other hand, if the material has virtually no range of plastic deformation, incipient cracking or even fracturing of the crystal will occur. The alterations in material depend on the rate of cooling. If it is too rapid, deformation will occur below the ductile-fracturing/brittle-fracturing transition temperature, causing the crystal to exhibit cracks. This phenomenon can be particularly troublesome in cases where the transition temperature is high: 1550°C in the case of $MgAl_2O_4$ spinel (Lewis, 1968). Quite often, the deformation takes the form of twinning, with fracturing occurring at the twinning joint, especially in the case of anisotropic materials.

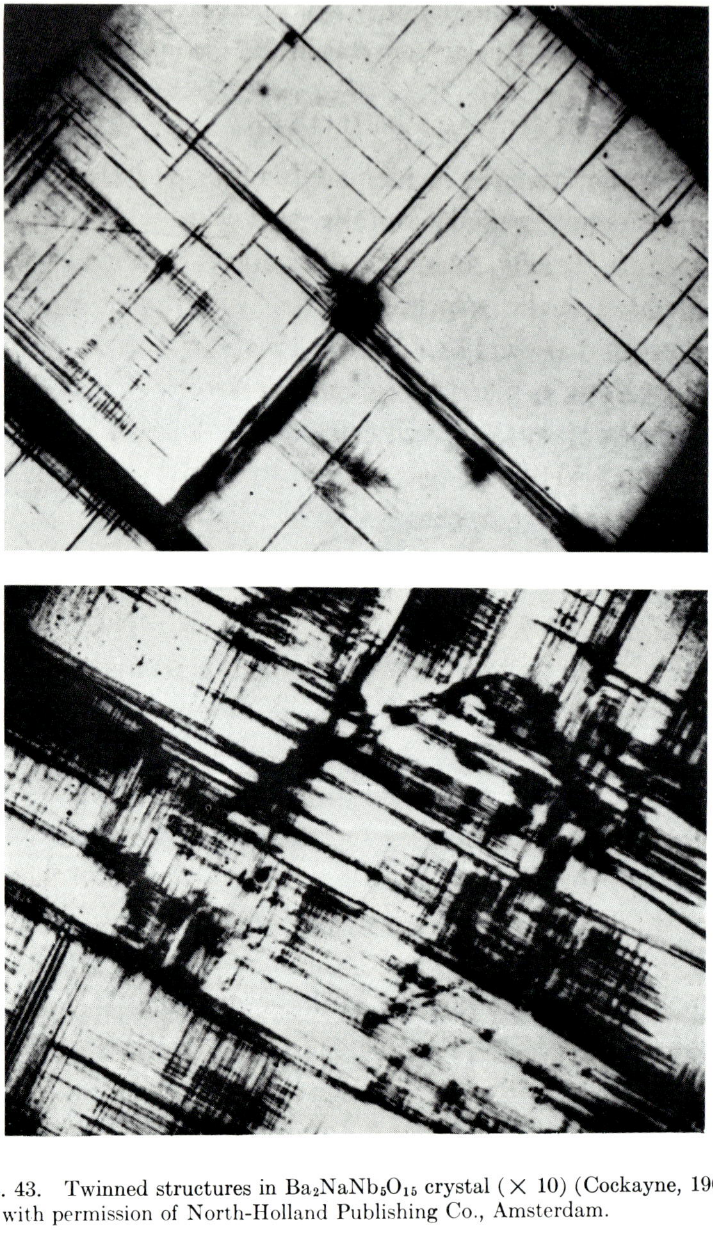

FIG. 43. Twinned structures in $Ba_2NaNb_5O_{15}$ crystal ($\times$ 10) (Cockayne, 1968). Used with permission of North-Holland Publishing Co., Amsterdam.

FIG. 44. Dislocations in a ruby crystals grown by zone floating (Revcolevschi, 1970).

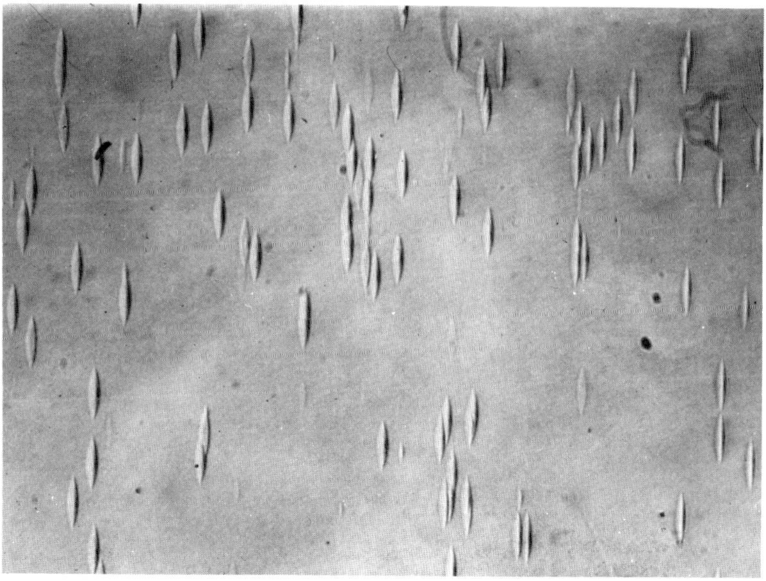

FIG. 45. Dislocations in a gallium oxide crystal grown by zone floating (Saurat *et al., C. R. Acad. Sci. Paris* **271**, 58, 1970).

The usual way of avoiding these alterations is to carry out the cooling slowly in annealing furnaces. Under optimum conditions, the density of the dislocations can be reduced to 10^2 cm^{-2} for alumina, spinel, and garnet. The temperature gradient should not exceed 50°C cm^{-1}. However, it is common to find dislocation densities of the order of 10^5 cm^{-2} in crystals grown at high temperatures.

There can be no doubt that a better understanding of the mechanical properties of oxides at high temperatures would greatly improve the results obtained.

c. *Chemical tensions.* These are caused by impurity segregations, by the effects of nonstoichiometry, and, more generally, by any local variation in the composition. They result in dislocations.

B. Defects Characteristic of Crystals Grown in Solution

Apart from the possibility of solvent inclusions, crystals grown in a solution are usually more perfect than those grown by other methods. There are no internal stresses, no mosaic structures, very little disorientation, and a small number of dislocations. This improved quality is mainly due to the temperatures and shallow temperature gradients used, which result in slow growth rates. Table XVIII gives an idea of the average rates achieved.

On the other hand, these slower rates restrict the use of solution techniques insofar as the production of large crystals on an industrial scale is concerned. Crystals of over 3 cm in size are exceptional with the flux method, whereas a size in the region of 10–20 cm. is possible by other methods (e.g., by melting).

TABLE XVIII

Method	Temperature (°C)	Linear growth rate (mm day^{-1})
Hydrothermal	650	0.1
Fluxed-melt	1200	1
Flame fusion	2100	450

1. *Morphology*

Crystal morphology depends both on the solvent and the temperature, but there are no hard and fast rules. Only a certain amount of data from observations are available. Thus zirconia prepared in a PbF_2 bath is obtained in the form of platelets (Anthony and Vutien, 1965), whereas only

FIG. 46. Striations in corundum crystal.

needles are obtained when it is grown in a sodium tetraborate melt (Harari *et al.*, 1967). Similarly, the corundum that crystallizes at the surface of a melt, where the evaporation technique is prevalent, is found in the form of fine platelets, whereas growth within the body of the melt, where the temperature gradient technique predominates, is more equidimensional. In the case of $MnLnO_3$ growth in Bi_2O_3, the higher the temperature (1400°C), the thicker the crystal.

There are many cases in which a laminar structure parallel to the growth face occurs (Fig. 46). This laminar structure is due mainly to external environmental changes, such as temperature fluctuations, rather than to internal deviations, such as dissolutions or impurity inclusions or incipient surface dislocation.

With the use of the slow cooling technique, nucleation can be limited to a few sites. A dendritic structure is often obtained. The preferential dissipation of the heat of crystallization at corners and edges, coupled with the constitutional supersaturation effects described above, may easily result in the formation of rapidly growing spikes that branch in a dendritic manner (Fig. 47).

Crystals very often exhibit twinning (Fig. 48). In the case of corundum (Wallace and White, 1967), the occurrence of twinning accounts for preferential nucleation sites perpendicular to the c axis. In all cases, the twinning operation involves a 180° rotation in the basic plane, equivalent to reflection in the (100) plane. The operation causes no discontinuities in planes of the ($h00$) or (hhl) type because they are common to both orienta-

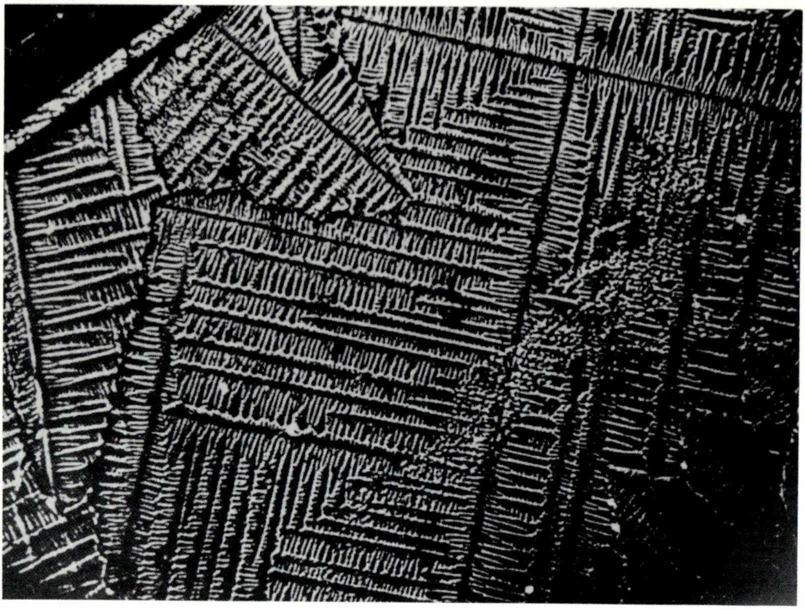

FIG. 47. The oriented growth ridges on the (111) face of NiO crystal (Hill and Wanklyn, 1968). Used with permission of North-Holland Publishing Co., Amsterdam.

tions. This twinning law differs from the law governing corundum spheres grown by the Verneuil method, corresponding to a rotation in the (012) plane. In the growth of barium titanate butterfly-twins, the persistence of undissolved powder is a necessary condition for platy growth. The occurrence of two twin planes in close proximity in some particles evidently led to the rapid growth of a "spine," with later development of the wings of the crystal.

In the case of hydrothermal synthesis, the origin of the nucleus and hence its degree of dislocation have a considerable effect on the morphology. In the case of rutile, for example, the (110) form is the prism and this requires a (100) growth rate twice as great as the (110) rate. Results are shown in Table XIX (Harvill and Roy, 1966).

TABLE XIX

No.	Crystal	Origin of nucleus	(001) ($\times 10^4$ cm hr^{-1})	(110) ($\times 10^4$ cm hr^{-1})	(100) ($\times 10^4$ cm hr^{-1})
1	TiO_2	Melt	10	7.1	3.7
2	TiO_2	Melt	7.5	7.8	7.1
3	TiO_2	Hydrothermal	0.7	4.3	3.9

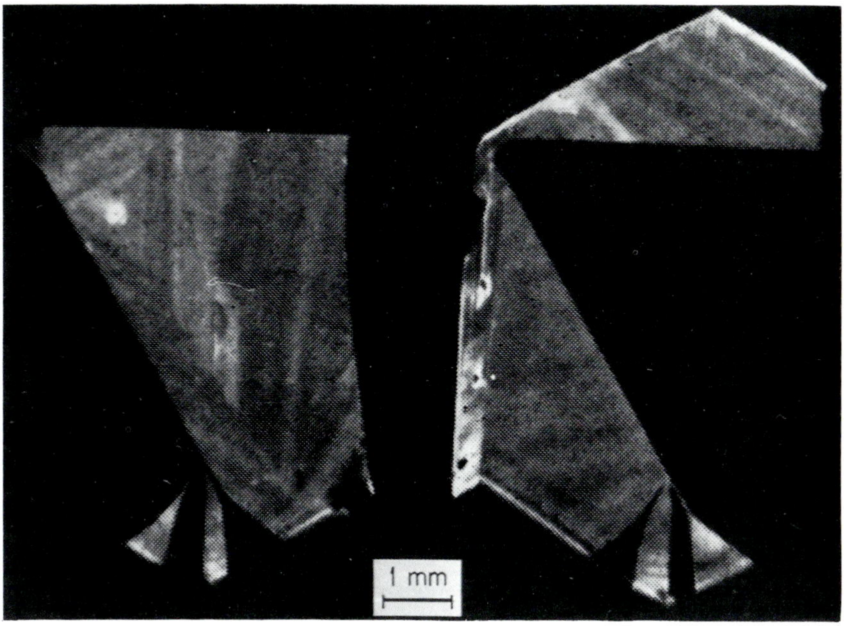

FIG. 48. Topograph showing multiple twinning at the termination of (110) boundaries (Wallace and White, 1967, used with permission of Pergamon Press).

If the rates remain constant, rutile No. 1 will be prism-shaped, whereas 2 and 3 may have either the (110) or (100) shape. In the case of crystals exhibiting very little dislocation, the (100) faces are very small. It has been shown that the nucleus dislocations propagate in the (110) direction.

2. *Dislocations*

As a rule, the degree of dislocation in crystals grown in a solution (melt or hydrothermal) is under 10^4 dislocations cm^{-2}. It is even possible to obtain crystals containing only a few dozen dislocations.

The crystalline perfection of crystals grown in solution is illustrated by a study of spinel grown at 1250°C in lead fluoride. The crystals are octahedral with (111) faces. The poorest contain 200 dislocations cm^{-2}, the best less than 50 dislocations cm^{-2}. Shallow trigonal pyramids can be observed on well-formed faces and are associated with a system of spiral dislocations (Fig. 49).

As in the case of beryllium oxide, the presence of screw dislocations is the cause of inversion twinning, and it has been shown that the presence of an axial screw dislocation, and hence inversion twinning, is related to environmental factors such as the composition and temperature of the melt. On the other hand, dislocations are often due to an impurity.

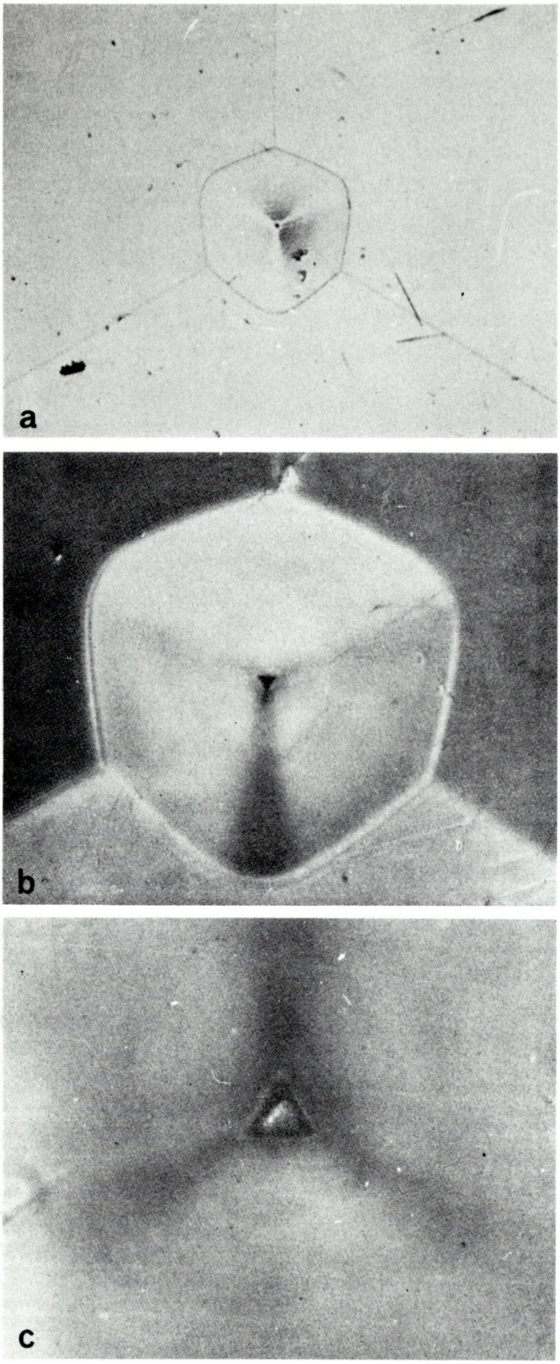

FIG. 49. Trigonal growth feature (Wood and White, 1968); (a) ×75; (b) ×200; (c) ×750. Used with permission of North-Holland Publishing Co., Amsterdam.

3. *Impurities*

The presence of impurities is unquestionably the major drawback of the solution growth method. It is inherent in the method itself, which causes a crystal to grow in an environment having a chemical composition very different from that of the crystal. No matter what improvements are made in the different techniques using this principle, it will never be possible to eliminate completely impurities due to contamination caused by the crucible material, or by the introduction of the solvent into the lattice and by impurites in the starting chemicals.

In the case of crystals grown by the hydrothermal method, the presence of water was revealed by infrared studies, though the amount would appear to be smaller than with melt techniques. The influence of impurities is greatest with the melt techniques, resulting in dislocations, growth sites, color centers, and nonhomogeneity of the crystals. The largest crystals grown in a melt frequently reveal zones of very great crystalline perfection separated by a zone of inclusions or nonhomogeneity.

Another type of impurity is the point defect, such as nonstoichiometry in compounds. This phenomenon causes color centers. For instance, in the case of rare-earth aluminates grown in a mixture of PbF_2 and PbO, brown crystals were obtained at the surface of the melt and green crystals at the bottom of the crucible. The addition of 1% of PbO_2 into the solution gives better oxygenation and produces brown crystals throughout the crucible. In the case of oxides, nonstoichiometry is closely related to the oxygen potential in the solution. Conversely, by operating under various degrees of oxygen activity, it is possible to obtain single crystals from substoichiometric or new-phase substances.

Thus, the Ti_nO_{2n-1} Magneli phases, with n varying from 4 to 10, as well as Ti_3O_5, have been obtained in a $Na_2B_4O_7$–B_2O_3 melt by carefully controlling the fugacity of the oxygen in the vapor phase (Bartholomew and White, 1970).

C. Example, Based on Corundum, of the Relationship between Dislocations and the Growth Method†

The Lang method was used to compare crystal quality and defect density in corundum crystals grown by four different methods (Fig. 50): (a) the Verneuil method, (b) the zone melting method, (c) the Czochralski method, and (d) the flux method. All the topographs obtained are transla-

† See Authier and Sauvage (1970).

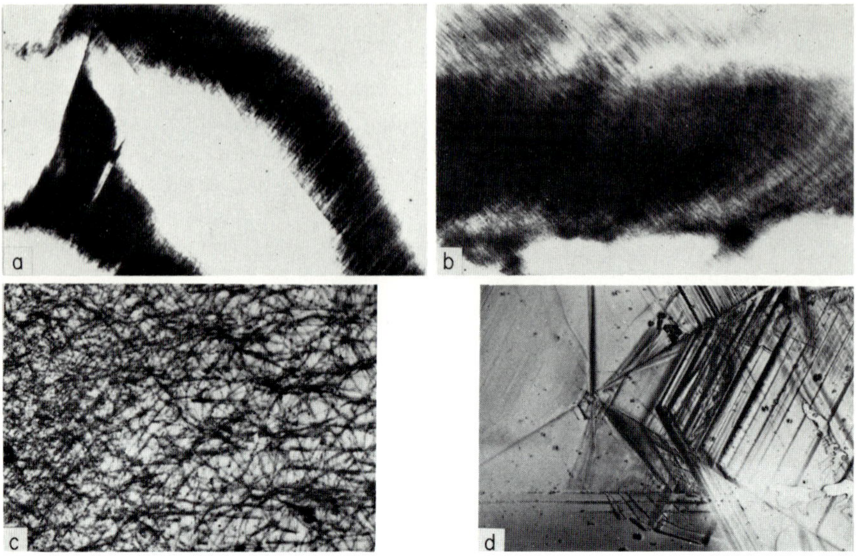

FIG. 50. Topographs of corundum crystals grown by (a) Verneuil method, (b) zone-melting method, (c) Czochralski method, (d) flux method (Authier and Sauvage, 1970).

tion topographs. The thickness of the plates ranges from 0.5 mm (for a crystal obtained by the flux method) to 1–2 mm, the radiation used being Mo $K\alpha_1$. Under such conditions, all the images obtained are direct images.

The series of topographs in Fig. 50 illustrates very different states of perfection. The topography taken with the crystal grown by the Verneuil method (Fig. 50a) exhibits very great variations in contrast, indicating that the crystal contains high internal stresses. The deformations correspond to disorientations of several minutes of arc. This is very commonly encountered in Verneuil crystals. The image observed on the topograph is essentially a Bragg contour. Subjoint walls can also be distinguished. Resolution of the dislocations is impossible, so that this corresponds to a high number of dislocations, in excess of 10^5 cm^{-2}.

The topograph in Fig. 50b shows a crystal obtained by zone melting. Crystal quality is better than that of the previous crystal, but one can still clearly see notable disorientation and a very high density of dislocations. The topographies also show segregations of impurities parallel to the growth front.

The topograph in Fig. 50c represents crystal sections sawn off a corundum bar obtained by the Czochralski drawing method. These sections and the bar from which they are obtained are fully single-crystalline. Maximum disorientations from any point to another on a section is no more than about 20″ of arc, and no subjoints can be seen. The density of dislocations ranges from 10^3 to 10^5 cm^{-2} at different places, and individual dislocations can be resolved without difficulty. Local fluctuations in the density and average orientation of the dislocation lines may be observed.

Essentially, the dislocations are of two types: (1) straight, parallel dislocations located in the symmetry planes (110); the topograph in Fig. 50, taken on sections perpendicular to the threefold axis show the three families of dislocations; (2) dislocations forming a three-dimensional Franck network, with meshes of varying slackness depending on the bar zones.

The peculiar arrangement of these dislocations shows that the actual density which can be measured on the topographs is much higher than can be deduced from the density of the pitting. For the straight dislocations in particular, the density of the pitting varies greatly with the orientation of the section.

The topograph in Fig. 50d shows a corundum plate obtained by the flux method. The excellent crystalline quality and the very low density of dislocations will be noted. Furthermore, the various growth sectors radiating from the first nucleus are underlined by contrasting along each growth front. It can be deduced from this that the growth was uneven and that there were slight variations in the composition of the melt.

The deformations associated with these growth fronts are anisotropic and are not visible when the planes of the reflectors used are perpendicular to them. All that can be seen are two growth steps decorated with impurities. The horizontal bands parallel to them become visible if one chooses a different family of reflecting planes.

REFERENCES

Anthony, A. M., and Vutien, L. (1965). *C. R. Acad. Sci. Paris* **260,** 1383.

Arend, H., Novak, J., and Coufova, P. (1966). *Rost. Kristall. Akad. Nauk. SSSR Inst. Kristallogr.* **7,** 281.

Arlett, R. H. *et al.* (1967). *J. Amer. Ceram. Soc.* **50,** 58.

Austerman, S. B. (1963). *J. Amer. Ceram. Soc.* **46,** 6.

Austerman, S. B. (1964). *J. Nucl. Mater.* **14,** 225.

Authier, A., and Sauvage, M. (1970). *Rev. Int. Hautes. Temp. Refract.* **7,** 94.

Bakradze, R. V., Kuznetsova, G. P., Baryshev, G. P., Selivanova, T. N., and Bychov, V. Z. (1968). *Izv. Akad. Nauk SSSR Neorgan. Mater.* **4,** 395.

Ballman, A. A., and Laudise, R. A. (1963). *In* "The Art and Science of Growing Crystals" (J. J. Gilman, ed.), p. 231–254. Wiley, New York.

Ballman, A. A., Porto, B. P., and Yarni, B. P. (1963). *J. Appl. Phys.* **34,** 3155.

Barber, W. A., and Sloan, C. L. (1961). *J. Phys. Chem.* **65**, 2026.
Bardsley, W., and Cockayne, B. (1967). *In* "Crystal Growth" (H. S. Peiser, ed.), p. 109. Pergamon, Oxford.
Barks, R. E., and Roy, D. M. (1967). *J. Phys. Chem. Solids Suppl.* **1**, 497.
Barns, R. L., Laudise, R. A., and Shields, R. M. (1963). *J. Phys. Chem.* **67**, 835.
Bartholomew, R. F., and White, W. B. (1970). *J. Crystal Growth* **6**, 249.
Basterfield, J., Prescott, M., and Cockayne, B. (1968). *J. Mater. Sci.* **3**, 33.
Boksha, S. S. (1968). *J. Crystal Growth* **3, 4**, 426.
Bridgman, P. W. (1914). *Proc. Amer. Acad. Arts Sci.* **49**, 625.
Brissot, J. J., and Belin, C. (1971). *J. Crystal Growth* **8**, 213.
Butcher, M. M., and White, E. A. D. (1965). *J. Amer. Ceram. Soc.* **48**, 492.
Charvat, F., Smith, J., and Nestor, O. (1967). *In* "Crystal Growth" (H. S. Peiser, ed.), p. 45. Pergamon, Oxford.
Chase, A. B., and Osmer, J. A. (1967). *J. Amer. Ceram. Soc.* **50**, 325.
Chase, J., and Van Ruyven, L. (1969). *J. Crystal Growth* **5**, 294.
Christensen, A. N. (1969). *Rev. Chim. Minéral.* **6**, 1187.
Christensen, A. N., and Rasmussen, S. E. (1963). *Acta. Chem. Scand.* **17**, 843.
Christensen, A. N., Broch, N. C., Von Heidenstam, O., and Nilsson, A. (1967). *Acta Chem. Scand.* **21**, 1046.
Class, W. (1968). *J. Crystal Growth* **3, 4**, 241.
Cobb, C. M., and Wallis, E. B. (1967). *US Govt. Res. Develop. Rep.* **67**, 115.
Cockayne, B. (1968). *J. Crystal. Growth* **3, 4**, 60.
Cockayne, B., and Gates, M. (1967). *J. Mater. Sci.* **2**, 118.
Cockayne, B., Chesswan, M., and Gasson, D. (1968). *J. Mater. Sci.* **3**, 224.
Cox, R., Revcolevschi, A., and Collongues, R. (1972). *J. Crystal Growth*, in press.
Dem'yanets, L. N., and Litvin, B. N. (1968). *Sov. Phys. Crystallogr.* **12**, 965.
Dem'yanets, L. N., Sidorov, P. M., Belov, N. V., and Pobedimskaya, E. A. (1966). *Vestn. Mosk. Univ. Ser. 4* **21**, 37.
Deutschbein, O., Grimouille, G., Pautrat, C., and Petit LeDu, G. (1966). *Rev. Appl. Phys.* **1**, 128.
Dhalenne, G., Rouchaud, J. C., Revel, G., Revcolevschi, A., and Collongues, R. (1971). *C. R. Acad. Sci. Paris* **272**, 538.
Drabble, J. (1968). *J. Crystal Growth* **3, 4**, 804.
Ervin, G., and Osborn, E. F. (1951). *J. Geol.* **59**, 4, 387.
Fay, H., and Brandle, C. (1967). *In* "Crystal Growth" (H. S. Peiser, ed.), p. 51. Pergamon, Oxford.
Field, W., and Wagner, R. (1968). *J. Crystal Growth* **3, 4**, 799.
Frank, F. C., Mullin, J. B., and Peiser, H. S. (1968). *Proc. Int. Conf. Crystal Growth, 2nd.* North-Holland Publ., Amsterdam.
Galtier, F., and Collongues, R. (1967). *C. R. Acad. Sci. Paris* **264**, 87.
Gilman, J. J., ed., (1963). "The Art and Science of Growing Crystals." Wiley, New York.
Grodkiewicz, W. H., and Nitti, J. J. (1966). *J. Amer. Ceram. Soc.* **49**, 576.
Grodkiewicz, W. H., Dearborn, E. F., and Van Uitert, L. G. (1967). *In* "Crystal Growth" (H. S. Peiser, ed.), p. 441–444. Pergamon, Oxford.
Guggenheim, H. (1963). *J. Appl. Phys.* **34**, 2482.
Harari, A., Thery, J., and Collongues, R. (1967). *Rev. Int. Hautes Temp. Refract.* **4**, 207.
Hart, P. B., and White, E. A. D. (1968). *J. Crystal Growth* **3**, 475.
Harvill, M. L., and Roy, R. (1967). *In* "Crystal Growth" (H. S. Peiser, ed.), p. 563–567. Pergamon, Oxford.

Hill, V. G., and Harker, R. I. (1968). *J. Electrochem. Soc.* **115,** 294.
Hill, G. J., and Wanklyn, B. M. (1968). *J. Crystal Growth* **3, 4,** 476.
Hill, V. G., and Zimmerman, K. G. (1968). *J. Electrochem. Soc.* **115,** 978.
Karan, C. (1954). *J. Chem. Phys.* **22,** 957.
Karan, C., and Skinner, B. J. (1953). *J. Chem. Phys.* **21,** 2225.
Kashkurov, K. F., Nikitichev, P. I., Osipov, V. V., Sizova, L. D., and Simonov, A. V. (1968). *Sov. Phys. Crystallogr.* **12,** 837.
Kennedy, G. C. (1950). *Amer. J. Sci.* **248,** 550.
Kestigian, M., Goldsmith, G. J., and Hopkins, M. (1964). "Thermal Imaging Techniques," p. 201. Plenum Press, New York.
Kharchenko, L. Yu, and Klevtsov, P. V. (1968). *Sov. Phys. Crystallogr.* **12,** 965.
Kharchenko, L. Yu, Klevtsov, P. V., and Solov'eva, L. P. (1967). *Dokl. Akad. Nauk SSSR* **176,** 826.
Klevtsov, P. V. (1965). *Sov. Phys. Crystallogr.* **10,** 370.
Klevtsov, P. V. (1968). *Inorgan. Mater.* **4,** 133.
Klevtsov, P. V., Kharchenko, L. Yu, and Klevtsova, R. F. (1967). *Dokl. Akad. Nauk SSSR* **176,** 575.
Kolb, E. D., and Laudise, R. A. (1966). *J. Amer. Ceram. Soc.* **49,** 302.
Kooy, G., and Couwenberg, H. J. (1962). *Philips Tech. Rev.* **23,** 161.
Kuz'mina, I. P., and Antonova, V. F. (1964). Crystal Growth IV "Nanka," p. 151.
Kuznetsov, V. A. (1968). *J. Crystal Growth* **3, 4,** 405.
Lacour, C., and Paulus, M. (1968). *J. Crystal Growth* **3, 4,** 814.
Laudise, R. A., and Ballman, A. A. (1958). *J. Amer. Chem. Soc.* **80,** 11, 2655.
Laudise, R. A., Linares, R. C., and Dearborn, E. F. (1962). *J. Appl. Phys. Suppl.* **33,** 1362.
Laurent, Y. (1969). *Rev. Chim. Minéral.* **6,** 1145–1186.
Le Bihan, R., and Vacherand, C. (1969). "Composés Minéraux Monocristallins," p. 147. Masson, Paris.
Levin, E., Robbins, C., and Murdie, H. Mc. (1964). *In* "Phase Diagrams for Ceramists" (Reser, ed.). Amer. Ceramic Soc., Columbus, Ohio.
Lewis, M. (1968). *Phil. Mag.* **17,** 481.
Liebertz, J. (1965). *Chem. Ing. Tech.* **37,** 830.
Linares, R. C. (1962a). *J. Amer. Ceram. Soc.* **45,** 307.
Linares, R. C. (1962b). *J. Appl. Phys.* **33,** 1747.
Linares, R. C. (1964). *J. Appl. Phys.* **35,** 433.
Linares, R. C. (1967). *J. Phys. Chem. Solids* **28,** 1285.
Linares, R. C. (1968). *J. Crystal Growth* **3, 4,** 443.
Litvin, B. N., Dem'yanets, L. N., and Garashina, L. S. (1966). "Growth of Crystals," Vol. 4, p. 134. Consultants Bur., New York.
Makram, H., Anthony, A. M., and Vutien, L. (1966). *Rev. Int. Hautes Temp. Refract.* **3,** 395.
Makram, H. (1969). "Composés monocristallins minéraux," p. 117–125. Masson, Paris.
Marais, M. (1969). "Composés monocristallins minéraux," p. 33–45. Masson, Paris.
Margotin, P. (1969). "Composés monocristallins minéraux," p. 47. Masson, Paris.
Marshall, D. J. (1967). *J. Mater. Sci.* **2,** 294.
Marshall, D. J., and Laudise, R. A. (1967a). *In* "Crystal Growth" (H. S. Peiser, ed.), pp. 557–561. Pergamon, Oxford.
Marshall, D. J., and Laudise, R. A. (1967b). *J. Crystal Growth* **1,** 88.
Michel, D., Perez Y Jorba, M., and Collongues, R. (1968). *C. R. Acad. Sci. Paris* **266,** 1602.

Naumova, I. I., and Anikin, I. N. (1966). *Zh. Neorgan. Khim.* **11,** 1746.
Nelson, D. F., and Remeika, J. P. (1964). *J. Appl. Phys.* **35,** 522.
Newkirk, H. W., and Smith, D. K. (1965). *Ann. Mineralog.* **50,** 44.
Nielsen, J. W., and Dearborn, E. F. (1958). *J. Phys. Chem. Solids* **5,** 202.
Nielsen, J. W., and Dearborn, E. F. (1960). *J. Phys. Chem.* **64,** 1762.
Pakter, A. (1968). *Kristallogr. Tech.* **3,** 51.
Parfitt, H., and Robertson, D. (1967). *J. Appl. Phys.* **38,** 1709.
Petit Le Du G., and Aubree, J. (1967). *Rev. Appl. Phys.* **2,** 264.
Petit Le Du, G. (1970). *Rev. Int. Hautes Temp. Refract.* **7,** 100.
Pohl, K. (1968). *Naturwissenschaften* **55,** 82.
Puttbach, R. C., Monchamp, R. R., and Nielsen, J. W. (1967). *In* "Crystal Growth" (H. S. Peiser, ed.). Pergamon, Oxford.
Remeika, J. P., and Marezio, M. (1966). *Appl. Phys. Lett.* **8,** 87.
Revcolevschi, A. (1970) Thesis, Paris.
Revcolevschi, A., and Collongues, R. (1968). *C. R. Acad. Sci. Paris* **266,** 1767.
Rogers, D. B., Gillson, J. L., and Gier, T. E. (1967). *Solid State Commun.* **5,** 263.
Roy, R., and Tuttle, O. (1956). *Phys. Chem. Earth* **1,** 138.
Roy, R., and White, W. B. (1968). *J. Crystal Growth* **3, 4,** 33.
Sasaki, H. (1964). *J. Phys. Soc. Japan* **19,** 2238.
Sato, M., and Hukudo, S. (1963). *Yogyo Kyogai Shi.* **71,** 5.
Saurat, M., Revcolevski, A., and Collongues, R. (1970). *C. R. Acad. Sci. Paris* **271,** 58.
Sawada, S., and Danielson, G. (1959). *Phys. Rev.* **113,** 1005.
Schieber, M. (1964a). *J. Amer. Ceram. Soc.* **47,** 537.
Schieber, M. (1964b). *J. Inorgan. Nucl. Chem.* **27,** 553.
Setoguchi, M., and Sakamoto, C. (1966). *Osako Jogyo Gijutsu Shikenjo Kiho* **17,** 169.
Shternberg, A. A., Lapsker, Ya E., and Kuznetsov, V. A. (1968). *Sov. Phys. Crystallogr.* **12,** 838.
Sleight, A. W., and Chamberland, B. L. (1968). *Inorg. Chem.* **7,** 1672.
Smith, S. H., and Elwell, D. (1968). *J. Crystal Growth* **3, 4,** 471.
Timofeeva, V. A. (1965). *Rost. Kristallov. Akad. Nauk. SSSR Inst. Kristallogr.* **6,** 86.
Tolksdorf, W. (1968). *J. Crystal Growth* **3, 4,** 463.
van Uitert, L. G., Grodkiewicz, W. H., and Dearborn, E. F. (1965). *J. Amer. Ceram. Soc.* **48,** 105.
Vieck, P., and Golay, M. J. (1953). *Phys. Rev.* **89,** 1297.
Viting, L. M., and Khomyakov, K. G. (1965). *Vestn. Mosk. Univ. Ser. Khim* **20,** 60.
Voronkova, V. I., Yanovskii, V. K., and Koptsik, V. A. (1967). *Dokl Akad. Nauk SSSR* **177,** 571.
Voronkova, V. I., Yanovskii, V. K., and Koptsik, V. A. (1968). *Neorgan. Mater.* **4,** 1727.
Vutien, L., and Anthony, A. M. (1969). "Composés Minéraux Monocristallins," p. 107. Masson, Paris.
Vutien, L., and Anthony, A. M. (1970). *Rev. Int. Hautes Temp. Refract.* **7,** 140.
Vutien, L., and Anthony, A. M. (1972). *J. Crystal Growth* **13, 14,** 601.
Walker, A. C., and Buehler, E. (1950). *Ind. Eng. Chem.* **42,** 1369.
Wallace, C. A., and White, E. A. (1967). *In* "Crystal Growth" (H. S. Peiser, ed.), p. 431. Pergamon, Oxford.
Wanklyn, B. (1968). *J. Crystal Growth* **2,** 251.
Webster, F. W., and White, E. A. (1969). *J. Crystal Growth* **5,** 167.
White, E. A. (1964a). *General Electric Co. J.* **31,** 43.
White, E. A. (1964b). *Hilger J.* **8,** 61.

White, E. A. (1965). *In* "Technique of Inorganic Chemistry" Vol. 4, pp. 31–64. Wiley, New York.

White, E. A., and Butcher, M. M. (1965). *J. Amer. Ceram. Soc.* **48**, 492.

Wilke, K. T. (1965). *Rost. Kristallov. Akad. Nauk SSSR Inst. Kristallogr.* **6**, 75.

Wood, J. D. C., and White, E. A. (1968). *J. Crystal Growth* **3, 4**, 480.

Yamaguchi, G., Yanagida, H., and Soejma, S. (1962). *Bull. Soc. Chem. Japan* **35**, 1789.

Zupp, R., Nielsen, J., and Vittorio, P. (1969). *J. Crystal Growth* **5**, 269.

Chemical Transport as a Preparative Procedure†

HARALD SCHÄFER

Institute of Inorganic Chemistry,
Münster University, Münster, Germany

† Translated by Barry E. Taylor, Brown University, Providence, Rhode Island.

Introduction

One of the preparative techniques used most often in solid state chemistry consists in heating the pulverized reactants together, often in pellet form. If one of the reactants is volatile or if the material to be synthesized possesses a noticeable decomposition pressure, then the firing is undertaken in closed tubes. The rate of the reaction is determined by volume diffusion, which as a rule takes place only slowly in the solid state. This means that relatively long times are necessary for complete reaction. This customary synthesis technique is not always the most convenient one, and there are various ways of influencing and promoting the course of the reaction (Schäfer, 1971), as follows:

(a) If one first combines the components through synthesis of coprecipitates, double salts, or solid solutions that give the desired product on heating under thermal decomposition, then the diffusion path can be shortened by an angstrom order of magnitude. At the same time, imperfections and large surface areas are produced in the original decomposition product. By this means, the synthesis of the final product is greatly promoted and reaction times are considerably shortened [see Schäfer (1971)].

(b) Melts are often good solvents, which in very small amounts can also be used as mineralizers.

(c) The hydrothermal synthesis procedure, particularly the recent variation with strong hydrogen halide acids as reaction medium, offers many possibilities (Rau and Rabenau, 1967, 1968; Rabenau and Rau, 1968, 1969).

(d) The promotion of surface diffusion through suitable gaseous chemisorbant molecules on the surface is not yet sufficiently developed for preparative purposes, but deserves consideration nevertheless (Schäfer, 1971).

(e) Frequently, reactions with solid reactants can be considerably improved by the use of chemical transport (Schäfer, 1962). This is the concern of the present chapter.

I. Fundamentals of Chemical Transport

First, the principle of chemical transport reactions in a cylindrical tube in a temperature gradient is discussed with the aid of the following equa-

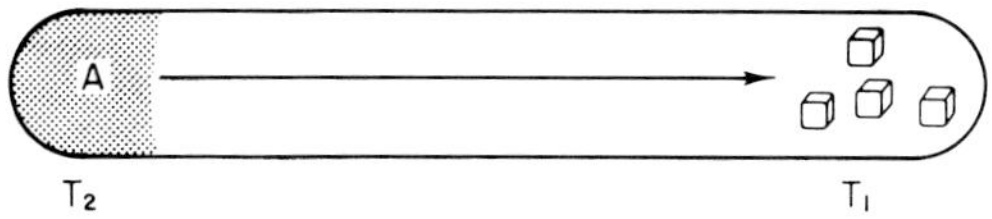

FIG. 1. Chemical transport in a cylindrical tube. Transport is from temperature T_2 to T_1.

tions (see Fig. 1):

$$A_s + B_g = AB_g$$

$$n_{AB} = (Dqt/sRT)(P_{AB(2)} - P_{AB(1)})$$

where n_{AB} is the number of moles of AB diffused, D is the diffusion coefficient, q is the tube cross section, s is the length of the diffusion path, t is the duration of the experiment, R is the gas constant, T is the absolute temperature of the diffusion path, and $P_{AB(i)}$ is the partial pressure of AB at a location with temperature T_i.

In the expression for n_{AB}, the first term on the right represents gas movement and the apparatus constant, and the second term represents heterogeneous reaction.

It is always true that $T_2 > T_1$. Furthermore, all transport equations are written so that the solid reactant is on the left-hand-side of the equation.

If the solid material A reacts with a gas B with the formation of a gaseous material AB and if the reaction is reversible, then a chemical transport may take place. If the observed reaction is *endothermic*, then the solid material A will be consumed at a higher temperature T_2, with the formation of gaseous AB, while after the migration of AB to the location with temperature T_1, the reverse reaction takes place, and A is deposited from the gas phase. *Exothermic* reactions transport the solid phase in the temperature gradient in the reverse direction, $T_1 \rightarrow T_2$.

Normally, for the transport reaction, the diffusion in the gas phase is rate-determining. This means that the important concentration gradient for the diffusion is determined by the existing heterogeneous equilibrium at the temperatures T_2 and T_1.

As an example, the following equation applies to the transport of a metal with an endothermic reaction:

$$Pt + O_2 = PtO_{2,g} \qquad \text{(endothermic)} \tag{1}$$

By this reaction, platinum can be transported via the gas phase in considerable volume at a temperature of 1200°C or higher. It is then deposited as well-formed crystals at lower temperature (Schäfer, 1962; Schäfer and Tebben, 1960).

Exothermic transport reactions in particular have been known for a long while through the work of van Arkel (1934) and van Arkel and de Boer (1925) as a means of purifying metals. For example, the metals Ti, Hf, Th, V, Nb, Ta, Cr, Cu, and Fe can be transported with iodine as a transporting agent, as gaseous iodide to a hot wire or to a heated quartz wall. A recent compilation of these procedures was made by Rolsten (1961). For instance, the following equation describes the transport of chromium (Rolsten, 1960):

$$Cr + I_2 = CrI_{2,g} \qquad \text{(exothermic)} \tag{2}$$

II. Criteria for the Choice of Transport Reactions

We consider next the case of chemical transport without subsequent reactions. The reacted solid phase at the starting location in the tube is therefore (aside from possible impurities) identical with the material deposited from the gas phase at the other end. The driving impetus is the concentration gradient caused by the temperature dependence of the heterogeneous equilibrium. The following factors are then essential in the choice of transport reactions:

(a) Obviously, chemical transport is only possible when, aside from the transported solid material, only gases participate in the reaction. Only in such a case are all components mobile throughout the gas phase.

(b) The reaction must be reversible. As a rule, this is true at the higher temperatures used because of the necessary volatility of the reactants.

(c) The equilibrium position of the transport reaction must not be extreme, that is, it must not lie too far to the reactant side or to the gaseous product side. Otherwise, the concentration gradient becomes too small.

One can see, by introducing plausible values in the diffusion equation ($q \approx 3$ cm^2, $s \approx 10$ cm), that useful preparative transport efficiency is to be expected if the factor $\Delta P_{AB}/\sum P$ is at least 10^{-4}. This also means that none of the partial pressures involved in the transport should be smaller than is given by this factor.

(d) Under the conditions given, the transport efficiency is a maximum if the concentration gradient (ΔP_{AB}) is a maximum. In this case, the free enthalpy of the transport reaction lies in the neighborhood of zero. However, this condition is not critical. [See Schäfer *et al.*, (1956) and Schäfer (1962).]

If the transport reaction is coupled with a subsequent reaction, then the necessary concentration gradient in the gas phase also prevails for the course of the reaction with isothermal heating. The important situation for preparative purposes can be described by the equations

$$\text{location 1:} \qquad A_s + B_g = AB_g \tag{3}$$

$$\text{location 2:} \qquad AB_g + C_s = AC_s + B_g \tag{4}$$

$$\text{total:} \qquad A_s + C_s = AC_s \tag{5}$$

In relation to Eqs. (3) and (4), the statement about the necessary value of $\Delta P/\sum P$ is valid.

With chemical transport, several heterogeneous equilibria can also work together. In the calculation of the necessary concentration gradient for the transport, one must then consider all equilibria (Schäfer, 1962).

III. Specifications of the Technique

Chemical transport reactions possess numerous areas of application, including the preparation of epitaxial layers, crystal-growing, compensation of tungsten vaporization in a filament lamp, detection of new gaseous compounds, and determination of thermodynamic data from transport inversion points; these will not be considered further. The emphasis here will be on the preparative application of chemical transport.

Most often, chemical transport is carried out in sealed cylindrical tubes made of glass or quartz glass. For preparative work, the loading of the reactants and the transport agent is no problem. The substance and the transporting agent, which is not volatile at room temperature (NH_4Cl, S, . . .), are sealed off under vacuum. If iodine is used as the transporting agent, the tube is cooled during the evacuation process. If transporting agents are used which are gaseous at room temperature (Cl_2, HCl, HBr, . . .), the tube loaded with reactants is flushed for a sufficient length of time with this gas and then sealed off. For further information, see Schäfer (1962).

The requirements are most exacting when the transport experiment is to give information about new molecules in the gas phase. This is outside the scope of the present considerations.

The heating of the transport tube in a temperature gradient can be accomplished in the natural temperature gradient of a simple tube furnace. Tube furnaces with two separate heating coils in a row are more favorable.

If transport tubes of ~20 mm diameter and 150–200 mm length are used and if the total pressure in the tube amounts to ≤ 1 atm at the experimental temperature, then the gas motion results primarily from diffusion. With greater pressures (and greater tube diameters), thermal convection is also important. This can significantly increase the transport efficiency, especially when the tube is heated in a tilted position with the hotter zone at the bottom.

Until now, the discussion has been limited to transport in sealed tubes. However, open flow arrangements can also be used. In such a manner, e.g., ZnS can be transported in a nitrogen stream charged with I_2 (Schäfer and Jagusch, 1966):

$$ZnS + I_{2,g} = ZnI_{2,g} + 0.5S_2 \tag{6}$$

This procedure is limited to those systems in which the reverse reaction—hence the deposit of the solid substance—is sufficiently rapid and complete. Related systems can differ considerably in this respect. Thus in closed transport tubes, the transport of Si by the following reactions takes place in a comparable manner (Schäfer and Morcher, 1957):

$$Si + SiCl_{4,g} = 2SiCl_{2,g} \tag{7}$$

$$Si + SiI_{4,g} = 2SiI_{2,g} \tag{8}$$

In the open flow system, however, only the iodine system is suitable for silicon transport. In the chlorine system, silicon deposition is retarded and one obtains instead the compounds Si_2Cl_6, Si_3Cl_8, . . ., Si_nCl_{2n+2} (Schäfer, 1953).

Perhaps the different behavior can be explained by the fact that silicon deposition in the iodide system can proceed not only kinetically by disproportionation (as in the chloride system), but also by direct decomposition into the elements.

IV. Transported Materials and Transporting Agents

Until now, chemical transport of representatives of the following classes have been observed (also see Table I):

1. Elements.
2. Fluorides, chlorides, bromides, iodides.
3. Oxides, sulfides, selenides, tellurides.
4. Nitrides, phosphides, arsenides, antimonides.
5. Silicides, germanides, stannides, borides.
6. Ternary oxides: spinels, titanates, niobates, tantalates, molybdates, tungstates, borates, germanates, phosphates.
7. Ternary and quaternary sulfides.

8. Oxyhalides, chalcogenohalogenides.
9. Silicidosulfides, silicidoselenides, silicidophosphides.

It is to be emphasized that, because of the reversibility of the process, the ability to be chemically transported also means, as a rule, that the solid substances can be synthesized from their components under transport conditions and can be obtained as crystals (see Section X).

The most important transporting agents are the following:

1. F_2, Cl_2, Br_2, I_2.
2. HF, HCl, HBr, HI.
3. O_2, S, H_2O, CO, CO_2.
4. $HgCl_2$, $AlCl_3$, $SiCl_4$, $NbCl_5$, ...

Of these, elemental fluorine (because of the problems with apparatus) is only used in exceptional cases, e.g., tungsten transport in the temperature gradient of a heated tungsten filament at about 3000°C (Schröder, 1967). On the other hand, the other halogens, especially iodine, are frequently appropriate.

In addition, new transporting agents are always being found. An interesting example is sulfur. A large number of sulfides, as well as oxides (e.g., Ta_2O_5), are transported at sulfur pressures of about 5–10 atm (Schäfer *et al.*, 1968). The underlying mechanism is as yet unknown.

V. Simultaneous Use of Several Transporting Gases

It has been known for a long time that the reaction of oxides with chlorine is considerably enhanced if carbon or carbon monoxide is introduced in addition. The oxygen of the chlorination reaction liberated initially is then bound as CO or CO_2. This means that there is a substantial displacement of the equilibrium position in favor of the chlorination product. On a similar basis, one can favorably influence the equilibrium position for transport processes.

Examples are given by the following reactions†:

$$Cr_2O_3 + 2Cl_2 + 0.5O_2 = 2CrO_2Cl_2 \tag{9}$$

$$2Ta_2O_5 + 4.5Cl_2 + 3CrCl_3 = 4TaOCl_3 + 3CrO_2Cl_2 \tag{10}$$

$$TiO_2 + 2I_2 + 0.5S_2 = TiI_4 + SO_2 \tag{11}$$

The oxides Cr_2O_3 and Ta_2O_5 are also transportable solely with Cl_2 at about 1000°C in a temperature gradient. The addition of O_2 or $CrCl_3$, respectively,

† For reactions (9)–(11), see Schäfer (1968), compare with Schäfer (1962) Chapter 3.1.4.3, and Nitsche (1967), respectively.

however, improves the equilibrium position and consequently the transport efficiency considerably. Determinative for this is the formation of the stable product CrO_2Cl_2.

Especially remarkable are the conditions of the TiO_2 transport. TiO_2 is not transportable with I_2 alone, because of the extremely unfavorable equilibrium position. The additional introduction of sulfur leads to the TiO_2 transport according to Eq. (11).

The use of several transporting agents does not always lead to results such as those just described; rather, more complex molecules can also be produced. Thus platinum is transportable with chlorine by an endothermic reaction (via $PtCl_{2,g}$), but the transport effect remains small and one needs temperatures of about 1000°C. However, if carbon monoxide is added, then the relatively stable compound $Pt(CO)_2Cl_2$ is produced and the following *exothermic* transport can be carried out at temperatures from 300 to 400°C (Schäfer, 1968; Schäfer and Wiese, 1971):

$$Pt + Cl_2 + 2CO = Pt(CO)_2Cl_{2,g} \qquad (12)$$

One can also consider the transport of tungsten. This element is not amenable to transport with iodine by the van Arkel method because gaseous tungsten iodides are much too unstable. In the presence of iodine and water, however, tungsten is transported from a 800°C zone into a 1000°C zone. The formation of gaseous oxyiodide WO_2I_2 is crucial; the transport equilibrium is as follows (Dettingmeijer *et al.*, 1969):

$$W + 2H_2O + 6I \quad (3I_2) = WO_2I_{2,g} + 4HI \qquad (13)$$

VI. The Amount of Transporting Agent

In principle, there is no upper limit for the amount of transporting agent. Reactions or crystal growth in supercritical media can be considered as transport processes at especially high pressure. However, these cases are not treated here.

For transport reactions in the narrower sense, one seldom uses transporting agent pressures greater than about 10 atm [e.g., transport with sulfur (Schäfer *et al.*, 1968)].

The transport tube is often loaded at room temperature and atmospheric pressure with gaseous transport agents (Cl_2, HCl). Iodine is normally used in the amount of a few mg ml^{-1} tube volume (1 mg ml^{-1} I_2 corresponds to a pressure of 0.1 atm at room temperature). It is especially interesting that, with small total pressures, very small amounts of transporting agent can in many cases cause a considerable transport effect.

This is true, e.g., for nickel transport with the help of hydrogen chloride (Schäfer and Etzel, 1959). If the reaction tube is carefully prepared with metallic nickel and nickel chloride, where the nickel is outgassed under high vacuum at 900°C and the nickel chloride is sublimed under high vacuum, then a chemical transport of nickel is observed after the sealing off of the tube and heating in a temperature gradient of 1000 → 600°C. This is caused by the small hydrogen content of the nickel and especially by residual water liberated from the walls of the quartz tube on heating. The water reacts with nickel chloride to form hydrogen chloride, so that all necessary components are present for the chemical transport according to

$$Ni + 2HCl = NiCl_{2,g} + H_2 \tag{14}$$

Only when extremely careful steps are adopted (Schäfer and Etzel, 1959), does no transport of nickel take place in the presence of nickel chloride.

Similar observations are made with Cu_2O (Schäfer and Etzel, 1957). This compound is transported by the following exothermic reaction even with only a few micrograms of HCl (600 → 900°C):

$$3Cu_2O + 6HCl = 2Cu_3Cl_{3,g} + 3H_2O \tag{15}$$

VII. Impurities in the Transported Solid Phase

A. Introduction of the Transporting Agent

One can expect at the outset that the amount of transporting agent taken up by the solid phase into the crystal lattice will not be large, because the concentration of the gaseous transporting agent is small compared to the concentration of the solvent in liquids or melts. Experiments performed with labeled iodine led to the following iodine content in the transported material in gram-atoms of iodine per mole of component:

GeGe	GaAs	ZnO	CuBr†
$< 10^{-7}$	5×10^{-6}	$<3 \times 10^{-6}$	∞
		ZnS	
		7×10^{-5}	
		ZnSe	
		3×10^{-4}	
		ZnTe	
		1.4×10^{-3}	

Exhaustive investigations concerning the introduction of iodine and the

† CuBr forms an entire series of solid solutions with CuI at 400°C. No transport experiment was carried out.

transport of CdS (Schäfer and Odenbach, 1966) have shown that the result is described by

$$2S^{2-} + CdI_{2,g} = 2I^- + \square(Cd^{2+}) + 2CdS \quad (16)$$

I^- therefore goes into an S^{2-} position.

It is seen from the given list of iodine contents that (1) the amount of iodine introduced into the crystal is small, (2) the size of the substituted anion has a considerable influence (sequence ZnO → ZnTe), and (3) the greatest influence is exerted by the charge and the bonding conditions (sequence GeGe → CuBr).

From this, it follows that the rate at which transporting agent is taken up can be influenced, and in particular, decreased, by the choice of the transporting agent.

B. Impurities from the Reaction Vessel

As a rule, transport tubes consist of quartz glass, and at lower transport temperatures, of Pyrex glass. In special cases, porcelain or platinum is used. If possible, a vessel material is used which is inert toward the transport reaction, and general rules concerning this cannot be given. With regard to the attack of the vessel, substitution of the transporting agent may be necessary. Thus, for example, Ta_2O_5 noticeably absorbs SiO_2 from the quartz wall when it is transported with $TaCl_5$ according to

$$Ta_2O_5 + 3TaCl_{5,g} = 5TaOCl_{3,g} \quad (17)$$

This is connected with the fact that SiO_2 is transported with $TaCl_5$ (Schäfer, 1962):

$$SiO_2 + 2TaCl_{5,g} = SiCl_{4,g} + 2TaOCl_{3,g} \quad (18)$$

On the other hand, if Ta_2O_5 is transported (by an unknown reaction) with sulfur, then the absorption of SiO_2 into the transported Ta_2O_5 is not detectable (Gruehn and Schäfer, 1969).

For the influence of the water content of the quartz glass, see Section X.A.

VIII. Complex Solid Phases

Until now, only elements or binary compounds have been cited as examples. In fact, the number of chemically transportable binary compounds has become very large. One can expect that the thermodynamic requirements for the transport reactions become greater as the number of solid-phase components increases. All of these components must be reversibly converted in the gas phase for chemical transport to occur. However, it has been shown that numerous ternary and even quaternary compounds are accessible to chemical transport. Some examples for the chemical transport of oxychlorides are as follows†:

$$AlOCl + NbCl_{5,g} = 0.5Al_2Cl_{6,g} + NbOCl_{3,g} \quad (19)$$

$$FeOCl + 2HCl = 0.5Fe_2Cl_{6,g} + H_2O \quad (20)$$

$$TiOCl + 2HCl = TiCl_{3,g} + H_2O \quad (21)$$

$$VOCl + 2VCl_{4,g} = VOCl_{3,g} + 2VCl_{3,g} \quad (22)$$

$$\left.\begin{aligned} 2CrOCl + CrCl_{4,g} + Cl_2 &= CrO_2Cl_2 + 2CrCl_{3,g} \\ 2CrOCl + 2Cl_2 &= CrO_2Cl_2 + CrCl_{4,g} \end{aligned}\right\} \quad (23)$$

$$NbOCl_2 + NbCl_{5,g} = NbOCl_{3,g} + NbCl_{4,g} \quad (24)$$

$$Nb_3O_7Cl + 4NbCl_{5,g} = 7NbOCl_{3,g} \quad (25)$$

Table I further illustrates ternary and quaternary compounds that as a rule can be obtained as well-formed crystals by chemical transport. In most cases, it is not the compounds listed, but rather their components that are placed in the transport tube. Synthesis and chemical transport therefore occur in the same step. This compilation lists many of the compounds in this class, but it is by no means exhaustive.

As a rule, the crucial heterogeneous equilibria for the transport are indicated without difficulty, but more complex equilibria may also be at hand [e.g., see Eq. (23)]. The correct description of the transport process then requires a detailed thermodynamic discussion, as was carried out, e.g., for the transport of boracites, $Me_3B_7O_{13}X$, with $HX + H_2O$ (Schmid, 1965) and for the transport of tungsten with $I_2 + H_2O$ (Dettingmeijer *et al.*, 1969). Such factors have not been considered up to now, e.g., for the transport of SbSI and the analogous Sb and Bi compounds.

It is further of interest that compounds that contain an alkali or alkaline earth element ($RbNb_4Cl_{11}$, $CsNb_4Cl_{11}$, and $CaNb_2O_6$) have now been transported. This is thermodynamically of interest: There are scarcely any binary compounds of these elements, whose chemical transport would be hoped for on the basis of the equilibrium position of the possible transport processes. By going to ternary (or higher) compounds, however, the situation can be shifted in the direction of nonextreme equilibrium positions, so that chemical transport is possible.

† For reactions (19)–(21), see Schäfer (1962); for (22), Schäfer and Wartenpfuhl (1961a); for (23), Schäfer and Wartenpfuhl (1961b); and for (24) and (25), Schäfer *et al.* (1961).

TABLE I

CHEMICALLY TRANSPORTED TERNARY AND QUATERNARY COMPOUNDS

Compound	Transporting agent	Reference
$RbNb_4Cl_{11}$, $CsNb_4Cl_{11}$	$NbCl_5$	Broll *et al.* (1969)
VCl_2Br	Br_2	McCarley *et al.* (1964)
VBr_2I	I_2	McCarley *et al.* (1964)
$TaOCl_2$	$TaCl_5$	Schäfer *et al.* (1961)
$MoOCl_2$	$MoCl_5$	Schäfer and Tillack (1964)
$WOBr_2$	$WOBr_4$	Tillack and Kaiser (1969)
$NbOI_2$	I_2	Schäfer and Gerken (1962)
$Me_3B_7O_{13}X$ (X = Cl, Br, I; Me = Fe, Co, Ni)	$HX + H_2O$	Schmid (1965)
NbS_2Cl_2 (NbY_2X_2, Y = S, Se; X = Cl, Br, I)	$NbCl_4$	Schäfer and Beckmann (1966)
SbSBr	Br_2	Klazer and Horák (1968)
BiSBr	Br_2	Kozáková *et al.* (1966)
SbSI	I_2	Kern (1962), Grigas (1968a)
SbSeBr	Br_2	Klazer and Horák (1968)
SbSeI	I_2	Nejezchleb and Horák (1968), Grigas (1968b)
BiTeBr	Br_2	Horák and Rodot (1968)
$CuCr_2Y_3X$ (X = Cl, Br; Y = S, Se)	Halogen	Miyatani *et al.* (1968)
$MeGeO_3$ (Me = Mn, Fe, Co)	HCl (NH_4Cl)	Royen and Forwerg (1962, 1963)
$MeTiO_3$ (Me = Mg, Ni)	Cl_2	Emmenegger (1968)
$MeNbO_4$ [Me = Fe(III), Cr(III)]	$Cl_2 + NbCl_5$	Emmenegger (1968); Emmenegger and Petermann (1968)
$FeTaO_4$	$Cl_2 + TaCl_5$	Emmenegger (1968)
$CrTaO_4$	Cl_2	Emmenegger (1968)
$MeMoO_4$ (Me = Mg, Mn, Co, Ni)	Cl_2	Emmenegger (1968)
$MeWO_4$ (Me = Mg, Mn, Fe, Co, Ni, Zn)	Cl_2	Emmenegger (1968)
Co_2SnO_4	Cl_2	Emmenegger (1968)
$MeCr_2O_4$ (Me = Mn, Co, Ni)	Cl_2	Emmenegger (1968)
$MeFe_2O_4$ (Me = Mn, Co, Ni)	HCl	Curtis and Wilkinson (1965)
$MeFe_2O_4$ (Me = Mg, Mn, Co, Ni)	HCl	Kleinert (1963, 1964)
$MeNb_2O_6$ (Me = Ca, Co, Ni)	Cl_2	Emmenegger and Petermann (1968)
$MeNb_2O_6$ (Me = Mg, Zn)	HCl	Emmenegger and Petermann (1968)
$MeNb_2O_6$ (Me = Fe, Mn)	$Cl_2 + NbCl_5$	Emmenegger and Petermann (1968)
$MeTa_2O_6$ (Me = Mg, Mn, Co)	Cl_2	Emmenegger (1968)
$NiTa_2O_6$	$Cl_2 + TaCl_5$	Emmenegger (1968)
ZrOS	I_2	Nitsche (1967)

TABLE I (continued)

Compound	Transporting agent	Reference
ZrSiS	I_2	Nitsche (1967)
$MeGeS_4$ (Me = Fe, Mn)	I_2	Nitsche (1967)
$MeIn_2S_4$ (Me = Zn, Cd, Hg)	I_2	Nitsche (1960); Nitsche *et al.* (1961)
$MeIn_2S_4$ (Me = Mn, Co)	I_2	Nitsche (1967)
$MeGa_2S_4$ (Me = Zn, Cd, Hg)	I_2	Nitsche *et al.* (1961)
$MeCr_2S_4$ (Me = Mn, Zn, Cd)	$AlCl_3$ (HCl)	Lutz and von Lovâsz (1968)
$CdCr_2S_4$	Cl_2	Berger and Pinch (1967)
$CoCr_2S_4$	HCl (NH_4Cl)	Nitsche (1967)
Cu_3MeY_4 (Me = Nb, Ta; Y = S, Se)	I_2	Nitsche and Wild (1967)
Cu_3NbS_4, Cu_3TaS_4	I_2	Nitsche (1967)
Cd_4SiS_6	I_2	Kaldis *et al.* (1967)
ZrSiSe	I_2	Nitsche (1967)
$MeIn_2Se_4$ (Me = Zn, Cd)	I_2	Nitsche (1960); Nitsche *et al.* (1961)
$MeGa_2Se_4$ (Me = Zn, Cd)	I_2	Nitsche *et al.* (1961)
$CdCr_2Se_4$	Br_2	Scholz (1966)
Cd_4SiSe_6	I_2	Kaldis *et al.* (1967)
$MeCu_2SiS_4$ (Me = Zn, Cd, Fe)	I_2	Nitsche *et al.* (1967)
$MeCu_2GeS_4$ (Me = Zn, Cd, Fe, Mn, Ni)	I_2	Nitsche *et al.* (1967)
$MeCu_2SnS_4$ (Me = Zn, Cd, Fe)	I_2	Nitsche *et al.* (1967)
$ZnCu_2GeSe_4$	I_2	Nitsche *et al.* (1967)
$ZnSiP_2$	I_2	Valov and Usakova (1968)
$CdSiP_2$	Cl_2, $SiCl_4$	Valov and Plecko (1967), (1968)

IX. Solid Solutions; Hägg–Magnéli Phases

Next, we consider the case where two substances are chemically transportable under comparable conditions (transporting agent, temperature) and are capable of forming a solid solution. Under such conditions, solid solutions are produced if the components are mutually transported. Examples for this are given in Table II.

There may be difficulties in the formation of homogeneous crystals: if one starts with a mechanical mixture of the binary compounds, then the crystals produced will not automatically have a homogeneous composition. Instead, preferential transport of one component could occur. By choice of a suitable transporting agent, possibly through simultaneous use of several transporting agents, such a selective transport may be suppressed. It is

TABLE II

EXAMPLES OF SOLID SOLUTION FORMATION BY CHEMICAL TRANSPORT[a]

Solid solution system	Transporting agent	Reference
Nb_3Cl_8–Nb_2Cl_8	$NbCl_5$	Schäfer and Dohmann (1959b)
ZnS–CdS	I_2	Davis and Lind (1968)
ZnS–MnS	I_2	Nitsche *et al.* (1961)
TiS_2–VS_2	I_2	Nitsche (1967)
TiS_2–NbS_2	I_2	Nitsche (1967)
TiS_2–TaS_2	I_2	Nitsche (1967)
GaAs–InAs	Cl_2	Sirrine (1964); Minden (1965)
2GaAs–$ZnSiAs_2$	Cl_2, I_2	Averhieva *et al.* (1967)
2GaAs–$ZnGeAs_2$	Cl_2, I_2	Averhieva *et al.* (1967)
SnO_2–SnS_2	I_2	Nitsche (1967)
TiS_2–$TiSe_2$	I_2	Nitsche (1967)
ZrS_2–$ZrSe_2$	I_2	Nitsche (1967)
$CdCr_2S_4$–$CdCr_2Se_4$	$I_2 + AlCl_3$	Pickardt (1970)
TiS_2–$TiTe_2$	I_2	Nitsche (1967)
GeSe–GeTe	I_2	Muir and Cashman (1968)
ZnSe–GaAs	I_2	Ku and Bodi (1968)
Cu_2GeSe_3–3GaAs	I_2, Cl_2	Averhieva *et al.* (1967)
GaP–GaAs	I_2	Pizzarello (1962); Ku (1963); Hull (1964)
	Cl_2	Holonyak *et al.* (1961)
	H_2O	Gottlieb (1965); Purohit (1968)
InP–InAs	I_2	Egorov *et al.* (1967)

[a] Frequently, the given solid solutions were synthesized so that the mechanical mixture of the binary compounds served as the reactant for transport. Sometimes, however, the previously prepared solid solution was used as reactant, too.

also suitable to press pellets from the finely ground mixture and to use these as reactant for the transport. Finally, the homogeneity of the solid solution can be remarkably improved by repeated transport (reversal of the transport gradient) in the same tube. Use was made of this in the investigation of the system Nb_3Cl_8–Nb_2Cl_8 (Schäfer and Dohmann, 1959b). This practice was also used for the synthesis of GaP–GaAs solid solutions (Hull, 1964).

It is not fundamentally necessary for the formation of solid solutions that each individual component under consideration be itself chemically transportable. Rather, it is sufficient if *one* component is transportable, while the second is present in the gas phase in some form. It will be deposited with the transport of the first component upon formation of the solid solution. For the production of homogeneous solid solutions, one should be as careful as possible that the concentration of the gaseous second component does not change too much, or one will repeat the transport several times in the sealed tube.

Solid solutions and Hägg–Magnéli phases are related, in so far as the homogeneous regions often can be resolved into narrow, closely adjacent phases with intrinsic structures. These Hägg–Magnéli phases may be thermodynamically only a little more stable than solid solutions of the same composition. If the mobility of the atoms is increased, as is the case with chemical transport, then the synthesis of the stable phases is facilitated.

Examples. The niobium oxides NbO, NbO_2, and Nb_2O_5 have been known for a long time. They can be obtained as crystals by chemical transport (Schäfer and Hüesker, 1962; Schäfer *et al.*, 1966). On the oxygen-rich side of the system, however, five further phases exist, as is shown by the phase diagram in Fig. 2, which was obtained at 1300°C through adjustment of the equilibrium with a H_2O/H_2 gas phase (Schäfer *et al.*, 1969). The system (also at 1300°C) required at least one day to reach equilibrium. By chemical transport, the phases B, E, F, and G were obtained at about 1000°C (Schäfer and Hüesker, 1962; Schäfer *et al.*, 1965, 1969; Gruehn and Norin, 1967), and phase D at about 1200°C as crystals (Gruehn and Norin, 1969).

The phase G (Nb_2O_5) possesses a narrow range of homogeneity. By proper choice of the transporting agent, it can be obtained as colorless, nearly stoichiometric crystals, as well as dark blue, oxygen-deficient crystals. The analogous situation is true for other Nb_2O_5 modifications (Schäfer *et al.*, 1966) and for TiO_2 (Schäfer, 1962).

In view of this, the following remarks should be made: Under conditions of equilibrium, the composition of the solid phase can be regulated with the help of the gas phase. If several closely adjacent solid phases exist, or if it is important to introduce exact compositions in a homogeneous region, then the chemical transport should be carried out in a buffered gas phase;

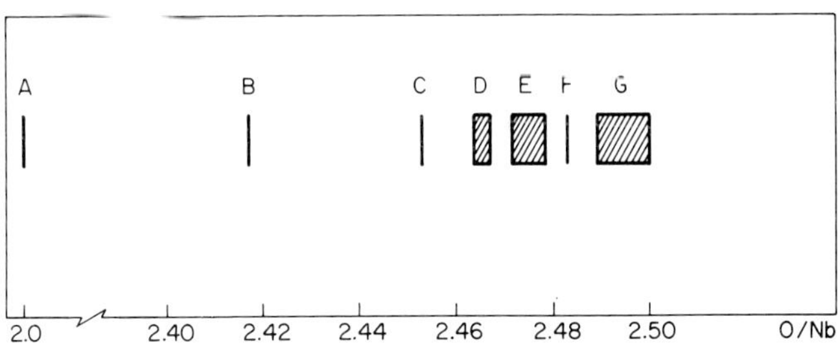

FIG. 2. Thermodynamically stable niobium oxide phases between NbO_2 and $NbO_{2.5}$ at 1300°C (Schäfer *et al.*, 1969).

e.g., consider the niobium pentoxide transport

$$Nb_2O_5 + 3NbCl_{5,g} = 5NbOCl_{3,g} \qquad (26)$$

A small Cl_2 pressure arises from the $NbCl_5$ decomposition,

$$NbCl_{5,g} = NbCl_{4,g} + 0.5Cl_2 \qquad (27)$$

and from this, an O_2 pressure results:

$$Nb_2O_5 + 3Cl_2 = 2NbOCl_{3,g} + 1.5O_2 \qquad (28)$$

$$P(O_2) = (Kp_{28})^{2/3}\, P^2(Cl_2)/P^{4/3}(NbOCl_3)$$

This oxygen pressure finally controls the stoichiometry of the NbO_x solid phase. With the help of the factor $P^2(Cl_2)/P^{4/3}(NbOCl_3)$, one can therefore exert a buffering effect. The situation is analogous to the bromide or iodide system, but with buffer regions displaced to smaller oxygen pressures.

Finally, one can also use buffer systems that do not take part in the transport. In such a manner, in the presence of a definite CO/CO_2 mixture as buffer the transport of Nb_2O_5 is possible with $NbCl_5$ according to Eq. (26).

X. Promotion of Solid State Reactions by a Transporting Gas Phase

A transporting gas phase can influence and promote the reaction of solid materials in various ways, as discussed in the following subsections. In none of these cases is one required to use the starting substances in stoichiometric quantities. Chemical analysis of the reaction product thus has more meaning than it could in normal solid state reactions.

A. Partial Transport

If one wishes to react stoichiometric amounts of Nb and Nb_2O_5 to form NbO, then this can occur in the absence of transporting agents at about 1500°C. The firing of the reactants in a quartz tube which contains a small amount of water (or hydrogen) is more suitable. Water oxidizes the niobium and the hydrogen produced reduces the niobium pentoxide; this happens mutually, until finally the conversion of the components to NbO is complete. This occurs after some hours of heating at 900°C (Schäfer and Hüesker, 1962). The system H_2/H_2O therefore acts as an oxygen carrier, without niobium being transported via the gas phase ("partial transport") (Schäfer, 1962).

If the reactants are separated, an excess of Nb_2O_5 (location 1) can be used. The reaction ceases if the reactant (location 2) that is not in excess is completely converted into NbO. At location 1, then, a mixture of NbO + NbO_2 is present.

In an analogous manner, NbO_2 can be prepared (Schäfer and Hüesker, 1962).

In this context, it should be pointed out that quartz glass normally gives up a significant amount of water if it is heated above 600°C. Often, the water liberated in this manner is of significance for the course of the reaction, although experimenters have not always taken this into account.

Partial oxygen transport can also be caused by a CO_2/CO gas phase.

For such a reaction, diffusion in the solid state always plays an important role. The gas phase produces the contact between the solid reactants. This has the result that the reaction takes place on the total granular surface, while for reactions without the help of a gas phase, the points of contact of the grains act as a bottleneck for the diffusion current.

B. Only One of the Solid Reactants Is Chemically Transportable

A typical example of this case is the reaction of niobium or tantalum with the silicon dioxide of the quartz wall. If the metal is heated at 1000°C in a quartz tube which is connected to a working high vacuum pump, then the thermodynamically expected reactions

$$11Nb + 3SiO_2 = 6NbO + Nb_5Si_3 \tag{29}$$

$$14Ta + 5SiO_2 = 2Ta_2O_5 + 5Ta_2Si \tag{30}$$

do not take place in appreciable amounts.

If, on the other hand, a quartz tube is used which is evacuated at room temperature and sealed under high vacuum, then an extensive conversion to the oxide and silicide (especially with niobium) takes place with successive tempering. Small amounts of water given off by the quartz wall act as a promotor for the reaction. The hydrogen formed by the reaction of the metal with water reacts with the quartz wall according to

$$SiO_2 + H_2 = SiO_g + H_2O \tag{31}$$

In this manner, SiO_2 from the reverse reaction is available to the entire surface of the metal; SiO_2 is transported to niobium (Schäfer and Dohmann, 1959a). An analogous reaction was observed with tantalum when $TaCl_5$ was present as transporting agent for SiO_2 (Schäfer, 1962; Schäfer *et al.*, 1964):

$$SiO_{2,s} + 2TaCl_{5,g} = SiCl_4 + 2TaOCl_{3,g} \tag{32}$$

Transport in the reverse direction can also take place if iodine is added as transporting agent. In this case, niobium and tantalum migrate to the quartz wall by means of the gaseous iodides, where the conversion to oxide and silicide takes place (Schäfer *et al.*, 1964; Schäfer and Fuhr, 1965). For further examples, see Schäfer (1962).

In these reactions, one component is heated in the "gaseous solution" of the other component. The volume diffusion in the nontransportable phase is still an important, possibly rate-determining step (see Section X.A).

C. The Reaction Product Is Chemically Transportable

We now consider the case where two substances are reacted together, and at least one is chemically transportable (or volatile). Moreover, the reaction product may be transportable.

This case is especially favorable because the formation of surface layers which hamper the reaction is avoided. The ability of all solid reactants to be transported is utilized in the preparation, e.g., of niobates, tantalates, and spinels, starting with the binary oxides.

The chemical transport of both solid reactants C and D and also the reaction product CD with the same transporting agent is only possible,

TABLE III

Reactions Where the Reaction Product Is Transportable[a]

Reaction	Transporting agent	Reference
$Zn_l + 0.5S_{2,g} = ZnS$	I_2	Schäfer (1962)
$Zn_l + 0.5Se_{2,g} = ZnSe$	I_2	Schäfer (1962)
$Cd_l + 0.5S_{2,g} = CdS$	I_2	Schäfer (1962)
$Cd_l + 0.5Se_{2,g} = CdSe$	I_2	Schäfer (1962)
$2Al_l + 1.5S_{2,g} = Al_2S_3$	I_2	Compare with Schäfer (1962) and Schäfer and Jagusch (1966)
$2Al_l + 1.5Se_{2,g} = Al_2Se_3$	I_2	Compare with Schäfer (1962) and Schäfer and Jagusch (1966)
$Ru_s + O_2 = RuO_2$	O_2	Schäfer *et al.* (1963)
$Ir_s + O_2 = IrO_2$	O_2	Schäfer and Heitland (1960)

[a] In these cases the formation of a surface layer is prevented.

if ΔG for the solid state reaction C + D = CD does not assume too large an absolute value (this is often the case).

Table III shows reactions between a liquid or a solid metal and a gas for which the reaction product is transported away in the temperature gradient.

The rates of the reactions in Table III are determined in the absence of transporting agents by the rate of diffusion in the sulfide, selenide, or oxide surface layer. However, if one makes use of a temperature gradient and transporting agents, then either the formation of the surface layer is completely avoided or the surface layer is kept thin by transporting the reaction product into the lower temperature region. The reaction product is deposited as well-formed crystals far from the starting material. For instance, if one reacts zinc or cadmium at 800°C with sulfur vapor, then the reaction scarcely takes place, because of the formation of a thin surface skin. On the other hand, in the presence of iodine, the reaction is complete within the same length of time at 800°C. A good synthetic possibility for aluminum sulfide is also given by this technique. One heats aluminum in a corundum boat at 800°C in the presence of sulfur vapor and a little iodine, in quartz tubes. At a cooler part of the tube (700°C), aluminum sulfide is deposited as large, colorless crystals.

XI. Purification and Separation of Materials

A. The Impurities Remain in a Residue

Great purifying effects can be obtained if the main component of the impure starting material is transported away while the impurities remain behind. This is true for endothermic as well as exothermic transport reactions. If, indeed, the impurities possess a noticeable vapor pressure (or if they are perceptibly attacked by the transport agent), then the temperature of the reacting solid phase should be kept as low as possible and the pure substance deposited in the hot zone—for example, on a hot filament—with the help of an exothermic transport reaction. This is the basis of the well-known van Arkel method for the purification of metals by transport with iodine (Rolsten, 1961). By this method, transition metals are liberated from their carbides, nitrides, and oxides. Only in exceptional cases are impurities of considerable amount simultaneously transported. This is true for the oxygen content in niobium (Rolsten, 1961), which can migrate via the gas phase in the form of $NbOI_3$; e.g., NbO transport with iodine (Schäfer and Hüesker, 1962).

B. The Impurities Go into the Gas Phase

It is possible for an impurity to remain in the gas phase as the main component is transported if the impurity forms a particularly stable compound with the transporting agent or if the formation of such a compound is achieved as a result of additions. Also, if in such cases the separation effect can be considerable, it is unfavorable in principle to work in closed reaction vessels: The impurity becomes more concentrated in the gas phase during the course of the transport process and the portion of the transported substance deposited last becomes more impure than the portion deposited first.

In this case, it is better to carry out the transport in an open flow system, assuming that the transported substance can be deposited from the gas flow sufficiently rapidly (see Section III). The gaseous impurity is carried away by the carrier gas. In some cases, a fractionation takes place.

Example. An N_2 flow was charged with I_2 at 40°C and passed over heated ZnS (T_3 = 800°C) forming there $ZnI_{2,g}$ and S_2. After leaving the ZnS space, the gas flow encountered a temperature gradient $T_3 \rightarrow T_2 \rightarrow T_1$. Here, ZnS was deposited by the reverse reaction. The ZnS crystals deposited at T_2 were practically colorless and as clear as glass, while those deposited at the lower temperature (T_1) were deeply colored yellow or honey-yellow. In this region, the impurities were undoubtedly concentrated (Schäfer and Jagusch, 1966). This technique deserves further consideration. The purifying effect can be multiplied by continual transport cycling.

C. Two Materials Are Transported in the Same Direction in a Temperature Gradient

If both of the transported materials are deposited as individual phases, then the following cases can be differentiated:

(a) Both crystal types E and F are normally not grown together. They can be separated by selection by. hand or on the basis of their specific weights.

Examples. Nb_2O_5 modifications (Schäfer *et al.*, 1966) or $MeNb_2O_6$ and $(Me, Nb)_{12}O_{29}$ (Emmenegger and Petermann, 1968) one beside the other.

(b) E and F can have different transport rates (in closed transport tubes).

Examples. Nb_3Sb, Nb_5Sb_4, and NbO are transported with iodine into the hotter zone of the tube. However, separations of Nb_3Sb/Nb_5Sb_4 and of Nb_3Sb/NbO occur because Nb_3Sb transports fastest (Schäfer and Fuhr, 1965). See also, for Nb/NbO, Schäfer and Hüesker (1962).

(c) In the temperature gradient (in spite of the same transport direction), a distinct succession of depositions can take place: $T_1 < T_2 < T_3$.

Example for exothermic transport. For the systems named under (b), the Nb_3Sb crystals always grow in the hottest part of the reaction space (T_3). Nb_5Sb_4 and NbO both migrate from T_1 in the direction of the hotter zone, but are deposited at the somewhat lower temperature T_2 compared with Nb_3Sb (Schäfer and Fuhr, 1965). See also Nb at T_3 and Nb_3Sb at T_2 (Schäfer and Fuhr, 1965) and Ta at T_3 and Ta_2Si at T_2 (Schäfer *et al.*, 1964).

Example for endothermic transport. Reactant at T_3, deposit of WO_2 at T_2 and of $W_{18}O_{49}$ at T_1 (Dettingmeijer *et al.*, 1969).

D. Two Substances Are Transported in Opposite Directions in a Temperature Gradient

If one of the substances is transported with an *exothermic* reaction while the other is transported with an *endothermic* reaction, then the separation is especially good and quite simple. The products of the reaction crystallize, separated from each other, at each end of a reaction tube lying in the temperature gradient. For examples, see Table IV.

TABLE IV

Separation by Means of Different Transport Directions

Product deposited at T_1	Product deposited at T_2	Temperature gradient T_1/T_2 (°C)	Transporting agent	Reference
CrTe	Cr	850/1025	I_2	Schäfer (1962)
Cu	Cu_2O	600/900	HCl	Schäfer and Etzel (1957)
CuO	Cu_2O	700/800	HCl	Schäfer and Jagusch (1966); Schäfer (1968)
$NbAs_2$	NbAs	850/1050	I_2	Schäfer and Fuhr (1965)
WO_2	W	800/1000	$I_2 + H_2O$	Dettingmeijer *et al.*, (1969)

Often, a metal A which is exothermically transportable by the iodide method forms a compound AB_x which is also chemically transportable with iodine. This is especially true when the compound concerned is not especially stable and is formed from the metal A and an element B that is gaseous at the transport temperature. Since the compound AB_x is derived from the elements exothermically, this frequently leads to a reversal in the transport direction compared to the direction of the transport of underlying metal.

Examples.

$$Cr + I_2 = CrI_{2,g} \qquad \text{(exothermic)} \tag{33}$$

$$CrTe + I_2 = CrI_{2,g} + 0.5Te_{2,g} \qquad \text{(endothermic)} \tag{34}$$

In this manner, AB_x can be easily separated from an excess of metal. It is expected that this principle can be used more frequently, since further examples are already known (Schäfer and Fuhr, 1965).

XII. Illustrations of Crystals

The crystals shown in Figs. 3–14 were prepared using normal transport procedures carried out at Münster. No special arrangements for crystal growth have been applied.

FIGS. 3–8.

3. $CsNb_4Br_{11}$, transported with $NbBr_5$. Crystal size about 0.2 mm (Broll *et al.*, 1969.)

4. ReO_2, transported with I_2. Diameter of the crystal block about 1 mm (Schäfer and Bode, 1967.)

5. TaS_2, transported with S. Crystal size $\geqq 10$ mm. (Schäfer *et al.*, 1968.)

6. RuO_2, transported with O_2. The thickest crystal is 6 mm long (Schäfer *et al.*, 1963).

7. Al_2S_3, transported with I_2. Crystal size up to about 10 mm. (Schäfer, 1962, Schäfer and Jagusch, 1966.)

8. SiO_2 (tridymite), transported with HF. Clear crystals up to 3 mm. (Gruehn, 1967.)

FIGS. 9–14.

9. CdO, transported with I_2. Crystals up to about 0.8 mm. (Schäfer and Fuhr, 1964.)

10. ReO_3, transported with I_2. Crystal size up to 1 mm. (Schäfer and Bode, 1967.)

11. $Ti_2Nb_{10}O_{29}$ (monoclinic), transported with $NbCl_5$. Length of the crystal needles is about 5 mm. (Gruehn *et al.*, 1967.)

12. Nb_3Sb, transported with I_2. Crystal size up to 3 mm. (Schäfer and Fuhr, 1965.)

13. NbO_2, transported with I_2. Crystal size 1 mm. (Schäfer and Hüesker, 1962.)

14. Nb_2O_5, in the *P* modification, transported with $NbCl_5$. Crystal size 1–2 mm. (Schäfer *et al.*, 1966.)

REFERENCES

Averhieva, G. K., Gorjunova, N. A., Nazarov, A., and Pročuchan, V. D. (1967). *Krist. Tech.* **2**, 517.

Berger, S. B., and Pinch, H. L. (1967). *J. Appl. Phys.* **38**, 949.

Broll, A., Simon, A., von Schnering, H. G., and Schäfer, H. (1969). *Z. Anorg. Allg. Chem.* **367**, 1.

Curtis, B. J., and Wilkinson, J. A. (1965). *J. Amer. Ceram. Soc.* **48**, 49.

Davis, E. A., and Lind, E. L. (1968). *J. Phys. Chem. Solids* **29**, 79.

Dettingmeijer, J. H., Tillack, J., and Schäfer, H. (1969). *Z. Anorg. Allg. Chem.* **369**, 161.

Egorov, L. A., Sachanov, P. V., Torbova, O. D., and Gordeev, I. V. (1967). *Izv. Akad. Nauk SSSR Neorg. Mater.* **3**, 881.

Emmenegger, F. (1968). *J. Crystal Growth* **3**, 135.

Emmenegger, F., and Petermann, A. (1968). *J. Crystal Growth* **2**, 33.

Gottlieb, G. E. (1965). *J. Electrochem. Soc.* **112**, 192.

Grigas, B. P. (1968a). *Fiz. Tekh. Poluprov.* **2**, 481.

Grigas, B. P. (1968b). *Fiz. Tekh. Poluprov.* **2**, 585.

Gruehn, R. (1967). Münster.

Gruehn, R., and Norin, R. (1967). *Z. Anorg. Allg. Chem.* **355**, 176.

Gruehn, R., and Norin, R. (1969). *Z. Anorg. Allg. Chem.* **367**, 209.

Gruehn, R., and Schäfer, H. (1969). Private communication.

Gruehn, R., Trenkel, M., and Schäfer, H. (1967).

Holonyak, N., Jillson, D. C., and Bevacqua, S. F. *In* "Metallurgy of Semiconductor Materials" (J. D. Schroeder, ed.). Wiley (Interscience), New York, 1961.

Horák, J., and Rodot, H. (1968). *C.R. Acad. Sci. Paris* **267**, 1427.

Hull, E. M. (1964). *J. Electrochem. Soc.* **111**, 1295.

Kaldis, E., Krausbauer, L., and Widmer, R. (1967). *J. Electrochem. Soc.* **114**, 1074.

Kern, R. (1962). *J. Phys. Chem. Solids* **23**, 249.

Klazar, J., and Horák, J. (1968). *Collect. Czech. Chem. Commun.* **33**, 973.

Kleinert, P. (1963). *Z. Chem.* **3**, 353.

Kleinert, P. (1964). *Z. Chem.* **4**, 434.

Kozáková, M., Horák, J., and Klikorka, J. (1966). *Z. Chem.* **6**, 431.

Ku, S. M. (1963). *J. Electrochem. Soc.* **110**, 991.

Ku, S. M., and Bodi, L. J. (1968). *J. Phys. Chem. Solids* **29**, 2077.

Lutz, H. D., and von Lovász, Cs. (1968). *Angew. Chem.* **80**, 562.

McCarley, R. E., Roddy, J. W., and Berry, K. O. (1964). *Inorg. Chem.* **3**, 50.

Minden, H. T. (1965). *J. Electrochem. Soc.* **112**, 300.

Miyatani, K., Wada, Y., and Okamoto, F. (1968). *J. Phys. Soc. Japan* **25**, 369.

Muir, J. A., and Cashman, R. J. (1968). *Phys. Rev.* **169**[2], 602.

Nejezchleb, K., and Horák, J. (1968). *Czech. J. Phys.* **B18**, 138.

Nitsche, R. (1960). *J. Phys. Chem. Solids* **17**, 163.

Nitsche, R. (1967). *Crystal Growth* **1967**, 215 (Suppl. for *J. Phys. Chem. Solids*).

Nitsche, R., and Wild, P. (1967). *J. Appl. Phys.* **38**, 5413.

Nitsche, R., Bölsterli, H. U., and Lichtensteiger, M. (1961). *J. Phys. Chem. Solids* **21**, 199.

Nitsche, R., Sargent, D. F., and Wild, P. (1967). *J. Cryst. Growth* **1**, 52.

Pickardt, J., Riedel, E., and Reuter, B. (1970). *Z. Anorg. Allg. Chem.* **373**, 15.

Pizzarello, F. A. (1962). *J. Electrochem. Soc.* **109**, 226.

Purohit, R. K. (1968). *J. Mater. Sci.* **3**, 330.

Rabenau, A., and Rau, H. (1968). *Naturwissenschaften* **55**, 336.

Rabenau, A., and Rau, H. (1969). *In* "Reactivity of Solids" (J. W. Mitchell, R. C. DeVries, R. W. Roberts, and P. Cannon, eds.), p. 763. Wiley, New York.
Rau, H., and Rabenau, A. (1967). *Solid State Commun.* **5,** 331.
Rau, H., and Rabenau, A. (1968). *J. Crystal Growth* **3,** 417.
Rolsten, R. F. (1960). *Z. Anorg. Allg. Chem.* **305,** 25.
Rolsten, R. F. (1961). "Iodide Metals and Metal Iodides." Wiley, New York.
Royen, P., and Forwerg, W. (1962). *Naturwissenschaften* **49,** 85.
Royen, P., and Forwerg, W. (1963). *Naturwissenschaften* **50,** 41.
Schäfer, H. (1953). *Z. Anorg. Allg. Chem.* **274,** 265.
Schäfer, H. (1962). "Chemische Transportreaktionen." Weinheim. (New York, London. 1964, Moskow 1964.)
Schäfer, H. (1968). *In* "Festschrift für Leo Brandt." Westdeutscher Verlag, Köln and Opladen.
Schäfer, H. (1971). *Angew. Chem.* **83,** 35; *Angew. Chem. Int. Ed. Engl.* **10,** 143.
Schäfer, H., and Beckmann, W. (1966). *Z. Anorg. Allg. Chem.* **347,** 225.
Schäfer, H., and Bode, M. (1967). Thesis of M. Bode, Münster.
Schäfer, H., and Dohmann, K. D. (1959a). *Z. Anorg. Allg. Chem.* **299,** 197.
Schäfer, H., and Dohmann, K. D. (1959b). *Z. Anorg. Allg. Chem.* **300,** 1.
Schäfer, H., and Etzel, K. (1957). *Z. Anorg. Allg. Chem.* **291,** 294.
Schäfer, H., and Etzel, K. (1959). *Z. Anorg. Allg. Chem.* **301,** 137.
Schäfer, H., and Fuhr, W. (1964). Dissertation of W. Fuhr, Münster.
Schäfer, H., and Fuhr, W. (1965). *J. Less-Common Metals* **8,** 375.
Schäfer, H., and Gerken, R. (1962). *Z. Anorg. Allg. Chem.* **317,** 105.
Schäfer, H., and Heitland, H.-J. (1960). *Z. Anorg. Allg. Chem.* **304,** 249.
Schäfer, H., and Hüesker, M. (1962). *Z. Anorg. Allg. Chem.* **317,** 321.
Schäfer, H., and Jagusch, W. (1966). Dissertation of W. Jagusch.
Schäfer, H., and Morcher, B. (1957). *Z. Anorg. Allg. Chem.* **290,** 279.
Schäfer, H., and Odenbach, H. (1966). *Z. Anorg. Allg. Chem.* **346,** 127.
Schäfer, H., and Tebben, A. (1960). *Z. Anorg. Allg. Chem.* **304,** 317.
Schäfer, H., and Tillack, J. (1964). *J. Less-Common Metals* **6,** 152.
Schäfer, H., and Wartenpfuhl, F. (1961a). *J. Less-Common Metals* **3,** 29.
Schäfer, H., and Wartenpfuhl, F. (1961b). *Z. Anorg. Allg. Chem.* **308,** 282.
Schäfer, H., and Wiese, U. (1971). *J. Less-Common Metals* **24,** 55.
Schäfer, H., Jacob, H., and Etzel, K. (1956). *Z. Anorg. Allg. Chem.* **286,** 42.
Schäfer, H., Sibbing, E., and Gerken, R. (1961). *Z. Anorg. Allg. Chem.* **307,** 163.
Schäfer, H., Schneidereit, G., and Gerhardt, W. (1963). *Z. Anorg. Allg. Chem.* **319,** 327.
Schäfer, H., Schibilla, E., Gerken, R., and Scholz, H. (1964). *J. Less-Common Metals* **6,** 239.
Schäfer, H., Gruehn, R., and Rönspiess, M. (1964/5). Investigations of chemical transport in buffered gas phase. Private communication.
Schäfer, H., Gruehn, R., Schulte, F., and Mertin, W. (1965). *Bull. Soc. Chim. Fr.* **1965,** 1161.
Schäfer, H., Gruehn, R., and Schulte, F. (1966). *Angew. Chem.* **78,** 28.
Schäfer, H., Wehmeier, F., and Trenkel, M. (1968). *J. Less-Common Metals* **16,** 290.
Schäfer, H., Bergner, D., and Gruehn, R. (1969). *Z. Anorg. Allg. Chem.* **365,** 31.
Schmid, H. (1965). *J. Phys. Chem. Solids* **26,** 973.
Scholz, H. (1966). Private communication.
Schröder, J. (1967). Compare with A. Rabenau, *Angew. Chem.* **79,** 43.
Sirrine, R. (1964). *J. Electrochem. Soc.* **111,** 750.

Tillack, J., and Kaiser, R. (1969). *Angew. Chem.* **81,** 149.
Valov, Ju. A., and Plecko, R. L. (1967). *Krist. Tech.* **2,** 535.
Valov, Ju. A., and Plecko, R. L. (1968). *Izv. Akad. Nauk SSSR Neorg. Mater.* **4,** 993.
Valov, Ju. A., and Usakova, T. N. (1968). *Izv. Akad. Nauk SSSR, Neorg. Mater.* **4,** 1054.
van Arkel, A. E., and de Boer, J. H. (1925). *Z. Anorg. Allg. Chem.* **148,** 345.
van Arkel, A. E. (1934). *Metallwirtschaft* **13,** 405.

Preparation of Transition Metal Compounds by Electrolytic Reduction of Fused Salts

AARON WOLD and DAVID BELLAVANCE

Department of Chemistry and Division of Engineering
Brown University, Providence, Rhode Island

I. Introduction

Electrolysis of fused salts was first used extensively by J.-L. Andrieux and his co-workers in order to obtain well-crystallized products for many transition metal compounds. Initially, transition metal oxides were dissolved in melts of alkali metal borates or mixtures of alkali metal borates and halides. Electrolysis of these melts, under controlled conditions, resulted in the formation of transition metals as well as a number of transition metal compounds. This work was also extended by other workers to include the electrolysis of other melt compositions, which resulted in the formation of oxides containing reduced oxidation states of the transition metals. In addition, transition metal sulfides, phosphides, silicides, borides, etc. were prepared, many for the first time, as large single crystals. Electrolysis of fused melt mixtures have also been used to prepare large, well-formed crystals of the alkali metal and rare earth tungsten bronzes as well as the alkali metal molybdenum bronzes. Ternary transition metal oxides, including the vanadium spinels, have also been prepared as single crystals by electrolysis.

This chapter will discuss the electrolysis of fused salts as a technique for the preparation of various transition metal compounds. The different electrolytic cells used will be discussed, as will the particular conditions required for the synthesis of several different compounds. In addition, this technique will be compared with other methods for the synthesis of the various phases considered. It will not be possible to treat all the reported compounds, but a sufficient number will be chosen so that the broad applicability of the method will be apparent. Compounds to be discussed include binary and ternary transition metal oxides, as well as a number of compositions that can be prepared from the combination of the transition metals with elements belonging to the families IIIB, IVB, VB, and VIB of the periodic table.

II. Preparation of Transition Metal Oxides

L. Andrieux (1929) indicated that it was possible to obtain a number of transition metal compounds by the electrolysis of melts containing transition metal oxides. In 1949 and 1950, Andrieux and Bozon reported on a number of vanadium spinels produced by electrolyzing melts of sodium tetraborate and sodium fluoride in which are dissolved the appropriate mixtures of transition metal oxides. The conditions used for electrolysis and the products obtained are given in Table I.

The products were extracted from the melt with dilute hydrochloric acid, and black, well-crystallized octahedra were obtained. Chemical and x-ray analysis and density determinations indicated that the products were spinels with the composition MV_2O_4 (M = Fe, Mn, Co, Zn, Mg).

Dodero (1952) and Dodero and Déportes (1956) have prepared a number of compositions corresponding to $NaMO_2$ (M = Fe, Co, Ni) by the electrolysis of sodium hydroxide melts contained in alumina crucibles. Electrodes of iron, cobalt, or nickel were used, depending on the desired composition of the final product. During electrolysis, the cathode was attacked, thereby providing the necessary transition metal ions required

TABLE I

ELECTROLYTIC CONDITIONS FOR THE FORMATION OF VANADIUM SPINEL

Melt composition	Voltage used (V)	Current used (A)	T (°C)	Time of electrolysis (hr)	Anode	Cathode	Product composition
$Na_2B_4O_7 + NaF + \frac{1}{12}V_2O_5 + \frac{1}{8}Fe_2O_3$	3	8	850	2	Fe crucible	Fe rod	FeV_2O_4
$Na_2B_4O_7 + \frac{1}{2}NaF + \frac{1}{7}V_2O_5 + \frac{1}{3}Mn_2O_3$	2.6	6	800	$2\frac{1}{2}$	C crucible	C rod	MnV_2O_4
$Na_2B_4O_7 + \frac{1}{2}NaF + \frac{1}{9}V_2O_5 + \frac{1}{2}CoO$	3	8	880	$3\frac{1}{2}$	C crucible	C rod	CoV_2O_4
$Na_2B_4O_7 + NaF + \frac{1}{3}V_2O_5 + ZnO$	3	15	850	$1\frac{1}{2}$	C crucible	C rod	ZnV_2O_4
$Na_2B_4O_7 + 1.5NaF + 2MgO + \frac{1}{2}NaVO_3$	3.5	15	950	$1\frac{1}{2}$	C rod	C crucible	MgV_2O_4

for compound formation. In addition to the sodium compounds, they were also able to prepare and characterize $KFeO_2$ and $LiFeO_2$.

Rogers *et al.* (1966) reported on the synthesis of members of the mixed spinel system $Co_{1+\delta}V_{2-\delta}O_4$ ($0.11 < \delta < 0.56$) by fused salt electrolysis. The starting materials used consisted of reagent-grade sodium tungstate, tungstic anhydride, cobaltous carbonate, and ammonium metavanadate. The flux components Na_2WO_4 and WO_3 were dried at 500°C for 12 hr before use and the other reagents were decomposed to their respective oxides, Co_3O_4 and V_2O_5, by slow ignition in air to 500°C. The desired molar proportion of these materials was mixed and fused for approximately 1 hr at 750°C in a porcelain crucible. The melt was ground to a coarse powder suitable for filling a platinum double cell as described by Kunnmann and Ferretti (1964). The platinum double cell used by Rogers *et al.* is shown in Fig. 1.

The cell was mounted in a slit tube furnace and the temperature raised slowly to the desired value. Initially, a platinum disk (21 cm²) was used

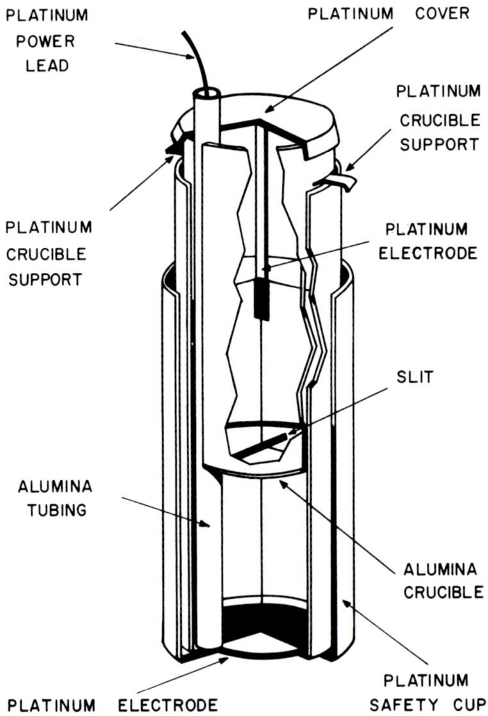

FIG. 1. The platinum double cell.

TABLE II

EXPERIMENTAL CONDITIONS FOR GROWTH OF SOME VANADIUM SPINELS BY FUSED SALT ELECTROLYSIS

Initial compositions of melt (mole fractions)				x in $(1-x)Na_2W_2O_7$–xNa_2WO_4	T (°C)	Product composition
Co_3O_4	V_2O_5	WO_3	Na_2WO_4			
0.055	0.165	0.156	0.624	0.75	900	$Co_{1.51}V_{1.49}O_4$
0.055	0.165	0.066	0.714	0.908	900	$Co_{1.38}V_{1.62}O_4$
0.055	0.165	0.033	0.747	0.956	900	$Co_{1.22}V_{1.78}O_4$
0.055	0.165	0.00	0.78	1.0	900	$Co_{1.11}V_{1.89}O_4$
0.055	0.165	0.14	0.64	0.781	800	$Co_{1.54}V_{1.46}O_4$
0.055	0.165	0.2	0.58	0.655	960	$Co_{1.29}V_{1.71}O_4$
0.076	0.230	0.0	0.694	1.0	900	$Co_{1.56}V_{1.44}O_4$

as the cathode; however, for the growth of larger crystals, a seed crystal was substituted as the cathode. Crystals as large as 1 cm on an edge were grown by using a crystal seed. The electrolytic reactions were carried out for a period of 7–10 days. It was found empirically that the optimum results could be obtained using current densities of about 300 mA cm^{-2} for the first three days, followed by electrolysis at 40 mA cm^{-2} for the remainder of the period. All reductions were carried out under an atmosphere of flowing nitrogen. The products were all carefully characterized by chemical and x-ray analysis as well as by resistivity and Seeback measurements. The results of these reactions are summarized in Table II.

As can be seen from the first four examples in Table II, when the solute concentrations and the temperature of electrolysis were held constant, more reduced products were formed with increasing basicity of the flux. It can also be seen that, whereas increasing the temperature favors the formation of more reduced phases, increasing the concentration of the solute results in a higher percentage of Co_2VO_4.

In general, electrolysis above 900°C was hampered by the inherent limitations of the equipment. In addition, for flux mixtures where $x < 0.75$, it was not possible to obtain crystalline samples. One of the end members of this series, CoV_2O_4, grown from a borate flux by Andrieux and Bozon (1949), could not be obtained from the polytungstate melt.

It should be noted that there are numerous problems associated with the crystal growth of vanadium spinels. These compounds have high melting points and appear to melt noncongruently. In addition, there is considerable difficulty in preparing vanadium compounds which possess a lower than normal valence state. Although Rogers *et al.* did not succeed in obtaining single crystals of the end-member spinels CoV_2O_4 and Co_2VO_4, the technique of fused salt electrolysis permitted the synthesis of a number of compounds with intermediate composition.

The reduction of TiO_2 or $CaTiO_3$ dissolved in a calcium chloride melt has been reported by several investigators (Bertaut and Blum, 1956a,b; Bright *et al.*, 1958). The electrolysis was carried out by Bright at 850°C for 10 min using a current of 14 A. Black, lustrous crystals formed at the cathode which on analysis were shown to have the composition $CaTi_2O_4$. It has not been possible to synthesize this compound except by electrolysis.

Electrolysis of fused melts has also been used to prepare single crystals of a number of unusual transition metal oxides. Single crystals of uranium (IV) oxide were deposited by Robins (1960, 1961) on a platinum cathode by the electrolysis of a solution of uranyl chloride in fused sodium chloride–potassium chloride eutectic at 840°C. A carbon rod served as the anode and the electrolysis time was 20 hr. The crystals obtained reached a size of about 3 mm across and the largest weighed approximately 50 mg.

Perloff and Wold (1967) also describe the preparation of large, well-

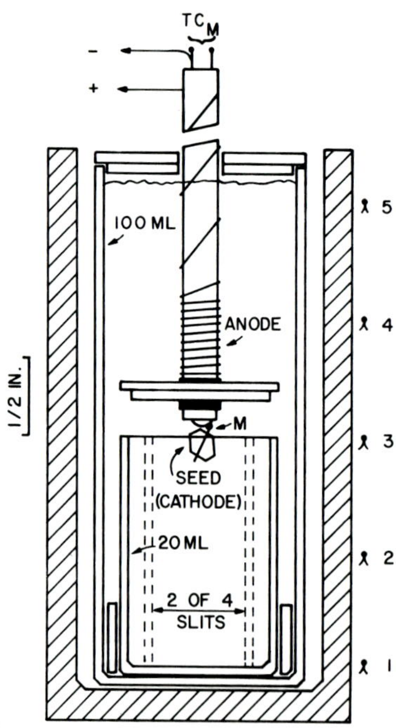

FIG. 2. Alumina double cell.

formed single crystals of molybdenum (IV) oxide. The growth was carried out under an argon atmosphere in the double cell shown in Fig. 2. The slablike crystals obtained were of dimensions 7 × 3 × 2 mm and were grown in a period of three days by electrolyzing a molar mixture of $1.40Na_2MoO_3$ and $1.0MoO_3$ at 675°C, 20 mA, and a cell potential of 0.75 V. The cell shown in Fig. 2 may be treated as two parts: the double cell assembly and the cell power unit. The double cell assembly consists of coaxial crucibles of impervious recrystallized alumina; the inner crucible has a volume of 20 ml and the outer of 100 ml. The crucibles are maintained in a coaxial position by means of a ring cut from a 50-ml alumina crucible. The cell power unit contains the electrode and a thermocouple for monitoring the temperatures. It consists of a long, double-bore alumina tube with a snug-fitting platinum sleeve at one end which is force-fitted into a hole drilled in the center of the 20-ml crucible cover. The tip of the rod is positioned to protrude through the bottom of the cover. The anode is a platinum wire fastened near the cover and wound in a helix. One bore of the tube carries a lead of platinum–10% rhodium which is welded to the platinum cathode wire at a point near the crystal to form a thermocouple junction.

During the electrolytic growth, the cell power unit is positioned with respect to the double cell assembly by seating the cover on the 20-ml crucible. In this way, a double cell is formed in which the cathode region is separated from the anode region. Ions can migrate freely between the regions through the slits of the inner crucible, but the oxygen which is liberated at the anode is excluded from the environment of the growing crystal.

The double cell shown in Fig. 2 eliminates temperature gradients in the melt, oxidation of crystals, and the problems associated with crystal recovery. As has been indicated, oxygen does not interfere with crystal growth and since argon may be introduced following the growth period, the melt can be lowered away from the electrode upon which the crystals have grown. By this technique, large, well-formed, strain-free single crystals of molybdenum (IV) oxide were obtained. It is also anticipated that this type of electrolysis apparatus can be used for the preparation of a large variety of binary and ternary transition metal compounds in which the transition element is present in a reduced state.

A number of oxide bronzes have also been prepared by electrolysis. The bronzes are considered to be nonstoichiometric crystalline phases of transition metal oxides, usually in ternary systems, although some binary phases qualify by virtue of their characteristic metallic or near-metallic conductivity and metallic lustre. In addition, the transition metal is present in an apparent oxidation state lower than the highest possible one. Banks and Wold (1968) have reviewed the properties and various synthetic methods used for the preparation of bronzes.

Samples of Na_xWO_3 can be prepared by electrolysis of a fused mixture of reagent-grade sodium tungstate and tungsten trioxide in a glazed ceramic or recrystallized alumina crucible. The electrolysis melts, with varying mole ratios of Na_2WO_4 to WO_3, result in different compositions of the bronzes. The electrolysis is carried out in the double cells of either Kunnmann and Ferretti (1964) or Perloff and Wold (1967). Crystal growth occurs at the cathode and current densities of about 20 mA cm^{-2} yield the best crystals.

Single crystals of Cu_xWO_3 were prepared (Conroy and Sienko, 1957) by the electrolytic reduction at 800°C of molten mixtures of $CuWO_4$ and WO_3. The value for x was varied from 0.26 to 0.77. Crystals of Li_xWO_3 were prepared (Sienko and Truong, 1961) by electrodeposition from fused mixtures of Li_2CO_3 and WO_3. The best crystals were obtained on the cathode under the following conditions: temperature, 780–850°C; cathode current density, 0.35 A cm^{-2}; time of electrolysis, 20–25 hr. Single crystals were obtained with the compositions $Li_{0.394\pm0.01}WO_3$, $Li_{0.377\pm0.01}WO_3$, and $Li_{0.365\pm0.01}WO_3$. These correspond, respectively, to starting ratios of Li_2CO_3:WO_3 of 1:1.5, 1:1.75, and 1:2.

Ostertag (1966) has utilized the method of electrolytic reduction in order to prepare single crystals of several rare earth tungsten bronzes.

Single crystals of Tm_xWO_3 approximately 2 mm in length were obtained from electrolysis of a fused mixture containing Tm_2O_3 and WO_3 at 1250°C. The electrodeposition was carried out for 20 hr using a crucible of recrystallized alumina, platinum electrodes, and a cathode current density between 0.15 and 0.20 A cm^{-2}. Crystals of Gd_xWO_3 (8 mm long) were prepared by a similar reduction at 1000°C. The melt was obtained from a mixture of $GdCl_3$ and WO_3. The electrolysis was started after an initial period of gas evolution (WO_2Cl_2).

Single crystals of molybdenum bronzes have also been grown by electrolytic reduction (Wold *et al.*, 1964) of alkali-metal molybdate–molybdenum (VI) oxide mixtures. The crystals can also be grown in the double cell shown in Fig. 2. Baker's reagent-grade sodium molybdate, potassium molybdate, and molybdic anhydride were dried for 12 hr at 500°C before use. These materials were mixed in the desired molar proportions and fused for approximately 1 hr at 650°C in a platinum crucible. The melt was removed and ground until it had the consistency of a coarse powder. The electrolytic cell was filled with this material. The sample was brought up slowly to the desired temperature and electrolysis begun, after waiting

TABLE III

MOLYBDENUM BRONZES

Formula	Initial flux composition	T (°C)	M^+ (wt%)	Mo^{+5} (wt%)	Total Mo (wt%)
$Na_{0.93}Mo_{5.9}O_{17}$	$3MoO_3 : Na_2MoO_4$	550	2.48	16.98	66.1
$Na_{0.87}Mo_{5.95}O_{17}$	$4MoO_3 : Na_2MoO_4$	550	2.33	14.83	66.16
$K_{0.28}Mo_{1.02}O_3$	$3.35MoO_3 : K_2MoO_4$	550	6.94	8.98	62.34
$K_{0.26}Mo_{1.01}O_3$	$3MoO_3 : K_2MoO_4$	560	6.48	10.12	62.62

a sufficient interval of time for the flux material to melt completely. Table III summarizes the conditions for the preparation of several alkali metal molybdenum bronzes. Initially, a platinum disk (4.5 cm^2) was used as the cathode; later, a seed crystal was substituted to obtain large crystals. All the electrolytic reductions were carried out for a period of 7–8 days at a current density of 10 mA cm^{-2} in air.

The sodium and potassium molybdenum bronze crystals were analyzed by standard procedures. The chemical analysis of the sodium molybdenum bronzes formed by the electrolytic reduction of $3MoO_3{:}Na_2MoO_4$ and $4MoO_3{:}Na_2MoO_4$ indicates a composition range from $Na_{0.93}Mo_{5.9}O_{17}$ to $Na_{0.87}Mo_{5.95}O_{17}$. It was pointed out by Wold *et al.* (1964) that the members of this series of sodium bronzes approached the composition $NaMo_6O_{17}$. From the chemical analysis of the potassium molybdenum bronzes, Wold *et al.* (1964) report the blue bronze to have a composition of $K_{0.28}MoO_3$ and the red bronze to have the composition $K_{0.26}MoO_3$. The ideal formulas of the blue and red bronzes as determined from x-ray crystallographic analysis (Graham and Wadsley, 1966; Stephenson and Wadsley, 1965) are $K_{0.30}MoO_3$ and $K_{0.35}MoO_3$, respectively. This analysis appears to be in good agreement with the reducing power analysis for the molybdenum. Since both the potassium and total molybdenum analyses are difficult to carry out to great accuracy, the crystallographically determined compositions are probably more accurate. The ideal compositions of $K_{0.35}MoO_3$ and $K_{0.30}MoO_3$ have been prepared by Bouchard *et al.* (1967) by the electrolytic reduction of molybdate melts.

The method of electrolytic reduction has been successfully used to prepare a considerable number of binary as well as ternary transition metal oxides. It is particularly useful when it is desired to obtain the transition metal in a lower valence state than its maximum value. In addition, it would not be possible to synthesize readily single crystals of such compounds unless electrolytic reduction were used.

III. Preparation of Transition Metal Sulfides

Weiss (1946) first prepared MoS_2 and WS_2 by electrolytic reduction. The conditions under which the electrolysis was carried out are given in Table IV.

The WS_2 formed as small, blue-gray crystals that were readily flattened on squeezing. The MoS_2 crystallized as hexagonal, blue-gray platelets, similar to natural molybdenite. X-ray examination was not carried out on the products. Chemical analysis of the products of the electrolyses gave the results shown in Table V. X-ray examination, however, was not carried out on these materials.

A more recent patent (Heinen, 1968) describes the synthesis of both molybdenum and tungsten sulfides by electrolysis of molten salts. For the

TABLE IV

ELECTROLYTIC PREPARATION OF MoS_2 AND WS_2

Composition of melt	Nature of electrode	T (°C)	Voltage (V)	Current (A)	Time (min)	Composition
$Na_2B_4O_7 + \left(\frac{1}{2}\right)NaF + \left(\frac{1}{3.8}\right)WO_3 + \left(\frac{1}{6}\right)Na_2SO_4$	C crucible cathode C anode	800	7.5	15	20	WS_2
$Na_2B_4O_7 + \left(\frac{1}{2}\right)NaF + \left(\frac{1}{4.75}\right)MoO_3 + \left(\frac{1}{10}\right)Na_2SO_4$	C crucible cathode C anode	800	7	15	25	MoS_2

TABLE V

CHEMICAL ANALYSIS OF MoS_2 AND WS_2

	I	II	III	IV	Calculated for MoS_2	Calculated for WS_2
S	39.5	39.7	25.3	25.6	40.0	25.81
Mo	59.6	59.5	—	—	60.0	—
W	—	—	73.9	74.0	—	74.19

synthesis of molybdenum (IV) sulfide, molybdenum (VI) oxide and sulfur were dissolved in a melt consisting of sodium tetraborate, sodium fluoride, potassium fluoride, and sodium carbonate. The concentration of molybdenum used was between 5–10 wt% of the total melt. The melt was contained in a graphite crucible which also served as an anode. The cathode consisted of either a graphite or molybdenum rod. The electrolysis was carried out at 1000°C at an initial cathode current density of 60 A cm^{-2}. No protective atmosphere was used and the cathode, with the desired product, was withdrawn from the melt and air-cooled. The excess electrolyte was dissolved in water and a typical yield was 80 g of MoS_2 per 100 A-hr of electrolysis. Crystals up to ½ in. in size were reported by the authors. X-ray analysis confirmed that the product was chiefly molybdenite. Chemical analysis indicated that the principal impurity present was approximately 2% carbon.

Tungsten (IV) sulfide was also prepared in a similar cell. However, the melt consisted of 10 wt% $KAlF_4$, 60 wt% NaF, 20 wt% WO_3, and 10 wt% S. Approximately 195 g of product was obtained after 100 A-hr of electrolysis. Analysis of the product indicated the same degree of purity as MoS_2.

A more elaborate cell was devised by Didchenko and Litz (1962) for the synthesis of CeS and ThS. The method, which may be applied to the synthesis of subsulfides of many other transition metals, is based on the electrolytic reduction of the sulfide of a metal in its higher oxidation state suspended or dissolved in a molten salt bath which consists of alkali metal chlorides and the appropriate metal chloride. The reduction is followed by vacuum distillation of the salts and electrolysis by-products at 1000°C. The electrolytic cell is shown in Fig. 3. The crucibles (55 mm o.d., 45 mm i.d., 90 mm high) were made from ATJ-grade graphite (National Carbon Co.). The cathode shaft and paddle were made of molybdenum and were

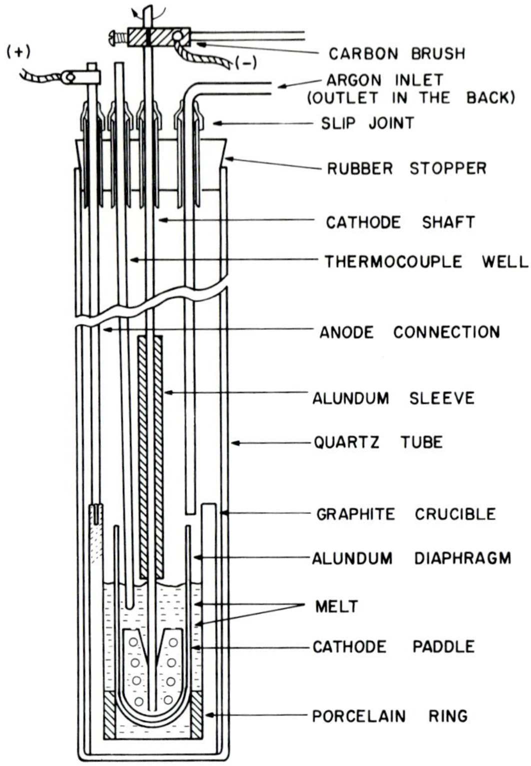

FIG. 3. Electrolytic cell for the preparation of rare earth sulfides.

protected from attack by chlorine produced at the anode. This was done by means of an alumina sleeve projecting above the melt. Electrical contact to the cathode was made by means of a spring-loaded carbon brush which permitted rotation of the cathode during the electrolysis. Alundum extraction thimbles (Norton, RA 98) of 34 mm diameter, 100 mm high were used as catholyte container and diaphragm.

For the preparation of cerium monosulfide, a mixture of 35 g of cerium sesquisulfide and 35 g of cerium (III) chloride were mixed together and put into the Alundum diaphragm cup. A sodium chloride–potassium chloride eutectic mixture was added to the bottom part of the cell until the porcelain support ring was covered. The Alundum cup then was fitted into the ring and more eutectic mixture added to the cell with constant tapping to make sure that both the anode and cathode compartments were filled with salt. The cell was lowered into the quartz envelope previously flushed with argon and closed, by the rubber stopper, with the cathode in the proper position. All these operations were performed in the dry box. The cell was then heated electrically to 800°C in a pot furnace while maintaining a slow flow of argon (2–3 liters min^{-1}) through the tube. The cathode was set in rotation at 60–80 rpm, with the blades 2–3 mm above the bottom of the Alundum cup, centered in such a manner that no scraping of the walls occurred. For the first 45 min a preelectrolysis voltage of 0.6–2.1 V was applied. The voltage was then raised to 5.0 V and maintained close to this value during the entire electrolysis. Sixteen ampere-hours were passed through the cell and then the stirring was continued for one additional hour without current. The cathode was then lifted above the melt and the cell allowed to cool under argon. The solidified melt was easily separated from the graphite crucible and the thimble was broken, exposing at the bottom a large, golden-brown button of cerium monosulfide embedded in the salts.

Much of the salt mixture could be removed mechanically. The remainder was distilled off under vacuum at 1000°C. The light-brown cake of CeS which remained was broken up into a crystalline powder of about 200 mesh. The powder was placed in a molybdenum crucible and heated inductively in a vacuum furnace. After first removing any remaining impurities by heating between 1000° and 1500°C, the pressure was reduced to 10^{-5} Torr and the sulfide heated to 1800–1900°C for 15–30 min. A bright golden ingot of CeS was produced on cooling down to room temperature.

In addition to CeS, Didchenko and Litz were able to prepare the following compounds by electrolytic reduction of fused salts: ThS, LaS, GdS, YbS, and US. The monosulfides obtained by this method are usually metal-rich and analyze as $MeS_{0.95\text{-}1.00}$. Oxysulfide forms quite readily, and even small quantities change the product to dark brown or black. For this reason, a rigid exclusion of moisture and oxygen from the electrolytic melt is necessary.

It is apparent from this section that electrolysis of fused salts can be used to prepare a number of unusual sulfides of the transition metals, and this method will undoubtedly be extended to include many other transition metal chalcogenides.

IV. Transition Metal Phosphides

The most common method used for the preparation of transition metal phosphides is the direct combination of the elements. However, in general, the products are not pure or homogeneous and any crystals produced are usually small and of poor quality. The various synthetic techniques for the synthesis of phosphides are summarized in the monograph by Aronsson *et al.* (1965). Of all the methods available, fused salt electrolysis produces large crystals, whereas the other techniques usually result in the formation of inhomogeneous powders.

The first electrolytic preparation of transition metal phosphides was carried out by Hartmann and his group (Hartmann *et al.*, 1931; Hartmann and Conrad, 1937).

Hartmann *et al.* (1931) obtained W_2P by the electrolysis of a fused bath of metaphosphoric acid and tungstic acid with a graphite anode and rotating nickel cathode. A fused salt mixture of $4NaPO_3$, $2LiPO_3$, and WO_3 was used by Hartmann and Orban (1936). When the current density was less than 0.0085 A cm^{-2} and the temperature kept between 470–520°C, an amorphous product, identified by chemical analysis as W_4P, was obtained. At temperature between 525 and 535°C, α-tungsten was the product and W_2P formed if the current densities were greater than 0.04 A cm^{-2}.

Andrieux and Chêne (1938a,b, 1939), Andrieux (1948), Chêne (1938, 1939, 1941) investigated extensively the formation of numerous transition metal phosphides by fused salt electrolysis. Among those claimed to have been prepared are Fe_3P, Fe_2P, FeP, Ni_3P, Ni_2P, Co_2P, CoP, CoP_2, Mn_2P, MnP, CrP, Mo_3P, MoP, W_2P, WP, V_2P, and VP. The electrolysis was carried out in a carbon crucible with a carbon rod as the other electrode. The temperatures of the electrolysis reaction varied from 500 to 1050°C, depending on the composition of the fused melt. The compositions of the melts used appear in Table VI.

The melts were leached with dilute hydrochloric acid and the products were characterized by classic chemical analysis, but x-ray diffraction studies were not done.

Wood (1953), Yocom (1958), Hsu *et al.* (1955, 1958), and Blaugher *et al.* (1965) further developed this method.

The simple phosphides prepared by Yocom (1958) are listed in Table VII. The compounds reported in Table VII were subjected to wet chemical analysis, spectrochemical analysis, and x-ray powder diffraction analysis. Wet chemical analysis accounted for the main constituents in all cases to about 99.99%, with no evidence of metallic impurities. The x-ray diffraction patterns corresponded with the structures previously reported in the literature. The compound W_3P was not previously identified, presumably because its preparation is difficult in the direct combination techniques used by previous workers (Schönberg, 1954).

TABLE VI

COMPOSITION OF CHENE'S MELTS

Phosphide	Phosphate	Metal-bearing constituent	Halides
Fe_2P	$2NaPO_3$	$\leq \frac{1}{2} Fe_2O_3$, $FeCl_2$	$\frac{1}{2}$ NaF
FeP	$2NaPO_3$	$\left(\frac{1}{8}-\frac{1}{10}\right) Fe_2O_3$	$\frac{1}{2}$ NaF
Ni_3P	$2NaPO_3$	$\left(\frac{1}{5}-\frac{1}{3}\right) NiO$	$\frac{1}{2}$ NaF
Ni_2P	$2NaPO_3$	$\frac{1}{10}$ NiO	$\frac{1}{2}$ NaF
Co_2P	$2NaPO_3$	$\left(\frac{1}{5}-\frac{1}{2.5}\right) CoO$	NaF
CoP	$2NaPO_3$	$\left(\frac{1}{15}-\frac{1}{12}\right) CoO$	$\frac{1}{2}$ NaF
CoP_2	$4NaPO_3$	$\left(\frac{1}{15}-\frac{1}{10}\right) CoO$	—
Mn_2P	P_2O_5	3MnO	(5–10)LiCl
MnP	$2NaPO_3$	($\leq$12MnO)	$\frac{1}{2}$ NaF
MnP	$2NaPO_3$	$2MnCl_2$	NaCl
MnP	$2NaPO_3$	$2MnCl_2$	KCl
CrP	$2NaPO_3$	$\left(\frac{1}{7.5}-\frac{1}{5}\right) Cr_2O_3$	$\frac{1}{2}$ NaF
Mo_3P	$2NaPO_3$	$\left(\frac{1}{6}-\frac{1}{3}\right) MoO_3$	$\frac{1}{5}$ NaCl
MoP	$2NaPO_3$	$\frac{1}{18} MoO_3$	$\frac{1}{5}$ NaCl
W_2P	$2NaPO_3$	$\left(\frac{1}{6}-\frac{1}{5}\right) WO_3$	$\frac{1}{2}$ NaCl
WP	$2NaPO_3$	$\left(\frac{1}{12}-\frac{1}{10}\right) WO_3$	$\frac{1}{10}$ NaCl

TABLE VI (continued)

Phosphide	Phosphate	Metal-bearing constituent	Halides
V_2P	$P_2O_5 \cdot 3Li_2O$	$\frac{1}{5} V_2O_5$	4LiF
V_2P	$P_2O_5 \cdot 3Li_2O$	$\frac{1}{2.5} V_2O_5$	$2LiF + MgF_2$
V_2P	$P_2O_5 \cdot 3MgO$	$\frac{1}{5} V_2O_5$	4LiF
VP	$P_2O_5 \cdot H_2O$	$\frac{1}{7.5} V_2O_5$	$(1\text{–}2)MgCl_2$, 2LiCl
VP	$P_2O_5 \cdot H_2O$	$\frac{1}{10} V_2O_5$	$MgCl_2 + 4LiCl$

Pure single crystals of FeP were prepared and characterized by Bellavance *et al.* (1969). The method used involved the technique of fused salt electrolysis followed by chemical transport of the products. Initially, iron monophosphide single crystals were grown by the electrolysis of sodium metaphosphate–iron (III) fluoride mixtures. The apparatus used for the electrolysis of these fused melts is shown in Fig. 4. The assembly is vertically mounted in a split tube furnace. A mullite tube, with water-cooled brass cap ends sealed by rubber O-rings, provides an enclosed chamber for electrolysis. Gas connection at either end allow the electrolysis to be carried out under an atmosphere of flowing argon. A carbon pedestal

TABLE VII

SOME TRANSITION METAL PHOSPHIDES

Compound	Structure	Flux used	Compound	Structure	Flux used
VP	*B* 8_1	$Na_4P_2O_7$	MnP	*B* 31	$Na_4P_2O_7$
Cr_3P	DO_e	$Na_4P_2O_7$	Fe_2P	*C* 22	$NaPO_3$
CrP	*B* 31	$Na_4P_2O_7$	FeP	*B* 31	$NaPO_3$
Mo_3P	DO_e	$NaPO_3$	Co_2P	*C* 23	$NaPO_3$
MoP	B_h	$NaPO_3$	Ni_3P	DO_e	$NaPO_3$
W_3P	DO_e	$NaPO_3$	Ni_2P	*C* 22	$NaPO_3$
WP	*B* 31	$NaPO_3$			

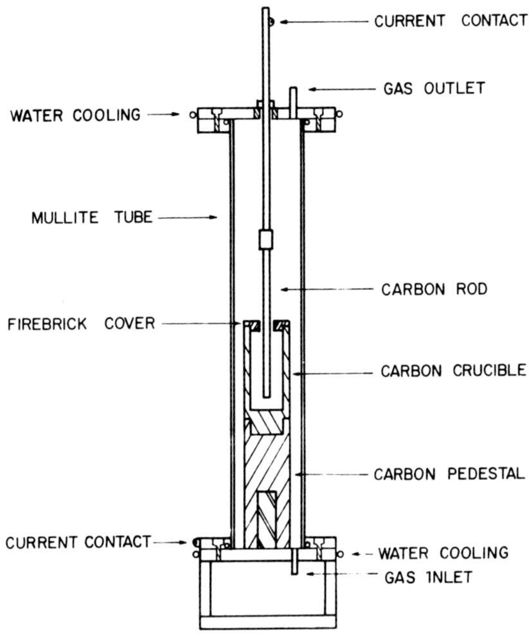

FIG. 4. Electrolytic cell for the preparation of phosphides.

centers the carbon crucible in the hot zone of the furnace. The crucible size varied from 1 in. i.d. by 5 in. deep to 1.5 in. i.d. by 4 in. deep. The crucible acts as the anode and a ¼-in. centered carbon rod serves as the cathode during electrolysis. The external anode lead wire is connected to the brass plate of the base, and the current is conducted through the pedestal to the crucible. The external cathode lead wire is attached to the ¼-in. stainless steel rod which supports the center carbon rod. The use of internal wire connection is prohibited by the corrosive nature of the atmosphere in the chamber during electrolysis. The thermocouple for the control of the furnace temperature is placed outside the reaction chamber. The thermocouple is calibrated with respect to a thermocouple placed within the reaction vessel in the absence of a melt.

The desired molar proportions of iron (III) fluoride and sodium metaphosphate were used as the melt components. The desired molar proportions of these materials were thoroughly mixed prior to placement in the crucible. The melt was heated to the reaction temperature at the rate of 75°C hr^{-1} under an atmosphere of flowing argon and allowed to equilibrate for 1 hr. The centered carbon rod was inserted to a depth of ¾ in. and the electrolysis carried out from 12 to 24 hr. A constant current source maintained the desired current. After electrolysis, the melt was cooled to room temperature at the rate of 75°C hr^{-1}. The product was recovered by leaching the melt in hot water or dilute hydrochloric acid. Free carbon was removed by flotation using methylene iodide.

The composition of the products obtained from the electrolysis of iron (III) fluoride and sodium metaphosphate was dependent on the phosphorus-to-iron ratio as well as the reaction temperature. The quality and size of the crystals depended on the current density and reaction time. Single crystals of FeP in the form of needles 1–2 mm in length were grown by the electrolysis of a melt containing a phosphorus-to-iron ratio of 11:1. The reaction temperature was 925°C and a current of 200 mA was used. Currents of less than 200 mA resulted in the growth of many small crystals. Excessive twinning was observed when currents of 400–600 mA were used. Previous investigators used a mixture of transition metal oxide and sodium metaphosphate to which sodium fluoride or carbonate was added in order to facilitate solution of the metal oxide. This method usually resulted in products of uncertain compositions that were also contaminated by transition metal oxide. In addition, the yields were low as a result of incomplete solution of the metal oxide in the phosphate melt, partial hydrolysis of the fluorides present, and incomplete reduction of the metal complexes present in the melt. However, the use of iron (III) fluoride is preferable since it is readily soluble in sodium metaphosphate. Slow heating under a flowing argon atmosphere permits the removal of moisture without the loss of appreciable fluoride. Low reduction potentials ($\sim$0.7 V) were sufficient to give the desired monophosphide. White phosphorus deposits, reported as a by-product in the earlier studies, were not observed to form in the electrolysis of the iron (III) fluoride–sodium metaphosphate melt.

In order to obtain sufficiently large crystals for electrical measurements, the products obtained from electrolysis were transported in silica tubes using iodine as the transport agent. The tubes for chemical transport were prepared in the manner previously described by Kershaw *et al.* (1967). The transport charge was a homogeneous powder of FeP which had been prepared electrolytically. All transport reactions were carried out in a two-zone furnace with the transport from the hot zone to the cool zone. Initially, the charge zone was set at 500°C and the growth zone at 800°C and the tube was heated for two days. This period of backtransport minimizes the number of nucleation sites. The temperature of the zones was then reset to 800°C for the charge zone and 540°C for the growth zone. The transport was carried out for periods of from one to two weeks. At the end of the transport period, the furnace was shut off and allowed to

TABLE VIII

X-RAY DATA FOR FeP

Sample	Unit cell dimensions (Å)		
	a	b	c
Electrolysis (single crystal)	5.19 ± .005	3.099 ± .005	5.79 ± .005
Electrolysis (powder)	5.188 ± .005	3.097 ± .004	5.789 ± .005
Chemical transport (single crystal)	5.196 ± .006	3.099 ± .005	5.792 ± .006
Chemical transport (powder)	5.189 ± .005	3.096 ± .004	5.786 ± .005
Rundqvist (1962)	5.19	3.099	5.79

cool to room temperature. The crystals were removed from the tube and washed with carbon tetrachloride and acetone to remove excess iodine.

The crystals of FeP were ground and dissolved in a mixture of hydrochloric and nitric acid (75:1). The solution was passed through a silver column and titrated with standardized ceric sulfate. Phosphorus was analyzed by reacting the phosphide with a quinquevalent vanadium sulfate solution and reoxidizing the quadrivalent vanadium produced with a standard solution of $KMnO_4$. Calculated analysis for FeP was: Fe, 64.33; P, 35.67; found: Fe, 64.17; P, 35.60.

The cell parameters for powder samples were determined with a Norelco diffractometer using monochromatic radiation (AMR-202 focusing monochromator) and a high-intensity copper source. The radiation used was λ(Cu $K\alpha$), 1.5405 Å. The values were confirmed by single-crystal analysis using precession techniques. The observed cell parameters are summarized in Table VIII.

Considerable detail has been given concerning the preparation of FeP by Bellavance *et al.* (1969). This was done because, first, the products obtained were of high purity, and second, it was desirable to show how the two techniques of electrolysis and chemical transport could be used together to eliminate the direct handling of elemental phosphorus.

Baker (1967) has also reported on the electrolytic preparation of monophosphides of titanium, manganese, and chromium from their respective oxides. The operating conditions established by Baker for repetitive, high-quality production of metal phosphides of the MP structure are presented in Table IX.

TABLE IX

CONDITIONS FOR METAL PHOSPHIDE DEPOSITION

Phosphide	T (°C)	Initial current density (A cm^{-2}) Anode	Initial current density (A cm^{-2}) Cathode	Cell voltage (V)	Electrolyte composition (wt%)
MnP	925	25–40	150–350	2.0–2.3	25 NaCl, 25 $Na_4P_2O_7$, 25 $NaPO_3$, 15 $Na_2B_4O_7$, 10 MnO_2
TiP	860	27–70	150–450	4.0–4.6	25 NaCl, 25 $Na_4P_2O_7$, 25 $NaPO_3$, 10 $Na_4B_2O_7$, 5 LiF, 10 TiO_2
CrP	1000	10–15	60–80	2.0–2.5	10 $Na_4P_2O_7$, 24 $NaPO_3$, 6 Cr_2O_3

An externally heated 3-in. i.d. by 6-in. deep graphite cell which served as the anode as well as a container for the electrolyte was used in these investigations, and a suspended, centrally positioned graphite rod served as the cathode. A closely fitting silicon carbide furnace liner minimized atmospheric oxidation of the crucible. Analysis of the crystals obtained is presented in Table X.

TABLE X

ANALYSIS OF METAL PHOSPHIDES

	Mn	P	Ti	Cr	C	O	Specific gravity[a]
MnP							
Actual	63.8	35.8	—	—	0.18	0.25	5.0 (27°C)
Theoretical	63.95	36.05	—	—	—	—	5.39 (21°C)
TiP							
Actual	—	38.4	58.9	—	0.4	0.44	3.90 (27°C)
Theoretical	—	39.3	60.7	—	—	—	3.95 (25°C)
CrP							
Actual	—	37.8	—	61.7	0.24	0.16	5.5 (27°C)
Theoretical	—	37.3	—	62.7	—	—	5.7 (18°C)

[a] Theoretical values from Handbook of Chemistry and Physics.

V. Transition Metal Arsenides and Antimonides

There has been relatively little work done on the preparation of the other transition metal pnyctides by electrolytic reduction of fused salts. Andrieux and Canaud (1944) have prepared two iron arsenides, FeAs and $FeAs_2$, by the electrolysis of melts consisting of iron (III) oxide or iron (III) chloride dissolved in alkali metal arsenates. They have also obtained compositions between FeAs and $FeAs_2$, depending upon the compositions of the melts and the conditions under which the electrolysis was carried out. These are summarized in Table XI.

The melts were contained in carbon crucibles which served as cathodes, and graphite rods acted as the anodes. After electrolysis, the melts were treated with dilute hydrochloric acid and the crystals that remained were treated with sodium peroxide. The arsenic was determined by iodometric analysis and the iron by gravimetric analysis (as Fe_2O_3). Unfortunately, x-ray diffraction studies were not done on the crystals.

Weiss (1946) obtained crystals of MoAs and WAs by the electrolysis of a melt of borax and sodium fluoride to which the appropriate required metal had been added. The conditions under which the electrolyses were carried out are summarized in Table XII. The product was separated from the melt and identification was made by chemical analysis. No identification using x-ray techniques was carried out on the samples. The

TABLE XI

ELECTROLYTIC CONDITIONS FOR THE PREPARATION OF IRON ARSENIDES

Composition of melt	T (°C)	Voltage (V)	Current (A)	Formula of product	% Fe	% As
$KAsO_3 + 2NaCl + \frac{1}{15}FeCl_3$	500	5–6	10	$FeAs_{2.24}$	24.7	74.5
$KAsO_3 + 2NaCl + \frac{1}{15}FeCl_3$	600	3–5	25	$FeAs_2$	27.6	72.0
$KAsO_3 + \frac{1}{3}NaCl + \frac{1}{10}FeCl_2$	550	14	20	$FeAs_2$	27.9	71.8
$\frac{1}{2}Na_4As_2O_7 + \frac{2}{3}NaCl + \frac{1}{6}FeCl_3$	700	8	20	$FeAs_2$	27.4	71.7
$\frac{1}{2}Na_4As_2O_7 + NaCl + \frac{1}{8}Fe_2O_3$	700	6	20	$FeAs_{1.64}$	31.05	68.4
$\frac{1}{2}Na_4As_2O_7 + 2NaCl + \frac{1}{6}Fe_2O_3$	700	5	25	$FeAs_{1.47}$	33.4	66.0
$\frac{1}{2}Na_4As_2O_2 + NaCl + \frac{1}{4}Fe_2O_3$	650	8	20	$FeAs_{1.34}$	35.9	63.9
$\frac{1}{2}Na_4As_2O_7 + 2NaCl + \frac{1}{2}Fe_2O_3$	700	7	25	$FeAs_{1.25}$	36.9	62.1
$\frac{1}{2}Na_4As_2O_7 + 2NaCl + \frac{1}{4}Fe_2O_3$	700	5	22	$FeAs_{1.13}$	39.4	60.2
$\frac{1}{2}Na_4As_2O_7 + 2NaCl + \frac{3}{4}Fe_2O_3$	700	6	20	FeAs	42.5	57.3
$\frac{1}{2}Na_4As_2O_7 + 2NaCl + Fe_2O_3$	800	6	20	FeAs	41.9	58.6

TABLE XII

ELECTROLYTIC PREPARATION OF MoAs AND WAs

Composition of melt	T (°C)	Voltage (V)	Current (A)	Time (min)	Ratio of MO_3/As_2O_3 in melt
$Na_2B_4O_7 + \frac{1}{2}NaF + \frac{1}{4.75}MoO_3 + \frac{2}{13}NaAsO_2$	800	7	15	20	2.73
$Na_2B_4O_7 + \frac{1}{2}NaF + \frac{1}{4.75}MoO_3 + \frac{1}{7}NaAsO_2$	800	7	15	19	2.95
$Na_2B_4O_7 + \frac{1}{2}NaF + \frac{1}{4.75}MoO_3 + \frac{2}{19}NaAsO_2$	800	7	15	18	3.37
$Na_2B_4O_7 + \frac{1}{2}NaF + \frac{1}{5.75}WO_3 + \frac{2}{13}NaAsO_3$	820–830	6	20	20	2.26
$Na_2B_4O_7 + \frac{1}{2}NaF + \frac{1}{4.6}WO_3 + \frac{2}{13}NaAsO_2$	820–830	6	20	22	2.82
$Na_2B_4O_7 + \frac{1}{2}NaF + \frac{1}{3.8}WO_3 + \frac{2}{13}NaAsO_3$	820–830	5	20	25	3.42

TABLE XIII

CHEMICAL ANALYSIS OF ELECTROLYTICALLY PREPARED MoAs

	I	II	% calculated for MoAs
As	43.2	43.5	43.86
Mo	55.5	55.6	56.14

results of the chemical analyses for MoAs and WAs are given in Tables XIII and XIV, respectively.

Weiss also prepared molybdenum and tungsten antimonides by the electrolysis of sodium tetraborate–sodium fluoride melts containing the transition metal oxide and antimony (III) oxide. The products varied in composition and this was a function of the original ratio of transition metal oxide to antimony oxide in the melt. The products were apparently also contaminated with free antimony. The technique of chemical transport appears to be a better way, in general, for preparing the transition metal antimonides.

TABLE XIV

CHEMICAL ANALYSIS OF ELECTROLYTICALLY PREPARED WAs

	I	II	Calculated for WAs
As	28.3	28.5	28.96
W	70.9	70.7	71.04

VI. Preparation of Transition Metal Carbides, Silicides, and Germanides

In general, it was shown by Andrieux and Weiss (1944) that it is more difficult to utilize the technique of electrolysis for the synthesis of carbides than for the silicides. In part, this is due to the relative ease of decomposition of many carbonates at elevated temperature. However, they reported that a mixture of lithium or sodium metaborate and lithium or sodium carbonate

TABLE XV

PREPARATION OF TRANSITION METAL CARBIDES

Composition of melt	T (°C)	Voltage (V)	Current (A)	Time (min)	Product
$2NaBO_2$, $3Na_2CO_3$, 4LiF, $\frac{1}{10}Fe_2O_3$	815	2	20	15	Fe_3C
$2NaBO_2$, $2Na_2CO_3$, 4.5LiF, $\frac{1}{6}WO_3$	780–800	2	20	30	WC
$2NaBO_2$, $2Na_2CO_3$, 4.5LiF, $\frac{1}{3}MoO_3$	780–800	2.5	20	20	Mo_2C

to which an alkali metal fluoride had been added makes an excellent melt for the synthesis of several carbides. Certain of these mixtures can be heated to 800°C without noticeable decomposition. The conditions for the preparation of several transition metal carbides are given in Table XV.

Considerably more work has been done on the synthesis of silicides by electrolysis of melts. Dodero (1934a) demonstrated that the electrolysis of a calcium silicate–calcium fluoride–calcium chloride melt resulted in the formation of $CaSi_2$. Dodero (1934b) also reported that the mixture of several grams of cerium oxide in the melt produced $CeSi_2$.

The electrolysis was carried out at a temperature of 1000°C for 2 hr with a current of 30 A. However, the product was contaminated with several per cent of free silicon. The preparation of $LaSi_2$ was carried out in a similar fashion. The use (1950) of a water-cooled aluminum cathode placed on the surface of the melt resulted in a product which was easier to isolate.

The electrolysis of melts containing Na_2SiF_6 or K_2SiF_6 and transition metal oxides or fluorides was also reported by Dodero (1939, 1952). He

TABLE XVI

PREPARATION OF TRANSITION METAL SILICIDES

Composition of melt	Cathode	Product obtained
$\frac{1}{10}TiO_2 + K_2SiF_6$	Iron	$TiSi_2$
$\frac{1}{5}Fe_2O_3 + Na_2SiF_6$	Iron	Fe_2Si
$\frac{1}{5}Na_2CrO_4 + Na_2SiF_6$	Iron	Cr_3Si
$\frac{1}{5}CrF_2 + Na_2SiF_6$	Iron	Cr_3Si
$\frac{1}{5}MnO_2 + K_2SiF_6$	Carbon	Mn_2Si
$\frac{1}{5}MnO_2 + K_2SiF_6$ or Na_2SiF_6	Iron (cooled electrode)	Mn_2Si

obtained the following silicides: Mn_2Si, Fe_2Si, Cr_3Si, and $TiSi_2$. The conditions under which the electrolysis was carried out are summarized in Table XVI. The current used was 20 A and the voltage was 6 V. The electrolysis was carried out for 20 min. The product formed at the cathode and was separated from the melt by crushing the boule and extracting the soluble salts with water.

Barbier-Andrieux (1955) has shown that it is possible to electrolyze melts consisting of sodium digermanate ($Na_2Ge_2O_5$), sodium fluoride, and either a transition metal oxide or fluoride. The following germanides were among those prepared: Ni_2Ge, Fe_2Ge, NiGe, $FeGe_2$, Mn_5Ge_3, Cr_3Ge, Mo_3Ge, CrGe, $\alpha MoGe_2$, Mo_2Ge_3, and Mo_3Ge_2. In addition, it was pointed out the technique of electrolysis is particularly useful for the preparation of the germanides that would be difficult to prepare by direct combination of the elements. This is particularly true for refractory metals such as Cr, Mo, and W. For several binary systems, products containing different ratios of transition metal to germanium were obtained by varying the ratio of transition metal to germanium in the original melt. Graphite crucibles were used to contain the melts and served as either the cathode or anode, depending upon the system. The other electrode used was usually a graphite rod. The phases prepared were characterized by both chemical analysis and x-ray diffraction.

VII. Transition Metal Borides

Andrieux (1929) published in 1929 his outstanding work on the preparation of borides by electrolysis of fused melts. By this method, which involved the electrolysis of a mixture of a borate and transition metal fluoride or alkaline earth fluoride and transition metal oxide, he was able to synthesize 17 borides, 10 of which were new. They had the following compositions: CaB_6, SrB_6, BaB_6, CeB_6, LaB_6, NdB_6, GdB_6, YB_6, ErB_6, YbB_6, ThB_6, Zr_3B_4, TiB_2, VB_2, Cr_3B_2, UB_4, and MnB.

Andrieux (1929c) was also able to prepare the diborides of tantalum and niobium, TaB_2 and NbB_2, by the electrolysis at about 950°C of a melt containing lithium tetraborate and lithium fluoride to which has been added either Ta_2O_5 or Nb_2O_5. A current of 24–25 A was passed through the melt for a period of 2 hr. The crucible and contents were then allowed to cool, and were crushed and treated with hydrochloric acid. Impurities of boron and calcium boride were attacked with nitric acid, leaving small crystals of either tantalum or niobium diboride. The products were characterized by chemical analysis.

In addition, Andrieux (1931, 1932) prepared the two cubic borides, ThB_6 and CeB_6, as well as the mixed borides. For these preparations, a carbon crucible acted as the cathode and a carbon rod as the anode. The electrolytic melt consisted of boron oxide, magnesium oxide, and magnesium fluoride, to which thorium and/or cerium oxides were added. The electrolysis was carried out at 1150°C and a current of 25 A was used for 2 hr. The ratio of thorium to cerium in the products was directly proportional to the corresponding ratio of the oxides in the melt.

Andrieux and Weiss (1948) also reported the preparation of MoB and WB. For molybdenum boride, the melt was kept at a temperature of 1000°C, and the current and voltage were 20 A and 5 V, respectively. MoB was the product of the melt containing less molybdenum oxide than represented by

$$Na_2B_4O_7 + NaF + \tfrac{1}{8}MoO_3$$

WB was also prepared from melts having the composition

$$Na_2B_4O_7 + NaF + \tfrac{1}{9}WO_3$$

Two uranium borides, UB_4 and UB_{12}, were also prepared by Andrieux and Blum (1949).

The best yield for the tetraboride was obtained under the conditions shown in Table XVII.

When the concentration of U_3O_6 in the melt was reduced from 1/20 to 1/40, a considerable quantity of UB_{12} was formed. The UB_4 impurity was removed by boiling either hydrochloric or hydrofluoric acid, since UB_{12} is not attacked by either of these acids.

Andrieux and Marion (1953) were able to prepare a number of chromium borides, including Cr_3B_4, CrB, Cr_5B_3, Cr_2B, and Cr_4B. These were obtained by varying the quantity of Cr_2O_3 in melts containing CaB_4O_7, B_2O_3, and CaF_2. Crystals of the products were examined by both x-ray and chemical analysis.

Marion (1957) was also able to prepare several nickel borides by elec-

TABLE XVII

CONDITIONS FOR ELECTROLYTIC PREPARATION OF UB_4

Composition of melt	T (°C)	Voltage (V)	Current (A)	Time (hr)	Cathode	Anode
$MgB_4O_7 + MgF_2 + \tfrac{1}{20}U_3O_8$	1100	12	23	1½	Carbon	Carbon

trolysis. Nickel oxide was dissolved in a melt containing potassium fluoroboride and potassium fluoride. The electrolysis was carried out at 900°C, and by varying the concentration of nickel oxide in the melt, NiB and NiB_2 were obtained. In addition, a third boride was prepared which had the same crystallographic parameters previously reported by Blok (1957) for Ni_3B.

Aléonard (1958) was able to extend the method developed by Marion and succeeded in preparing several borides of iron and cobalt. They included CoB and Co_2B, as well as FeB and Fe_2B. The products were identified by x-ray analysis. Aléonard (1960) also prepared borides of molybdenum and tungsten. The different borides could be prepared by varying the amount of molybdenum or tungsten oxide dissolved in the flux. It was observed that if the atomic ratio of Mo/B and W/B in the melt was less than 3–5%, the boride M_2B_5 was formed, but with some excess boron. If the ratios are 4–14% for Mo and 5–20% for W, the borides obtained were M_2B_5 and MB. Finally, if the ratio of W/B was greater than 20% and Mo/B greater than 14%, the free metal itself was formed.

The conditions used by Aléonard for the synthesis of borides can be summarized by describing the conditions for the preparation of the manganese borides (Aléonard, 1961). The melt consisted of 50 g of potassium fluoborate with varying amounts of potassium fluoride and Mn_3O_4. The electrolysis was carried out in a graphite crucible at 950°C under an argon atmosphere. The crucible served as the anode and a molybdenum rod attached to a graphite electrode served as the cathode. The duration of the electrolysis was 1 hr, after which the product was separated by the usual procedures. Two borides, MnB and Mn_3B_4, were magnetic and could be identified by x-ray analysis. In addition, two new borides, MnB_{12} (or MnB_{10}) and MnB_4, were prepared. The latter compounds were not characterized fully.

It should be noted that whereas electrolysis of melts is still the best method for the preparation of single crystals of many of the borides, there is considerable question as to the purity of the products. Almost all of the preparations utilized graphite crucibles and electrodes, and it is well known that carbon can substitute for boron to form borocarbides.

VIII. Summary

The technique of fused salt electrolysis has been applied successfully to the preparation of many transition metal compounds. Several of these compounds can be prepared as single crystals only by this technique. In addition, a variety of new transition metal compounds have been synthesized by this technique. However, fused salt electrolysis has not been widely used as a preparative technique, nor has there been any systematic investigation of all the variables involved in this method. Some of the areas which need further development are discussed in this section.

In general, little is known concerning the thermodynamics of fused salt electrolysis. Although tables of reduction potentials at room temperature exist for many ions, there has been no investigation of reduction potentials at elevated temperatures. Furthermore, there is little understanding of the nature of the reacting species or complexes in the melt at the elevated temperatures required for electrolysis.

One of the most important experimental factors, the composition of the fused salt flux, has not been well investigated. Almost all investigators have based their flux compositions on those used by Andrieux and Chêne (1938a). The mixture metal oxide–alkali borate (and/or phosphate, sulfite, etc.)–alkali halide (or carbonate) is the most common flux composition. Alkali borates have been used in the melt composition even when the desired product was an oxide (Andrieux and Bozon, 1949a, 1950), sulfide (Heinen *et al.*, 1968; Weiss, 1946), or phosphide (Baker, 1967). Recently, Bellavance *et al.* (1969) used a simpler flux composed of a transition metal fluoride and sodium metaphosphate for the preparation of some transition metal phosphides. Rogers *et al.* (1966) studied the effect of the acidity of the electrolytic flux on the reaction products. By varying the acidity of the melt, they were able to prepare a unique mixed-spinel system. However, in general, there have been no other investigations of the relationship between the acidity of the electrolytic flux and the reaction product and very little innovation in the composition of the electrolytic flux.

In most systems, the problem of material for the crucible and electrodes has been resolved. The crucible must be able to withstand high temperatures, be impermeable to the melt constituents, and remain inert to reaction with both the melt constituents and the product. Crucibles of recrystallized alumina have been used in the preparation of the oxide compounds. Platinum and/or carbon rods have served as electrodes. Carbon crucibles and electrodes are preferred in the preparation of the sulfides, carbides, phosphides, arsenides, silicides, and germanides. However, no suitable crucible material has been found for the preparation of the borides. Carbon crucibles have been used previously, but it has been observed that carbon will enter into the reaction to yield borocarbides.

Both Rogers *et al.* (1966) and Bellavance *et al.* (1969) have observed that the quality of the crystals in the product is dependent upon the applied current. In both studies, the optimum current densities were determined empirically. No other investigators, however, have attempted to optimize conditions for crystal growth. There is also little understanding of the factors governing crystal growth for this technique.

REFERENCES

Aléonard, S. (1958). *Bull. Soc. Chim. Fr.* **1958,** 827.

Aléonard, S. (1960). *Bull. Soc. Chim. Fr.* **1960,** 653

Aléonard, S. (1961). *C. R. Acad. Sci. Paris* **253,** 1213.

Andrieux, J.-L. (1948). *Rev. Met.* **45,** 49

Andrieux, J.-L., and Blum, P. (1949). *C. R. Acad. Sci. Paris* **229,** 210.

Andrieux, J.-L., and Bozon, H. (1949). *C. R. Acad. Sci. Paris* **228,** 565.

Andrieux, J.-L., and Bozon, H. (1950). *C. R. Acad. Sci. Paris* **230,** 952.

Andrieux, J.-L., and Canaud, A. (1944). *C. R. Acad. Sci. Paris* **218,** 711.

Andrieux, J.-L., and Chêne, M. (1938a). *C. R. Acad. Sci. Paris* **206,** 661.

Andrieux, J.-L., and Chêne, M. (1938b). *C. R. Acad. Sci. Paris* **206,** 1562.

Andrieux, J.-L., and Chêne, M. (1939). *C. R. Acad. Sci. Paris* **209,** 672.

Andrieux, J.-L., and Marion, S. (1953). *C. R. Acad. Sci. Paris* **236,** 805.

Andrieux, J.-L., and Weiss, G. (1944). *C. R. Acad. Sci. Paris* **219,** 550

Andrieux, J.-L., and Weiss, G. (1948). *Bull. Soc. Chim. Fr.* **1948,** 598.

Andrieux, L. (1929a). Thesis, Paris.

Andrieux, L. (1929b). *Ann. de Chim.* **12**(10), 423.

Andrieux, L. (1929c). *C. R. Acad. Sci. Paris* **189,** 1279

Andrieux, L. (1931). Séance du 21 Décembre.

Andrieux, L. (1932). *C. R. Acad. Sci. Paris* **194,** 720.

Aronsson, B., Lundstrom, T., and Rundqvist, S. (1965). "Borides, Silicides and Phosphides." Wiley, New York.

Baker, Jr., D. H. (1967). *Trans. Met. Soc. A.I.M.E.* **239,** 755.

Banks, E., and Wold, A. (1968). "Preparative Inorganic Reactions," Vol. 4, p. 237. Wiley (Interscience), New York.

Barbier-Andrieux, M.-J. (1955). *Ann. de Chim.* **10**(12), 754.

Bellavance, D., Vlasse, M., Morris, B., and Wold, A. (1969). *J. Solid State Chem.* **1,** 82.

Bertaut, E. F., and Blum, P. (1956a). *J. Phys. Radium* **17,** 175.

Bertaut, E. F., and Blum, P. (1956b). *Acta Crystallogr.* **9,** 121.

Blaugher, R. D., Hulm, J. K., and Yocom, P. N. (1965). *J. Phys. Chem. Solids* **26,** 2037.

Blok, N. I., Korlova, M. N., Lashko, N. F., and Shpunt, K. J. A. (1957). *Dokl. Akad. Nauk S.S.S.R.* **113,** N04, 811.

Bouchard, Jr., G. H., Perlstein, J., and Sienko, M. J. (1967). *Inorg. Chem.* **6,** 1682.

Bright, N. F. H., Rowland, J. F., and Wurm, J. G. (1958). *Can. J. Chem.* **36,** 492.

Chêne, M. (1938). *C. R. Acad. Sci. Paris* **207,** 571.

Chêne, M. (1939). *C. R. Acad. Sci. Paris* **208,** 1144.

Chêne, M. (1941). *Ann. Chim.* **15,** 266.

Conroy, L. E., and Sienko, M. J. (1957). *J. Amer. Chem. Soc.* **79,** 4048.

Didchenko, R., and Litz, L. M. (1962). *J. Electrochem. Soc.* **109**(3), 247.

Dodero, M. (1934a). *C. R. Acad. Sci. Paris* **198,** 1593.

Dodero, M. (1934b). *C. R. Acad. Sci. Paris* **199,** 566.

Dodero, M. (1939). *C. R. Acad. Sci. Paris* **208,** 799.

Dodero, M. (1950). *Bull. Soc. Chim. Fr.* **1950,** 545.

Dodero, M. (1952a). *J. Chim. Phys.* **49,** C11.
Dodero, M. (1952b). *J. Chim. Phys.* **49,** C210.
Dodero, M., and Déportes, C. (1956). *C. R. Acad. Sci.* **242,** 2939.
Graham, J., and Wadsley, A. D. (1966). *Acta Crystallogr.* **20,** 93.
Handbook of Chemistry and Physics, 45th ed., The Chemical Rubber Co., Cleveland, Ohio.
Hartmann, H., and Orban, J. (1936). *Z. Anorg. Allg. Chem.* **226,** 257.
Hartmann, H., and Conrad, U. (1937). *Z. Anorg. Allg. Chem.* **233,** 313.
Hartmann, H., Ebert, F., and Bretschneider, O. (1931). *Z. Anorg. Allg. Chem.* **198,** 116.
Heinen, H. J., Baker, Jr., D. H., and Gomes, J. M. (1968, March 26). U. S. Patent 3,375,180.
Hsu, S. S., Yocom, P. N., and Cheng, T. C. C. (1955). Interim Rep., Contract N6ori-071(50) Project NR 052-341, December 1953–July 1955.
Hsu, S. S., Yocom, P. N., Cheng, T. C. C., Oldham, K. B., Myers, C. E., Gingerich, K., Travaglini, C. H., Bailar, Jr., J. C., Laitinen, H. A., and Swann, Jr., S. (1958). Final Rep., Contract N6ori-071(50), Project NR 052-341, July 1953–December 1958.
Kershaw, R., Vlasse, M., and Wold, A. (1967). *Inorg. Chem.* **6,** 1599.
Kunnmann, W., and Ferretti, A. (1964). *Rev. Sci. Instrum.* **35,** 465.
Marion, S. (1957). *Bull. Soc. Chim. Fr.* **1957,** 522.
Ostertag, W. (1966). *Inorg. Chem.* **5**(5), 758.
Perloff, D. S., and Wold, A. (1967). *Crystal Growth Suppl. J. Phys. Chem. Solids* **28,** 361 *Proc. Int. Conf. Crystal. Growth, Boston, June, 1966.*
Robins, R. G. (1960). *J. Nucl. Mater.* **2**(2), 189.
Robins, R. G. (1961). *J. Nucl. Mater.* **3**(3), 294.
Rogers, D. B., Ferretti, A., and Kunnmann, W. (1966). *J. Phys. Chem. Solids* **27,** 1445.
Rundqvist, S. (1962). *Acta Chem. Scand.* **16,** 287.
Schönberg, N. (1954). *Acta Chem. Scand.* **8,** 226.
Sienko, M. J., and Truong, T. B. N. (1961). *J. Amer. Chem. Soc.* **83,** 3939.
Stephenson, N. C., and Wadsley, A. D. (1965). *Acta Crystallogr.* **18,** 241.
Weiss, G. (1946). *Ann. Chim.* **1**(12), 446.
Wold, A., Kunnmann, W., Arnott, R. J., and Ferretti, A. (1964). *Inorg. Chem.* **3,** 545.
Wood, D. W. (1953). Ph.D. Thesis, Univ. of Illinois, Urbana, Illinois.
Yocom, P. N. (1958). *Dissertation Abstracts* **18,** 1974. (1961). U. S. Dept. of Commerce, Office of Tech. Services P. B. 147,079.

Preparation of Single Crystals of III–V Compounds

MICHEL DUGUE, JEAN FRANÇOIS GOULLIN, PIERRE MERENDA, and MICHEL MOULIN

Thomson-CSF Laboratories, Corbeville, France

I. Physical Properties and Applications

Before analyzing the various aspects of the preparation of III–V compounds and discussing the growth of the most important ones, it is necessary to recall the history of these compounds and to sum up their physical properties and their applications.

It was Welker who demonstrated in 1952 that the III–V compounds have particularly interesting properties and that they are nearly all semiconductors. He was able to predict important new characteristics of these semiconductors and prove them experimentally with InSb. After the first exploratory research, nearly all of the work was limited to InSb, partly because it was the easiest compound to grow as relatively pure single crystals.

The very high electron mobility and the ionic component in the crystal bonding were the particular features that attracted interest. From a technical point of view, the interesting properties were a high mobility (for Hall devices) and an energy gap larger than that of silicon (for devices working at high temperatures).

However, very soon, the particular properties of the band structure caused a marked change in this situation. The invention of the semiconductor laser and the discovery of the Gunn effect shifted interest to GaAs, GaP, and compounds or alloys which can emit visible or microwave radiation.

In this section, only the most important features of the III–V compounds are presented. Further information can be found in several books (Hilsum and Rose-Innes, 1961; Madelung, 1964; Willardson and Beer, 1966, 1967, 1968) and review articles. Only compounds of the elements of groups IIIb and Vb are considered.

A. Crystal Structure

The phosphides, arsenides, and antimonides of boron, aluminum, gallium, and indium have the zincblende structure and the nitrides of these elements have the wurtzite lattice, except for boron. The remaining known compounds (InBi, TlSb, TlBi) are metallic. We note here the importance of the absence of an inversion center in the zincblende structure, which means that the [111] axis is a polar axis in this structure. This fact leads to the special properties of the III–V compounds.

B. Band Structure

The conduction band consists of a number of subbands which have minima at $k = 0$ and along the [111] and [100] axes. It is difficult to predict which of these minima will be the lowest. Kane's theory suggests that the band is not parabolic, but is nearly isotropic if the minimum of the lowest subband is at $k = 0$ and if the other subbands lie high enough so that they are not occupied by electrons. The valence band consists of three subbands: two degenerate at $k = 0$, and the other lying lower. The band maximum is not exactly at $k = 0$, but is nearly along the [111] axis. If the lowest minimum of the conduction band is along the [111] axis, the gap is direct, and if it is along the [100] axis, the gap is indirect. The width of this gap decreases with increasing atomic weight for the elements of the fourth column of the periodic chart.

C. Transport Properties

Those III–V compounds that have a direct gap have a low electron effective mass and a much higher electron mobility than those of fourth-column elements of the periodic chart. The mobility of these compounds is limited by polar scattering owing to their appreciable ionicity. For electrons and holes in compounds with an indirect gap, the effective mass is much larger and acoustic scattering, which leads to a mobility proportional to $(m^*)^{-5/2}$, becomes relatively more important than polar scattering, which gives a $(m^*)^{-3/2}$ mobility dependence.

In contrast to the acoustic scattering, no relaxation time can be defined for the polar scattering. Roughly speaking, the electron mobility increases with increasing atomic weight.

D. Impurities and Defects

In semiconductors of the fourth group of the periodic chart, foreign atoms from group III elements are built into the lattice substitutionally as acceptors, and foreign atoms from group V elements as donors. If this picture is extended to the III–V compounds, one suspects that the most important acceptors will be foreign group II atoms, which can be located substitutionally in the sublattice of the trivalent atoms. Similarly, the main donors will be atoms of group VI in the sublattice of the pentavalent atoms. The situation is often more complex. The ionization energies of these donors or acceptors are not always small compared to the width of the gap and such foreign atoms play a more important role as traps for electrons and holes and as recombination centers.

Atoms of group IV may be donors if they are built into the anion sublattice, or acceptors if they are in the cation sublattice. Therefore, they often show an amphoteric behavior which is very useful in the realization of p–n junctions.

It is rather well established that the crystals of the most important III–V compounds have a low deviation from stoichiometry. On the other hand, this apparently complete stoichiometry does not exclude the possibility of lattice defects. In these compounds, donors, acceptors pairs, vacancies, or more complex point defects are found. For instance, the possibility of such defects explains some changes of the physical properties of III–V compounds after heat treatment, annealing and quenching. These complex point defects act generally as traps or recombination centers.

E. Mixed Crystals; Ternary Compounds

Qualitative criteria for mixed crystal formation is that the atomic size and the polarization of both components must not be too different. The III–V mixed crystal systems which have been investigated rather intensively are $(In_xGa_{1-x})P$, $In(As_{1-x}P_x)$, $(In_{1-x}Ga_x)As$, $Ga(As_{1-x}P_x)$, $(In_{1-x}Ga_x)Sb$, $(Ga_{1-x}Al_x)Sb$, and $(Ga_{1-x}Al_x)As$. In each of these cases the lattice parameter is related linearly to the composition x. This is not always true for the band gap. If the components do not have the same band structure, i.e., one has a direct gap and the other an indirect gap, there is a change of slope in the variation of gap with composition: this is true with $Ga(As_xP_{1-x})$. On one side the gap is direct (GaAs), and on the other side it is indirect (GaP). For $In(As_{1-x})P_x$, in which the components have the same structure, the variation of gap versus composition is nearly linear.

For the other compounds, the situation is often more complicated. The mobility of the charge carriers is not very reduced in mixed crystals by the statistical distribution of the atoms of the components. The situation is different for the movement of phonons. Consequently, the thermal conductivity is lower in mixed crystals than in the components.

F. Applications

Most of the applications of the III–V compounds are now based on the particular properties of their band structure. Germanium and silicon research is more technically advanced, and these materials are used for diodes, transistors, and integrated circuits.

InSb, InAs, and now GaAs, for which the electronic mobility is very high, are used in the manufacture of Hall or magnetoresistance devices, mixers, choppers, and switches for brushless motors. GaAs, which has a very much higher gap than InSb and InAs, is the best material available for high temperatures.

Another application is the development of infrared or visible light emitters based on radiative recombination either between the conduction and the valence bands for direct-energy-gap compounds, or between one band and impurity levels for indirect-energy-gap compounds. The radiative efficiency is the highest with direct-energy-gap compounds. The main compounds used in this field are: GaAs, in infrared ($\lambda = 1\ \mu m$), in coherent (laser), or incoherent light; and GaP, in visible (red or green, according to the impurity levels introduced in the material), always in incoherent light. Ternary compounds (for instance, $Al_xGa_{1-x}As$ or $GaAs_xP_{1-x}$) allow the range of emitted radiation to be increased, and the production of visible radiation (red) with theoretical efficiency much greater than that possible with GaP, which is at present limited by technological problems.

A few watts of monochromatic light can at present be obtained with GaAs diode lasers at liquid nitrogen temperature with an efficiency of 30%. In incoherent light, the power is 500 mW at room temperature with an efficiency of 20%. With a GaP diode, the efficiency is only 6% for green and 3–7% for red, and the power is of the order of some milliwatts at room temperature. Studies on solid solutions are in constant progress, for instance, Ga–As–P gives 1% in yellow and Ga–Al–As 1–2% in red at room temperature.

For the metallurgist, several problems appear in the preparation of electroluminescent materials. First, to obtain good device efficiency, it is necessary to avoid impurities or point defects acting as traps or recombination centers. Materials prepared at low temperature by vapor-phase epitaxy and above all by liquid-phase epitaxy are always more suitable than bulk materials which have been prepared at higher temperature. Injection across forward-biased *p–n* junctions is the easiest way of exciting luminescence and the only one suitable from a technical point of view. Therefore it is necessary for the metallurgist to obtain *n*-type materials and *p*-type materials. Since the concentration of the impurities is high enough ($10^{17}\ cm^{-3}$ to $5 \times 10^{18}\ cm^{-3}$), one must avoid precipitation and segregation. At present, diode production is sometimes carried out during epitaxy—in particular, by using the amphoteric behavior of a group IV element (for instance, Si in GaAs).

A third important application concerns GaAs. The discovery of the Gunn effect in 1964 increased interest in GaAs for the production of microwave devices. The current versus electric field curve has a negative slope for an electric field larger than $3\ kV\ cm^{-1}$ if the electron concentration is smaller than $10^{16}\ cm^{-3}$ for good mobility *n*-type GaAs. Roughly speaking, this effect is due to the transfer of electrons from the primary minima of the conduction band, where the effective mass is small, to the secondary minima in this band, where the effective mass is large. This means that the number of fast electrons decreases, and so the current decreases when the electric field increases, a fact leading to instabilities. GaAs is the only III–V compound showing this effect at room temperature.

Studies concerning ternary compounds such as $In_xGa_{1-x}Sb$ or $InAs_xP_{1-x}$ have been carried out.

Gallium arsenide devices employing the Gunn effect produce an X-band power ranging from 1 mW to a few watts with efficiency reaching 5%. The metallurgical problem is rather difficult: the active layer of GaAs used for Gunn effect is deposited by epitaxy on low-resistivity substrates, their thickness varying from 1 to 100 μm. Since the mobility must be as high as possible to increase the efficiency, it is necessary to eliminate traps and compensation. Impurity profiles must be very well controlled to avoid resistivity variations, which disturb the field distribution. Moreover, a poor interface between layer and substrate causes a decrease in efficiency. The suitable concentration range extends from 5×10^{15} to 10^{14} cm^{-3}. Gallium arsenide is also used to make other microwave devices: Schottky mixers, varactors, and field-effect transistors.

The properties of III–V compounds are listed in Table I.

TABLE I

Compound	Lattice constants (Å)	Density g cm^{-3}	Melting point (°C)	Energy gap at room temperature (eV)	Mobility cm^2 V^{-1} sec^{-1} μ_e	μ_h
BN	3.515	2.25	—	—	—	—
BP	4.537	2.89	3000	—	—	—
BAs	4.777	—	—	—	—	—
AlN	3.111, 4.978	3.26	—	6 (direct)	—	—
AlP	5.4625	2.85	1900	2.4 (indirect)	—	—
AlAs	5.662	3.79	1600	2.10 (indirect)	—	—
AlSb	6.135	4.218	1065	1.62 (indirect)	200	550
GaN	3.189, 5.185	6.1	1500	3.39 (direct)	—	—
GaP	5.4505	4.13	1467	2.26 (indirect)	300	100
GaAs	5.6532	5.316	1238	1.38 (direct)	9300	400
GaSb	6.0954	5.619	706	0.7 (direct)	4000	1400
InN	3.53, 5.69	6.88	—	1.98 (direct)	—	—
InP	5.8687	4.787	1062	1.29 (direct)	4600	150
InAs	6.0584	5.66	942	0.35 (direct)	33,000	460
InSb	6.4788	5.775	530	0.17 (direct)	78,000	750

II. Preparation of Large Single Crystals

A. Vapor Phase

The first single crystals of III–V compounds were obtained by vapor-phase techniques (evaporation, deposition, or chemical transport). The relatively small size of crystals grown in this manner still permitted the study of physical, chemical, and electronic properties of most III–V compounds.

For the preparation of large single crystals of III–V compounds, vapor-phase techniques are now seldom used, and nothing more will be said here except to note that chemical vapor-phase transport has become the most important method in the case of thin-layer growth and will be considered in Section III.

B. Growth from a Stoichiometric Melt

1. *The Czochralski method*

a. Introduction. The Czochralski (1917) method is a process for single-crystal pulling starting with a small, oriented seed. The seed is put in contact with the melt, which is heated to a temperature slightly higher than the melting point of the material to be crystallized. The heat loss by conduction is determined by the thermal conductivities of the liquid and solid, and the temperature gradient through both phases. As growth temperatures increase, losses by radiation become more and more important. These losses are the cause of the difference ΔT between the freezing point and the melting temperature. The mass growth rate (hence diameter) is controlled by two parameters, namely, ΔT and the pulling speed (linear growth rate).

The crucible and the seed holder can be rotated in reverse directions in order to provide a better homogeneity of liquid concentration and temperature. The pulling apparatus is rather simple and consists of a crucible, a heating system (radiofrequency coil or resistor), a mechanical setup for pulling and rotating the seed holder, and a mechanical system for performing these actions on the crucible. The crucible and the seed are in a chamber under an inert-gas atmosphere.

b. Pulling under pressure. The Czochralski process has been intensively developed and used for germanium and silicon pulling. In the case of III–V compounds, the saturation vapor pressure over the melt can be high (1–50 atm for As and P compounds). Control of this pressure is then

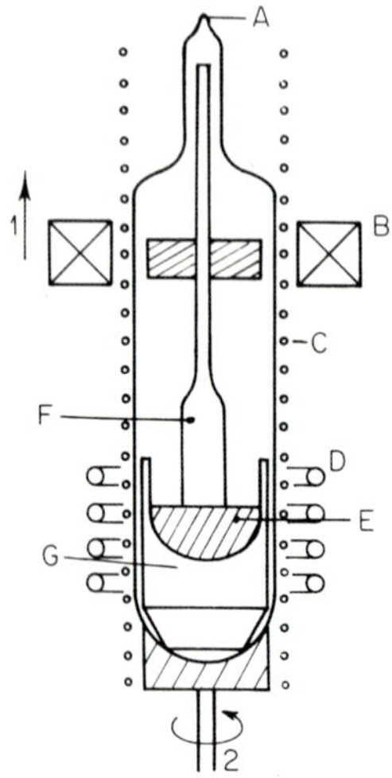

FIG. 1. Controlled-vapor-pressure pulling. A, Sealed ampoule; B, magnetic seed holder; C, vapor-pressure temperature control; D, rf coils; E, melt; F, crystal; G, crucible; (1) pulling direction; (2) rotation.

needed during the entire growth. The Czochralski method as applied to III–V compounds was developed by Gremmelmaier and Madelung (1953). The main difference from the classical method is the use of a tight chamber, the walls of which are temperature-regulated to control the vapor pressure of the volatile component over the melt (see Fig. 1).

c. Liquid encapsulation. A very interesting improvement of the pulling technique was first suggested by Mullin *et al.* (1965) and led to appreciable simplification for the growth of compounds with high-vapor-pressure dissociation. At the surface of the melt, a substance of a lower density is melted. This substance is chosen to be gas-tight and chemically inactive, and an inert-gas atmosphere is used to prevent the loss of the volatile component of the material under preparation, which is thus encapsulated. The pressure of inert gas must obviously be equal to or slightly greater than the saturation pressure over the melt. A typical apparatus is shown in Fig. 2.

A big problem, of course, is to find the correct substance for encapsulation, one having suitable values for the following parameters: density, melting point (lower than the material melting point), chemical reactivity (versus material and crucible), and wetting. If the encapsulant wets the crucible and the solid crystal, the loss of volatile species is minimum, and evaporation from the solid is also prevented. To date, B_2O_3 is the only substance giving satisfactory results in III–V compound preparation.

d. Gallium arsenide pulling. Pulling is carried out from a stoichiometric melt at about 1260°C, corresponding to a dissociation pressure of 0.9 atm.

A good control of this pressure is obtained by the use of a vacuum-sealed quartz chamber whose walls are maintained at a temperature of 680–700°C (Gremmelmaier and Madelung, 1953; Gremmelmaier, 1956).

The required quantity of arsenic comes from a solid arsenic "reservoir" or from the melt; in the latter case, the melt is previously made somewhat rich in arsenic. Seed rotation and pulling can be done by magnetic coupling. Several different systems have been studied by Moody and Kolm (1958) and Willardson and Goering (1962).

The crystalline quality of crystals grown in this way is strongly dependent on the quality of the mechanical moving systems, which must be completely free of vibrations.

The pulling speed of GaAs single crystals varies with the growth orientation, the required diameter of ingot (10–25 mm), and the desired grade of homogeneity in the case of doping. Pulling rates generally range between 1 and 4 cm hr^{-1}. Seed rotation speeds are between 10 and 20 rpm.

Growth can occur on orientations (110) (Ellis, 1959) and (111) and $(\overline{111})$ (Moody *et al.*, 1960; Steineman and Zimmerli, 1963), the latter being preferable for the growth of dislocation-free single crystals.

e. Liquid encapsulation growth of gallium phosphide. Gallium phosphide melts congruently at 1467°C under a dissociation pressure of 35 atm; the encapsulation technique is thus of particular interest (Bass *et al.*, 1968; Mullin *et al.*, 1968). A water-cooled, gas-tight crystallization chamber is used; it is made of stainless steel and can bear pressures as high as 200

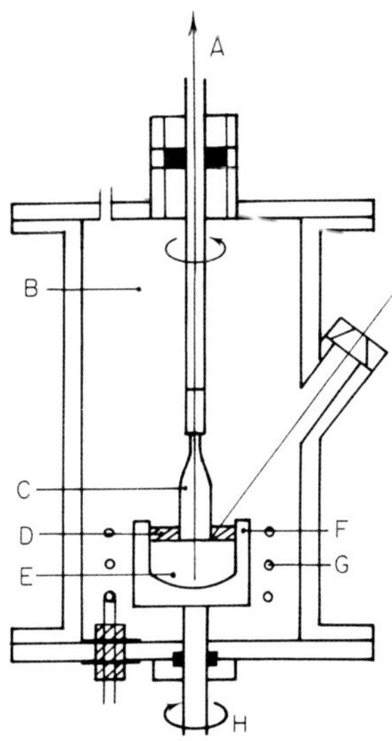

FIG. 2. Liquid encapsulation pulling. A, Pulling rod; B, inert-gas pressure; C, crystal; D, B_2O_3; E, melt; F, crucible; G, rf coil; H, crucible rotation.

atm. It is necessary to put the heating element inside the chamber and to use leak-tight current connections. Furthermore, the effect of pressure on the crystal-holder rotating gasket must be balanced by a reverse pressure to preserve the quality of pulling.

A typical operation would run as follows:

1. A load of gallium phosphide and boric oxide is put in a graphite (or vitreous carbon) crucible.
2. The chamber is evacuated and refilled with argon or helium under a pressure of 50 atm.
3. Heating is started at 300°C, boric oxide melts, and encapsulation is achieved before melting of GaP.
4. Seed is set down and dipped in the melted gallium phosphide through the B_2O_3.
5. Then, the crystal is pulled up in the classical way.

The whole operation can be made automatic and controlled and monitored through a closed-circuit television set.

Single crystals of lengths between 3 and 7 cm and diameters between a few and 20 mm have been grown with a growth speed of about 2 cm hr^{-1}. The boron concentration of these crystals is greater than 200 ppm due to the contamination by the encapsulant. Nonetheless, the liquid encapsulation pulling technique is now becoming the major industrial process for the preparation of single-crystal GaP.

InAs and InP have also been grown by the liquid encapsulation technique (Mullin *et al.*, 1968).

2. *The Bridgman–Stockbarger techniques*

a. Introduction. These techniques are also based on the solidification of a stoichiometric melt. An oriented solidification of the melt is obtained by using either a relative displacement of the melt and a thermal gradient at

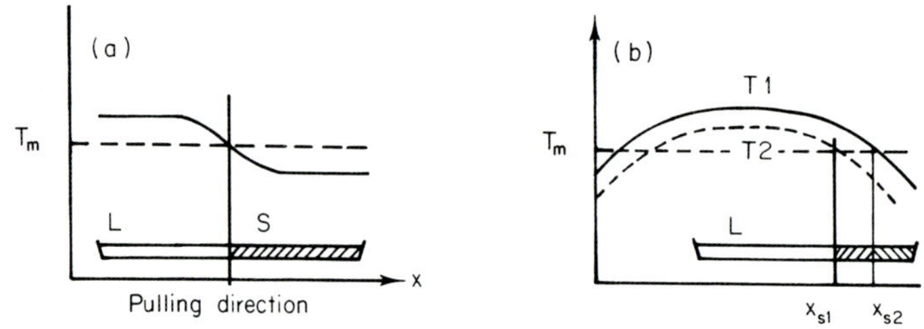

FIG. 3. The (a) Stockbarger and (b) Bridgman techniques. L, Liquid; S, solid.

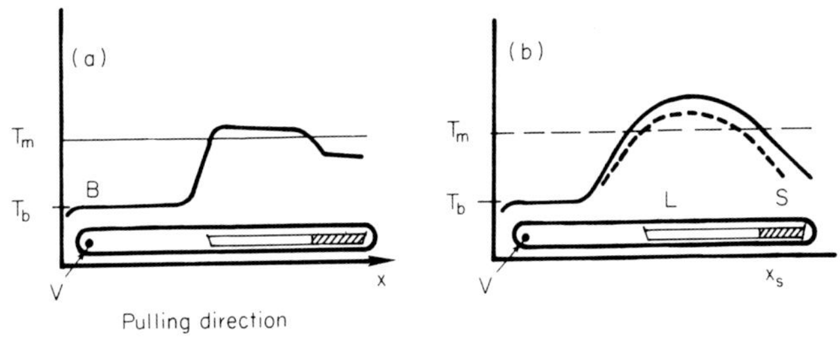

FIG. 4. Horizontal (a) Stockbarger and (b) Bridgman techniques. B, Boiler; V, volatile element; L, liquid; S, solid.

a controlled speed (Stockbarger, 1936) (see Fig. 3a), or a programmed cooling along a thermal gradient over a static melt (Bridgman, 1925) (see Fig. 3b). Both techniques are used with a vertical tube or a horizontal boat, and crystallization can be started from a seed. If there is no seed, the shape of the tube or boat end is an important parameter in starting a single crystal.

Horizontal systems, either Bridgman- or Stockbarger-type, have been used for the preparation of III–V compounds (Weisberg *et al.*, 1959). As in the Czochralski pulling method, dissociation pressure control is necessary: An extremity of the ampoule contains an excess of the volatile constituent and during the whole operation stays in a cold zone of the furnace with a separate thermal regulation which controls the partial vapor pressure (Fig. 4).

b. Horizontal zone melting. This technique is rather similar to the Stockbarger technique, but the thermal profile is such that only a small part of the total load is melted (see Fig. 5). This is of interest in the case of doping.

If a volatile dopant is used the small size of the melted zone minimizes

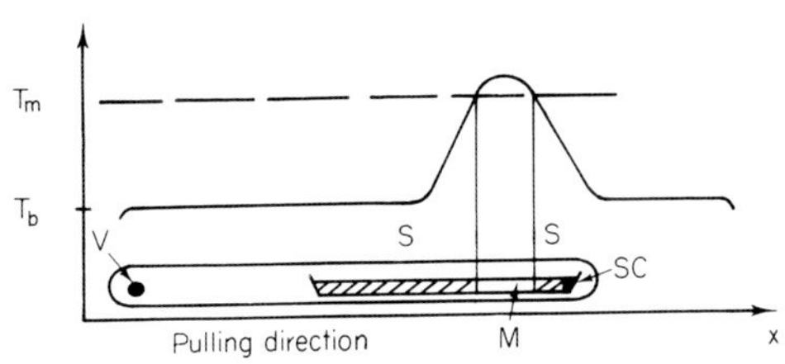

FIG. 5. Horizontal zone melting. S, Solid; V, volatile element; M, molten zone; SC, single-crystal seed.

the loss of dopant through the vapor phase and, moreover, the loss is kept constant during the whole operation, since the volume of the melted zone does not change. Thus, the concentration profiles obtained by zone melting are generally "flatter" than with the Bridgman technique. In most cases, single-crystal growth is ensured by starting on a seed.

3. *Floating zone*

In this case, the melted zone is maintained between the two parts of a vertical ingot by the combined effects of surface tension and magnetic induction. Absence of a crucible allows the crystallization of reactive III–V compounds (i.e., those of aluminum).

However, if a volatile constituent is present, one has to face the same problems as in the Czochralski technique under pressure.

4. *Gallium Arsenide*

The gradient cooling technique was developed for the preparation of large gallium arsenide single crystals and is now often preferred to the Czochralski technique, the principal reason being a much better homogeneity of dopant concentrations.

Important parameters are:

1. The arsenic pressure over the melt (1–2 atm).
2. The thermal gradient at the liquid–solid interface (50°C cm^{-1}).
3. The shape of the isotherm surfaces, which must be flat or slightly convex toward the liquid.
4. The cooling rate.

There are several possibilities for the cooling system. A good solution consists in utilizing a furnace made of several independent, small heating elements and separately programming their temperatures. It is then possible to modify the temperature at a point in the furnace without changing the thermal gradient over the liquid–solid interface.

The cold zone for the arsenic pressure control must be far enough from the crystallization zone to be thermally completely independent. Growth speeds are about 1 or 2 cm hr^{-1}.

In the case of zone-melting growth of doped GaAs single crystals, the starting load is a polycrystalline ingot containing a certain amount of the required dopant. One has to determine the apparent segregation coefficient, which depends on k, the equilibrium separation coefficient at the temperature of crystallization, and the growth rate. The influence of growth rate is very important in the case of small segregation coefficients.

Doping of gallium arsenide is generally carried out with the following elements: semiinsulating—Cr, Fe, or O_2; *n*-type—Si, Sn, Se, Te; *p*-type—Cd, Zn.

The higher resistivities are obtained with chromium, 10^8 ohm-cm; the lower ones with selenium and tellurium, 10^{-3} ohm-cm.

The doping agent is incorporated in the polycrystal in a separate operation before growth, except for oxygen, which is introduced through the vapor phase during growth.

Single crystals with a section larger than 2 cm^2 and length greater than 15 cm are grown in cylindrical or trapezoidal boats. However, the contact between the crystal and the boat makes it difficult to lower the dislocation concentration to zero, and the best crystals available have between 100 and 10,000 dislocations cm^{-2}.

Both techniques, Bridgman and zone melting, have also been used for the single-crystal preparation of GaP (Davis, 1960; Frosch and Derick, 1961; Gershenzon *et al.*, 1961); InP (Lawson and Nielson, 1958; Weisberg *et al.*, 1959; Guire and Weiser, 1959); and InAs (Harman *et al.*, 1956; Hilsum and Rose-Innes, 1961; Effer, 1961).

C. Growth from a Nonstoichiometric Melt

1. *Solution growth*

In contrast to the previously described techniques, in which solidification occurs from a stoichiometric melt at high temperatures, the solid is here obtained by a controlled precipitation from a solution of the compound to be prepared, the solvent being a constituent of the required material or another simple element. The temperature of solidification of a compound AB from its solution in A or B is always lower than the melting point of AB. Crystals can be grown at relatively low temperatures; the theoretical lower limit, depending on the shape of the phase diagram, is either the melting temperature of constituent A or B or the formation temperature of a eutectic which causes the appearance of a second solid phase. This second phase is rather difficult to eliminate by annealing and quenching, and the quality of final crystals is rather poor.

Solution growth in a solvent other than one of the constituents of the compound is often hazardous, due to lack of knowledge of the ternary phase diagrams.

Low temperatures of growth give a low concentration of native lattice defects and low contamination.

However, it is important to note that the lower the temperature, the richer the concentration of solvent in the solution, and the smaller are the diffusion coefficients, which implies very low growth rates because of the constitutional supercooling (1 mm day^{-1} for gallium phosphide, for example).

If not completely avoided by a sufficiently low growth rate or a strong liquid–solid interface thermal gradient or both, constitutional supercooling leads to poor crystalline quality and a drastic lack of homogeneity for impurity concentrations.

2. *Traveling solvent method*

An interesting technique is the traveling solvent method (TSM), which is a combination of solution growth and zone melting (Fig. 6). A polycrystalline ingot and a seed crystal are placed in a boat, separated from each other by a thin zone of solvent. The temperature profile is flat except for a peak which at the start is over the solvent zone. The boat or the furnace are then moved so that the solvent zone travels along the whole ingot. Horizontal and vertical systems can be used (Broder and Wolff, 1963).

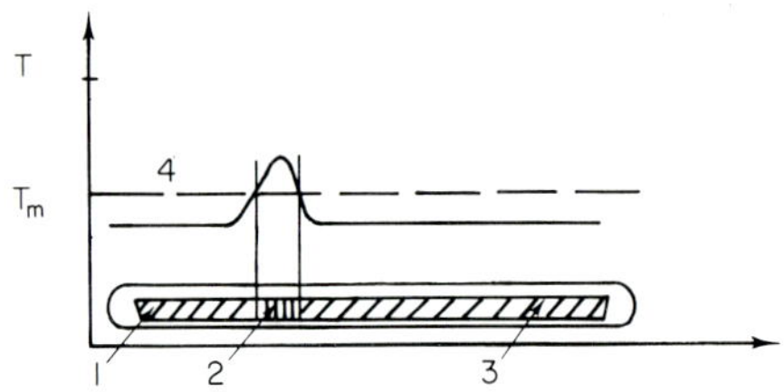

FIG. 6. Traveling solvent method. (1) Single-crystal AB seed; (2) AB–solvent; (3) polycrystalline AB; (4) liquid solution zone.

3. *Solution growth of gallium phosphide*

The growth of gallium phosphide is achieved in a gallium-rich solution Ga_xP_{1-x} ($0.8 < x < 0.9$). The temperature of the appearance of solid is about 850°C and the pressure of phosphorus in equilibrium with the liquid solution is close to 1 atm.

The saturation process can be done in two ways: dissolution of polycrystalline GaP in liquid gallium, or saturation of the gallium by the phosphorus vapor pressure. In both cases, the homogeneity of the solution is an important parameter of the starting growth conditions. The growth ampoule, which is sealed under vacuum (1 μm Hg), requires a phosphorus reservoir at controlled temperature to keep a constant degree of saturation of the solution during the whole time of the growth. The crystallization is achieved by the Bridgman–Stockbarger method in a vertical furnace with a pulling rate as slow as 1 mm day^{-1}. Great care must be observed in temperature and pulling control to avoid instabilities. This method was initiated by Vieland and Skalski (1960) and has been developed by Rodot *et al.* (1968).

In the present state of technology, it is possible to grow large-grain polycrystalline ingots of very high purity by this technique.

Modified solution-growth methods have been studied to control the nucleation process in the early steps of the growth. For this purpose, observation of the growth interface would permit the use of the seeding technique.

Of advantage with this method is the simplicity of the technology for the crystallization of high-vapor-pressure-dissociation compounds, because of the low growth temperature.

The slow rate of pulling, which is the main difficulty of this technique, must be considered with regard to the very high purity and the homogeneity of doping obtained in these crystals.

The solution growth of III–V compounds has also been carried out with success for GaAs, InP, and InAs.

III. Growth of Thin Layers

The processing of bulk materials for the preparation of electronic devices generally presents many difficulties and often does not allow one to obtain the required qualities. Therefore, the direct preparation of thin layers is preferred for many applications and this problem was given considerable attention in the case of III–V compounds.

In most cases, thin layers are grown epitaxially on bulk substrates. However, single crystals are not required for some applications and the preparation of polycrystalline layers is also of interest.

A. Vapor-Phase Epitaxy

Vapor-phase epitaxy is one of the most powerful methods of growth applicable to semiconductors. The crystalline and electrical quality of the depots can be very good and, in the case of III–V compounds, superior to that of bulk-grown crystals. Moreover, it is sometimes possible to grow by this method materials that cannot be grown by other methods (for example, refractory materials). Finally, it is a very flexible method which lends itself especially well to device manufacturing.

1. *Choice and preparation of substrates*

Epitaxial layers are deposited either on single-crystal bulk substrates of the same nature (this will be referred to as homoepitaxy) or on substrates of a different nature (heteroepitaxy).

In the second case, the substrate can be another III–V compound with lattice parameters not too different from those of the deposited material. Other materials are also used, provided one can find, for a specific orientation, a good agreement between the lattice parameters of both materials. The interatomic distances in the planar unit cells (or multiples of order two or three) must not differ more than a few per cent. Thermal expansion mismatch must also be as low as possible; GaAs, for example, has been prepared by heteroepitaxy on Al_2O_3, $MgAl_2O_4$, BeO, and ThO_2 (Manasevit, *et al.*, 1968) or on germanium (Kontrimas and Blakeslee, 1968; Schulze, 1966; Bobb, 1966a).

In every case, the preparation of the substrate surface is a very important stage which strongly influences the quality of the interface between the substrates and the deposit. Crystalline defects and an anomalous variation of chemical composition are often due to an incorrect surface treatment before starting epitaxy.

The substrate must be chosen as perfect as possible. It must have a low concentration of dislocations and other crystalline defects, and homogeneous distributions of those defects present and of impurities.

Cutting and polishing processes giving flat surfaces with few perturbations must be developed. No amorphous layer, dislocations, strains, or mechanical damage can be allowed after polishing. Furthermore, it is necessary to avoid any chemical pollution of the surface. This can be done by several processes: mechanical, chemical, or mechanochemical polishing. However, it is not possible to give a rule applicable for all the III–V compounds.

When mechanochemical polishing can be done, it is surely the best method. This polishing involves a chemical etching which does not perturb the lattice, in conjunction with a mechanical action which regulates the etching and leads to a more planar surface.

The particular case of gallium arsenide requires the following extra considerations.

Substrate materials must possess a low dislocation density, lower than 10^4 dislocations cm^{-2}, and be free from impurity segregation. Substrate quality can be tested by metallography, x-ray topography, ion analysis, and other methods (Huber and Champier, 1970).

After cutting, a mechanochemical polishing in a bromine–methanol bath etches away the perturbated layer. A final etch in the mixture

($3H_2SO_4$:$1H_2O_2$:$1H_2O$) is performed just before entering the reactor. Particular care is given to the whole rinsing process. The substrate is first rinsed in filtered deionized water under ultrasonic agitation, and then in acetone, methanol, or isopropylol, and dryed under pure hydrogen.

These considerations apply to the preparation of substrates for liquid-phase epitaxy as well as for vapor-phase epitaxy.

2. *Principle of Vapor-Phase Transport*

We will briefly sum up the main features of vapor-phase transport. In considering vapor-phase transport, we have first to deal with a solid–gas thermodynamic equilibrium: it is thus necessary that the material to be deposited can exist in the vapor phase or in equilibrium with a gaseous mixture including its components or chemical compounds of them. This condition must be realized over a range of temperatures and pressures compatible with available technology. Two cases are possible:

a. Physical transport. The compound exists in the vapor state, dissociated or not:

$$\mathrm{AB(solid)} \rightleftarrows \mathrm{AB(vapor)}: \qquad P_{\mathrm{AB}} = F(T) \tag{1}$$

$$\mathrm{AB(solid)} \rightleftarrows (1/y)\mathrm{A}_y\mathrm{(vapor)} + (1/z)\mathrm{B}_z\mathrm{(vapor)}:$$
$$(P_{\mathrm{A}_y})^{1/y}(P_{\mathrm{B}_z})^{1/z} = K_P(T) \tag{2}$$

where P_i is the partial pressure of component i, and T is the absolute temperature.

This type of transport is seldom applied to III–V compounds, because of the low partial pressures of the compounds of group III elements. Examples can be found in close-space epitaxy or the flash evaporation processes.

b. Chemical transport. The compound is transported by means of a gaseous mixture of chemical compounds of the components. The deposition reaction is a thermochemical equilibrium reaction, such as

$$\mathrm{AB_s} + (m/x)\mathrm{C_g} + (n/y)\mathrm{D_g} \rightleftarrows (1/x)\mathrm{A}_x\mathrm{C}_{m,\mathrm{g}} + (1/y)\mathrm{B}_y\mathrm{D}_{n,\mathrm{g}} \tag{3}$$
$$(P_{\mathrm{A}_x\mathrm{C}_m})^{1/\mathrm{x}}(P_{\mathrm{B}_y\mathrm{D}_n})^{1/\mathrm{y}}/(P_{\mathrm{C}})^{m/x}(P_{\mathrm{D}})^{n/y} = K_{\mathrm{P}}(T)$$

C and D are gaseous reagents that we will call transporting agents.

This type of transport is directly applicable to most of the III–V compounds, the group V element often being transported by means of its vapor (AS_4, P_4, etc.).

The operation can be divided into three stages:

1. Constitution of the vapor phase: The equilibria in (1), (2), or (3) are completely or partially displaced toward the right. This occurs at a temperature T_S referred to as the source temperature. Sometimes, T_S is room temperature. This happens when hydrides of group V elements are used as starting products (ASH_3, PH_3, SbH_3, NH_3).
2. Transport: The transport of gaseous species from the source zone to the deposition zone is due to diffusion, natural convection, or forced convection.
3. Epitaxial deposit: In the deposition zone, the equilibrium is displaced to the left at a temperature T_D referred to as the deposition temperature. Growth takes place on the substrate due to a surface-catalyzed reaction.

Thus the transport process is achieved through a shift of physical or chemical equilibrium due to the variation of the equilibrium constant K_P with respect to temperature. Deposition occurs if $K_P(T_D) < K_P(T_S)$. An endothermic deposition reaction leads to $T_S < T_D$, and an exothermic reaction to the opposite situation.

3. *Choice of the transport agent*

This choice is controlled by considerations developed by Jeffes (1968) with respect to vapor-phase transport.

(a) The chosen transport agent associated with the component species of the III–V compound must give compounds which are gaseous in a practicable range of temperature.

(b) The variation of enthalpy during the transport reaction must be large enough to provide a sufficient variation with the temperature of the equilibrium constant of the reaction K_P. These two variations are related by the Van't Hoff equation

$$d(\ln K_P)/d(1/T) = -\Delta H_0/R$$

(c) To ensure a reasonable yield of transport, a reaction constant K_P not too different from the one at the deposition conditions is preferable. This implies that ΔG_0 is not too far from zero, for

$$K_P = \exp - (G_0/RT)$$

Taking into account these last two conditions, the expression for the free energy,

$$\Delta G_0 = \Delta H_0 - T\,\Delta S_0$$

TABLE II

		Transport agents			
Group	Element	Halides	Hydrides	Vapor of the element	Oxide
V	N	—	NH_3	N_2	—
	P	PCl_3/Br_3	PH_3	P_2	—
	As	$AsCl_3/Br_3$	AsH_3	As_4, As_2	—
	Sb	$SbCl_3/Br_3$	SbH_3	Sb	—
III	B	BBr_3	B_2H_4	—	—
	Al	$AlCl_3$	—	—	—
	Ga	$GaCl_3$, GaCl	—	—	Ga_2O
	In	$InCl_3$	—	In	—

gives

$$\Delta S_0 \approx \Delta H_0/T$$

(d) If possible, the partial pressures of the constituents of the gaseous III–V compound should be almost identical, to make it easier to obtain a stoichiometric deposit and to avoid a great loss of material.

(e) Finally, the growth of the right compound will be possible only if the chemical potentials of both constituents (III and V) are equal to each other in the source and deposit conditions.

These rules applied to III V compounds lead to the choice of possible transport agents listed in Table II.

4. *Closed-tube system*

Closed-tube transport was the first vapor-phase technique developed for the preparation of III–V compounds.

It was used for growing either small single crystals or epitaxial layers, but today does not play an important part in III–V compound preparation. However, mainly for the sake of completeness, it will be described here briefly.

In this technique, the transport of the volatile compounds is generally ensured only by diffusion of the gaseous species. However, natural convection sometimes also plays a part in the transport.

The process is quite simple, and was very well described by Schäfer (1962). A tube (generally made of quartz) contains at one end, the source material, and at the other end (if desired), the substrate for epitaxy

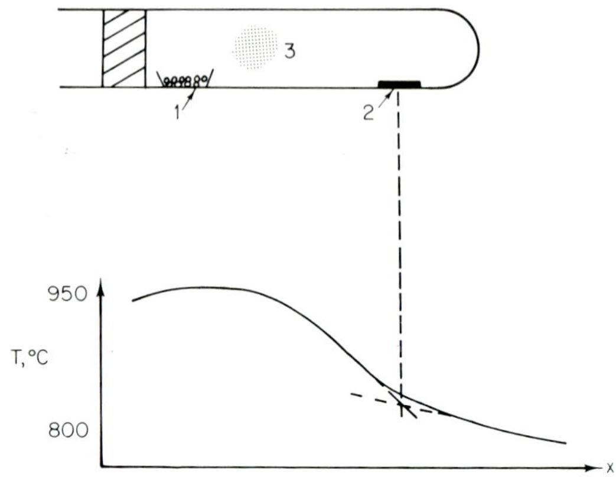

FIG. 7. Transport of GaP by iodine in a closed tube. (1) Polycrystalline GaP; (2) GaAs substrate; (3) I_2.

(Fig. 7); a known quantity of the transport agent is also included. Sealing is done under vacuum or a vapor pressure of the transporting agent, if necessary. A polycrystalline powder of the material under preparation is often used as source material. The tube is then introduced into a two-zone furnace so that the source is at the chosen temperature T_S and the other end is at the deposition temperature T_D. A useful refinement of the process consists in starting the operation with a reverse temperature gradient in order to etch the wall in the deposition zone and so avoid parasitic nucleation. It is also helpful to etch the substrate prior to deposit in the case of homoepitaxy.

The following compounds were thus prepared: GaP, InP, GaAs, InAs (Antell and Effer, 1959), BP, GaAs, GaP, $GaAs_{1-x}P_x$ (Angeli, 1966; Nitsche, 1967), and GaP (Alferov *et al.*, 1966).

The use of this technique, however, is severely limited by its numerous and important disadvantages:

1. Continuous recycling of impurities, which does not allow one to prepare very pure compounds.
2. No possible control of the stoichiometry of the deposit throughout the whole operation.
3. Difficulty in controlling the thickness of the deposit.
4. A general lack of flexibility of operation in the case of doping, for example.

These factors have led in most cases to the use of open-tube techniques.

5. *Open-tube processes*

a. General considerations. Vapor-phase epitaxy in an open tube is now the industrial technique used for the preparation of high-quality III–V compounds. This implies the preparation of bulk materials of a lower perfection which are then used as substrates. In the open-tube process, transport is achieved by a stream of gas flowing through the tube, the carrier gas being inert or simply not involved in the chemical reactions. The mass transport of materials is directly related to the gas flow speed, and the influence of diffusion and natural convection can be made negligible with respect to the forced convection of the vapor phase.

On the other hand, hydrodynamic problems play an important part and must carefully be taken into account in the design of epitaxy apparatus. The main requirements for the epitaxial deposit of III–V compounds are the following:

1. Preparation of thin layers (5000 Å to 100 μm) of great crystalline perfection.
2. Electrical characteristics as close as possible to the theoretical optimum.
3. Control of doping during growth and even preparation of multilayer structures.
4. Preparation of refractory compounds (nitrides and some phosphides) for which other processes have failed.

These various conditions demand a very accurate control of the vapor-phase composition, which implies:

1. A good knowledge of the thermochemistry of the system and a fair temperature regulation in the whole apparatus.
2. Control of the gas-flow parameters (input quantity, speed, and pressure).
3. Control of nucleation; deposits on the walls of the tube upstream of the substrate greatly affect the supersaturation of the vapor phase.

We pointed out earlier what transport agents and compounds can be used to achieve the transport.

If the compound is stable at room temperature, it is introduced in the apparatus either from a gas tank (e.g., AsH_3) or from a gas saturator (bubbler for liquids, e.g., $H_2/AsCl_3$). If the compound is stable only at high temperature, it is synthesized in the apparatus by the reaction of a source product and a gaseous transport agent (e.g., $Ga + HCl \rightleftarrows GaCl + \frac{1}{2}H_2$).

b. Choice of the carrier gas. If a carrier gas is used, an inert gas is the best solution, the criteria for the choice being purity and facility of manipulation. If the carrier gas takes part in the chemical reaction, this makes the chemical problem somewhat more complex. However, some advantages can be found in this situation. A typical example is the use of hydrogen; its reducing properties allow an ultimate elimination of oxygen, and moreover, its reactivity with halogens is used to control the thermochemistry of the system when a halogen is the transporting agent. Above all, hydrogen is available at very high purity from purification by diffusion through palladium. One cannot insist enough on the importance of the purity of the carrier gas, a great amount of which flows through the reactor during the whole operation. As a matter of fact, though it has been rarely pointed out in publications, proper account of this factor was one of the keys in the preparation of high-purity gallium arsenide by the vapor-phase technique.

c. Gallium arsenide. This is by far the most well known of the III–V compounds. This is a logical consequence of the tremendous amount of effort that went into its preparation, which was induced by the numerous applications of this material in the field of electronics. We will consider the preparation of gallium arsenide in some detail, since it is a typical example of the vapor-phase epitaxy of a III–V compound.

The problem here is the simultaneous transport of gallium and arsenic.

Arsenic is readily obtainable in the vapor phase and the range of choices is rather large; As_2 and As_4 exist at sufficiently high pressures over a wide temperature range that numerous volatile compounds are available (e.g., AsH_3, $AsCl_3$). Gallium, however, has a very low vapor pressure in the usual range of temperature ($T_{EB} = 1983°C$), and it is necessary to achieve transport through a gaseous compound. The thermodynamic considerations given before practically limit the choices to halides (GaCl, GaI) and the volatile oxide Ga_2O.

Three main processes of vapor transport will be taken into consideration.

(1) Transport from a source of GaAs. In this case, the starting material is polycrystalline or single-crystal GaAs; crushed or bulk samples can be used as well. By this technique, Fig. 8, it was possible to grow epitaxial layers with good crystalline properties and with an impurity content lower than that of the starting material (for example, a concentration of 10^{15} cm^{-3} is obtainable from a source with impurity concentration of 10^{16} cm^{-3}). This is due to the fact that the separation coefficient between the solid and the vapor phase is not the same at the source and on the substrate. Dilution in the carrier gas can also have a favorable effect.

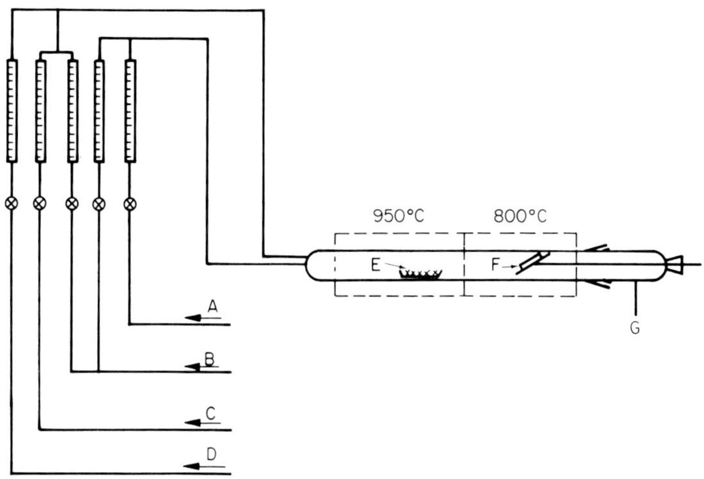

FIG. 8. Transport from a bulk source by HCl or H_2O. A, Pure HCl or H_2 saturated by H_2O; B, pure H_2; C, dopant; D, HCl for etching; E, polycrystalline GaAs; F, seed; G, to waste.

In the search for high purity, however, the impurity content in the source material is a severe limitation.

(*2*) *Water as a transport agent.* Here, gallium is transported through its oxide Ga_2O and arsenic through its own vapor As_2 or As_4. In most cases, the carrier gas is hyperpure hydrogen which is mixed with water vapor by passing it through a container full of ice (temperature between −30°C and 0°C).

The main reactions involved are the following:

$$GaAs + \tfrac{1}{2}H_2O \rightleftarrows \tfrac{1}{2}Ga_2O + \tfrac{1}{2}H_2 + \tfrac{1}{2}As_2$$

$$2As_2 \rightleftarrows As_4$$

The range of suitable source temperatures is 600–1050°C. The temperature difference between the source and the deposit zones varies from 50 to 100°C.

The partial pressure of water must be low enough (a few millibars) to avoid the formation of the oxide Ga_2O_3, which is solid in this temperature range. This technique has been developed by Michelitsh *et al.* (1964) and Lawley (1966).

(*3*) *Transport of GaAs by halides.* Calling X the chosen halogen, this type of transport involves the formation of the monohalide GaX at the source by the reaction

$$GaAs + X \rightleftarrows GaX + \tfrac{1}{4}As_4 \tag{4}$$

or

$$GaAs + HX \rightleftarrows GaX + \tfrac{1}{4}As_4 + \tfrac{1}{2}H_2 \tag{5}$$

In the deposition zone, the following reaction also takes place

$$3GaX \rightleftarrows GaX_3 + 2Ga$$

which, associated with the reaction

$$Ga + \tfrac{1}{4}As_4 \rightleftarrows GaAs \tag{6}$$

gives

$$3GaX + \tfrac{1}{2}As_4 \rightleftarrows 2GaAs + GaX_3$$

In every case, arsenic is transported by its own vapor. The suitable transporting agents are HCl, Cl_2, HI, and I_2 (under pressures varying between 10^{-2} and 10^{-5} atm). Hydrogen is the most common carrier gas (flow rates range from 100 to 500 $cm^3\ min^{-1}$). Deposition temperatures lie between 650 and 900°C in the case of HCl or Cl_2; the use of HI or I_2 lowers these limits by about 100°C. Temperature differences between source and deposit zones vary from 50 to 100°C. More details will be found in the papers of Ewing and Greene (1964), Hoyt and Haisty (1966), Moest (1966), Effer (1965), and Mehal and Gronin (1966).

Transport by iodine is of particular interest; the low growth temperature gives less crystalline defects in thermal equilibrium in the layer and reduces the risk of contamination. Unfortunately, very pure iodine is not easily available and the heaviness of its vapor leads to a problem with regard to the homogeneity of the gas stream.

(4) The Ga–AsCl₃ process. This technique is the most frequently used and best-known of all the vapor-phase epitaxy processes for the preparation of gallium arsenide. Though a little more complicated with regard to thermochemistry, it allows the growth of the purest GaAs obtainable by vapor-phase transport.

Here, high-purity gallium (99.9999%) is transported by HCl through the monochloride GaCl (Fig. 9). The single source of hydrochloric acid and arsenic is arsenic trichloride, which is now available at high purity (99.9999%). This trichloride is purified by distillation. The carrier gas can be either argon (Taylor, 1967) or hydrogen (Effer, 1965; Knight *et al.*, 1965; Bobb *et al.*, 1966b; Wolfe *et al.*, 1968). The latter is preferred, due to its higher purity, notwithstanding the fact it takes part in the chemical reactions and makes the system a little more complex.

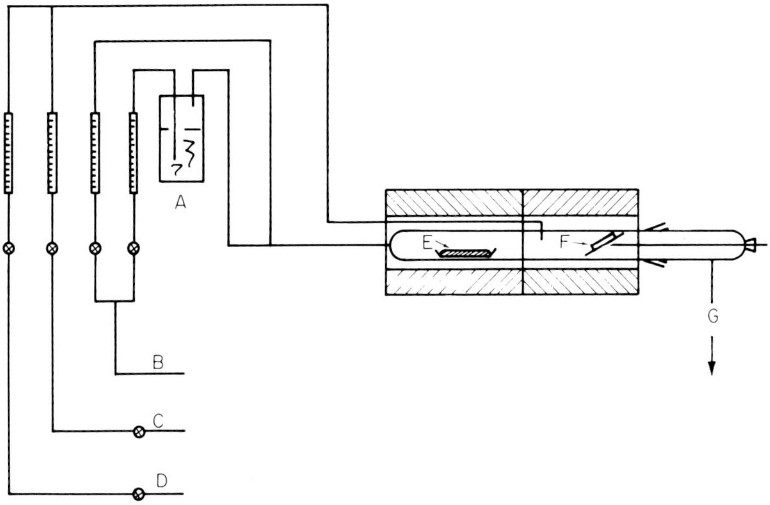

FIG. 9. The Ga–$AsCl_3$ process. A, $AsCl_3$; B, pure palladium-diffused H_2; C, HCl for etching; D, gaseous dopant (H_2Si); E, Ga boat ($T_S \approx 850°C$); F, substrate ($T_D \approx 950°C$); G, to waste.

The operation can be briefly described as follows: part of the carrier gas (50–300 $cm^3\ min^{-1}$) goes through a bubbler containing arsenic trichloride and held at a constant temperature (10–20°C). The rest of the carrier gas bypasses the bubbler. The mixture then enters the reactor itself (a quartz tube a few centimeters in diameter) and in the hot zone, passes over a boat of gallium; farther downstream, in a colder zone, is located the substrate, on which growth occurs using the gallium chloride and the arsenic present in the vapor phase. It is important to note that the growth takes place by the intermediary of a surface-catalyzed reaction, and under suitable conditions, no material is deposited on the reactor walls in the substrate zone. Finally, one or more additional gas lines enter the reactor tube: for admission of a doping agent, and/or for addition of hydrochloric acid in the substrate zone in order to perform a chemical etch before starting the growth.

The main chemical reactions involved are as follows, depending on the nature of the carrier gas.

In the case of hydrogen:

$$2AsCl_3 + 3H_2 \rightleftarrows 6HCl + \tfrac{1}{2}As_4$$

$$As_4 \rightleftarrows 2As_2$$

$$Ga + HCl \rightleftarrows GaCl + \tfrac{1}{2}H_2$$

$$3GaCl + \tfrac{1}{2}As_4 \rightleftarrows 2GaAs + GaCl_3$$

In the case of argon:

$$AsCl_3 \rightleftarrows \tfrac{1}{4}As_4 + \tfrac{3}{2}Cl_2$$

$$Ga + \tfrac{1}{2}Cl_2 \rightleftarrows GaCl$$

$$3GaCl_3 + \tfrac{1}{2}As_4 \rightleftarrows 2GaAs + GaCl_3$$

Some authors have mentioned the fact that $GaCl_2$ should also be taken into account in this case.

Though it does not seem to be necessary, the deposition takes place in a zone where the vapor phase is not at equilibrium, and the supersaturation is quite high. Many authors use a thermal gradient (3–10°C cm^{-1}) in the deposition zone in order to obtain layers of constant thickness over the whole substrate.

Typical conditions with the use of hydrogen are the following:

Temperature of the source (gallium boat)	800–850°C
Temperature of substrate (100)	750–800°C
Temperature of $AsCl_3$ (bubbler)	10–20°C
Hydrogen flow rate (bubbler)	50–100 cm^3 min^{-1}
Hydrogen flow rate (bypass)	50–100 cm^3 min^{-1}
Thermal gradient (substrate zone)	3–7°C cm^{-1}

The conditions are slightly different for different orientations of the substrate, (110), (111)*A*, and (111)*B*. In the case of the (100) orientation, the growth rate would be about 20 μm hr^{-1}.

A typical vapor-phase epitaxy includes the following steps:

1. Saturation of the gallium. During this period (about 15 hr for 25 g of gallium at the indicated conditions), the gallium is saturated with arsenic (the solubility of arsenic in gallium is 4 atom % at 850°C) coming from the vapor phase. It is of great importance that the gallium be completely saturated before the beginning of the growth process; otherwise, droplets of gallium are deposited on the substrate, causing important defects (pyramids) due to a vapor–liquid–solid growth process (Joyce and Mullin, 1966).
2. Introduction of the substrate in the deposition zone.
3. Etching of the substrate *in situ*. This can be done by addition of HCl in the vapor phase in the deposition zone or by raising the temperature of the substrate for a few minutes.
4. Flushing of the reactor by hydrogen, to flush away the products of the etching of the substrate.
5. Epitaxial deposition.
6. Cessation of the $AsCl_3$ flow, and cooling of the wafer under pure hydrogen flow.

In the search for high-purity gallium arsenide, emphasis is put on the following factors: (1) the absolute necessity of a very pure carrier gas; (2) the need for purity of the starting materials; and (3) the requirements that the whole apparatus be leak-proof and that dead stream zones be absent, to allow efficient cleaning by a stream of pure gas. The use of synthetic quartz is recommended for the hot parts. Between runs, cleaning can be performed by etching with the HCl produced by the cracking of $AsCl_3$.

(5) *Transport of constituent species by separate fluxes.* In this technique, often referred to as the "synthesis technique," arsenic and gallium are carried by two separate streams of gas in a reaction chamber. This chamber is held at the deposition temperature and the point of arrival of the two gas fluxes is situated a little upstream of the substrate.

Bringing the arsenic and gallium in separately allows greater control of the composition of the vapor phase and a better theoretical approach to the thermochemistry of the system, the parameters having been clearly separated. In addition, the reaction chamber volume can be minimized, and thus "memory effects," which often are a big problem in the doping processes, are lowered. Although there are several possible shapes for the reactor (cylinders with coaxial inputs, T-shaped or Y-shaped), the main

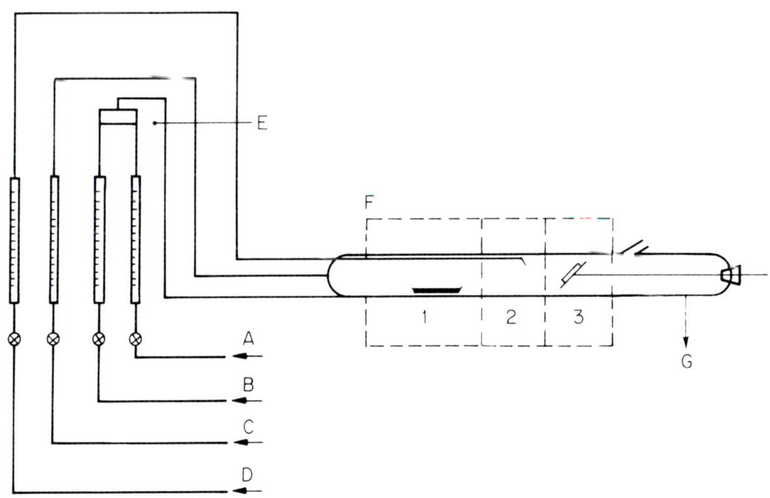

FIG. 10. Separate-flux system using HCl and AsH_3. A, AsH_3; B, gaseous dopant; C, HCl–H_2 mixture; D, H_2 or HCl for etching; E, mixing bulb; three-zone furnace: (1) Ga, T = 850°C; (2) reaction zone, T = 900°C; (3) substrate, T = 700°C; G, to waste.

features of the apparatus are the same. In the case of GaAs, the requirements are as follows:

1. Space for the synthesis of GaCl (temperature about 850°C) controlled by $Ga + HCl \rightleftarrows GaCl + \frac{1}{2}H_2$.
2. Space for the evaporation of arsenic (400–500°C) (Rubenstein and Myers, 1966; Michelitsh, 1965; Conrad *et al.*, 1967; Shaw, 1968a,b) or an input line of AsH_3 (Fig. 10) (Tietjen and Amick, 1966; Enstrom and Peterson, 1967).
3. A reaction zone where the mixing is achieved (900°C).
4. A deposition zone, isothermal or with a slight gradient (deposition temperature, 550–750°C).

An extra line is often used to bring gaseous HCl directly into the reaction zone—after the mixing stage, the supersaturation is rather high and it is necessary to shift the equilibrium by the addition of HCl to avoid deposits on the walls. This extra line can also be used for an etching *in situ* of the substrate before starting epitaxy. This type of apparatus is very versatile and, with only a few modification, is able to solve most of the problems of vapor-phase III–V compound synthesis. However, compared to the $AsCl_3/H_2$ method, the starting materials (HCl, AsH_3, As) are not available in very high purity. A solution is to use $AsCl_3$ as the source of HCl. This could be done by cracking $AsCl_3$ in presence of hydrogen using the reaction $AsCl_3 + \frac{3}{2}H_2 \rightleftarrows 3HCl + \frac{1}{4}As_4$ ($T > 800°C$), arsenic and unreacted $AsCl_3$ being eliminated in a cold trap (Fig. 11).

Finally, it is also possible to use $GaCl_3$ and arsenic as source materials, hydrogen being the chemically active carrier gas.

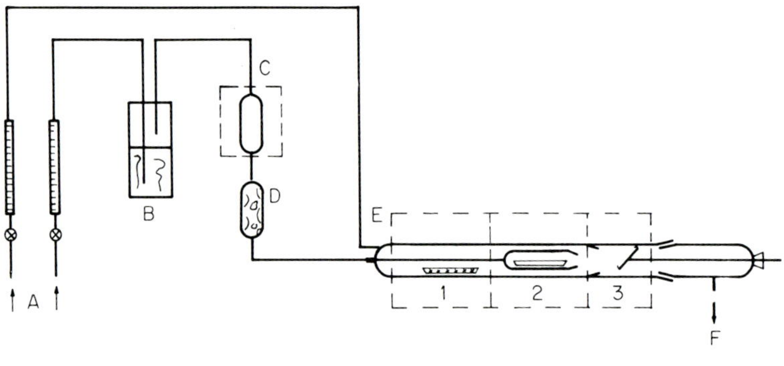

FIG. 11. Separate-flux system using $AsCl_3$ for HCl source. A, H_2; B, $AsCl_3$; C, $AsCl_3$-reducing furnace; D, cold trap; E, three-zone furnace: (1) As, $T = 450°C$; (2) Ga, $T = 900°C$; (3) substrate, $T = 700°C$; F, to waste.

(*6*) *Problems of doping.* Epitaxial layers always contain a certain amount of impurities which are not intentionally added: residual impurities of the vapor phase, or those coming from the contamination of the tube walls. Besides, etching the substrate before starting epitaxy often causes the impurities of the substrate to contaminate the beginning of the epitaxial layers. In most applications, the electrical carrier profile (and thus the impurity profile) is strictly defined. This makes evident the necessity of minimizing the first two sources of impurities and using great care in the etching process and at the start of growth. The etching of the substrate that is often necessary should be as brief as possible and followed by a flush of carrier gas.

Intentional doping can be achieved either by introduction of dopant (Sn, Si) in the gallium boat, or by introduction in the reaction chamber of a gaseous compound of the dopant (S_2, SH_2, SeH_2, etc.). This second technique is preferable with regard to the control of doping level during growth and for the realization of scheduled profiles. However, obtaining such profiles is always a difficult problem, and this delicate process is often to be determined in an empirical way. Important parameters to be taken into consideration are the ratio of gallium and arsenic concentrations in the vapor phase (Hasegawa and Saito, 1968) and the orientation of the substrate (Moest, 1966; Joyce, 1968). The range of intentional doping levels covers from 10^{14} cm^{-3} to 10^{18} cm^{-3} for the usual n-type dopants (S, Se, Te, Si, Sn); p-type layers of high concentration (from 10^{17} to 5×10^{18} cm^{-3}) were also grow using Zn or Cd as dopants. Preparation of semiinsulating epitaxial GaAs doped with Fe and Cr was also reported.

d. Gallium phosphide. The general considerations given for the epitaxial deposition of gallium arsenide are also applicable in the case of gallium phosphide. However, replacing arsenic by phosphorus leads to some modifications, mainly with regard to the source and deposition temperatures and the nature of the substrate. Gallium arsenide is often used as substrate material, single-crystal GaP only recently becoming available and still very expensive. In this case, in order to minimize the lattice mismatch, a gradual transition is made between GaAs and GaP through the ternary compound $GaAs_xP_{1-x}$, letting x go down slowly from one to zero. The more frequent methods will now be discussed.

(*1*) *Transport from gallium phosphide.* This can be done with HCl (Van Der Does, 1969) or H_2O (Luther, 1969); the carrier gas is hydrogen and typical operating conditions are: source temperature, 1000–1150°C; deposition temperature, 1000–1100°C; H_2 flow, 150–300 cm^3 min^{-1}; the growth rates range from 10 to 20 μm hr^{-1}.

(*2*) *Transport by the $PCl_3/Ga/H_2$ process.* This process is similar to the $AsCl_3/Ga/H_2$ process previously described for gallium arsenide, hydrogen being the carrier gas. Taylor *et al.* (1968), Taylor (1969) and Luther (1968) reported the growth of gallium phosphide on gallium arsenide substrates; orientations were (111)*A*, (111)*B*, (100), and (110). Zinc, sulfur, selenium, or tellurium were used as dopants.

(*3*) *Transport by separate fluxes.* This method is of particular interest in the case of GaP. It is easier to monitor a graded transition between the GaAs substrate and the GaP deposit. Graded region thickness can be extended over 10–20 μm (Richman and Tietjen, 1967). The present authors have obtained the growth of doped layers (sulfur, selenium, tellurium) and also semiinsulating layers using iron as dopant, the resistivity being as high as 10^{10} ohm-cm.

In addition, this technique allows a very high growth rate (3 μm min^{-1}) due to the high supersaturations one can produce in the reaction zone.

Temperatures are typically 950°C in the zone of the gallium boat and about 800°C in the deposition zone.

(*4*) *Transport by organometallic compounds.* Thomas (1969) reports the epitaxial deposit of gallium phosphide by the thermal decomposition of a gas mixture of gallium triethyl and phosphorous triethyl. The deposition temperature is below 500°C, but, of course, the growth rate is quite low, about 1500 Å hr^{-1}.

e. Other compounds. Some particular details will now be given on the vapor-phase transport or epitaxy of other III–V compounds.

(*1*) *Boron phosphide.* So far as we know, boron phosphide has not yet been prepared as a single-crystal layer. However, using the reaction

$$BCl_3 + PH_3 \rightleftarrows BP + 3HCl$$

or

$$PCl_3 + BCl_3 + 3H_2 \rightleftarrows BP + 6HCl$$

it was possible to obtain polycrystalline layers of acceptable purity.

(*2*) *Aluminum nitride.* Epitaxial deposits of AlN were obtained on substrates of Al, N, and SiC by the separate-flux process (Chu, 1967); Noreika and Ing (1968), using a similar process, prepared films of AlN on Si substrates; these films, however, possessed a mosaic structure. The reaction

$$AlCl_3 + NH_3 \rightleftarrows AlN + 3HCl \qquad (1000 < T < 1300°C)$$

and the pyrolysis of the complex $AlCl_3$–NH_3 were used. Doped layers were grown in the presence of gaseous H_2Se (n-type) or mercury (p-type).

The main difficulties come from the relatively high deposition temperature (over 1200°C), the lack of suitable substrates, and a marked tendency of this compound for the formation of nitrogen vacancies.

(*3*) *Aluminum phosphide.* Large-area single-crystal layers of AlP were prepared by Richman (1968). The substrates were silicon or sapphire and the reaction was

$$3AlCl + 2PH_3 \rightleftarrows 2AlP + 3AlCl_3 + 3H_2$$

Aluminum monochloride was synthesized from aluminum and hydrochloric acid at 1450°C in the reaction tube. Deposition occurred at about 1250°C. Platelets and needles were prepared in the same way. The main difficulty, besides the high temperatures, is the reactivity of the aluminum chloride, which makes it necessary that the reaction tube be made of alumina. Furthermore, deposits depart from stoichiometry due to an excess of phosphorus.

(*4*) *Aluminum arsenide.* Epitaxial deposits of GaAlAs were obtained (Tietjen *et al.*, 1969a) by the separate-flux process. Aluminum monochloride and arsine are used as transport agents. Hydrogen is the carrier gas. Once more, the problem is the reactivity of AlCl with the walls of the reaction chambers. Alumina and graphite-coated quartz tubes have been successfully used.

Pure AlAs, which is very hygroscopic, is not interesting in this context.

(*5*) *Indium nitride.* The dissociation temperature of InN is about 500°C. This makes it difficult to use the vapor-phase technique for its preparation. Indium halides and NH_3 as transport agents give a very poor reaction yield, due to the low pressure of indium halides below 500°C. Polycrystalline films were grown on quartz substrates using oxygen to transport indium (Tietjen *et al.*, 1969b), but the low deposition temperature reduces the surface mobility of the reagents and the layer is a mixture of indium and indium nitride.

(*6*) *Indium phosphide.* The methods previously described for GaAs epitaxy are suitable for indium phosphide. Good results with regard to the crystalline quality and carrier concentration (n-type, 10^{15} carriers cm^{-3}) have been reported by Tietjen *et al.* (1968, 1969a,b) and Joyce and Williams (1970).

Homoepitaxy on substrates grown by the encapsulation pulling technique, as well as heteroepitaxy on an InAs substrate, are also possible. In the latter case, a graded phosphorus-concentration region is necessary to achieve a good crystalline quality.

The deposition of InP involves the reactions

$$3InCl + 2PH_3 \rightleftarrows InCl_3 + 2InP + 3H_2$$

$$3InCl + 3P_2 \rightleftarrows InCl_3 + 2InP$$

following the use of pure phosphorus or phosphine as transport agent.

Indium monochloride is synthesized by the reaction of HCl or a mixture of PCl_3 and H_2 on liquid indium between 800 and 1000°C.

Deposition temperatures are below 800°C and can be as low as 400°C.

(*7*) *Indium arsenide.* Indium arsenide epitaxial layers were grown by the same techniques as used for gallium arsenide. It can also be grown by the halide metal process ($AsCl_3$–In) and by the separate-flux method with arsenic and indium monochloride as sources. MacCarthy (1966, 1967) and Cronin *et al.* (1966) reported the preparation of InAs layers on semi-insulating GaAs substrates. Deposits of both types were grown, selenium and zinc being the respective dopants. The maximum purity was about 10^{15} carriers cm^{-3}. The great tendency of indium vacancies to appear makes the electrical characteristics of the deposition very sensitive to the composition of the vapor phase.

(*8*) *Indium antimonide.* Two main difficulties arise when the vapor-phase transport of InSb is attempted. The first one, common to all antimony compounds, is the low vapor pressure of antimony at the melting point of InSb ($7 \cdot 10^{-5}$ Torr). The use of chlorides is not very convenient, because the dissociation reaction is endothermic and would lead to a need for complicated apparatus, InCl and antimony chloride flowing in opposite directions. The only practical way is to transport antimony as stibine (SbH_3); however, even this is quite difficult due to the instability of stibine.

The second difficulty here is the low melting point (425°C) of InSb, creating the same problems found in the case of indium nitride: low reaction yield due to the low pressure of InCl at the deposition temperature, and low surface mobility of the reagents.

For these reasons, single-crystal epitaxial deposition has not yet been obtained (Tietjen, 1969a).

(*9*) *Gallium nitride.* Recently, effort has been expended on the preparation of GaN due to its promising possibilities for electroluminescence (forbidden gap width and direct transition mode) (Maruska and Tietjen, 1969; Faulkner *et al.*, 1970).

A separate-flux process is generally used with the starting gaseous products NH_3 and GaCl, the latter being synthesized by HCl flowing over a gallium boat at 900°C. The carrier gas is pure hydrogen and the reaction is

$$3GaCl + 2NH_3 \rightleftarrows 2GaN + GaCl_3 + 3H_2$$

Deposition temperature is 825°C. Layers grown in this manner contain a large number of nitrogen vacancies and, if undoped, are *n*-type, the concentration of carriers being over 10^{19} cm^{-3}. With germanium dopant, *p*-type layers were prepared with poor reproducibility.

(*10*) *Gallium antimonide.* For the reasons given with regard to InSb, epitaxial layers of GaSb are rather hard to prepare. However, single-crystal layers on GaAs substrates were grown at 650°C using stibine and GaCl as source products in a separate-flux apparatus (Tietjen *et al.*, 1968, 1969).

6. *Close-spaced epitaxy.*

This method is as follows. Two wafers of a III–V compound, one the source and the other the substrate, are fixed together by means of spacers, the distance between them ranging from 0.5 to 2 mm.

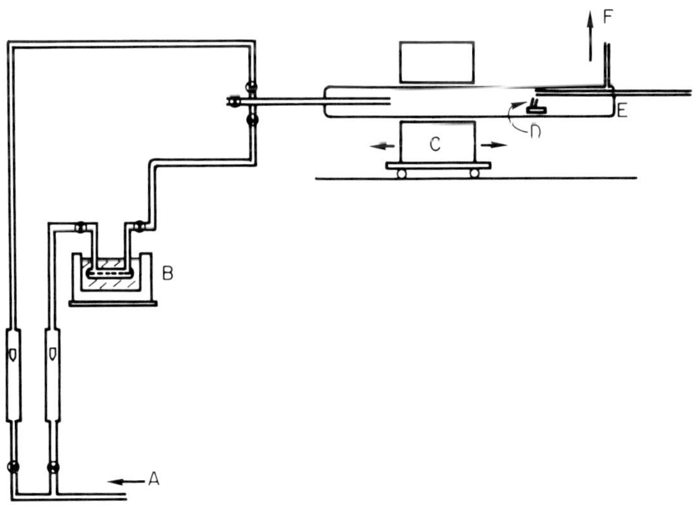

FIG. 12. Close-spaced system. A, Pure H_2; B, H_2O source (−10°C); C, furnace; D, seed holder; E, thermocouple; F, H_2 outlet.

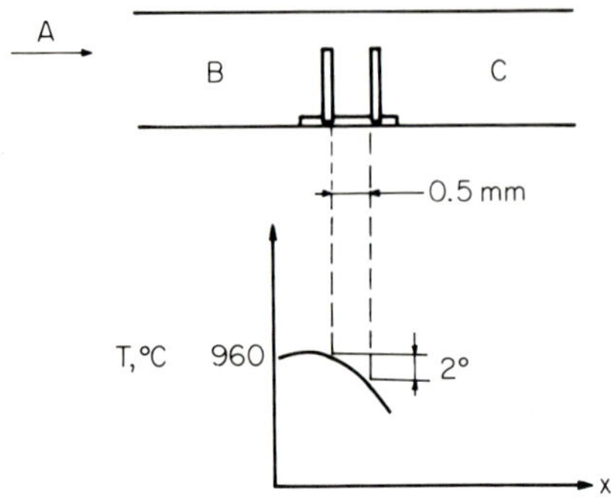

FIG. 13. Close-spaced system, seed holder and temperature profile. A, Pure $H_2 + H_2O$ vapor; B, polycrystalline GaP; C, GaAs single-crystal substrate.

A very steep temperature gradient (100°C cm^{-1}) is established across the assembly, with the source at the higher temperature. The apparatus, which is very simple, can be flushed with an inert gas or with a carrier gas plus a transport agent (see Figs. 12 and 13).

A purely physical transport can be obtained without any transport agent at very slow rates. Adding a transport agent (H_2O, HCl) results in considerably higher growth rates.

Hoss *et al.* (1968) and Merenda (1968) have described the preparation of GaP, GaAs, and GaAsP by this method. Doping of the deposit can be achieved, for example, by $ZnCl_2$.

This process seems to be more promising for preparing doped crystals than undoped ones, because of the great similarity between source and deposit with regard to purity.

One must note that the source is not necessarily a single crystal, but can be a sintered platelet of the powdered mixture of III–V compound plus dopant.

B. Liquid-Phase Epitaxy

1. *General considerations*

Liquid-phase epitaxy is a very simple technique from the theoretical point of view. As in the solution-growth method, deposition is achieved by the use of the properties of the liquid–solid phase diagram. Crystallization occurs in a liquid binary mixture of the group III and group V constituents. The whole operation takes place in a relatively small region of the phase diagram, richer in one of the two components. It is also possible to use another element as solvent, the doping agent, for example. At any rate, the growth temperature is much lower than the melting point of the compound under preparation. This is obviously advantageous with regard to contamination from the crucible and the number of thermally activated defects.

Furthermore, by proper choice of the solvent, it is possible to minimize the tendency for departure from stoichiometry of some compounds. Gallium arsenide, for example, when grown in gallium solution, does not have gallium vacancies, as is the case with the vapor-phase technique if the concentration of the gaseous mixture is not correct.

Two main processes are available, as follows.

a. Variation of the solubility of the III–V compound in the liquid phase by modification of the temperature. A bath of one of the constituents of the compound is saturated with the other constituent at a temperature below the melting point. At equilibrium, the representative point of the phase diagram stands on the liquidus curve. A substrate is then introduced in the bath, and in the case of homoepitaxy, no dissolution occurs.

In the case of heteroepitaxy, a material for the substrate has to be found which is weakly reactive with respect to the bath; this is necessary to ensure a rapid growth of a protective layer of the compound to be prepared. After introduction of the substrate, the temperature is slowly lowered, the solubility of the solute goes down, and epitaxy starts. If the bath is isothermal with respect to volume, nucleation occurs only on the substrate (minimization of free energy), but this will be true only when the cooling rate is slow enough. This process of liquid-phase epitaxy is by far the more usual.

b. Deposition by addition of solute. This process is isothermal. Deposition is caused by addition to a bath in equilibrium (with respect to concentration and temperature), a certain amount of the solute or of the compound to be grown. This can be done by the vapor-phase technique (VLS or by addition of the solid compound), concentrations in the bath being fixed by the temperature.

The excess of solute is absorbed by the epitaxial growth. This is of importance in the case of the solution growth of bulk crystal, for it can be operated as a continuous process. In every operation of liquid-phase epitaxy, the quality of the crystals as well as the purity of the deposits are strongly dependent on the phenomenon of constitutional supercooling (Andre and Le Duc, 1968; Minden, 1970).

In this type of growth, epitaxial layers show a lamellar or cellular structure, and the solvent can be trapped in the crystal. Residual impurity concentration can be abnormally high due to departures from the theoretical separation coefficient.

The maximum growth rate is given by

$$V = G[K/(1 - K)]D(1/mC_s)$$

where V is the growth rate, K is the ratio of the concentrations of the minority constituent of the bath in the solid and in the liquid phases, D is the diffusion coefficient of the same species in the liquid, m is the slope of the liquidus curve, and C_s is the concentration of the minority constituent in the solid.

An interesting study of the growth rate of crystals from solution will be found in a paper by Tiller and Kang (1968).

2. *Choice of the solvent*

a. Epitaxy of pure materials. Here, the solvent is by necessity one constituent of the III–V compound. Choice will be guided by the following considerations:

1. Shape of the phase diagram.
2. Vapor pressure over the melt.
3. Purity of the purest available solvent.
4. Separation coefficient of the impurity: some solvents act as a getter (e.g., gallium).
5. Reactivity with respect to the possible crucible materials.

In practice, compounds of gallium and indium are grown in gallium or indium solutions, respectively, though epitaxy in the group V constituent can be of interest. For a better approach to stoichiometry, gallium antimonide is grown in antimony solution. The case of aluminum compounds is not so easy; the reactivity of aluminum to the quartz crucible makes it necessary to use alumina, graphite, or vitreous carbon.

b. Epitaxy of doped materials. If the required level of dopant concentration is not very high, that is, far enough from the limit solubility criterion, the choice of solvent is the same as previously listed, and the doping agent is added to the bath. If highly doped materials are to be prepared, the doping agent can sometimes be chosen as solvent; during the growth, a certain amount of the solvent is incorporated into the crystal and the concentration obtained can be quite close to the limit solubility. For example, GaAs can be grown in tin solution when heavily doped layers are wanted.

3. *Technique of liquid-phase epitaxy*

The various practical techniques of liquid-phase epitaxy essentially differ from each other by the way the substrate comes in contact with the solution.

a. Tilting apparatus. Here, the substrate is fixed at the end of a boat placed in a tilting furnace (Fig. 14). At the other end of the boat are the solvent and the solute. During the saturation, the furnace is tilted so that the substrate is in the higher end of the boat. When the bath is correctly saturated, the furnace is set back to the horizontal position and the bath spreads over the substrate. The cooling is then started. Growth can be stopped at any time by tilting the furnace again. This technique is very simple; however, the thermal distribution in the furnace is difficult to keep constant during the tilt. The substrate is at high temperature during the saturation, and the wetting of the substrate by the solution makes it difficult to completely stop growth on the whole surface of the substrate.

This technique was initiated by Nelson (1963) and developed mainly for GaAs by numerous workers (Bolger *et al.*, 1966–1967; Kang and Greene, 1967; Goodwin *et al.*, 1968; Harris and Snyder, 1969; Hicks and Manley, 1969).

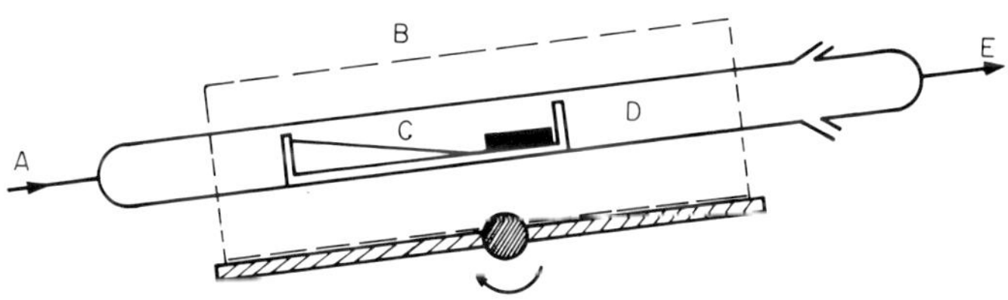

FIG. 14. Tilting system. A, Pure H_2; B, furnace; C, melt; D, substrate; E, to waste.

b. Rotating crucible. In this case, the furnace is fixed. Inside a tube, a chamber in the form of a parallelipiped is set up so that it can be rotated around an axis through rotating seals. During saturation, the substrate is fixed on the upper wall inside the inner chamber and the bath lies at the bottom. Growth is started by rotating the crucible by 180° and cooling (Fig. 15).

This technique was initiated by Andre and Le Duc (1968) and is generally used when a strong thermal gradient is needed at the growth interface. The cessation of growth is generally better than in the previous tilting furnace method.

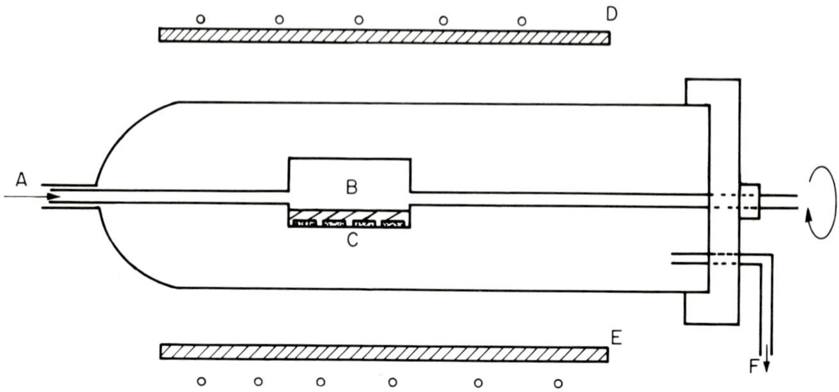

FIG. 15. Liquid-phase epitaxy, rotating system. A, Pure H_2; B, Ga-rich melt; C, substrates; D, hot zone; E, cold zone; F, H_2 to waste.

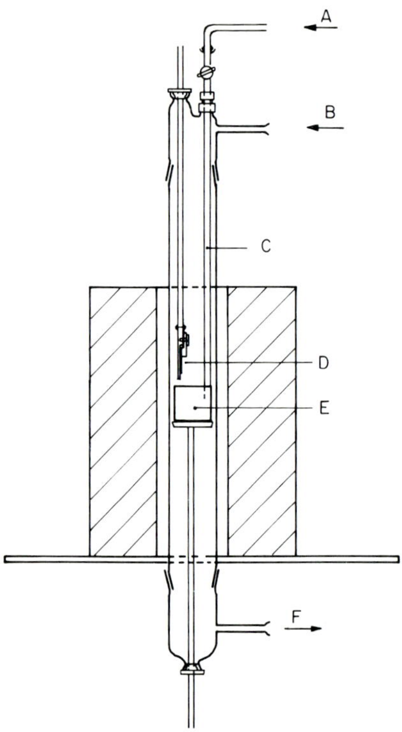

FIG. 16. Liquid-phase epitaxy, dipping system. A, Pure H_2; B, pure H_2; C, doping line; D, seed holder; E, melt; F, H_2 to waste.

c. Dipping process. This process differs from those previously described in that it uses a vertical geometry. The melt is placed in a crucible located in the hot zone of a furnace. The substrate is fixed on a seed holder which can move up and down (Fig. 16). When the melt is saturated, contact is established between the substrate, which has been kept in a cold zone, and the solution by moving the seed holder downward and immersing the seed. At the end of the epitaxy run, the seed can be easily removed from the solution by pulling the seed holder up (Williams and Blacknall, 1967; Shih *et al.*, 1968a,b).

4. *Gallium arsenide*

Here, as for vapor-phase epitaxy, we will go into detail concerning the liquid-phase epitaxy of gallium arsenide. First, we will deal with the deposition of pure GaAs, and then consider the different doping processes.

Pure GaAs has been deposited by the three processes described: tipping furnace, rotating crucible, and dipping.

In order to lower the deposition temperature and avoid gallium vacancies, deposits have all been made in the low arsenic-concentration range (10 atom %; see the phase diagram in Fig. 17).

Gallium-rich melts (10–50 g) are saturated at a temperature lower than 900°C and generally baked under pure, palladium-diffused hydrogen for several hours, which allows volatile impurities to be removed.

After immersion of the seed, which is generally cut in the (100)*B* or (111)*B* orientation, the melt is cooled down, rates ranging from 1°C min^{-1} to 1°C hr^{-1}; however, higher cooling rates are allowed when a steep tem-

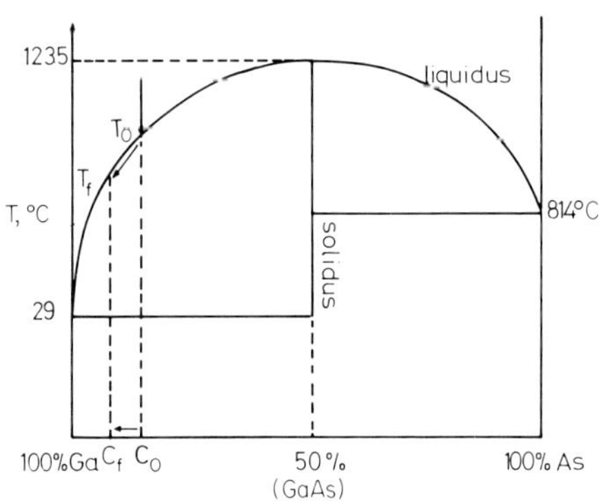

Fig. 17. Liquid-phase epitaxy, GaAs phase diagram.

perature gradient is located at the solid–liquid interface (tipping or rotating crucible) and very slow cooling is not necessary to avoid supercooling.

Typical growth rates are 10–20 μm hr^{-1}.

If the seed can be extracted easily from the melt (dipping, rotating crucible), the end of the run takes place at any temperature, allowing a good control of the grown layer thickness. If not, one is compelled to cool the melt to room temperature. Gallium stuck on the wafer can be removed by a hot HCl treatment.

Saturation of the melt can be accomplished by several different processes.

1. Use of pure arsenic. This is rarely used because of the difficulty in properly dissolving As in Ga (high vapor pressure) and the scarcity of very pure As.
2. Use of polycrystalline or bulk single-crystal undoped GaAs. This process is the more commonly used. Polycrystalline or bulk single-crystal GaAs is now available in reasonable purity (10^{16} impurities cm^{-3}), and dissolves sufficiently in Ga melts.
3. Saturation of the melt by a stream of the mixture $AsCl_3/H_2$, very similar to that used in vapor-phase epitaxy (Goodwin *et al.*, 1968).

In all cases, high-purity, palladium-diffused hydrogen is passed over the melt at flow rates of 20–200 cm^3 min^{-1} during the whole run. This is possible because the arsenic pressure over the melt is low enough in the temperature and concentration ranges used.

Because of the high segregation coefficient of impurities between the Ga-rich melt and the solid, high-purity gallium arsenide of total impurity content less than 10^{15} cm^{-3} can be deposited by these methods.

Of course, great attention must be given to the gallium purity (99.9999%, at least) and to the preparation of the boat containing the melt.

The boat and the glassware of the system are made of high-purity silica and are carefully washed with electronic grade chemical reagents (HNO_3–HCl, HF), then rinsed in deionized water and dried before use.

The electrical properties of pure GaAs grown by such processes are excellent: low free-carrier concentration, $10^{12} < n < 10^{14}$ cm^{-3}; high mobilities, $\mu(300°K) > 8000$ cm^2 V^{-1} sec^{-1}, $\mu(77°K) > 100{,}000$ cm^2 V^{-1} sec^{-1}.

a. Doping processes. Doping is done in GaAs for many different purposes. Most of them require the use of pure material doped with only one type of impurity. The techniques that we have just described are suitable for doing this provided a few modifications are made.

If the dopant is not too volatile and if a controlled impurity profile is not required, a certain amount of dopant can be directly added to the melt before saturation. Silicon and tin are often used in this way.

If the dopant is volatile (S, Se, Te, Zn), the melt can be enriched at any time during the run through the vapor phase. This allows a partial control of the impurity profile.

Finally, it is rather easy in the dipping process to drop solid or liquid dopants into the melt though a special tubing leading to the top of the crucible.

Whatever process is chosen, one must note that doping in LPE is far less flexible than in VPE. In particular, it is impossible to make an undoped layer after a doped one without changing the melt.

b. High doping levels. As previously mentioned, one of the best ways to get high doping levels is to use a solvent different from the group III and group V elements involved in the III–V compound to be grown. Both Sn and Zn can be used to grow n_{++} and p_{++} gallium arsenide doped in the range 10^{19}–10^{20} free carriers cm^{-3}.

This technique is often used to make ohmic contacts by LPE. The regrown layers must be as thin as possible and some authors have described a specially designed horizontal apparatus in which the graphite boat possesses a drawer at its bottom, allowing the seed and the melt to be in contact as briefly as possible.

c. Conclusion. Comparing GaAs LPE with VPE, we find the following:

1. The purity of LPE-grown GaAs is slightly higher than that of VPE-grown GaAs.
2. Fewer Ga vacancies are found in LPE-grown GaAs.
3. Good interfaces are obtained more easily by LPE than by VPE.
4. Growth rates and surface finish are less well-controlled in LPE than in VPE. (it is still difficult to grow very thin layers—$< 10\ \mu m$—by LPE).
5. Finally, doping is less flexible in the case of LPE.

To sum up, growing good GaAs is easier by LPE than by VPE, but VPE is far more flexible and better adapted to device-making, except in the case of electroluminescent devices, where very few Ga vacancies are wanted—LPE is then always prefered.

5. *Gallium phosphide*

Liquid-phase epitaxy is well-suited to growing GaP for electroluminescent devices, which is its main application.

Undoped GaP is not very useful in itself, and it is necessary to dope it with selected impurities, the matrix being as pure as possible.

The different methods described for GaAs are applicable to GaP at somewhat higher temperatures (900–1150°C, 2–10 atom % phosphorus). However, new considerations arise from the fact that the vapor pressures of phosphorus and of certain dopants (e.g., zinc) over the melt are high enough if the starting temperature is above 1100°C.

When n-type material is needed, epitaxy can be carried out at 900°C because the solubility of dopants (Te, Se, Si) in Ga is sufficient for this. To grow good, uncompensated material, low cooling rates are chosen (60–10°C hr^{-1}) and the apparatus is flushed with pure hydrogen (Lorentz and Pilkuhn, 1966).

On the other hand, p-type material, doped with zinc or ZnO, is generally deposited in a closed system because of high vapor pressures, especially for ZnO doping, which requires high temperatures ($T > 1100$°C) due to the low solubility of oxygen in gallium (Foster and Scardefield, 1969).

Cooling rates allowed here are higher than for n-type epitaxy (500–180°C hr^{-1}). More details will be found in papers by Saul *et al.* (1969), Saul and Hackett (1970), and Shih *et al.* (1968a).

It is sometimes interesting to grow pn junctions during the same epitaxy run, as is done, for example, with GaP doped with sulfur and then counter-doped with zinc for the manufacture of green-light-emitting diodes. The dipping process is well suited for this process, sulfur being put in the melt before saturation and zinc then dropped in later during the cooling period (Shih *et al.*, 1968b; Shih and Petit, 1968).

Gallium phosphide grown by LPE seems to be far better for making electroluminescent devices than GaP grown by any other method. Vapor growth is often used to provides single-crystal substrates for LPE or for the manufacture of ternary compounds, which are more difficult to grow by an LPE process.

Finally, we note that work has also been done on platelet growth in systems which are very similar to LPE systems, but which are unseeded.

6. *Other compounds*

Little information is available on the synthesis of binary compounds other than GaAs or GaP by LPE. However, the method is applicable to almost any of these compounds using the group III element as solvent. Group V elements can also be used as solvents if vapor pressures are not to high (As, Sb).

a. GaN. GaN can be prepared in a Ga-rich melt, NH_3 being used to saturate the solution.

b. GaSb. The synthesis of GaSb in Sb-rich melts (75 atom %) has been achieved by Reed *et al.* (1966).

c. AlP. The solubility of AlP in Al has been studied by Beer (1969) in the temperature range 900–1200°C. The LPE of AlP is certainly an interesting achievement.

Finally, as will be noted later, LPE has been used for the preparation of ternary III–V compounds.

C. Vapor–Liquid–Solid Processes

This process is a particular case of LPE, saturation being achieved through a vapor phase and deposition occurring at constant temperature by a variation of the total mass of the system. In fact, it is very similar to the process described for bulk solution growth, the differences being as follows:

1. The cause of growth is the use of a vapor pressure of the volatile component slightly higher than the equilibrium pressure. The system tends to return to the equilibrium by consuming the volatile component, which is achieved depositing the compound.
2. The use of a seed crystal.
3. The small mass of the system, due to the fact that only thin layers are wanted.

This method has not yet been used very much, but is interesting because of the ability to grow single-crystal layers at constant temperature.

D. Vacuum Decomposition of Thin Layers

Thin layers of III–V compounds have not very often been deposited under vacuum except for InAs or InSb, because of the difficulty of making single crystals by other methods and the problems of stoichiometry. Nevertheless, a certain amount of work has been done on methods of growth under vacuum, the attraction being the ability to grow very thin layers (<5000 Å).

Four different methods have been used:

1. Vacuum evaporation by the "three-crucible method."
2. Flash evaporation.
3. Sputtering.
4. Electron beam recrystallization under vacuum.

We will not describe these methods further; they are by now quite well known for the deposition of many materials in thin layers, giving deposits that are generally highly oriented polycrystalline layers.

True homoepitaxy or heteroepitaxy on a well-chosen substrate can sometimes be achieved by adjusting the temperature of the seed wafer. However, the electrical properties are often poor, except for InSb and InAs, whose mobilities, for example, have been found to be close to bulk material mobilities.

Three-crucible methods can be used when the two components have relatively high vapor pressures (InSb, InAs), with reasonable growth rates (Juhasz and Anderson, 1965; Holloway *et al.*, 1966; Sharma and Jain, 1968). When a group III element has a low vapor pressure, the technique can still be used, but growth rates are low and it is more difficult to control the stoichiometry of the deposit. Both GaAs (Steinberg and Scruggs, 1966; Davey and Pankey, 1968a) and GaP (Davey and Pankey, 1968b) have been deposited by the method.

Flash evaporation (Light *et al.*, 1968) and sputtering (Molnar *et al.*, 1964) seem to be more suitable methods for growing compounds containing a slightly volatile group III element such as Ga.

Finally, electron beam recrystallization can be a complementary method to the previous one, making it possible to turn highly polycrystalline layers into large, single-crystalline areas. Studies have been done on InSb by Davis and Wieder (1966) and Teede (1967).

IV. III–V Ternary Compounds

A. Introduction

Early in the research on III–V compounds, interest arose in ternary solid solutions both for theoretical reasons and because of their physical properties. Applied research on electroluminescence and volume field effects has led to the need for adjustable band-gap semiconductors.

B. Physical Properties of III–V Ternary Solid Solutions

1. *Phase diagrams*

The ternary systems investigated are those for which the width of the solid solubility region for the two binary compounds is large.

The other phase diagrams are unknown, although a few have been partially studied. Most of the known ternary diagrams are of the pseudobinary type $A_{1-x}B_xC$, where x is the mole fraction of substitutional B atoms

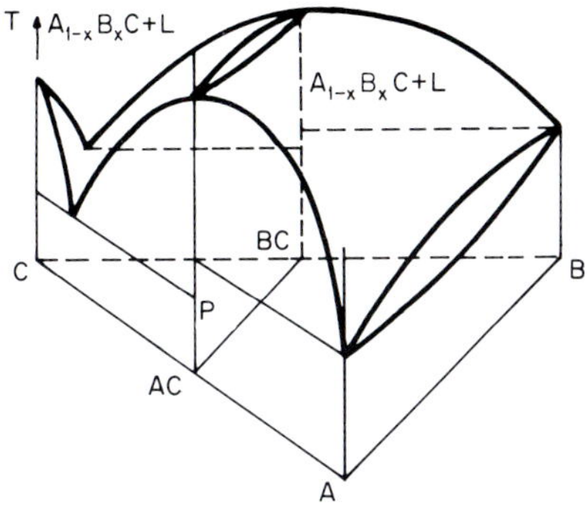

FIG. 18. Schematic ternary phase diagram: binary and pseudobinary compounds.

in the AC compound, or in other terms, the mole fraction of BC compounds in the AC–BC solid solution. In this particular case, the ternary solution is considered perfectly stoichiometric, $[C]/([A] + [B]) = 1$ for every $x, 0 < x < 1$.

The main interest in this type of diagram is in the indication of the domain of solubility of the solid solution versus temperature and the segregation coefficient of B or A in the ternary compound in the whole range of composition (Fig. 18).

If the solution-growth of the compound is to be investigated, it is necessary to settle on the most general phase diagram, or at least particular sections of it, such as P' and P'' in Fig. 19 (u = cte, $0 < x < 1$) or $Q(x$ = cte, $0 < u < 1)$, where u is the mole fraction of C in the liquid phase.

The first attempts to calculate III–V ternary phase diagrams were restricted to the regular solution approach (Wagner, 1958; Schottky and

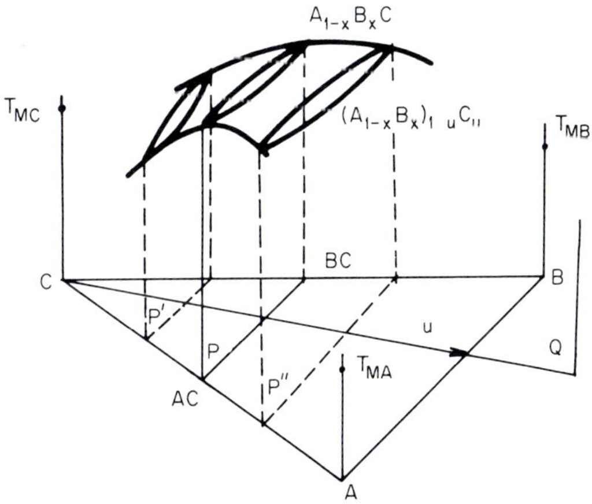

FIG. 19. Schematic ternary phase diagram: pseudobinary and ternary compounds.

Bever, 1958; Vieland, 1963; Thurmond, 1965; Arthur, 1967; Darken, 1967).

Theoretical determinations of such systems are described by Stringfellow and Green (1969), Antypas and James (1970), Panish (1970), and Steininger (1970) in the case of total miscibility of the two binary compounds.

Mathematical treatments require knowledge of temperature, entropy of melting, and activity coefficients for elements and binary compounds present in the system.

The calculations give the interaction parameters and chemical potentials versus the molar fraction of the elements in the solution. Good agreement between computed and experimental determinations has been obtained.

More thermodynamic data on elements and compounds are needed to extend the use of such models to permit a systematic determination of ternary phase diagrams.

2. *Band structure*

The most important property of a ternary solid solution is the possibility for a continuously controlled variation of band-gap structure with the chemical composition x over the solubility range. Data on the energy gap and band shape of some interesting III–V ternary solutions are listed in Table III. The other effects of the chemical composition on the band structure are the direct–indirect transition for radiative recombination of carriers and the temperature and pressure dependences of the band gap.

TABLE III

Solid solution	Method of growth[a]	Investigated composition range	Composition x for direct–indirect transition	E_g (energy gap) (eV)
$In_{1-x}Ga_xP$	VP	$0 \leq x \leq 1$	0.76	2.26
$Ga_{1-x}Al_xAs$	VPE–LPE	$0 < x < 0.6$	0.30–0.36	1.9
$InAs_{1-x}P_x$	VPE	$0 \leq x \leq 1$	Direct	—
$GaAs_{1-x}Sb_x$	VPE	$0 < x < 0.12$	Direct	—
$GaAs_{1-x}P_x$	VPE–LPE	$0 \leq x \leq 1$	0.44	~1.95
$In_{1-x}Ga_xSb$	LPE	—	Direct	—
$InAs_{1-x}Sb_x$	VPE	$x > 0.95$	Direct	—
$Ga_{1-x}In_xAs$	VPE	$x > 0.67$	Direct	—

[a] VP, Vapor phase; VPE, vapor-phase epitaxy; LPE, liquid-phase epitaxy.

C. Metallurgy of III–V Ternary Solid Solutions

a. Large crystals. The more frequent techniques here are melt and vapor growth. Melt growth gives a typical concentration gradient with the Bridgman–Stöckbarger technique (normal freezing) which is of great interest for investigations of the physical properties of solid solutions in a wide range of composition. Growth of InSb–InAs has been achieved by Kudman and Ekstrom (1968) and solution growth of Ga–InP by Rodot *et al.* (1969).

Vapor-phase transport in a sealed ampoule is a very simple method, but the single crystals obtained are small. It has been used for Al–As–Ga by Black and Ku (1966) and for BGaAs by Ku (1966).

In these experiments, the starting material is a mixture of the two compounds of suitable composition and the transport reaction is achieved by halogens. Graded-composition crystals are also grown by this method.

b. Epitaxial layers. Most of the devices using ternary solid solutions need semiconductor materials with precise forbidden band gaps. Due to this fact, the chemical composition is determined and must stay constant or have a controlled variation. Epitaxial growth methods are particularly interesting for this purpose, mostly with regard to homogeneity of composition and the possibility of doping during the growth of the layers.

(*1*) *Vapor-phase transport in an open tube.* The basic chemical transport mechanism is the same for ternary and binary compounds, and the thermochemical reactions are developed in Section III.A. The process, which consists in simultaneous transport of the two compounds, is now preferably achieved in separate-flux reactors. The active chemical species are obtained in separate chambers and the ternary solid solution is synthesized in a common area of the reactor.

The chemical composition and the stoichiometry are adjusted by using the flow rates of the vector gas in each chamber and are both controlled.

The chemical transport is performed by halogenides of group III elements in the presence of group V elements, or by hydrides. In most of the experiments, only chlorine or iodine is used, depending on the formula of the ternary compound. If the substitutional element in the $A_{1-x}B_xC$ ternary compound is a group III element (e.g., $Ga_{1-x}Al_xAs$, $In_{1-x}Ga_xP$), the reaction is of the type shown in Fig. 20a, and for a group V substitutional element (e.g., $InAs_{1-x}P_x$, $GaAs_{1-x}P_x$), the reaction is of the type shown in Fig. 20b. Most of the III–V ternary compounds have been grown successfully by vapor-phase epitaxy: In–GaP by Maruska and Tietjen (1969) and Tietjen *et al.* (1969a,b); In–AsP by Folberth (1955), Folberth and Weiss (1955),

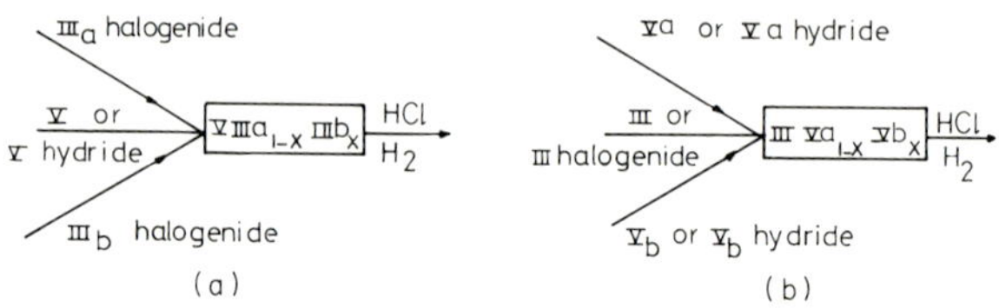

FIG. 20. Reactions for the growth of epitaxial layers of ternary solid solutions by tube. (a) Group III substitutional element, (b) group V substitutional element.

Weiss (1956), Bowers (1959), and Tietjen *et al.* (1968); Ga–AsSb by Tietjen *et al.* (1968) and Berka *et al.* (1969); Ga–AlAs by Ku (1965), Rubenstein (1965), Tietjen (1967), and Berka (1969); Ga–AsP by Tietjen *et al.* (1967), Ewing and Smith (1968), and Holonyak *et al.* (1962); and In–GaAs by Tietjen *et al.* (1968).

(*2*) *Liquid-phase epitaxy.* Solution growth of epitaxial layers has been performed for a number of III–V ternary compounds. Experimental determinations of the phase diagrams are available in specific cases (InAs–InSb), (InAs–GaAs), (GaAs–AlAs), (GaAs–GaP). Pseudobinary diagrams give values for the segregation coefficient of A or B in the $A_{1-x}B_xC$ solid solution which are strongly different from unity: $10^{-4} < k = c_S/c_L < 10^{-1}$; this causes strong variations of x during the growth of the layer.

According to Tiller (1963), constitutional supercooling will not take place for

$$V \leq GD_L k_0 / mC_{S0}(1 - k_0)$$

where V is the growth rate, G is the thermal gradient at the growth interface, k_0 is the segregation coefficient, C_{S0} the average concentration of A or B in solid solution, $m = dT/dC$ is the slope of the liquidus curve at the growth temperature, and D_L is the diffusion coefficient of A or B in the liquid.

For rich solutions and isothermal or low-gradient experiments, slow rates of deposition must be used. In this technique, it is necessary to calibrate the melt composition before growth starts, to have a good control of the solid composition and the associated energy gap.

The most important work on ternary III–V solution epitaxy has been done with Al–Ga–As (Ku, 1966; Rupprecht *et al.*, 1966; Kressel *et al.*, 1968; Ilegems *et al.*, 1968; Panish *et al.*, 1968; Merenda, 1969), In–GaP (White *et al.*, 1968; Hakki *et al.*, 1970), and InAsSb (Antypas and James, 1970). Recent developments in high efficiency III–V ternary photocathodes increase the interest in liquid epitaxy of GaAs–Sb, InSb–P, AlAs–P (James *et al.*, 1970) and new discoveries in this field are coming rapidly.

REFERENCES

Alferov, Z. H. I., Korol'kov, M. K., Trukan, M. K., and Chashchin, S. P. (1966). *Sov. Phys. Solid-State* **7,** 2370–2374.

Andre, E., and Le Duc, J. M. (1968). *Mater. Res. Bull.* **6,** 1–6.

Angeli, M. (1966). French Government Rept. DGRST 65.34.096.

Antell, G. R., and Effer, D. (1959). *J. Electrochem. Soc.* **106,** 509.

Antypas, G. A., and James, L. W. (1970). *J. Appl. Phys.* **41,** 2165.

Arthur, J. R. (1967). *J. Phys. Chem. Solids* **28,** 2257.

Bass, S. J., and Olivier, P. E. (1968). *J. Crystal Growth* **3, 4,** 286.

Beer, S. Z. (1969). *J. Electrochem. Soc.* **116,** 263.

Berka, L., Briggs, T., Millard, M., and Jolly, W. (1969). *J. Inorg. Nucl. Chem.* **14,** 190.

Black, J. F., and Ku, S. M. (1966). *J. Electrochem. Soc.* **113,** 249.

Bobb, L. C., Holloway, H., and Maxwell, K. H. (1966a). *J. Appl. Phys.* **37,** 4687.

Bobb, L. C., Holloway, H., and Maxwell, K. H. (1966b). *J. Phys. Chem. Solids* **27,** 1679–1685.

Bolger, D. E., Franks, J., Gordon, J., and Whitaker, J. (1966–1967). *Proc. Int. Symp. GaAs*, IPPS Conf. Ser. Inst. of Phys. and the Phys. Soc., pp. 16–22.

Bowers, R. *et al.* (1959). *J. Appl. Phys.* **30,** 1010.

Bridgmann, P. W. (1925). *Proc. Amer. Acad. Sci.* **60,** 305.

Broder, J. D., and Wolff, G. A. (1963). *J. Electrochem. Soc.* **110,** 1150.

Chu, T., Ing, D. W., and Noreika, A. J. (1967). *Solid State Electron.* **10,** 1023–1027.

Conrad, R. W., Reynolds, R. A., and Jeffcoat, M. W. (1967). *Solid State Electron.* **10,** 507.

Cronin, G. R., Larrabee, G. B., and Osborne, J. F. (1966). *J. Electrochem. Soc.* **113,** 293.

Czochralski, J. (1917). *J. Phys. Chem.* **92,** 219.

Darken, L. S. (1967). *Trans. AIME* **239,** 80.

Davey, J. E., and Pankey, T. (1968a). *J. Appl. Phys.* **39,** 1941.

Davey, J. E., and Pankey, T. (1968b). *Appl. Phys. Lett.* **12,** 38.

Davis, N. M., and Wieder, H. H. (1966). *Proc. Laser Beams Soc., April,* 385.

Davis, R. E. (1960). "Properties of Elemental and Compound Semiconductors." Wiley (Interscience), New York.

Effer, D. (1961). *J. Electrochem. Soc.* **108,** 357.

Effer, D. (1965). *J. Electrochem. Soc.* **112,** 1020.

Ellis, S. G. (1959). *J. Appl. Phys.* **30,** 947.

Enstrom, R. E., and Peterson, C. (1967). *Trans. Met. Soc. AIME* **239,** 413.

Ewing, R. E., and Greene, P. E. (1964). *J. Electrochem. Soc.* **111,** 1266.

Ewing, R. E., and Smith, D. K. (1968). *J. Appl. Phys.* **39,** 5943.

Faulkner, K. P., Wickenden, D. K., Isherwood, B. J., Richards, B. P., and Scobey, I. N. (1970). *J. Mater. Sci.* **5,** 308.

Folberth, O. G. (1955). *Z. Naturforsch.* **10a** 502.

Folberth, O. G., and Weiss, H. (1955). *Z. Naturforsch.* **10a,** 615.

Foster, L. M., and Scardefield, J. (1969). *J. Electrochem. Soc.* **116,** 494.

Frosch, C. J., and Derick, L. (1961). *J. Electrochem. Soc.* **108,** 251.

Gershenzon, M., Frosch, C. J., and Mikulyak, R. M. (1961). *J. Electrochem. Soc.* **108,** 53C.

Goodwin, A. R., Gordon, J., and Dolsson, C. D. (1968). *Brit. J. Appl. Phys. Ser. 2* **1,** 115.

Gottlieb, G. E. (1965). *J. Electrochem. Soc.* **112,** 192.

Gremmelmaier, R. (1956). *Z. Naturforsch.* **11a,** 511.

Gremmelmaier, R., and Madelung, O. (1953). *Z. Naturforsch.* **8a,** 333.

Guire, R. J., and Weiser, K. (1959). U.S. Patent 2871100.
Hakki, B. W., Jayavaman, A., and Kim, C. K. (1970). *J. Appl. Phys.* **41,** 5291.
Harman, T. C., Goering, H. L., and Beer, A. C. (1956). *Phys. Rev.* **104,** 1562.
Harris, J. S., and Snyder, W. L. (1969). *Solid State Electron.* **12,** 337.
Hasegawa, F., and Saito, T. (1968). *Japan J. Appl. Phys.* **7,** 1125.
Hicks, H. G. B., and Manley, D. F. (1969). *Solid State Commun.* **7,** 1463.
Hilsum, C., and Rose-Innes, A. C. (1961). "Semiconducting III–V Compounds." Pergamon, New York.
Holloway, H., Richards, J. L., Bobb, L. C., Perry, J., Jr., and Zimmerman, E. (1966). *J. Appl. Phys.* **37,** 4694.
Holonyak, N., Jr. *et al.* (1962). "Metallurgy of Semiconductor Materials." Wiley (Interscience), New York.
Hoss, P. A., Murray, L. A., and Rivera, J. J. (1968). *J. Electrochem. Soc.* **115,** 0553.
Hoyt, P. L., and Haisty, R. W. (1966). *J. Electrochem. Soc.* **113,** 296.
Huber, A., and Champier, G. (1970). *Proc. Int. Sym. AsGa, 3rd, IPPS, Aachen.*
Ilegems, M., and Pearson, G. L. (1968). *Proc. Symp. AsGa, 2nd, IPPS, Dallas,* pp. 3–10.
James, L. W., Antypas, G. A., Uebbing, J. J., Edgecumbe, J., and Bell, R. L. (1970). *Proc. Symp. AsGa, 3rd, IPPS, Aachen.*
Jeffes, J. H. E. (1968). *J. Crystal Growth* **3,** 13.
Joyce, B. A. (1968). *J. Crystal Growth* **3, 4,** 43.
Joyce, B. A., and Mullin, J. B. (1966). *Solid State Commun.* **4,** 463.
Joyce, B. D., and Williams, E. W. (1970). *Proc. Symp. AsGa, 3rd, IPPS, Aachen,* p. 57.
Juhasz, C., and Anderson, J. C. (1965). *Vacuum Conf. 3rd* **6G,** 333.
Kang, C. S., and Greene, P. E. (1967). *Appl. Phys. Lett.* **11,** 171.
Knight, J. R., Effer, D., and Evans, P. R. (1965). *Solid State Electron.* **8,** 178–180.
Kontrimas, R., and Blakeslee, A. E. (1968). *Electrochem. Technol.* **6,** 78.
Kressel, H., Dunse, J., Nelson, H., and Hawrylo, F. (1968). *J. Appl. Phys.* **39,** 4337.
Ku, S. M. (1965). *J. Electrochem. Soc.* **110,** 991.
Ku, S. M. (1966). *J. Electrochem. Soc.* **113,** 813.
Kudman, I., and Ekstrom, L. (1968). *J. Appl. Phys.* **39,** 3385.
Lawley, K. L. (1966). *J. Electrochem. Soc.* **113,** 240.
Lawson, W. D., and Nielson, S. (1958). "Preparation of Single Crystals," p. 17. Butterworth, London.
Light, T. B., Hull, E. M., and Gereth, R. (1968). *J. Electrochem. Soc.* **115,** 857.
Lorentz, M. R., and Pilkuhn, M. (1966). *J. Appl. Phys.* **37,** 4094.
Luther, L. C. (1969). *J. Electrochem. Soc.* **116,** 374.
Luther, L. C., and Roccasecca, D. D. (1968). *J. Electrochem. Soc.* **115,** 850.
Mac Carthy, J. P. (1966). *Solid State Commun.* **5,** 605.
Mac Carthy, J. P. (1967). *Solid State Electron.* **10,** 649.
Madelung, O. (1964). "Physics of III–V Compounds." Wiley, New York.
Manasevit, H. M. (1968). *Appl. Phys. Lett.* **12,** 156.
Maruska, H. P., and Tietjen, J. J. (1969). *Appl. Phys. Lett.* **15,** 327.
Mehal, E. W., and Gronin, G. R. (1966). *Electrochem. Technol.* **4,** 540.
Merenda, P. (1968). French Government Rept. DGRST 66.00.420.
Merenda, P. (1969). Thomson-CSF unpublished paper. Orsay, France.
Michelitsh, M. (1965). *J. Electrochem. Soc.* **112,** 747.
Michelitsh, M., Kappallo, W., and Hellbardt, G. (1964). *J. Electrochem. Soc.* **111,** 1248.
Minden, H. T. (1970). *J. Crystal Growth* **6,** 228.
Moest, R. R. (1966). *J. Electrochem. Soc.* **113,** 141.
Molnar, B., Flood, J. J., and Francombe, M. (1964). *J. Apply. Phys.* **35**(12), 3554.

Moody, P. L., and Kolm, C. (1958). *Rev. Sci. Instrum.* **29,** 1144.

Moody, P. L., Gatos, H. C., and Lavine, M. C. (1960). *J. Appl. Phys.* **31,** 1696.

Mullin, J. B., Straughan, B. W., and Brickell, W. S. (1965). *J. Phys. Chem. Solids* **26,** 782.

Mullin, J. B., Heritage, R. J., Holliday, C. H., and Straughan, B. W. (1968). *J. Crystal Growth* **3, 4,** 281–285

Nelson, H. (1963). *RCA Rev.* **24,** 603.

Nitsche, R. (1967). *J. Phys. Chem. Solids Suppl.* **1,** 215.

Noreika, A. J., and Ing, D. W. (1968). *J. Appl. Phys.* **39,** 5578.

Panish, M. B., and Ilegems, M. (1970). *Proc. Symp. AsGa, 3rd, IPPS, Aachen.*

Reed, F. T., Baxter, R. D., and Miller, S. E. (1966). *J. Electrochem. Soc.* **113,** 713.

Richman, D. (1968). *J. Electrochem. Soc.* **115,** 845.

Richman, D., and Tietjen, J. J. (1967). *Trans. Met. Soc. AIME* **239,** 418.

Rodot, H., Hruby, A., and Scheider, M. (1968). *J. Crystal Growth* **3, 4,** 305.

Rodot, H., Florak, J., Rouy, G., and Bourneix, J. (1969). *C. R. Acad. Sci. Paris* **269B,** 381–384.

Rubenstein, M. (1965). *J. Electrochem. Soc.* **112,** 426.

Rubenstein, M., and Myers, E. (1966). *J. Electrochem. Soc.* **113,** 365.

Rupprecht, H., Woodall, J. M., Konneath, K., and Petit, D. G. (1966). *Appl. Phys. Lett.* **9,** 221, 3.

Saul, R. H., and Hackett, W. H., Jr. (1970). *J. Electrochem. Soc.* **117,** 921.

Saul, R. H., Armstrong, J., and Hackett, W. H., Jr. (1969). *Appl. Phys. Lett.* **15,** 229.

Schäfer, H. (1962). "Chemische Transportreaktionen." Verlag Chemie Gybh, Weinheim, Germany.

Schottky, W. F., and Bever, M. B. (1958). *Acta Metall.* **6,** 320.

Schulze, R. G. (1966). *J. Appl. Phys.* **37,** 4295.

Sharma, S. K., and Jain, V. K. (1968). *Solid State Electron.* **11,** 423.

Shaw, D. W. (1968a). *J. Electrochem. Soc.* **115,** 405.

Shaw, D. W. (1968b). *J. Electrochem. Soc.* **115,** 777.

Shih, K. K., and Petit, G. D. (1968). *J. Appl. Phys.* **39,** 5025.

Shih, K. K., Lorentz, M. R., and Foster, L. M. (1968a). *J. Appl. Phys.* **39,** 2747.

Shih, K. K., Woodall, J. M., Blum, S. E., and Foster, L. M. (1968b). *J. Appl. Phys.* **39,** 2962.

Steinberg, R. E., and Scruggs, D. M. (1966). *J. Appl. Phys.* **37,** 4526.

Steinemann, A., and Zimmerli, U. (1963). *Solid State Electron.* **6,** 597.

Steininger, J. (1970). *J. Apply. Phys.* **41,** 6, 2713.

Stockbarger, D. C. (1936). *Rev. Sci. Instrum.* **7,** 133.

Stringfellow, G. B., and Greene, P. E. (1969). *J. Phys. Chem. Solids* **30,** 1779.

Taylor, R. C. (1967). *J. Electrochem. Soc.* **114,** 410.

Taylor, R. C. (1969). *J. Electrochem. Soc.* **116,** 383.

Taylor, R. C., Woods, J. F., and Lorenz, M. R. (1968). *J. Appl. Phys.* **39,** 5404.

Teede, N. F. (1967). *Solid State Electron.* **10,** 1069.

Thomas, R. W. (1969). *J. Electrochem. Soc.* **116,** 1149.

Thurmond, C. D. (1965). *J. Phys. Chem. Solids* **27,** 785.

Tietjen, J. J., and Amick, J. A. (1966). *J. Electrochem. Soc.* **113,** 724.

Tietjen, J. J. *et al.* (1967). *Trans. Met. Soc. AIME* **239,** 385.

Tietjen, J. J., Clough, R., and Richman, D. (1968). Vapor Phase Growth Technique and System for Several III–V Semiconductors. RCA Rept. on NASA contract NAS 12–538 by Clearinghouse N. 69–33320.

Tietjen, J. J., Clough, K., Dreeben, A. B., Enstrom, R., and Richman, D. (1969a). Vapor Phase Growth Technique and System for Several III–V Semiconductors. RCA Rept. on NASA contract NAS 12–538. Clearinghouse N. 69–33262.

Tietjen, J. J., Maruska, H. P., and Clough, R. B. (1969b). *J. Electrochem. Soc.* **116**, 492.

Tiller, W. A. (1963). *In* "The Art and Science of Growing Crystals" (J. J. Gilman, ed.) p. 276. Wiley, New York.

Tiller, W. A., and Kang, C. (1968). *J. Crystal Growth* **2**, 345.

Van Der Does De Bye, J. A. W., and Peters, R. C. (1969). *Philips Res. Rept.* **24**, 210.

Vieland, L. J. (1963). *Acta Met.* **11**, 137.

Vieland, L. J., and Skalski, S. (1960). *Metall. Elemental Compound Semiconduct.* **12**, 303.

Wagner, C. (1958). *Acta Met.* **6**, 309.

Weisberg, L. R., Rosi, F. D., and Herkart, P. G. (1959). "Properties of Elemental and Compound Semiconductors," p. 25. Wiley (Interscience), New York.

Weiss, H. (1956). *Z. Naturforsch.* **11a**, 430.

Welker, H. (1952). *Z. Naturforsch.* **7a**, 744.

White, A. M., Williams, E. W., and Porteous, P. (1968). *Phys. Stat. Sol.* **30**, K125.

Williams, E. W., and Blacknall, D. M. (1967). *Trans. Met. Soc. AIME* **239**, 387.

Willardson, R. K., and Beer, A. C. eds. (1966). "Semiconductors and Semimetals. Vols. 1, 2. Physics of III–V Compounds," Academic Press, New York.

Willardson, R. K., and Beer, A. C., eds. (1967). "Semiconductors and Semimetals. Vol. 3. Optical Properties of III–V Compounds." Academic Press, New York.

Willardson, R. K., and Beer, A. C., eds. (1968). "Semiconductors and Semimetals. Vol. 4. Physics of III–V Compounds." Academic Press, New York.

Willardson, R. K., and Goering, H. L. (1962). "Preparation of III–V Compounds," Vol. 1. Van Nostrand–Reinhold, Princeton, New Jersey.

Wolfe, C. H., Foyt, A. G., and Lindley, W. T. (1968). *Electrochem. Technol.* **6**, 208.

Preparation of CdS Single Crystals by Epitaxial Growth on Germanium Substrates

H. J. A. VAN DIJK

Philips Research Laboratories
N. V. Philips' Gloeilampenfabrieken
Eindhoven, The Netherlands

I. Introduction

Cadmium sulfide is probably the most intensively studied of the II–VI compounds. Because many physical properties are related to crystal structure and crystal perfection, there is a need for homogeneous single crystals of CdS. This chapter describes a method for the preparation of single-crystalline CdS platelets of thicknesses up to 0.5 mm and 20 mm diameter. The crystals are of the hexagonal wurzite type and mostly grown in the [$10\bar{1}3$] direction.

The epitaxial growth of CdS and other II–VI compounds has been reported by a number of authors. A great variety of substrates have been used for this purpose. Weinstein and Wolff (1967) and Curtis and Brunner (1970) investigated the epitaxial deposition of CdS on GaAs substrates. Heyraud and Capella (1968) and Wilcox and Holt (1969) used ionic crystals such as NaCl, KCl, and CaF_2 as substrates, whereas Aven and Garwacki (1963) deposited CdS on other II–VI compounds such as ZnSe. Van Dijk and Goorissen (1967) grew CdS and CdSe on germanium substrates and Aitkhozhin *et al.* (1969) reported the epitaxial growth of CdS on mica and other substrates. For epitaxial growth from the vapor phase, the material to be grown must be transported from a solid or liquid source to the seed. In the case of CdS, the source material usually consists of polycrystalline solid CdS that has to be volatilized. This can be done by evaporation in vacuum (Wilcox and Holt, 1969), sublimation in an inert gas stream (Caveney, 1968), or use of a gas transport reaction (Weinstein and Wolff, 1967; Curtis and Brunner, 1970; Heyraud and Capella, 1968). Hydrogen and iodine are often used (Schäfer, 1962) as the transporting agent for CdS, and the material transport occurs by displacement of the equilibrium in a temperature gradient of the reactions

$$CdS_s + H_{2,g} = H_2S_g + Cd_g \tag{1}$$

$$CdS_s + 2HX_g = CdX_{2,g} + H_2S_g \qquad X = Cl, Br, I \tag{2}$$

II. Experimental

We used purified hydrogen as the transporting agent in an open flow system. The experiments were carried out in a horizontal silica tube of 35 mm inside diameter. The furnace consisted of a three-zone, kanthal-wound silica tube surrounded by another silica tube that was spray-coated with SnO_2 as a reflection shield (Fig. 1a). The gas flow in the system could be established in either direction (Fig. 1b). The source material consisted of about 10 g of dense, luminescent grade CdS powder contained in a silica boat and located in the source zone of the system. Doped material

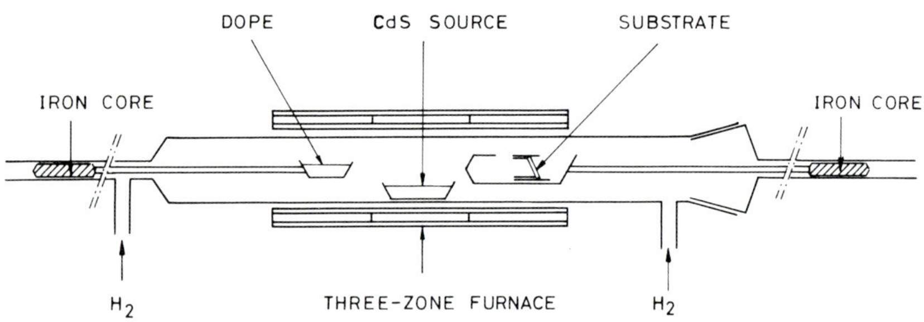

FIG. 1a. Apparatus for the epitaxial growth of CdS.

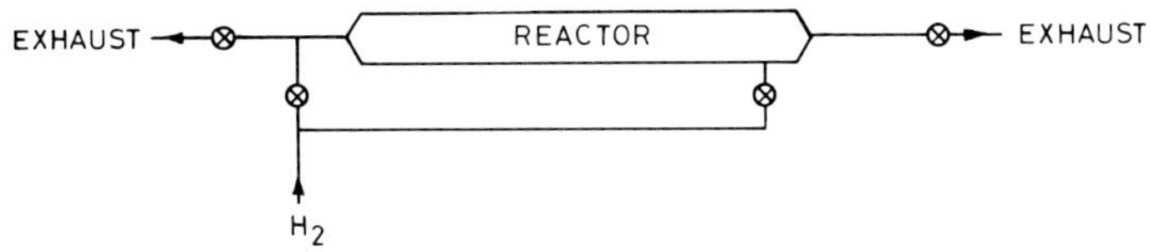

FIG. 1b. Flow sheet of the apparatus.

could be grown by placing an additional boat containing the dopant (as a rule, In_2S_3) in the dope zone of the reactor. The substrates were cut and lapped germanium slices of about 20 mm diameter.

Before use, the substrates were cleaned ultrasonically in water and acetone and etched for 15 min in a solution of 0.5 g $KMnO_4$ in 25 ml hydrofluoric acid (48%). While etching, air was bubbled through the etchant. After etching, the germanium slice was rinsed in deionized water, spin-dried, and placed on the substrate holder. Both the substrate holder and the dope container could be moved by means of a magnet. The bakeout and growth procedure for a doped CdS crystal is as follows.

A gas stream of 200 ml min^{-1}, palladium-diffused hydrogen is established in the direction toward the substrate. The substrate is located in the substrate zone, which is heated to 760°C. After 15 min, the furnace is cooled to 250°C and the substrate is withdrawn from the substrate zone of the furnace. The gas flow is reversed and the dope container is moved into the dope zone, which is heated to 760°C. After a 15-min bakeout, the furnace is cooled to 250°C and the dope container is withdrawn from the dope zone. The three furnace zones are now heated to 700, 860, and 660°C for the dope, source, and substrate zones, respectively. The CdS source is baked out under these conditions for 30 min. After cooling, the gas stream is reversed and the temperatures of the three zones are reestablished. As soon as a CdS deposit occurs just out of the furnace, the substrate is moved into the substrate zone. Since the substrate is cold when moved into position, the condensation of cadmium on the substrate must be avoided. This is done by a quartz shield in front of the substrate, on which the cadmium condenses in the initial stage of growth. It appeared to be necessary to move the substrate cold into position, because otherwise etching occurs. Under the conditions mentioned, a growth rate of about 1 μm min^{-1} was observed and smooth CdS crystals were grown up to thicknesses of about 0.3 mm. Thicker crystals were grown by increasing the deposition rate, which can be done by increasing the gas flow rate, the CdS–H_2 contact area, and the source temperature. The smoothness of the surface under these conditions is less than under those mentioned above.

After the deposition is complete, the substrate covered with CdS is withdrawn from the furnace, which is then cooled. The CdS crystal is separated from the germanium seed by dissolving the latter anodically or by selective chemical etching. To apply the anodic separation method successfully, the following requirement must be fulfilled. Since germanium is a semiconductor, only *p*-type and heavily doped *n*-type germanium can be dissolved anodically without strong illumination of the surface (Holmes, 1962; Gereth and Cowher, 1968). Therefore the seed crystals were always *p*-type (resistivity 1 ohm-cm) or heavily doped *n*-type (resistivity 0.005 ohm-cm). The CdS–Ge structure was cemented on a glass plate with beeswax, the sulfide side facing the glass plate. Since the back of the wafer is also covered with CdS, this is removed by etching in diluted HCl or by lapping. A platinum strip serving as the anode contact is pressed on the germanium by a perspex clamp and the wafer is immersed completely in the electrolyte. The electrolyte acts as the cooling agent of the contact area during dissolution. The cathode is made from a platinum gauze. The electrolyte consists of a 0.3 *M* solution of potassium ferricyanide in one liter of 1 *N* potassium hydroxide. In this electrolyte, a current density of about 0.5 A cm^{-2} can be drawn at low voltages (less than 10 V). The etch rate of the *p*-type germanium samples is about 3 $\mu m\ min^{-1}$. Since the resistivity of the CdS is much higher than that of germanium, the current in the circuit drops off as soon as the germanium has been dissolved. All that is left is a small germanium spot underneath the platinum contact. The remaining CdS crystal can be removed from the glass plate by melting or dissolving the beeswax in hot carbon tetrachloride.

High-resistivity germanium can be removed chemically in the etchant, but the etching time is then much longer (etch rate $\approx 0.1\ \mu m\ min^{-1}$). Crystals of 20 mm diameter can be handled safely provided they are at least 40 μm thick.

In the same way as described earlier, CdSe, CdTe, and ZnSe could be deposited on germanium substrates.

III. Properties of the CdS Crystals

Spectrochemical analysis of a typical sample revealed 10 ppm of silicon originating from the tube wall and 8 ppm of silver, which was found also in the source material. The germanium concentration was found to be only 2 ppm and it is likely that this contaminant is only present in the former Ge–CdS contact surface. No other contaminants in concentrations above 2 ppm were detectable. The resistivity of the as-grown material was 200 ohm-cm, which is low for undoped CdS. However, we were able to increase the dark resistivity to 10^{10}–10^{11} ohm-cm by annealing the samples in H_2S at 700°C for a few hours, indicating that the low resistivity originates from sulfur vacancies.

IV. Epitaxial Relations

The crystal perfection was examined by anomalous transmission of x rays. No deviations from the crystallographic orientation could be found that exceeded the limit of detection of this technique, which is 1 min of arc, so no small-angle boundaries were present and only the hexagonal structure could be detected.

A very distinct crystallographic relation between the grown CdS and the germanium seed was observed:

$$[111]\ \text{Ge} \parallel [0001]\ \text{CdS}$$

$$[110]\ \text{Ge} \parallel [2\bar{1}\bar{1}0]\ \text{CdS}$$

For [111]-oriented seeds, the *c* axis of the CdS is perpendicular to the substrate surface. A highly textured CdS surface resulted if this seed orientation was used. This may be due to the fact that the (0001) CdS plane is not stable under the growth conditions. On [110]-oriented seeds, there are two equivalent [111] directions that make an angle of 55° with the substrate surface. On precisely oriented seeds, we observed a strong twinning of the CdS crystal. The *c* axis of each part of this twin makes an angle of 55° with the seed surface. Twinning could be eliminated by misorienting the seed 1–5°. For this reason, most of the crystals were grown on substrate material cut 1–5° off the (110) plane toward the nearest [111] direction. On these seeds, the twinning is suppressed and the CdS *c* axis is oriented along the [111] Ge direction that makes the greatest angle with the seed surface. On [100]-oriented seeds, the four [111] directions are equivalent. We did not succeed in growing single crystals of CdS on these seeds. The grains of the polycrystalline films were oriented randomly along one of the four [111] directions. No influence of seed misorientation on this behavior was observed.

REFERENCES

Aitkhozhin, S. A. *et al.* (1969). *Sov. Phys. Crystallogr.* **14**, 309.
Aven, H., and Garwacki, W. (1963). *J. Electrochem. Soc.* **110**, 401.
Caveney, R. J. (1968). *J. Crystal Growth* **2**, 85.

Curtis, B. J., and Brunner, H. (1970). *J. Crystal Growth* **6,** 269.
Gereth, R., and Cowher, M. E. (1968). *J. Electrochem. Soc.* **115,** 645.
Heyraud, J. C., and Capella, L. (1968). *J. Crystal Growth* **2,** 405.
Holmes, P. J. (1962). "The Electrochemistry of Semiconductors," p. 181. Academic Press, New York and London.
Schäfer, H. (1962). "Chemische Transportreaktionen," pp. 53–54. Verlag Chemie, Weinheim.
van Dijk, H. J. A., and Goorissen, J. (1967). *J. Phys. Chem. Solids Suppl.* No. 1, 531.
Weinstein, M., and Wolff, G. A. (1967). "Crystal Growth" (H. S. Peiser, ed.), p. 537. Pergamon, Oxford.
Wilcox, D. M., and Holt, D. B. (1969). *J. Mater. Sci.* **4,** 672.

Oriented Eutectic Crystallization

W. ALBERS

Philips Research Laboratories
N.V. Philips' Gloeilampenfabrieken
Eindhoven, The Netherlands

I. Introduction

Composite materials are defined as heterogeneous materials consisting of two or more solid phases with one of the phases having at least one dimension ranging between about 10^{-6} and 10^{-2} cm. The interest in composite materials arises from the fact that their properties can be essentially different from those of the constituent phases, particularly if these are geometrically oriented to one another in some regular, periodic, and/or anisotropic way. For example, composite materials can offer advanced possibilities as reinforced materials for the construction and further development of supersonic aircraft, space vehicles, high-pressure tanks, etc., for which strength properties are required that cannot be provided by the existing homogeneous materials. A special class in this area are composites consisting of a ductile matrix (e.g., metal or plastic) enclosing a large amount of parallel whiskers, elongated thin crystals, fibers, or coated thin wires of a strong but brittle material (such as SiC, Al_2O_3, metal borides, glass fibers, etc.). Whisker or fiber lengths between 10^{-2} and 1 cm and diameters ranging from 10^{-5} to 10^{-2} cm may be considered as typical. If, for instance, whiskers with undamaged surfaces are embedded parallel to each other in an attaching, nonaggressive, ductile metal, then the resulting artificial composite can carry heavy loads parallel to the whiskers (Broutman and Krock, 1967).

A great variety of strong whiskers can be grown, e.g., from the gaseous phase. However, the problems of sorting, aligning, and embedding the whiskers without damaging their surface are numerous. Therefore it may be advantageous to grow the aligned whiskers *in situ* in the matrix in a one-step process, e.g., by means of a eutectic reaction (Straumanis and Brakss, 1935; Kraft and Albright, 1961; Salkind and Lemkey, 1967). This has the additional advantage that the whiskers and the matrix are then more or less in chemical equilibrium, which promotes the perfection of the whisker surface, even at higher temperatures.

Composite materials are equally promising in a number of fields in addition to those that employ their mechanical properties. At present, most metallic permanent magnets are made of alnico alloys, obtained by unidirectional solid-state decomposition under cooling conditions in an applied magentic field (De Vos, 1966). A representative example in electronics is the rod-type composite obtained by unidirectional solidification of a homogeneous eutectic melt of the quasibinary NiSb–InSb system. This composite consists of parallel, metallic NiSb rods (length $>5 \times 10^{-3}$ cm, diameter $\sim 10^{-4}$ cm) in a matrix of semiconducting InSb. The positive magnetoresistance is strongly anisotropic, and extremely large if the direction of the metallic rods, the electric current, and the applied magnetic field are mutually perpendicular to one another (Weiss and Wilhelm, 1963).

During the last decade, increased interest has been shown in the *in situ* growth of oriented composite materials. The field has become extensive and a number of review papers have appeared (Tiller, 1958; Winegard, 1961; Chadwick, 1963; Kraft, 1966; Kerr and Winegard, 1966; Hunt, 1968; Durand, 1969; Hogan *et al.*, 1970).

Oriented growth *in situ* can be accomplished by eutectic, monotectic, and eutectoid reactions (Sections II, IV.A, and IV.B, respectively) or by discontinuous precipitation from supersaturated solid solutions (Section IV.C). The growth of composite materials by means of continuous precipitation is not considered in this chapter.

II. Unidirectional Eutectic Solidification

The growth *in situ* of composite materials from a eutectic melt has the disadvantage that one must work with the eutectic systems provided by nature. The constituent phases cannot be chosen arbitrarily. Nevertheless, eutectic growth *in situ* covers a wide field, because eutectic reactions are

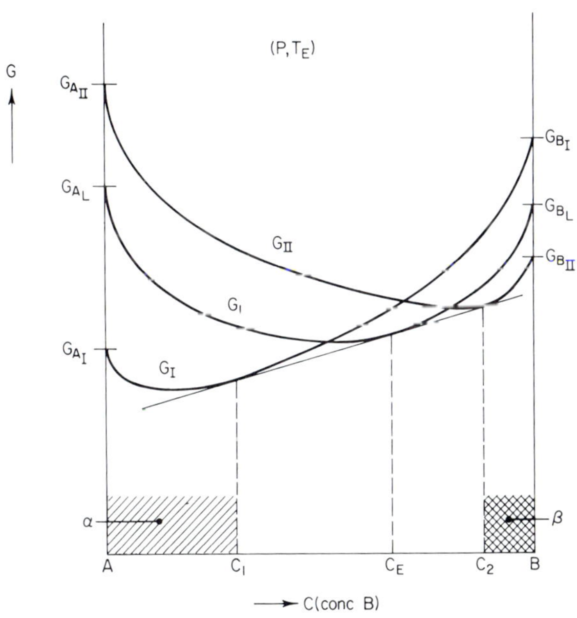

FIG. 1. The eutectic equilibrium. G_{A_I} is the Gibbs free energy of the crystalline component A with space group I, etc. Phase α of composition C_1 and phase β with composition C_2 coexist with the eutectic liquid of composition C_E at the eutectic temperature T_E and pressure P.

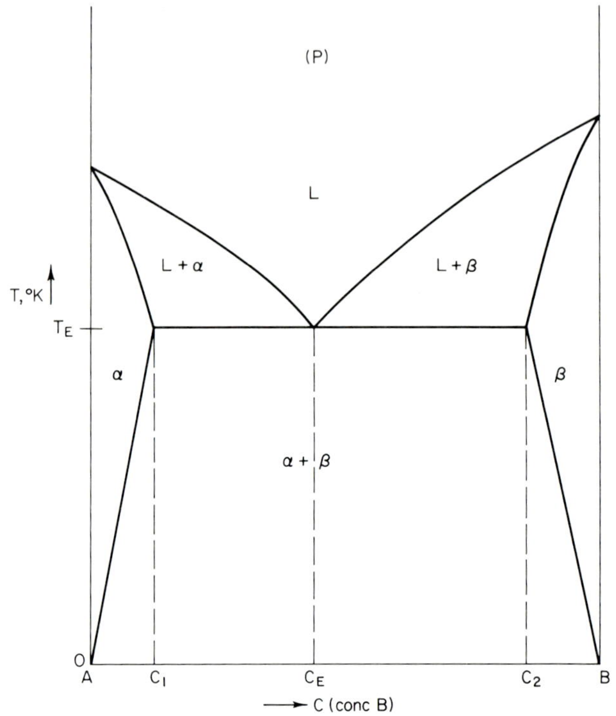

FIG. 2. Phase diagram of the binary A–B system. A and B are completely miscible in the liquid state and exhibit a miscibility gap in the solid state.

very common. This means that the components in many cases are miscible in the liquid state, but not in the solid state. A general argument for this, apart from reasons of free energy, is the following. One of the conditions for miscibility is that both components shall exhibit very strict symmetry relations (Landau, 1937; Landau and Lifshitz, 1959; Haas, 1965), one of them being that the symmetry elements shall be the same. The required symmetry condition is always obeyed when the components are in the normal liquid state (isotropic liquids), but it is rarely obeyed in the solid state. As a consequence the situation shown schematically in Fig. 1 can be representative of a large number of binary systems. This figure shows the Gibbs free energy G versus composition curves for the binary A–B system at a given pressure P and at the eutectic temperature T_E. In the liquid state, A and B are completely miscible (curve G_L). The stable configuration (minimum G at P and T_E) of solid A is a crystal of space group I, and of solid B is a crystal of space group II. As a consequence, $G_{AI} < G_{AII}$

and $G_{BII} < G_{BI}$. These space groups generally do not obey the above-mentioned symmetry requirements. Figure 1 shows diagrammatically the case where A and B are completely miscible in the solid state if they have the same symmetry. From the tangent construction, it follows that at T_E, the eutectic liquid of composition C_E coexists with two solid phases, i.e., α (A saturated with B) and β (B saturated with A). Consequently, for $T < T_E$, the eutectic reaction occurs (Fig. 2).

The fixed eutectic composition is another restriction, which, however, is less severe, since it is possible to grow regular structures from off-eutectic melts as well (Section III).

A. Regular Structures

The eutectic solidification has been the most widely studied. The composites that can be obtained by this process exhibit a large diversity of micromorphologies. Many efforts to classify the eutectic composites have been reported in the literature. One of the proposals bases the classification of eutectics on the type of growth of the two constituent single-crystalline phases (Hunt and Jackson, 1966).

Single crystals may grow with a nonfaceted (nf) or a faceted (f) shape. It has been argued that nf growth (the solidification front does not exhibit crystalline facets) occurs if the entropy of melting ΔS_F has a low value, roughly $\Delta S_F < 4$ cal deg^{-1} $mole^{-1}$,† while faceted growth occurs at higher values of ΔS_F (Jackson, 1958a,b). Most metals have a low ΔS_F and grow almost isotropically from the melt. The same is true of a number of organic materials such as CBr_4 and C_2Cl_6, and salts such as AgCl and CuCl. The corresponding nf–nf eutectics generally tend to grow in a coupled mode (see next paragraph) with rods or lamellae perpendicular to the solid–liquid interface while, under fixed growth conditions, the interrod distance or the interlamellar spacing has a definite value.‡ The resulting composite has a regular structure. Regularity is generally completely lost in the case of f–f eutectics. The f–nf eutectics are the intermediate case.

Whether the two phases grow coupled in an edgewise manner, i.e., joined together, or not depends on the relative magnitudes of the relevant interfacial tensions (Chadwick, 1965a,b). Consider a eutectic, lamellar,

† $\Delta S_F < 2R/\xi$, where R is the gas constant and ξ is a crystallographic factor which is smaller and almost equal to one (Hunt and Jackson, 1966).

‡ The fact that the solid–liquid interface between solid A and liquid A is nonfaceted does not necessarily imply that the interface between solid A and the eutectic melt $A + B$ is equally nonfaceted. This point is not discussed here.

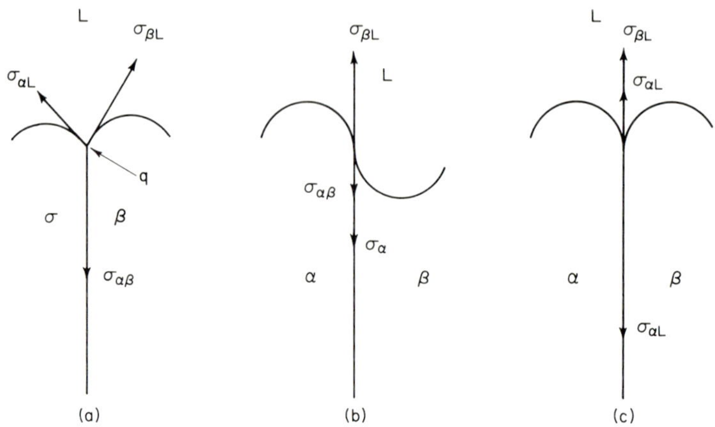

FIG. 3. Diagrammatic representation of the solid–liquid interface for coupled non-faceted, lamellar, two-phase growth: (a) $\sigma_{\beta L} - \sigma_{\alpha L} < \sigma_{\alpha\beta} < \sigma_{\beta L} + \sigma_{\alpha L}$; (b) $\sigma_{\alpha\beta} = \sigma_{\beta L} - \sigma_{\alpha L}$; (c) $\sigma_{\alpha\beta} = \sigma_{\beta L} + \sigma_{\alpha L}$.

two-phase composite, α–β, growing perpendicular to the solid–liquid interface from a eutectic liquid L with, for reasons of simplicity, isotropic interfacial tensions between the phases at the solidification front, $\sigma_{\alpha \mathrm{L}}$, $\sigma_{\beta \mathrm{L}}$, and $\sigma_{\alpha\beta}$, respectively. Then the solid phases can grow in a coupled mode (i.e., with a fixed orientation between the α–L, β–L, and α–β interfaces, which requires surface tension equilibrium) if

$$\sigma_{\beta \mathrm{L}} - \sigma_{\alpha \mathrm{L}} \leq \sigma_{\alpha\beta} \leq \sigma_{\beta \mathrm{L}} + \sigma_{\alpha \mathrm{L}} \tag{1}$$

for the case $\sigma_{\beta \mathrm{L}} > \sigma_{\alpha \mathrm{L}}$. The directions of the α–L and β–L solid–liquid surfaces at the three-phase point q adapt themselves to the surface tension equilibrium. Figure 3a represents the case where $\sigma_{\beta \mathrm{L}} - \sigma_{\alpha \mathrm{L}} < \sigma_{\alpha\beta} < \sigma_{\beta \mathrm{L}} + \sigma_{\alpha \mathrm{L}}$ and Figs. 3b and 3c show the extreme cases of Eq. (1) with $\sigma_{\alpha\beta} = \sigma_{\beta \mathrm{L}} - \sigma_{\alpha \mathrm{L}}$ and $\sigma_{\alpha\beta} = \sigma_{\beta \mathrm{L}} + \sigma_{\alpha \mathrm{L}}$, respectively. If $\sigma_{\alpha\beta}$ has a relatively high value, such that $\sigma_{\alpha\beta} > \sigma_{\beta \mathrm{L}} + \sigma_{\alpha \mathrm{L}}$, then the system tends to noncoupled growth, the growing α and β lamellae (or rods, etc.) being separated by the liquid phase.

Examples of regular, coupled-grown, lamellar eutectics of the nf–nf type are shown in Fig. 4. The lamellae are parallel to the growth direction. Figure 4a shows the metallic Pb–Sn eutectic, Fig. 4b the inorganic AgCl–CuCl salt eutectic, and Fig. 4c the organic C_2Cl_6–CBr_4 eutectic. Figure 5 represents a regular, rod-type eutectic.

Many eutectic composites grown *in situ* have been reported to show a unique crystallographic orientation relationship both between the constituent phases and their mating planes (Straumanis and Brakss, 1935; Hogan *et al.*, 1970). In a cross section perpendicular to the growth direc-

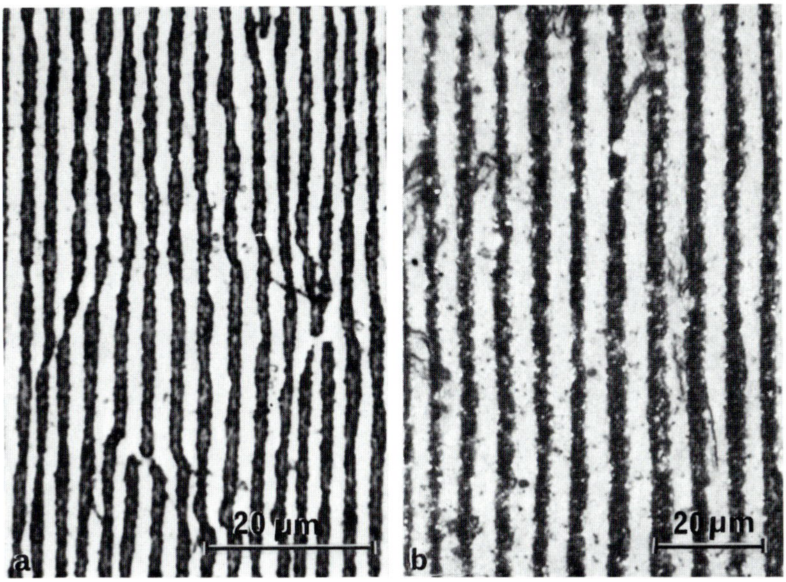

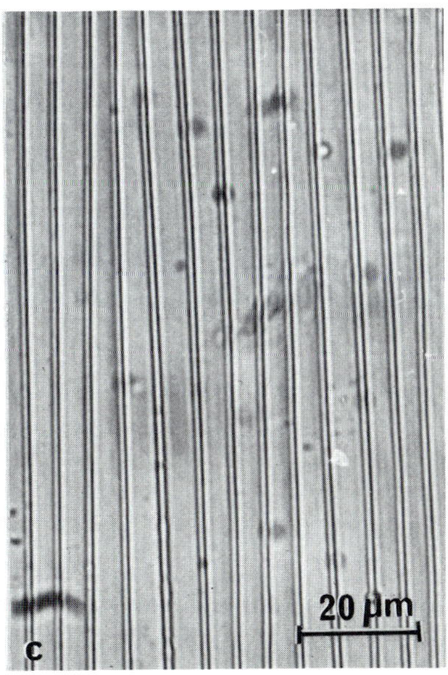

FIG. 4. Section parallel to the growth direction (bottom to top) of lamellar eutectics. (a) The metallic Pb–Sn eutectic ($R = 1.1 \times 10^{-3}$ cm sec^{-1}; 2140 ×). (b) The inorganic AgCl–CuCl eutectic ($R = 1.4 \times 10^{-4}$ cm sec^{-1}; 685 ×). (c) The organic CBr_4–C_2Cl_6 eutectic ($R = 1.7 \times 10^{-3}$ cm sec^{-1}; 685 ×).

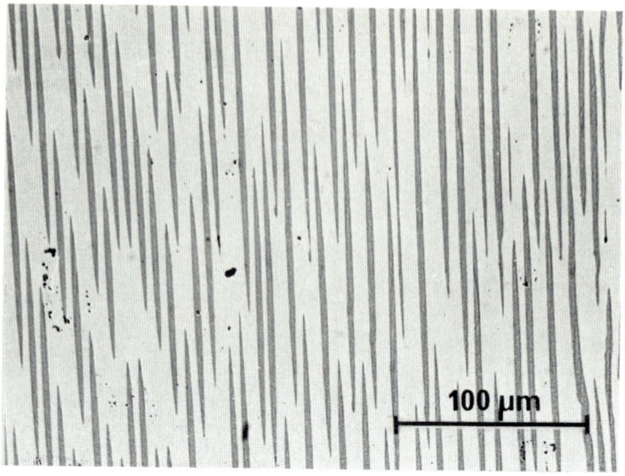

FIG. 5. Section parallel to the growth direction (bottom to top) of the rod-type MnSb–Sb eutectic. The MnSb needles are sectioned at a small angle and are much longer than shown here (230 ×).

tion, the lamellar structure proves to be less regular than would be suggested by Fig. 4. Lamellar misfit is manifest as grain boundaries and subgrain boundaries (Section II.D).

B. THE EUTECTIC REACTION PRODUCTS

Concerning the chemical stability of composites grown *in situ*, e.g., eutectic composites, three points may be mentioned. First, the constituent phases in the final composite are (more or less) in chemical equilibrium and are therefore mutually nonaggressive. Second, during growth, geometric configurations may be favored which exhibit a low interfacial α–β energy. As a consequence, the driving force for the spheroidization of the constituent phases at higher temperatures may be relatively small. This has been indirectly demonstrated experimentally. For example, the coarsening of the microstructure of the aligned lamellar Al–$CuAl_2$ eutectic has been studied by quantitative metallographic and x-ray diffraction techniques, and it was found that the specimens retained a very high degree of anisotropy even after prolonged heat treatments at a temperature less than 8°K below the eutectic temperature. The crystallographic orientation relationship between the phases remained constant (Graham and Kraft, 1966). Finally, attention should be drawn to a general property of the eutectic phase diagram: From general thermodynamic considerations, it follows that the solubility of B in A and of A in B decreases with decreasing temperature below T_E and is zero at $T = 0°K$ (see the solubility lines in Fig. 2). As a consequence, after solidification at T_E (or somewhat below T_E due to supercooling), it is to be expected that during cooling to room temperature, there will be solid state precipitation of β from α and α from β. The β precipitation will occur on the adjacent β lamellae if the interface spacing is small and/or the diffusion coefficient of B in A is high. In this case, the lamellae will remain free of precipitates. In the opposite case, β may nucleate inside the bulk of the α lamellae, as has been shown very clearly in the extreme case of the Cd cells of Fig. 23. The same holds for α precipitation in β.

This extra solid state precipitation deserves attention when considering the physical properties of the final composite. For instance, in the case of a composite with semiconducting constituents, it may bring about a considerable change of the charge carrier concentration and hence lead to untransparent inconstancy of the physical properties of the composite in temperature ranges where the precipitation–dissolution rates of the precipitates become significant (Albers, 1965, 1967).

The foregoing remarks are also applicable *mutatis mutandis* to monotectic and to eutectoid composites.

C. Lamellar Eutectic Growth

One of the earliest mathematical analyses of composite formation *in situ* by the decomposition of a single-phase material into two phases on cooling was related to a eutectoid transformation (Zener, 1946). A similar approach was subsequently used to discuss the eutectic decomposition (Tiller, 1958; Hunt and Jackson, 1966; Jackson and Hunt, 1966). A relatively simple case of eutectic crystallization is the growth from a binary eutectic melt of a two-phase lamellar composite, the two phases α and β growing coupled in an edgewise manner while the macroscopic solid–liquid interface is flat and the α–L and β–L interfaces are nonfaceted, the interfacial energies being independent of orientation. Under steady-state conditions, the lamellae then tend to grow parallel to one another and perpendicular to the solid–liquid interface (Fig. 6). The solidification front can be forced to be macroscopically flat by imposing a unidirectional heat flow during crystallization. The ratio of the widths of the α and β lamellae, $d\alpha/d\beta$, is such that the average composition $\bar{C}$ of the composite equals C_E. In detail, the situation is as follows (Fig. 6). The solidification front is, to a good approximation, isothermal so that the total supercooling ΔT_{tot} of the liquid phase along the entire solidification front has a fixed value (i.e., it has the same value in front of the α as at the β lamellae) on the assumption

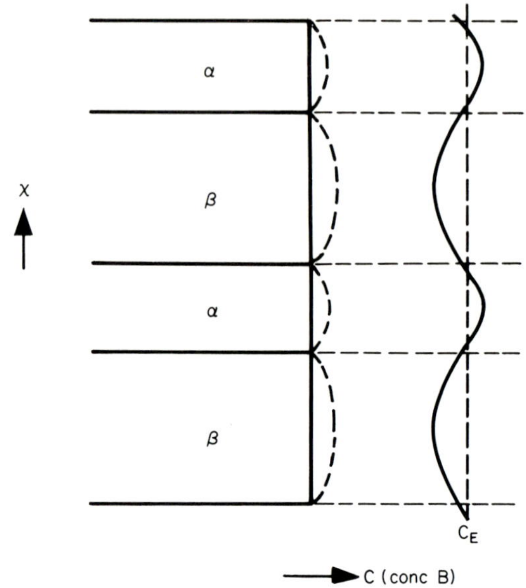

FIG. 6. Unidirectional lamellar crystallization of a binary eutectic melt A–B. The B concentration in the liquid at the solidification front as a function of x, i.e., the direction along the solidification front, is given by the solid curve. The dotted curves represent the curvature of the solid–liquid interface. (After Hunt and Jackson, 1966.)

that the diffusion of heat occurs rapidly compared with the diffusion of matter. Under ordinary circumstances, ΔT_{tot} is of the order of 1°K. The kinetic supercooling, the driving force for crystallization, is considered to be small with respect to the concentration and the curvature supercooling, and is not taken into account. This simplification is at least reasonable for metallic systems (Hunt and Chilton, 1963; Kramer and Tiller, 1965). Growth is made to proceed at a controlled velocity by passing through the system a temperature gradient at a fixed rate, e.g., by lowering the reaction vessel vertically through a two-zone furnace so that no (turbulent) convection occurs. Due to the presence of the temperature gradient, thermal diffusion (Soret effect) influences the concentration profile of the liquid near the solid–liquid interface (Jesse, 1969). For small gradients ($\sim$10°K cm^{-1}), this influence is small and will be neglected here. Furthermore, it is assumed that during growth at the solidification front, the solid phases and the liquid are very close to the equilibrium situation. Then, the three phases α, β, and L coexist where the α–β interfaces and the solidification front intersect and the concentration of L will be very close to the eutectic composition C_E, according to the Gibbs phase rule. When growth proceeds in front of the α lamellae, B will be expelled into the liquid and A will similarly be rejected in front of the β lamellae. The greater part of the expelled B will be transported to the β lamellae and the expelled A to the α lamellae by lateral diffusion in a thin liquid layer along the solid–liquid interface (x direction, see Fig. 6). As a result, the lamellae keep growing in the direction perpendicular to the solidification front in pace with the applied rate of the temperature gradient up to values of 1 cm sec^{-1}, until the heat flow becomes the rate-determining factor.

The lateral concentration pattern at the solidification front during steady-state growth is represented schematically in Fig. 6 by the wavy line showing an average concentration near C_{E} with a small excess in B in front of the α lamellae and a small excess of A in front of the β lamellae. Under usual experimental conditions (growth rates between 3×10^{-4} and 3×10^{-3} cm sec^{-1}), these excesses are of the order of 1%. The corresponding concentration supercooling ΔT_{conc} is largest in front of the center of the lamellae (solid curves in Fig. 7a). By extrapolation of the

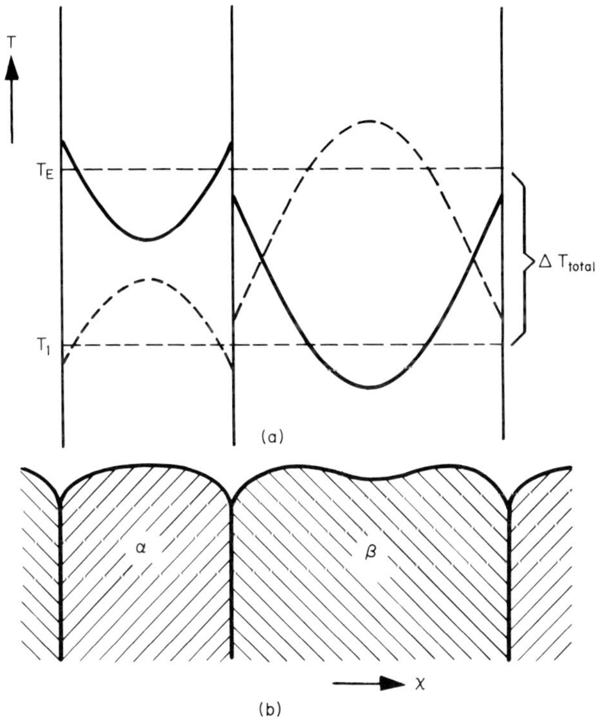

FIG. 7. Supercooling at the composite solidification front. T_{E} is the eutectic temperature for the flat solid–liquid interface. T_1 is the actual temperature at the interface. The solid curves and the dashed curves represent the concentration supercooling and the curvature supercooling, respectively. (After Hunt and Jackson, 1966.)

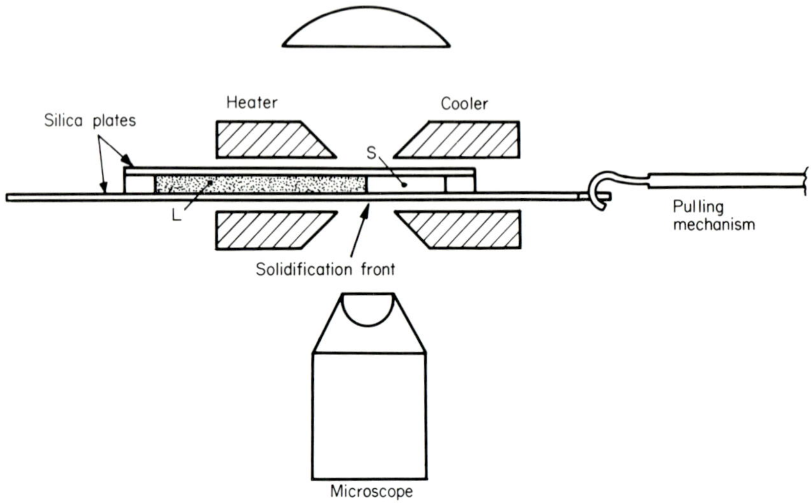

FIG. 8. Apparatus for observing the composite solidification front. L, liquid; S, solid composite.

liquidus below T_E, ΔT_{conc} can be derived from the phase diagram (Fig. 2). In order to obey the condition that at a given growth rate, ΔT_{tot} shall have the same value along the entire solidification front, the α–L interfaces and the β–L interfaces are curved on a microscopic scale (dashed curves in Fig. 6) such that the corresponding curvature supercooling ΔT_{curv} equals $\Delta T_{tot} - \Delta T_{conc}$. The corresponding lamella–liquid interfaces can be derived from ΔT_{curv} via the Gibbs–Thomson equation. Typical results are given in Fig. 7b. These types of microshapes of the solid–liquid interface for coupled and nonfaceted growth are in very good agreement with the experimentally observed solid–liquid interface shapes of growing lamellar eutectics observed in thin, flat cells in the microscope designed as in Fig. 8 (Jackson and Hunt, 1966).

The width d of the lamellae, or the lamellar period $\lambda = (d_\alpha + d_\beta)/2$, is a function of the growth conditions. It has been experimentally found for many eutectic systems that the period λ is inversely proportional to the square root of the imposed growth rate R [see Eq. (4)]. This relationship can be derived theoretically from the model under discussion, adding the postulate that at a given growth rate, the steady-state lamellar width has been adapted such that ΔT_{tot} has a minimum value. For the average concentration supercooling, the expression

$$\overline{\Delta T}_{conc} = K_1 \lambda R \tag{2}$$

can be derived (Hunt and Jackson, 1966; Jackson and Hunt, 1966), where K_1 is a constant containing the diffusion coefficient in the liquid phase at the solidification front, the difference between C_E and the compositions of the α and β lamellae, and the slopes of the liquidi in the phase diagram of the system A–B. Equation (2) formulates the fact that the amount of rejected B in front of the α lamellae and the amount of A in front of the β lamellae per time unit are proportional to R, while the final steady-state excesses of B in front of the α lamellae and of A at the β lamellae (see Fig. 6) also increase with increasing λ.

Via the Gibbs–Thomson equation, the following expression has been derived for the average curvature supercooling:

$$\overline{\Delta T}_{\text{curv}} = K_2/\lambda \tag{3}$$

where K_2 is a constant containing entropy of fusion and surface energy terms. Substituting Eqs. (2) and (3) in $\Delta T_{\text{tot}} = \overline{\Delta T}_{\text{conc}} + \overline{\Delta T}_{\text{curv}}$ and with the extremum condition $\delta(\Delta T_{\text{tot}})/\delta\lambda = 0$, the following relationship is obtained:

$$\lambda^2 R = K_3 \tag{4}$$

the constant $K_3 = K_2/K_1$ generally being of the order of magnitude of 5×10^{-11} cm^3 sec^{-1} according both to theory and experiment.

The change in λ on increasing R may be achieved by the movement of lamellar faults (i.e., sites or planes of misfit between the lamellae) or by growth of new α and β lamellae at the center of the β–L and α–L interfaces in order to decrease λ. The latter mechanism can be observed in thin cells (Fig. 8) where lamellar faults are absent (Hunt and Jackson, 1966). The process can be described as follows. Assume that a regular lamellar eutectic grows with a given R and the corresponding λ and minimum ΔT_{tot}. With R increasing and λ remaining constant, $\overline{\Delta T}_{\text{conc}}$ will increase [Eq. (2)], while ΔT_{curv} remains the same [Eq. (3)] and thus the value of the average concentration supercooling $\overline{\Delta T}_{\text{conc}}$ will approach the value of ΔT_{tot}. This means that the actual value of ΔT_{conc} in front of the centers of the lamellae may become considerably larger than ΔT_{tot}. As a consequence, central concave dents will arise (see β lamella in Fig. 7b). Supersaturation of A in the β dents or of B in the α dents can occur due to the fact that lateral diffusion is locally hampered, and as a consequence, a new α lamella can grow out in the β dent, seeded by a neighboring α lamella, or similarly, a new β lamella can grow from the α dent, seeded by a neighboring β lamella.

Similar discussions have been given, *mutatis mutandis*, with respect to rod-type eutectics (Hunt and Jackson, 1966; Jackson and Hunt, 1966), where, on solidification of the eutectic melt, parallel rods of one phase grow perpendicular to the flat solid–liquid interface in a matrix consisting of the other phase. Here, too, a relationship

$$\omega^2 R = \text{const} \tag{5}$$

has been obtained (where ω is the distance between the neighboring rod centers), which was first verified for the Al–Al_3Ni eutectic (Lemkey *et al.*, 1965).

Equations (4) and (5) have been experimentally confirmed for many eutectic systems over a wide range of growth rates, R ranging from 10^{-5} cm sec^{-1} to higher than 2×10^{-1} cm sec^{-1}, and the corresponding λ ranging from about 10^{-3} to 10^{-5} cm. High-speed solidification data have recently been reported for the eutectic composite systems Pb–Sn (Cline and Livingston, 1969), Ag–Cu, Zn–Al, Al–$CuAl_2$, Al–Al_3Ni, and Mg–γ-(MgAl) (Livingston *et al.*, 1970). In these cases, directional solidification was achieved by driving 10-cm-long samples in thin graphite crucibles of diameter ~0.6 cm, from a vertical furnace downward through a fitting hole in a water-cooled copper toroid.

The upper limit of regular, unidirectional, high-speed solidification may be determined by unfavorable heat transfer conditions which result in a concave solid–liquid interface. The lower limit of regular low-speed solidification is set by the occurrence of irregular degenerate structures, the nature of which is not yet fully understood.

D. Eutectic Grains

Though the regularity of coupled-grown nf–nf eutectics is striking, it must be emphasized that the perfect lamellar microstructures obtained by unidirectional solidification (e.g., by the Bridgman technique) are not as a rule faultlessly reflected on a centimeter scale, particularly in sections perpendicular to the growth direction (Kraft and Albright, 1961; Kraft, 1962; Chadwick, 1963; Hogan *et al.*, 1970).

After nucleation, many eutectic grains grow more or less perpendicular to the solidification front. In a lamellar eutectic grain, the single-crystalline lamellae are mutually parallel, while more or less fixed crystallographic relationships can exist between the lamellae and their mating planes (for a more rigorous definition, see Hogan *et al.*, 1970). The relative crystallographic relationships between α, β, and the α–β interface differ in various grains. After growth of a few centimeters, only grains of a certain preferred crystallographic orientation relationship survive, due to a competitive growth mechanism. These selected grains are parallel to the growth direction, but the rotation about the growth axis is not controlled and perpendicular to it they are initially randomly orientated to one another. On continued growth, one single grain of 1–10 cm cross section may finally result. These large single grains, however, are not perfectly regular on a microscale. The crystallographic relationship between the interface and the constituent phases is fixed only within a few angular degrees. Moreover, the eutectic grain consists of a large number of subgrains of about 10^{-3}–10^{-2} cm dimension in transverse section. The subgrain boundaries are formed by transverse misfits (kinks) in the parallel lamellae. Though subgrain boundaries occur in practically all observed cases, it is possible that they should not be considered as a definite characteristic of the lamellar structure. Perfect lamellar grains, free of mismatch surfaces, have been found in the lamellar Sn–Cd eutectic (Gruzleski and Winegard, 1968). Finally, in a number of systems, the eutectic grains do not exhibit a fixed crystal orientation with respect to external lines of reference, but show a constant and continuous twist about an axis which is almost parallel to the growth direction (Al–Zn; Al–$CuAl_2$; Al–Ag_2Al) (Double *et al.*, 1968; Double and Hellawell, 1969). This twist may amount to 30° cm^{-1} or even considerably more (Pb–Sn) (Lemkey, 1970).

E. Growth of Oriented Eutectic Films

Apparatus conventionally used for the growth of large single-phase crystals from the melt can generally also be used for the multiphase growth of bulky specimens. Both horizontal and vertical equipment have been successfully operated; Czochralski and Bridgman techniques as well as floating zone techniques are appropriate under given circumstances.

Thin composite films can be grown by withdrawing a wire loop or a substrate from a eutectic melt or by growing in a thin, flat cell (width 10–100 μm). The latter method is especially instructive for studies of the solidification front if the solid phases of the composite are transparent, enabling the microshape of the solidification front to be microscopically observed in transmitted light (Hunt and Jackson, 1966; see Fig. 8).

Additional information can be obtained by using a eutectic with one or two birefringent phases. By this artifice, lamellae of different crystallographic orientations are made to show up in different colors in the microscope, using crossed polarizers, which greatly facilitates the study of the crystallographic orientation relationships between the constituent phases, and the investigation of the mechanism of the competitive growth of eutectic grains (Van Suchtelen, 1970).

The growth of thin films of composites by withdrawing a centimeter-size loop of wire from the eutectic melt is possible if the wettability between melt and wire, the surface tension, the elasticity, the temperature, etc. have favorable values. In this way, films have been prepared of the lamellar eutectics Sn–Pb, Al–$AlAg_2$, Al–Zn (Takahashi and Kazato, 1956), Al–Al_2Cu (Takahashi, 1960), Cd-Zn and CuCl–$PbCl_2$ (Albers and Van Hoof, 1969). For the eutectic films Sn–Pb (Takahashi and Ashinuma, 1958/1959) and Al-Al_2Cu (Takahashi, 1960), it has been reported that the lamellae were monocrystalline and that in a given grain, similar lamellae had the same crystal orientations. For the Cd-Zn eutectic film, the lamellae tend to grow parallel to the growth direction (deviations up to 20 angular degrees, Fig. 9) and, for films with a thickness less than 5×10^{-3} cm, at the same time perpendicular to the film surfaces (Fig. 10). Similar behavior was found when thin films of the Cd–Zn eutectic were grown by withdrawing a solid substrate (e.g., a silica slide) from the melt. In this case also, the lamellae tend to grow parallel to the growth direction and perpendicular to the substrate surface, as is shown in Fig. 11 (Albers and Van Hoof, 1969).

F. Identification of the Phases by Vapor Etching

The constituent phases of a solid eutectic (monotectic or eutectoid) can be identified in various ways: e.g., by etching with specific chemical solutions, by x-ray analysis, or by means of the electron probe microanalyzer. In some cases, however, if the equilibrium vapor pressures and the diffusion coefficients have suitable values, the phases can be identified by means of a simple chemical vapor etching technique that requires no specific know-how or sophisticated apparatus (Albers and Verberkt, 1970). The principle will be illustrated on the basis of an A–B system with two compounds, AB and AB_2, and a eutectic in between (Fig. 12a). If the AB–AB_2 eutectic composite has been annealed at temperature T_1, then at equilibrium the AB phase and the AB_2 phase have the compositions C_4 and C_5, respectively. The partial vapor pressure of B, P_B, in the coexisting vapor phase at T_1 as a function of the gross concentration of the condensed phases is shown schematically in Fig. 12b. The equilibrium vapor pressures at the AB(A) boundary (AB saturated with A) and at the AB(B) boundary generally differ by many orders of magnitude. For example, in the semiconducting compound SnS at 730°C, the equilibrium partial vapor pressures of S_2 for SnS(Sn) and SnS(S) are 10^{-6} and 10^2 Torr, respectively (Rau, 1965). When the AB(B)–AB_2(A) composite (with equilibrium partial vapor pressure $P_B = P_1$) is annealed at temperature T_1 with a large amount of a mixture of A and AB of the arbitrary composition C_2 between C_3 and

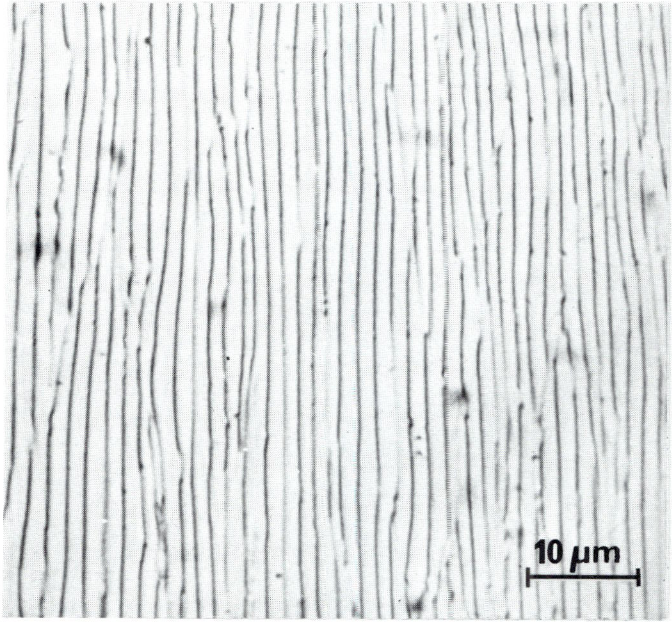

FIG. 9. The Cd–Zn eutectic film grown in a loop (longitudinal, bottom to top, 1480 ×).

FIG. 10. Top. The Cd–Zn eutectic film grown in a loop (transverse section, 2400 ×).

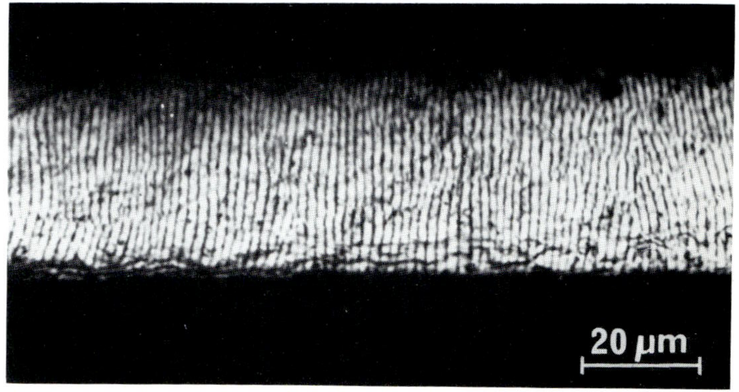

FIG. 11. Bottom. The Cd–Zn eutectic film grown on a silica substrate (cross section, substrate at bottom, 760 ×).

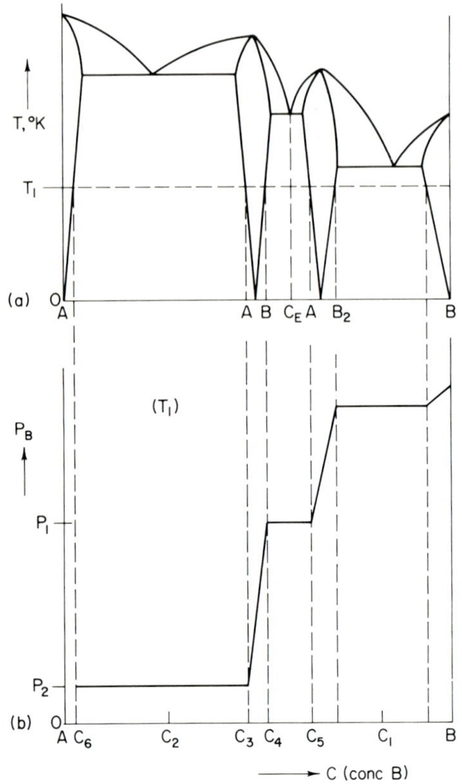

FIG. 12. The binary A–B system including two compounds. (a) The temperature–composition projection; the gas line has been omitted. (b) Diagrammatic dependence of the equilibrium B vapor pressure P_B on the gross composition of the condensed phases. (Albers and Verberkt, 1970.) Reproduced by permission of the publishers for the Journal of Materials Science, **5**, 24 (1970).

C_6, then at equilibrium, $P_B = P_2$. As a consequence the AB phase remains intact while its composition has been shifted from C_4 to C_3. The AB_2 phase, however, will be destroyed and converted to AB(A). If a flat, polished sample of the composite and relatively short annealing times are used so that no equilibrium can be reached, then at the surface, the AB_2 phase will be only partly destroyed and will transform in areas of nucleated AB in a matrix of AB_2. This effect can easily be observed by optical means.

This isothermal substitutional method has been demonstrated on the SnSe–$SnSe_2$ lamellar eutectic composite (Fig. 13). A polished sample was annealed at 500°C with a relatively large amount of a Sn + SnSe powder of a gross composition of 27 atom % Se in an evacuated silica tube for 2 hr. The result is shown in Fig. 14; the white phase is the (undisturbed) SnSe phase, while the originally black phase (after treatment, attacked and speckled black and white) is the $SnSe_2$ phase.

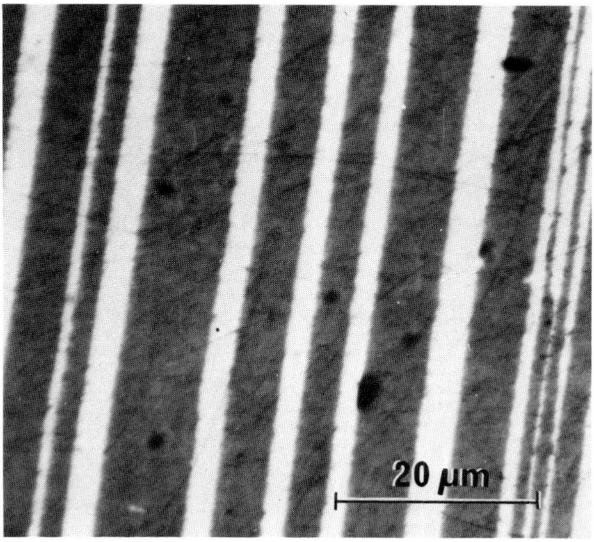

FIG. 13. Section of the lamellar SnSe–$SnSe_2$ composite before vapor-etching (1380 ×; Albers and Verberkt, 1970). Reproduced by permission of the publishers for the Journal of Materials Science, **5,** 24 (1970).

FIG. 14. The surface of the SnSe–$SnSe_2$ composite after annealing with a mixture of Sn and SnSe. (a) The white regions represent SnSe (focused on the white area; 1240 ×). (b) The speckled areas represent the original $SnSe_2$ regions (focused on the speckled area; 1240 ×). (Albers and Verberkt, 1970.) Reproduced by permission of the publishers for the Journal of Materials Science, **5,** 24 (1970).

G. The Mutual Connection of Similar Phases

It may be of interest for some applications to remark that with the procedure discussed in the preceding section, it is possible in principle to interconnect the similar phases. Take a slice of the eutectic lamellar composite AB(B)–AB_2(A) (Fig. 15a) and anneal it at a given temperature with a large amount of a powder mixture of A + AB (composition C_2 in Fig. 12), such that the entire surface is converted into a layer of composition AB(A), as shown in Fig. 15b. Then, anneal the sample at the same temperature with a powder mixture of AB + AB_2. As a consequence, the composition of the surface layer shifts from composition AB(A) to composition AB(B), interconnecting the AB(B) lamellae (Fig. 15c). Along the same lines, the AB_2(A) lamellae can be interconnected with an AB_2(A) surface layer by first annealing the sample with a powder mixture of AB_2 + B (composition C_1 in Fig. 12) and afterward, with a powder mixture AB + AB_2.

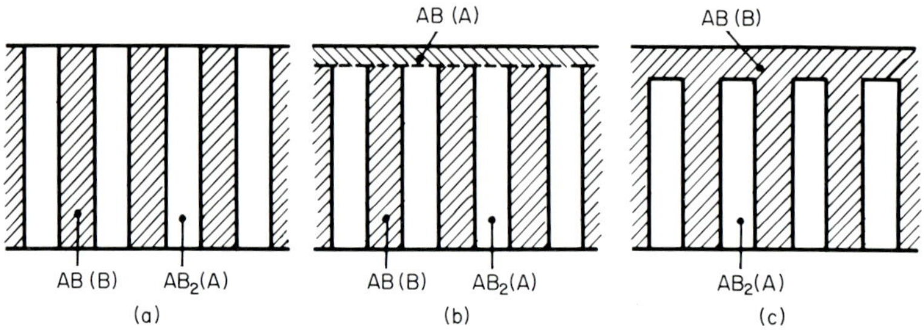

Fig. 15. The interconnection of similar phases. (a) Original eutectic sample; surface at top. (b) After equilibration against A + AB powder. (c) After subsequent equilibration against AB + AB_2 powder.

III. Off-Eutectic Lamellar Growth

Composite crystallization from a liquid phase of exact eutectic composition is in practice almost impossible because the relevant eutectic composition is a function of the growth rate, and consequently is rarely known exactly.

Suppose the solid curves of Gibbs free energy versus composition in Fig. 1 show diagrammatically the eutectic situation for the case that the solid–liquid interfaces are flat. However, if the solid phases are curved, then the corresponding Gibbs-free-energy curves shift to higher values with respect to the G_L curve, proportional to the interface energy and the curvature (Gibbs–Thomson equation). As a result, the eutectic temperaature has a lower value. This shift of the G curves is generally asymmetric because the interface energies, the entropies of fusion, and the curvatures of the constituent phases have different values. As a consequence, not only the eutectic temperature, but also the relevant eutectic equilibrium composition will shift. For the systems under discussion, the deviations ΔC_E [$= C_E$(flat) $- C_E$(curved)] can amount to values of 1 atom % for curvature radius r between 10^{-5} and 10^{-4} cm.

According to Eq. (4), λ and hence r, changes with R and therefore ΔC_E is a function of R. For the same reasons, the eutectic concentrations of phase diagrams reported in literature may be uncertain up to 1 atom % because the curvatures of the solid phases coexisting with the liquid phase have not been taken into account.†

In practice, however, the uncertainty in C_E is not a serious drawback. Experiments have shown that under definite conditions, regular composites can be grown *in situ* from off-eutectic melts as well. As a consequence, the ratio $d\alpha/d\beta$ is not confined only to the eutectic ratio, but can be varied. This gives the *in situ* growth of composites from the melt a wider applicability.

The experiments on regular growth from off-eutectic Pb–Sn melts with concentrations ranging from 13 to 26 atom % Pb (C_E = 26 atom % Pb) were very convincing in this respect (Mollard and Flemings, 1967a,b). Regular structures could be grown even at 13 atom % lead if high-temperature gradients G and low growth velocities were applied. The experiments suggested that the maximum deviation from C_E at which regular growth can occur is proportional to the ratio G/R. Such a relation has been theoretically derived starting from a simple constitutional supercooling criterion as follows (Mollard and Flemings, 1967a).

The same model and the same growth conditions are considered as those mentioned for eutectic growth, with the exception that the concentration of the off-eutectic melt C_i is poor in B with respect to the eutectic concentration C_E. If growth proceeds under steady-state conditions, then the ratio $d\alpha/d\beta$ is such that the average composition of the growing composite equals C_i. The average concentration of the liquid phase at the solidification front, however, is close to C_E again (three-phase equilibrium). As a consequence, in front of the growing composite, a B-rich boundary layer l exists in the liquid phase of composition $C_i \leq C \leq C_E$ (Fig. 16, solid curve). At the

† Another reason for a shift in C_E is the fact that the R-dependent kinetic undercoolings of the α and the β phases can have different values. This effect is not discussed here.

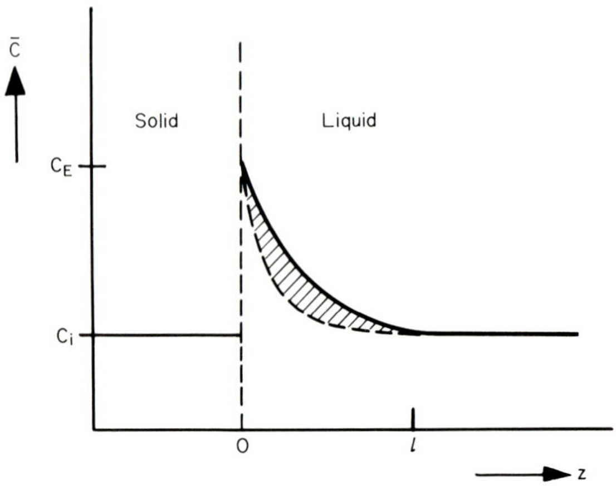

FIG. 16. Off-eutectic lamellar growth. The average B concentration $\bar{C}$ is a function of the distance from the solidification front z. C_i is the weighed-in concentration, C_E the eutectic concentration. The solid and the dashed lines represent the steady-state situations at R_1 and R_2, respectively ($R_2 > R_1$).

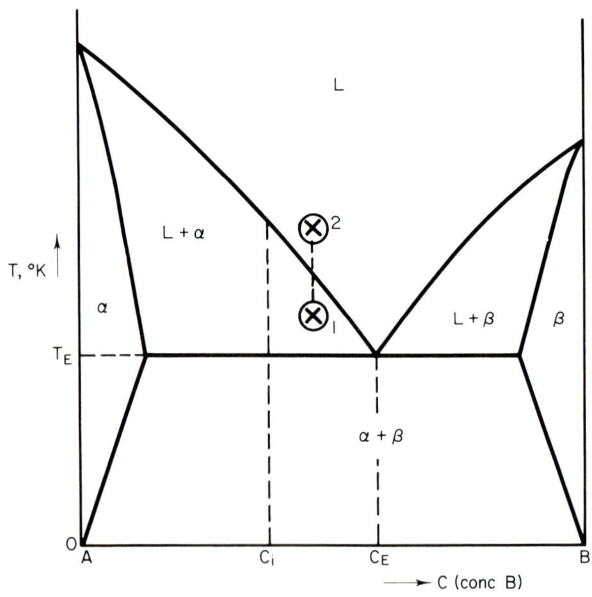

FIG. 17. Phase diagram of the binary A–B system. (1) Unstable melt, (2) stable melt.

same time, a temperature gradient G is present, i.e., in the boundary layer $T > T_E - \Delta T_{total}$. Two situations may arise in the boundary layer: the liquid is unstable (point 1 in Fig. 17) or stable (point 2 in Fig. 17). The latter case favors regular composite growth and occurs if, for $0 \leq z \leq l$,

$$G > m_\alpha \, dC/dz \tag{6}$$

where m_α is the slope of the liquidus at the A-side of the phase diagram and z is the distance perpendicular to the solidification front (Figs. 16 and 17). During steady-state growth, the amount of B rejected at the solidification front equals the amount of B transported via the concentration gradient in the liquid boundary layer l. Thus

$$(C_E - C_i)R = -D(dC/dz)_{z=0} \tag{7}$$

Combination of Eqs. (6) and (7) gives the condition for the growth of a regular composite from an off-eutectic melt:

$$G/R > -(m_\alpha/D)(C_E - C_i) \tag{8}$$

Figure 18 illustrates Eq. (8), including the B-rich compositions with respect to C_E. It shows a G/R versus concentration region of regular growth corresponding to structures shown in Figs. 4 and 5. Outside this region, cellular or dendritic growth should occur; in many cases, in combination with regular composite growth, as shown in Fig. 23. Figure 18 is qualitatively in good agreement with the off-eutectic experiments mentioned previously except at low values of G/R, where the extent of the regular

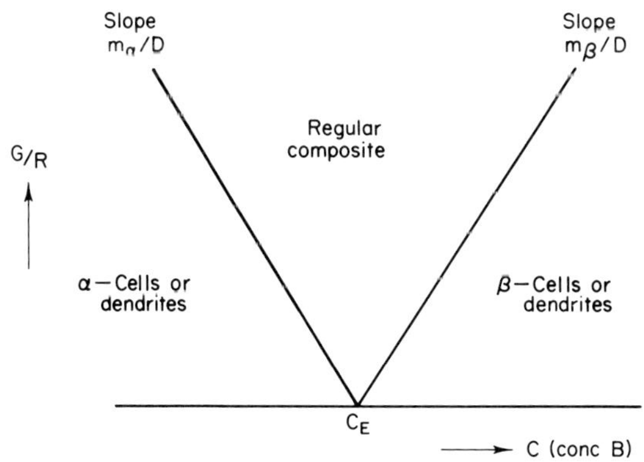

FIG. 18. The G/R versus concentration region of regular composite growth. G is the temperature gradient, R the growth rate, C_E the eutectic concentration, and m_α and m_β the slopes of the liquidi in the A–B phase diagram.

range found experimentally is considerably larger than that exhibited by Fig. 18. Efforts are being made to give the experimental results a more rigorous theoretical background, based on a perturbation analysis of the solid–liquid interface (Hurle and Jakeman, 1968; Cline, 1968) or on the competitive growth velocity between the regular composite and a dendritic structure (Jackson, 1968). The problem is still under discussion (Hunt, 1968; Cline and Livingston, 1969; Sekerka, 1969; Cline and Tarshis, 1969; Hunt *et al.*, 1970).

From Eq. (7), it follows that increase of R requires a steeper concentration gradient in the boundary layer in the liquid to obtain a new steady-state situation (see dashed line in Fig. 16), i.e., the excess of B in the hatched area has to be removed from the liquid phase. In the case that no extra β lamellae are created, this can be achieved by temporary broadening of the growing β lamellae. Analogously, the α lamellae will broaden if R decreases. As a consequence, extra geometric elements can be built into the composite by programming the fluctuations in R. This has been shown experimentally during growth in thin quartz cells of, among others, the CBr_4–C_2Cl_6 eutectic (Jackson and Hunt, 1966) and the AgCl–CuCl salt eutectic (see Fig. 19; Van Suchtelen, 1969). If dR/dt, ΔR (= total variation in R), and $C_E - C_i$ have large values, then one solidifying phase may overflow the other. This banding phenomenon is shown for the rod-type MnSb–Sb composite. During growth from an off-eutectic melt, R was fluctuating and the rods pinched off, the banding period b being $b = R/\nu$, where ν is the frequency of the fluctuation of R (Fig. 20).

IV. Composite Growth Other Than Eutectic

A. Monotectic Solidification

The monotectic decomposition differs from the eutectic decomposition in that one of the reaction products is a liquid and solidifies at a temperature (considerably) below the monotectic temperature. The unidirectional solidification of monotectic alloys has been studied less than eutectic solidification, but a few examples can be mentioned.

The phase diagram of the Bi–Se system shows a monotectic point at 618°C and 72 atom % Se. Unidirectional cooling of the monotectic liquid resulted in rod-shaped particles of a Se-rich liquid phase parallel to the growth direction, embedded in a single-crystal matrix of Bi_2Se_3. This liquid phase solidifies 400°K below the monotectic temperature. For

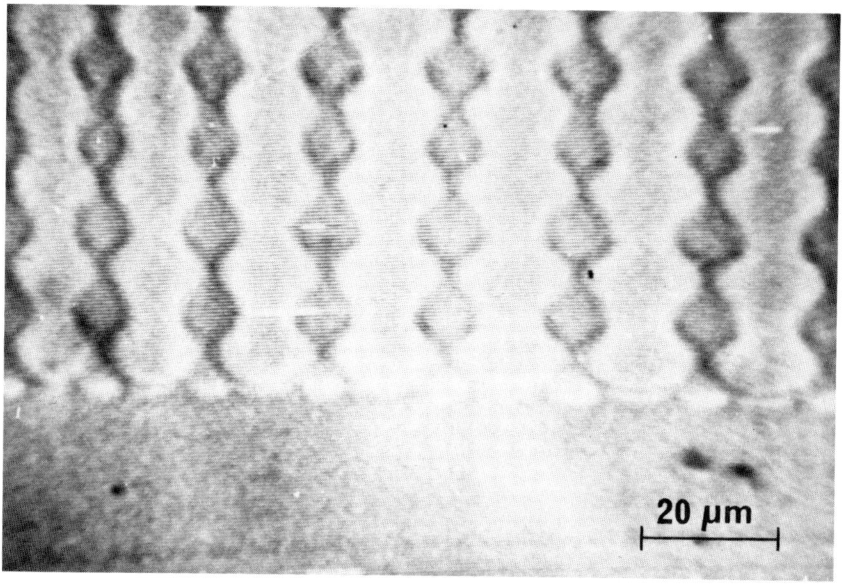

FIG. 19. Modulation of the lamellae of the AgCl–CuCl eutectic (900 ×; Van Suchtelen, 1969).

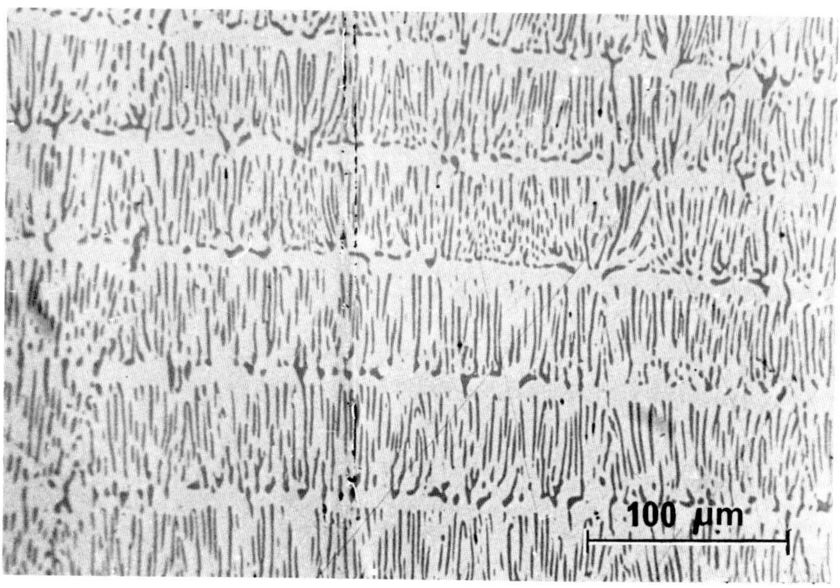

FIG. 20. Longitudinal section of the rod-type MnSb–Sb composite showing banding; Sb is the continuous phase. Growth direction from bottom to top (275 ×).

growth rates between 3×10^{-5} and 3×10^{-4} cm sec^{-1}, a value $\omega^2 R = 10^{-10}$ cm^3 sec^{-1} has been found, though the relation $\omega^{2.5}R$ = const gives a somewhat closer fit with the experimental data (Knight *et al.*, 1963).

Similarly, from the monotectic Sb–S melt, an aligned microstructure has been obtained: S-rich rods in a single-crystal matrix of Sb. The spacing ω decreased on increasing R (Lemkey and Ford, 1965).

The Cu–Pb system exhibits a monotectic point at 14.7 atom % Pb. High unidirectional growth rates, up to 2×10^{-1} cm sec^{-1}, yielded a composite of aligned continuous rods of liquid Pb, parallel to the growth direction in a solid copper matrix (Livingston and Cline, 1969). The relationship (5) was obeyed, the value of the constant being 2×10^{-8} cm^3 sec^{-1}. This value is roughly two to three orders of magnitude higher than has usually been found for eutectic solidification (Section II).

B. Eutectoid Disproportionation

Aligned lamellar composite structures have recently been obtained by discontinuous decomposition of eutectoid compositions under unidirectional cooling conditions (Carpay, 1970). The temperature gradient was about 100°K cm^{-1} and the pulling rate varied between 10^{-5} and 3.5×10^{-3} cm sec^{-1}. The resulting bar (length 10 cm, diameter 0.4 cm) consisted of large, elongated grains of millimeter size (Co_2Si–Co, NiIn–Ni_2In_3, Cu_9Al_4–Cu) or centimeter size (Cu–In system), in which the lamellae were oriented parallel to the applied temperature gradient (Fig. 21). Recently, the aligned lamellar Cu_9Al_4–Cu eutectoid has also been prepared by Livingston (1970).

For the NiIn–Ni_2In_3 eutectoid, regular structures could be obtained up to pulling rates of 8.5×10^{-4} cm sec^{-1}. At this pulling rate, the interphase spacing was equal to 3×10^{-5} cm. For the Cu–In system, the eutectoid reaction β-phase (20.2 atom % In) $\rightarrow$ δ-phase (29 atom % In) + Cu (11 atom % In) aligned lamellar structures were obtained between 10^{-5} and 3.5×10^{-4} cm sec^{-1}. At the higher pulling rates, the alignment of the lamellae was lost and the interlamellar spacing was inconstant.

The eutectoid reaction $Co_3Si \rightarrow Co_2Si + Co$ resulted in large, regular lamellar grains at all pulling rates applied. At 3.5×10^{-3} cm sec^{-1}, $\lambda = 8 \times 10^{-6}$ cm.

A relationship $\lambda^4 R = 1.6 \times 10^{-23}$ cm^5 sec^{-1} has been found; this in contrast to the eutectic relationship [Eq. (4)] (Carpay and van den Boomgaard, 1971).

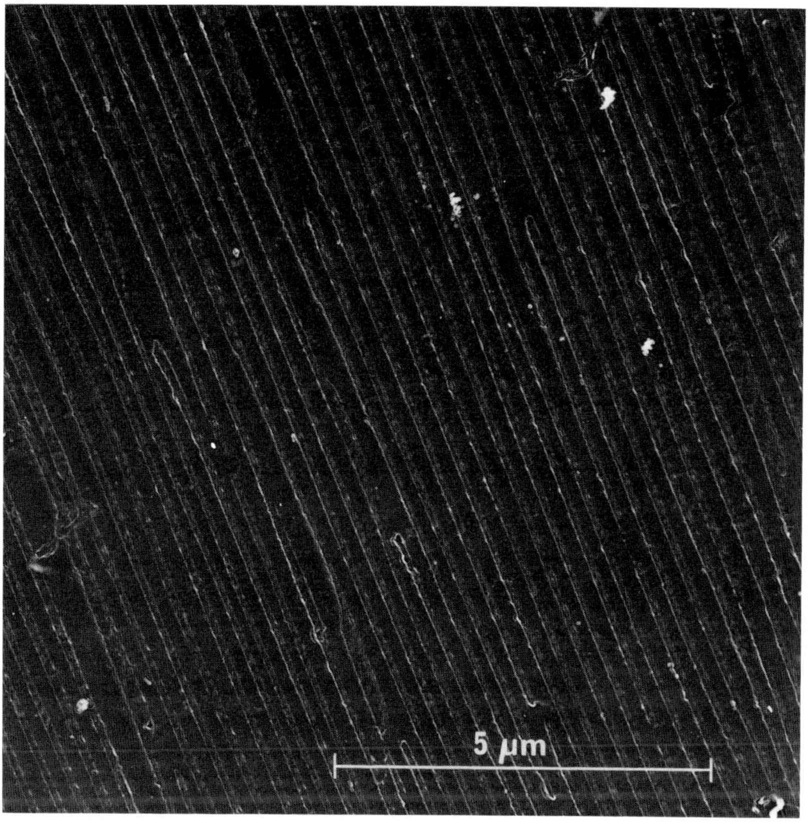

FIG. 21. Electron microscope replica of a transverse section of the Cu (11 atom % In) δ-phase (29 atom % In) lamellar eutectoid (10,000 ×; Carpay, 1970, used with permission of Pergamon Press).

C. Regular Growth from Supersaturated Solid Solutions

Regular composite structures can also be grown *in situ* by decomposition from supersaturated solid solutions (Carpay, 1970). As an example, the precipitation of solid Sn in a Pb matrix has been reported. A solid solution of 16.6 atom % Sn in Pb was used which becomes supersaturated below 148°C. The precipitation reaction is discontinuous. An aligned structure with Sn lamellae or Sn platelets parallel to the temperature gradient could be obtained only at pulling rates less than 1.4×10^{-4} cm sec^{-1} at a temperature gradient of about 100°K cm^{-1}.

V. Growth of Complex Regular Structures

A. Combination of Disproportionation Reactions

Complex composite materials can be grown *in situ* by combination of the growth modes discussed previously so that at a higher temperature, the homogeneous solid or liquid starting material decomposes into two phases, while at least one of these phases disproportionates again at a lower temperature. In this way, complex composites can be obtained that consist either of more than two phases, or of two phases, with one of the phases exhibiting more than one geometric shape. The latter case occurs if the A–B system has a eutectic melt at T_E between the solid element A and the solid compound A_3B, while at $T_{E'} < T_E$, the compound A_3B disproportionates eutectoidically into A and A_2B. As an example, the binary Co–Si system can be mentioned, which has a eutectic at 23 atom % Si and 1195°C. On directional cooling of the homogeneous eutectic melt, a lamellar composite can be obtained consisting of alternating Co and Co_3Si lamellae parallel to the growth direction. At 1170°C, the Co_3Si phase decomposes eutectoidically into alternating Co and Co_2Si lamellae with a smaller interphase spacing than the primary Co–Co_3Si composite (Fig. 22) (van den Boomgaard and Wolff, 1972).

A similar example is the aligned complex (Cd–Zn)–(Cd–Zn) composite (Fig. 23) (Jesse and Giller, 1970). This composite was grown *in situ* from an off-eutectic melt in the Cd-rich cellular/dendritic region (Section III). On unidirectional cooling, single crystalline Cd cells, saturated with Zn, solidify separated by a regular Cd–Zn eutectic structure. At lower temperatures, solid state separation takes place. In the intermediate eutectic, Zn precipitates from the Cd lamellae on the adjacent Zn lamellae. In the Cd cells, however, due to the great width, extra Zn nucleates as discrete platelets.

B. Ternary Eutectics

Three-phase composite materials can also be grown from a eutectic melt. For instance, the ternary eutectic Sn–Pb–Cd melt solidifies with three phases. The resulting structure is such that Sn and Cd lamellae are separated by Pb lamellae: PbSnPbCdPbSnPbCd ··· (Kerr *et al.*, 1964, 1965).

C. Nonparallel Directional Growth

Parallel alignment of rods or lamellae as discussed in the preceding sections has been achieved by means of a macroscopically flat reaction front.

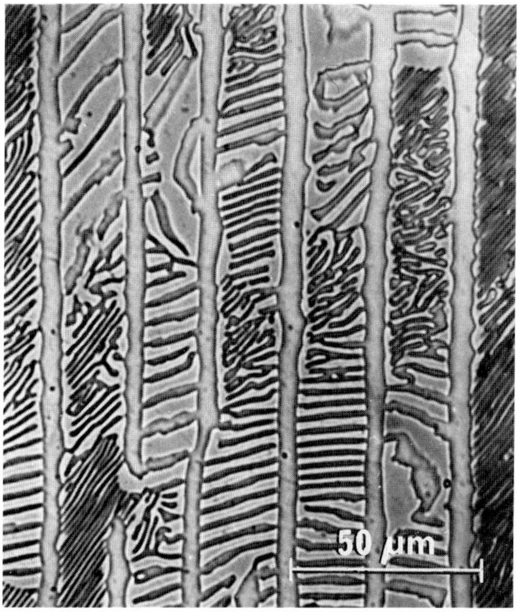

FIG. 22. Longitudinal section of the complex Co–(Co–Co_2Si) composite (500 ×).

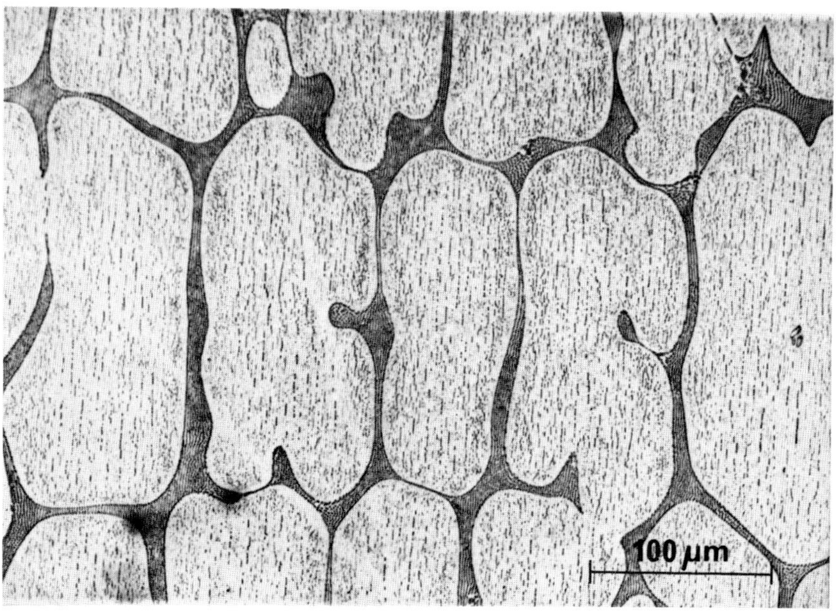

FIG. 23. Transverse section of the aligned complex (Cd–Zn)–(Cd–Zn) composite. The large, white, speckled cells consist of a Cd matrix with Zn inclusions (247 ×; Jesse and Giller, 1970).

Other, more complicated geometries can be obtained by modifying the shape of the reaction front. Two examples will be given for eutectic solidification, the first being colony formation.

If an impurity is present in a eutectic melt, then, due to extra constitutional supercooling effects (arising from the rejection of impurities from the growing solid), the solid–liquid interface between the eutectic melt and the solid eutectic may become cellular (Weart and Mack, 1958; Kraft and Albright, 1961; Chadwick, 1963). In a representative case, the cellular period is of the order of 100 times the eutectic period. As a consequence, the resulting eutectic, growing perpendicular to the cellular solidification front, will exhibit a fanlike superstructure known as a colony (Fig. 24).

The second example concerns radial directional growth. When a molten liquid column of eutectic composition is constrained to solidify from the outside to the center axis, then fibers or lamellae may grow perpendicular to the column axis such that a radial configuration will occur as is shown in Fig. 25. This figure represents a transverse section of the Fe–Cr_7C_3 com-

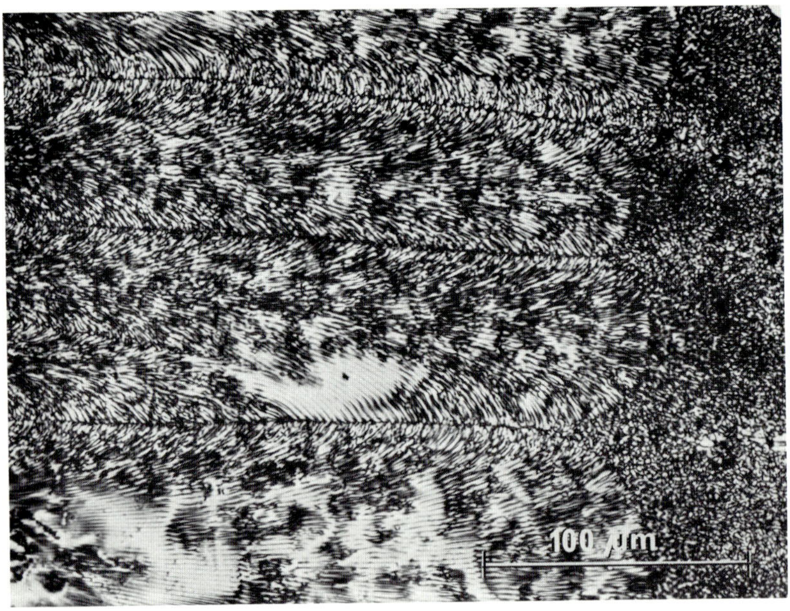

FIG. 24. Colony formation. During growth from left to right, the solid–liquid interface was cellularly curved, forcing the solidifying lamellae into a fanlike arrangement. At right: quenched liquid ($PbCl_2$–CuCl eutectic; longitudinal section; 360 ×).

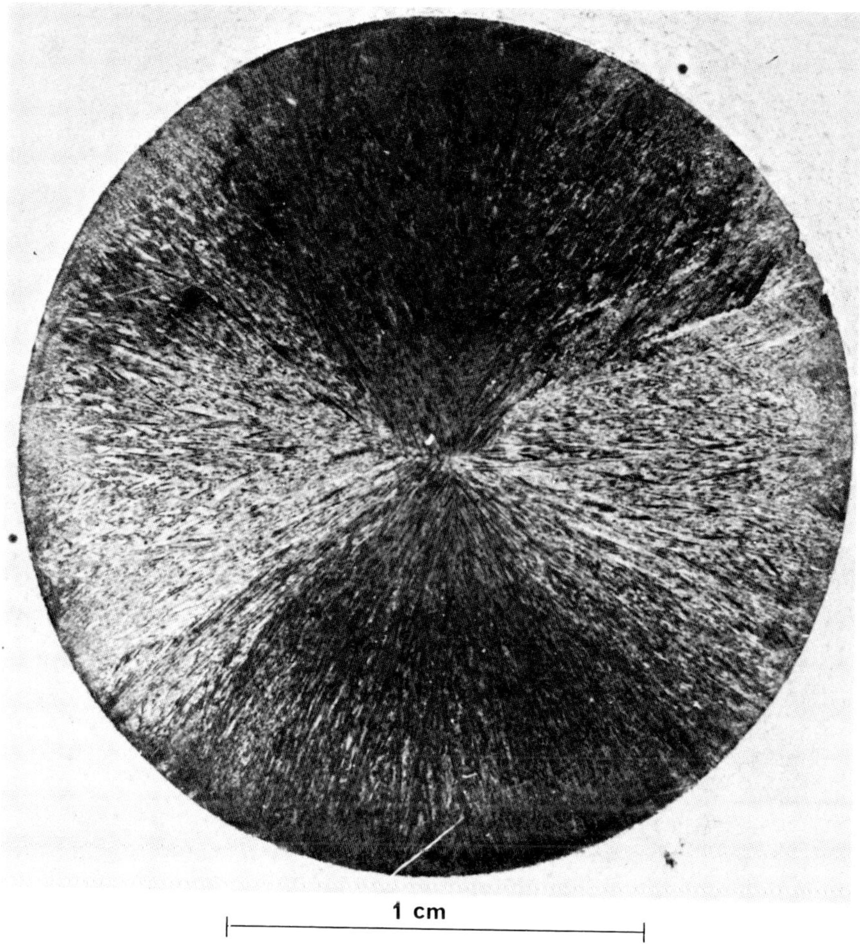

FIG. 25. Radially oriented crystallization of Cr_7C_3 rods in an Fe matrix (transverse section; 5.7 ×; van den Boomgaard and Wolff, 1972). Used with permission of North Holland Publishing Co., Amsterdam.

posite with radially oriented Cr_7C_3 rods in an Fe matrix (van den Boomgaard and Wolff, 1972).

VI. Final Remarks

Though considerable experimental and theoretical improvement has been made, particularly during the past decade, in the field of coupled nf–nf isotropic composite growth, many problems remain to be solved. For instance, which is the general model for eutectic growth (Section II; see Hunt, 1968; Lesoult and Turpin, 1969)? What are the incontestable criteria for off-eutectic growth (Section III)? How are we to get rid of the subgrains in lamellar eutectics? What is the nature of the crystallographic twist in grains (Section II)? How can a lamellar composite be transformed into a rod-type structure and *vice versa*?

Besides the study of coupled growth with anisotropic solid–liquid interface energy, the study of the crystallization of faceted eutectics and the growth *in situ* of n-phase composites ($n > 2$) is only in a preliminary state. Progress on these topics can be expected in the near future.

REFERENCES

Albers, W. (1965). *J. Chem. Phys.* **43,** 4401.

Albers, W. (1967). *In* "Physics and Chemistry of II–VI Compounds" (M. Aven and J. S. Prener, eds.), Chapter 4. North Holland Publ., Amsterdam.

Albers, W., and van Hoof, L. (1969). Unpublished.

Albers, W., and Verberkt, J. (1970). *J. Mater. Sci.* **5,** 24.

Broutman, J. L., and Krock, R. H. (1967). "Modern Composite Materials." Addison-Wesley, Reading, Massachusetts.

Carpay, F. M. A. (1970). *Acta Met.* **18,** 747.

Carpay, F. M. A., and van den Boomgaard, J. (1971). *Acta Met.* **19,** 1279.

Chadwick, G. A. (1963). *Progr. Mater. Sci.* **12,** 97.

Chadwick, G. A. (1965a). *Brit. J. Appl. Phys.* **16,** 1095.

Chadwick, G. A. (1965b). *J. Aust. Inst. Metals* **10,** 178.

Chilton, J. P., and Winegard, W. C. (1961). *J. Inst. Metals* **89,** 162.

Cline, H. E. (1968). *Trans. Met. Soc. AIME* **242,** 1613.

Cline, H. E., and Livingston, J. D. (1969). *Trans. Met. Soc. AIME* **245,** 1987.

Cline, H. E., and Tarshis, L. A. (1969). *Trans. Met. Soc. AIME* **245,** 422.

Double, D. D., Truelove, P., and Hellawell, A. (1968). *J. Crystal Growth* **2,** 191.

Double, D. D., and Hellawell, A. (1969). *Phil. Mag.* **19,** 1299.

Durand, F. (1969). *Bull. Soc. Fr. Minéral. Crystallogr.* **92,** 571.

Graham, L. D., and Kraft, R. W. (1966). *Trans. Met. Soc. AIME* **236,** 94.

Gruzleski, J. E., and Winegard, W. C. (1968). *J. Inst. Metals* **96,** 301.

Haas, C. (1965). *J. Phys. Chem. Solids* **26,** 1225; *Phys. Rev.* **140,** A863.

Hogan, L. M., Kraft, R. W., and Lemkey, F. D. (1971). *In* "Advances in Materials Research" Vol. 5, (H. Herman, ed.). Wiley (Interscience), New York.

Hunt, J. D. (1968). *J. Crystal Growth* **3, 4,** 82.

Hunt, J. D., and Chilton, J. P. (1963). *J. Inst. Metals* **92,** 21.

Hunt, J. D., and Jackson, K. A. (1966). *Trans. Met. Soc. AIME* **236,** 843.

Hunt, J. D., Hurle, D. T. J., Jackson, K. A., and Jakeman, E. (1970). *Met. Trans.* **1,** 318.

Hurle, D. T. J., and Jakeman, E. (1968). *J. Crystal Growth* **3, 4,** 574.

Jackson, K. A. (1958a). *In* "Liquid Metals and Solidification," p. 174. ASM, Cleveland, Ohio.

Jackson, K. A. (1958b). *In* "Growth and Perfection of Crystals" (R. H. Doremus, D. Turnbull, and B. W. Roberts, eds.), p. 319. Wiley, New York.
Jackson, K. A. (1968). *Trans. Met. Soc. AIME* **242,** 1275.
Jackson, K. A., and Hunt, J. D. (1966). *Trans. Met. Soc. AIME* **236,** 1129.
Jesse, R. E. (1969). *J. Crystal Growth* **5,** 132.
Jesse, R. E., and Giller, H. F. J. I. (1970). Communicated at the Spring Meeting of AMS of AIME, May 11–14, Las Vegas, Nevada.
Kerr, H. W., and Winegard, W. C. (1966). *J. Metals* **18,** 563.
Kerr, H. W., Plumtree, A., and Winegard, W. C. (1964). *J. Inst. Metals* **93,** 63.
Kerr, H. W., Bell, J. A., and Winegard, W. C. (1965). *J. Aust. Inst. Metals* **10,** 64.
Knight, R. J., Che-Yu Li, and Spencer, C. W. (1963). *Trans. Met. Soc. AIME* **227,** 18.
Kraft, R. W. (1962). *Trans. Met. Soc. AIME* **224,** 65.
Kraft, R. W. (1966). *J. Metals* **18,** 192.
Kraft, R. W., and Albright, D. L. (1961). *Trans. Met. Soc. AIME* **221,** 95.
Kramer, J. J., and Tiller, W. A. (1965). *J. Chem. Phys.* **42,** 257.
Landau, L. D. (1937). *Phys. Z. Sowjetunion* **11,** 26.
Landau, L. D., and Lifshitz, E. M. (1959). "Statistical Physics." Pergamon Press, Oxford.
Lemkey, F. D. (1970). Private communication.
Lemkey, F. D., and Ford, J. A. (1965). *J. Metals* **17,** 91.
Lemkey, F. D., Hertzberg, R. W., and Ford, J. A. (1965). *Trans. Met. Soc. AIME* **233,** 334.
Lesoult, G., and Turpin, M. (1969). *Mem. Sci. Rev. Met.* **66,** 619.
Livingston, J. D. (1970). *J. Mater. Sci.* **5,** 951.
Livingston, J. D., and Cline, H. E. (1969). *Trans. Met. Soc. AIME* **245,** 351.
Livingston, J. D., Cline, H. E., Koch, E. F., and Russel, R. R. (1970). *Acta Met.* **18,** 399.
Mollard, F. R., and Flemings, M. C. (1967a). *Trans. Met. Soc. AIME* **239,** 1526.
Mollard, F. R., and Flemings, M. C. (1967b). *Trans. Met. Soc. AIME* **239,** 1534.
Rau, H. (1965). *Ber. Bunsenges. Phys. Chem.* **69,** 731.
Rutter, J. W. (1958). *In* "Liquid Metals and Solidification," p. 243, ASM, Cleveland, Ohio.
Salkind, M., and Lemkey, F. D. (1967). *Sci. Technol.* **63,** 52.
Sekerka, R. F. (1969). *Bull. Soc. Fr. Minéral. Cristallogr.* **92,** 540.
Straumanis, M., and Brakss, N. (1935). *Z. Phys. Chem.* **B30,** 117.
Takahashi, N. (1960). *J. Appl. Phys.* **31,** 1287.
Takahashi, N., and Ashinuma, K. (1958/1959). *J. Inst. Metals* **87,** 19.
Takahashi, N., and Kazato, K. (1956) *C. R. Acad. Sci. Paris* **243,** 1408.
Tiller, W. A. (1958). *In* "Liquid Metals and Solidification," p. 276. ASM, Cleveland, Ohio.
van den Boomgaard, J., and Wolff, L. R. (1972). *J. Cyrst. Growth* **15,** 11.
van Suchtelen, J. (1969). To be published.
van Suchtelen, J. (1970). Communicated at the *Symp. Composites Grown in Situ, April 13–14, Eindhoven.*
De Vos, K. J. (1966). The Relationship between Microstructure and Magnetic Properties of Alnico Alloys. Thesis, Technical University, Eindhoven, Netherlands.
Weart, H. W., and Mack, J. D. (1958). *Trans. Met. Soc. AIME* **212,** 664.
Weiss, H., and Wilhelm, M. (1963). *Z. Phys.* **176,** 399.
Winegard, W. C. (1961). *Metals Rev.* **34,** 57.
Zener, C. (1946). *AIME Trans.* **167,** 550.

Preparative Methods for Solid Fluorine Compounds

R. BOUGON and J. EHRETSMANN

Centre d'Études Nucléaires de Saclay
Gif-sur-Yvette, France

J. PORTIER and A. TRESSAUD

Service de Chimie Minérale Structurale associé au CNRS
Faculté des Sciences de Bordeaux,
Talence, France

I. Introduction

Fluorine is an exceptional element in that it has a very high electronegativity. It has a high reactivity, which explains the specific nature of the experimental methods used in the field of solid state chemistry.

The main difficulties that these methods have to overcome are due to the physicochemical properties of fluorine and of its derivatives: a strong corrosive action on almost all materials, low resistance to hydrolysis, high volatility, etc.

The magnetic, electrical, and optical properties of solid inorganic fluorides have been widely studied by physicists: It is, in effect, easier to draw up theories for fluorides than for oxides or sulfides because of, for example, electron localization brought about by the high electronegativity of fluorine. Furthermore, although solid fluorides do not yet have many important applications in industry, some developments can be predicted in certain fields (magnetooptic, electrochemical, etc.). This new interest in solid fluorides thus justifies a review of the methods which can be used to prepare them.

The aim of this chapter is not to give an exhaustive list of methods of preparing fluorine compounds; several works have appeared which have already done this (Kirk and Othmer, 1966; Sharpe, 1959; Peacock, 1960). Here, an attempt will be made to show the most important methods by describing some simple practical processes; at the same time, care will be taken to draw attention to the most common pitfalls in this field of preparative chemistry.

II. General Fluorine Technology

The preparation of inorganic fluorides generally calls for the use of "fluorine lines." These are made up essentially of a primary and perhaps a secondary pumping unit (according to the nature of the degassing required), vacuum gauges, manometers, and tubing and joints for linking the various containers of products to be studied. Often, accessory equipment is also connected for analytical purposes.

As far as the choice of materials is concerned, and in comparison with more conventional chemical manipulations, the advantages of using glass (low price, transparency) are more than offset in fluorine chemistry by certain disadvantages such as a marked outgassing effect or the impossibility of using this material when hydrofluoric acid is present in any appreciable amount. More recently developed materials such as Teflon (polytetrafluorethylene), Kel F (polymerized monochloro-trifluoroethylene) and FEP (fluorinated ethylene propylene copolymer) prove invaluable for this type of chemistry; for example, with Kel F and FEP, it is possible to follow reactions visually or to observe changes in liquid levels, etc. These materials are flexible and can be used to make tubes through

TABLE I

RECOMMENDED MATERIALS FOR EQUIPMENT COMPONENTS[a]

Component	Gaseous fluorine service	Liquid fluorine service
Lines, fittings, and flanges	Nickel Monel Copper Brass Stainless steel, 300 series Aluminum 2017, 2024, 5052, 6061 Mild steel (low pressure)	Monel Stainless steel, 300 series Copper Aluminum 2017, 2024, 5052
Storage tanks	Stainless steel, 300 series Aluminum 6061 Mild steel (low pressure)	Monel Stainless steel, 300 series Aluminum 6161
Valve bodies	Stainless steel, 300 series Bronze Brass	Monel Stainless steel, 300 series Bronze
Valve seats	Copper Aluminum 1100 Stainless steel, 300 series Brass	Copper Aluminum 1100 Monel
Valve plugs	Stainless steel, 300 series Monel	Stainless steel, 300 series Monel
Valve packing	Tetrafluoroethylene polymer	Tetrafluoroethylene polymer
Valve bellows	Stainless steel, 300 series Monel Bronze	Stainless steel, 300 series Monel Bronze
Gaskets	Aluminum 1100 Copper Tetrafluoroethylene polymer Lead Red rubber (<5 psig) Neoprene (<5 psig)	Aluminum 1100 Copper Tetrafluoroethylene

[a] See Kirk and Othmer (1966).

which liquids can be transferred from one container to another. However, the chemical resistance is not always sufficient; Kel F, for example, can undergo swelling and can even react with compounds such as ClF_5, MoF_6, VF_5, etc. These materials also have the disadvantage of proving satisfactory only in a relatively narrow temperature range; above about 150°C, it is therefore necessary to use metallic equipment. Among the metals and alloys which can be employed for fluorine chemistry work, the most common are aluminum, copper, stainless steel,Monel, Inconel, and nickel. This last-named metal is very suitable in particular for use at higher temperatures, of the order of 600–650°C. A list of various materials which may be used is given in Table I.

A very complete description of modern equipment suitable for handling fluorides has been given by Canterford *et al.* (1967) and Canterford and O'Donnell (1968). These authors have given a detailed description of Kel F equipment of the type used at the Argonne National Laboratory, as well as of a commonly used fluorine line. Hyman and Katz (1965) have also given information about the handling of these active fluorides, and we will go into no further detail here.

III. Reactions in Solution

Two types of reaction will be considered: reactions in aqueous hydrofluoric acid solution, and reactions in nonaqueous solvents.

A. Reactions in Hydrofluoric Acid Solutions

1. *Low-pressure reactions*

This method of synthesis is no doubt the oldest. It consists in reacting together an aqueous solution of hydrofluoric acid with a solid or another liquid. By this method, one can obtain anhydrous or hydrated fluorides, oxyfluorides, or hydroxyfluorides.

We will not go into detail on this well-known method. Certain difficulties can be noted, however, in the case where well-defined compounds are required. The ions F^-, OH^-, and O^{2-} have very similar ionic radii and are therefore fairly easily interchangeable in a crystal lattice. Two examples will be given here to illustrate this phenomenon of substitution.

Hassel and Nilssen (1929) believed they had prepared cubic bismuth fluoride BiF_3 using a solution method. Aurivillius (1955) showed that BiF_3 is in fact orthorhombic and that the compound prepared by Hassel and Nilssen was an oxyfluoride. Portier *et al.* (1970) have shown that the composition of this compound was close to $BiO_{0.65}F_{1.70}$.

The exchange between F^- and OH^- is also fairly easy. Gladney and Street (1968) claimed that $RbFeF_4$ prepared in aqueous solution was tetragonal: Tressaud *et al.* (1969) have shown that it was in fact orthorhombic and that the compound which Gladney and Street had prepared was the hydroxyfluoride $RbFeF_{4-x}(OH)_x$.

Therefore it is of great importance, after syntheses in solution, to dry the products in a stream of gaseous hydrogen fluoride. A careful analysis should then be carried out, preferably of all the components, although this is difficult in the case of oxygen and hydroxyl groups. If their concentrations are obtained by difference, the water content of the product should be determined by a method such as that due to Karl Fisher; this avoids any confusion between residual surface water (if drying is inadequate) and oxide or hydroxyl groups. Furthermore, use should be made of infrared spectroscopy, which is a powerful tool for detecting qualitatively the presence of water molecules and hydroxyl groups in the products.

2. *Reaction at high pressure; hydrofluorothermal syntheses*

In a hydrofluorothermal synthesis, a solid product (oxide, fluoride, etc.) is made to react with an aqueous hydrofluoric acid solution. The reacting products are placed in a thin gold tube (thickness, 0.1–0.2 mm). This sealed tube is then subjected to a pressure either by a liquid or by a gas. Generally, one uses an autoclave of the same type as those used in conventional hydrothermal syntheses (Laudise, 1962). The thin gold tube is crushed by the pressure, which is thus transmitted to the solid substances. The pressures and temperatures which can be obtained depend on the apparatus; the most commonly used experimental conditions are pressures of 5–6 kbar and temperatures close to 500°C.

This method can be used for preparing hydroxyfluorides. Forsberg (1957), for example, prepared $In(OH)F_2$ by the action of hydrofluoric acid on $InF_3 \cdot 3H_2O$ at 200°C. Rault *et al.* (1970) prepared the hydroxyfluorinated pyrochlore $M_8[(OH)_{1-x}F_x)]_{24} \cdot 3H_2O$ [M = Cr(III), Ga(III)] by the action of hydrofluoric acid at 300°C and 3.5 kbar on the oxides Ga_2O_3 and Cr_2O_3.

A very interesting variant of this method was used by Wold (1969). A metal, the oxide of the same metal, and a solution of hydrofluoric acid were made to interact:

$$\delta x\mathrm{M} + (y - \delta)\mathrm{M}_x\mathrm{O}_y \xrightarrow[\text{under pressure}]{\text{HF solution}} y\mathrm{M}_x\mathrm{O}_{y-\delta}\mathrm{F}_\delta$$

(M is the metal, M_xO_y is an oxide of the same metal, δ can have any value between 0 and y.) In this case, hydrofluoric acid has two roles: one is to provide fluorine for the sample, the other is to transmit the pressure (Fig. 1). By this method, Bayard (1969) was able to prepare single crystals

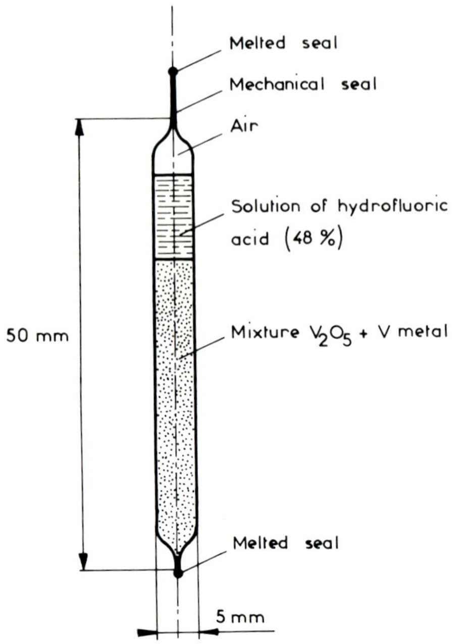

FIG. 1. Gold tube used for hydrofluorothermal synthesis (Wold, 1969).

having the formula

$$V_2O_{5-\delta}F_\delta, \qquad 0 < \delta < 0.40$$

This method devised by Wold was adopted by Sleight (1969) for preparing oxyfluorides having the composition $MoO_{3-x}F_x$ ($x = 0.74$–0.97) and $WO_{3-x}F_x$ ($x = 0.17$–0.66). A pressure of 3 kbar and a temperature of 700°C were used.

B. Reactions in Nonaqueous Solvents

In this case, two methods can be distinguished, depending on whether the solvent is, or is not, the fluorinating agent.

1. *The solvent acts as fluorinating agent*

Examples of this type of reaction are fluorinations with solvents such as BrF_3, BrF_5, ClF_5, SbF_5, and IF_5. For these reactions, more modern materials, in particular fluorinated polymers such as Kel F or FEP, usually replace metals for the construction of the equipment.

One of the best solvents for preparing many inorganic compounds is

BrF_3. From conductivity measurements in liquid BrF_3 and from acid–base reaction studies, it is thought that the ionization equilibrium in this solvent is

$$2BrF_3 \rightleftarrows BrF_2^+ + BrF_4^-$$

In order to carry out a fluorination, the oxide or halide is treated with the solvent and the excess liquid evacuated. Certain reactions can be followed by conductivity measurements.

Whitney *et al.* (1964) have carried out solvolysis reactions in liquid ClF_3 and BrF_5; they were able in this way to prepare the tetrafluorochlorates and hexafluorobromates of potassium, rubidium, and cesium using the following reactions:

$$MF + ClF_3 \rightarrow MClF_4$$

$$MF + BrF_5 \rightarrow MBrF_6$$

where M = K, Rb, or Cs. These authors followed the amount of interhalogen compound introduced by means of a calibrated tube. The interhalogen was then distilled over into the nickel reactor containing the alkali fluoride; after removal of this reactor from the fluorine line, it is placed on an external agitator and heated. The excess halide is then distilled off and the product weighed in order to determine the amount of conversion.

The work of Christe and Sawodny (1969) on the preparation of the complex $Cl_2F^+ \cdot AsF_6^-$ must also be mentioned. The complex is obtained by the action of ClF on AsF_5. Because two compounds having high vapor pressures are involved, the following procedure is adopted: AsF_5 and ClF are condensed at $-195°C$ in a U trap made of FEP. The mixture is then reheated slowly until the products melt and a reaction occurs; when the pressure reaches 760 Torr, the mixture is once again cooled. This procedure is repeated several times until reaction is complete. The excess ClF is then distilled off at $-78°C$.

2. *The solvent is not the fluorinating agent*

To give an example, we will consider again the work already mentioned of Christe and Sawodny (1969). In order to obtain a reaction between SbF_5 and ClF, which have very different volatilities, they poured SbF_5 into a Kel F trap in a dry glove box; the trap was fitted with a device for stirring the liquid. After HF was introduced by condensation at $-196°C$, the trap was reheated to room temperature so as to dissolve all the SbF_5. Then, ClF was added and the mixture successively heated and cooled as before. This method can be used mainly in order to ensure that two reactants having different volatilities come into contact in the condensed form.

The same process can be used to obtain interaction between ClF_5 and SbF_5. Christe and Pilipovich (1969) used either a common solvent (HF), or an excess of ClF_5, so as to conserve a liquid phase at the temperature of the reaction.

For their work on the synthesis of uranium(V) complexes and despite the fact that these can be prepared by dry methods, Sturgeon *et al.* (1965) used HF as solvent in order to obtain, after concentration and evaporation of the solution, single crystals for x-ray studies. The technique used here is as follows. The containers in which the reaction occurs are made of Kel F tubes. At the top of the tubes, some Teflon "wool" is placed for filtration of the solutions. The products to be reacted are introduced in a dry glove box into the tubes, which are then connected to a vacuum line. The HF is condensed using liquid nitrogen and the tube is then reheated to room temperature and stirred. The hydrofluoric solution is cooled to about 10°C and decanted through the "wool" plug into a second Kel F tube. The HF is evaporated at room temperature and the crystals formed are dried on filter paper.

Some other reactions using $NOUF_6$ dissolved in anhydrous HF may be mentioned (Geichman *et al.*, 1963). The uranium(V) complex is made to react with volatile Lewis acids. The acids are condensed in the solution, which is then reheated to room temperature. In this way, it is possible, using BF_3 or VF_5, to prepare $NOBF_4$ or $NOVF_5$, respectively.

For the sake of completeness, mention may be made of reactions in molten salts. Various syntheses of fluorides can be carried out in this way. The most commonly used salts are the chlorides, which generally have low melting points. This type of reaction is mainly used for obtaining fluorides in the form of single crystals (Garton and Wanklyn, 1967). The advantage of this method is that the fluoride formed is protected from the atmosphere of the furnace by the melt.

IV. Gas–Solid Reactions

The fluorides of many elements, in particular those of the transition metals, are frequently hygroscopic and generally have a marked tendency to form hydrates or hydroxyfluorides. Because of the difficulties of obtaining anhydrous products with methods involving aqueous hydrofluoric acid solutions, it is necessary in these cases to operate under strictly anhydrous conditions.

A. Action of Fluorinating Gases

Fluorination by fluorine and by gaseous hydrofluoric acid constitute the main methods for synthesizing fluorine compounds. Since a few years ago, however, the use of fluorinating gases such as halogen fluorides has made it possible to carry out both very vigorous and more gentle reactions.

1. *Reaction using low-pressure fluorine*

Fluorine has ceased to be a dangerous laboratory curiosity and, if handled in small amounts at pressures not exceeding 5 atm, it can be used practically without danger. With the equipment now available, it is possible to build relatively simple and cheap fluorination units. Figure 2 shows an apparatus for fluorinating solid or liquid products up to 600°C (Grannec, 1970).

Attention should be drawn to some of the characteristics of these techniques. The fluorine cylinder is located in a well-ventilated independent structure and is placed in a box packed with glass wool. All the tubing is made of nickel. A reserve volume makes it possible to carry out reactions

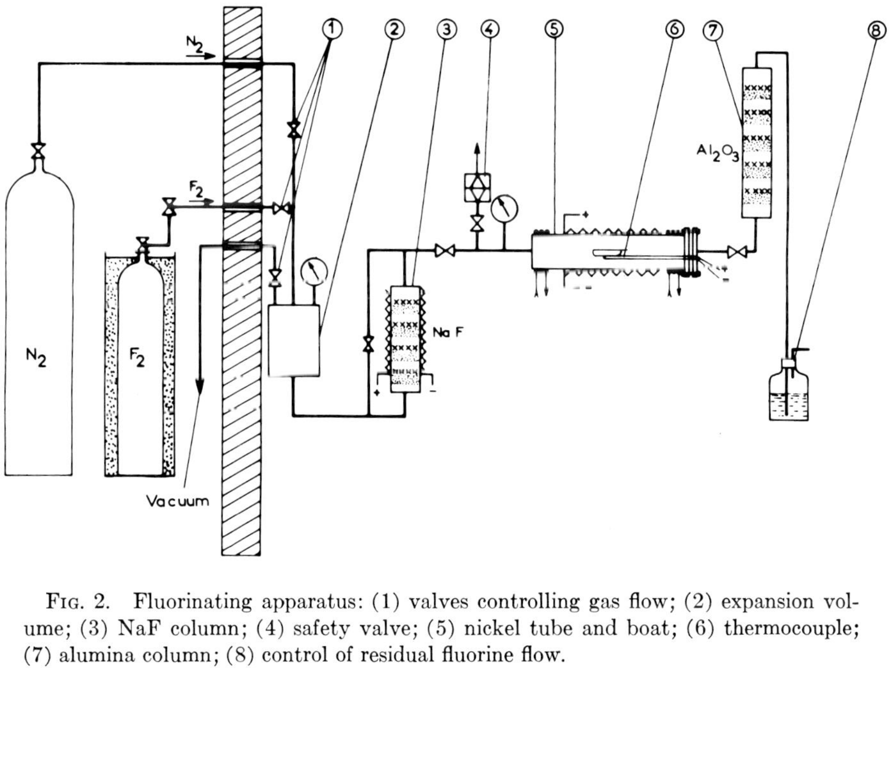

Fig. 2. Fluorinating apparatus: (1) valves controlling gas flow; (2) expansion volume; (3) NaF column; (4) safety valve; (5) nickel tube and boat; (6) thermocouple; (7) alumina column; (8) control of residual fluorine flow.

TABLE II

ACTION OF FLUORINE ON METALS

Metal	Experimental conditions (°C)	Compound formed	References
Sc	300–600	ScF_3	Jack (1959)
Ti	250	TiF_4	Kirk and Othmer (1966)
V	300–350	VF_5	Clark and Emeleus (1957); Trevorrow *et al.* (1957); Trevorrow (1958b)
Nb	300–500	NbF_5	Clark and Emeleus (1958;) Fairbrother and Frith (1951); Fairbrother *et al.* (1954)
Ta	300–500	TaF_5	Clark and Emeleus (1958); Fairbrother and Frith (1951); Fairbrother *et al.* (1954)
Cr	300–600	CrF_3, CrF_4, CrF_5	Wartenberg (1941); Clark and Sadana (1964)
Mo	400	MoF_5, MoF_6	Barber and Cady (1956); Edwards *et al.* (1962); Iwasaki *et al.* (1962); O'Donnell (1956)
W	300	WF_6	Barber and Cady (1956); Garton and Wanklyn (1967); Iwasaki *et al.* (1962); O'Donnell (1956)
Mn	300–600	MnF_3, MnF_4	Hepworth and Jack (1957); Nyholm and Sharpe (1952); Roesky *et al.* (1965)
Tc		TcF_5, TcF_6	Edwards *et al.* (1963)
Re	300	ReF_6	Aynsley *et al.* (1950)
Fe	300–600	FeF_3	Hepworth *et al.* (1957)
Co	300–600	CoF_3	Hepworth *et al.* (1957); Nyholm and Sharpe (1952)
Ni	700	NiF_2	
Ru	300	RuF_5	Ruff and Bahlau (1918); Holloway and Peacock (1963)
Rh	400	RhF_4, RhF_5, RhF_6	Holloway *et al.* (1965)
Pd	300–600	PdF_3	Hepworth *et al.* (1957); Nyholm and Sharpe (1952)
Ir	350–380	IrF_5	Bartlett and Rao (1965)
Pt	700	PtF_5, PtF_6	Bartlett and Lohmann (1964); Beveridge and Clark (1967)
Cu	500	CuF_2	Billy and Haendler (1957)
Ag	<80	AgF	Palmer (1954)
	150–200	AgF_2	Gruner and Klemm (1937)
Au	300–600	AuF_3	Jack (1959)

TABLE III

ACTION OF FLUORINE ON HALIDES AND OXYCOMPOUNDS

Initial product	Experimental conditions (°C)	Compound formed	References
Sc_2O_3	450	ScF_3	Zalkin and Templeton (1953)
Y_2O_3	400–500	YF_3	Zalkin and Templeton (1953)
Ln_2O_3	400–500	LnF_3	Zalkin and Templeton (1953)
HfO_2	300	HfF_4	Sense *et al.* (1954); Clark and Emeleus (1958); Trevorrow (1958b)
V_2O_5	475	VOF_3	Trevorrow (1958a); Haendler *et al.* (1954)
Nb_2O_5, $NbCl_5$	300–500	NbF_5	Clark and Emeleus (1958), Fairbrother and Frith (1951) Faifbrother *et al.* (1954)
Ta_2O_5, $TaCl_5$	300–500	TaF_5	Clark and Emeleus (1958), Fairbrother and Frith (1951), Fairbrother *et al.* (1954)
$Cr + CrCl_3 + CrF_3$	300	CrF_4–CrF_5	Wartenberg (1942); Wartenberg (1941)
CrO_3	—	CrF_5	Wartenberg (1941)
$Mo(CO)_6$	$-70 \rightarrow Mo_2F_9$ decomp 170	MoF_4	Peacock (1957)
Mn_2O_3, $MnCl_3$	300–500	MnF_3	Hepworth and Jack (1957); Nyholm and Sharpe (1952)
Fe_2O_3, $FeCl_3$	300–500	FeF_3	
Co_2O_3, $XoCl_2$	300–500	CoF_3	
$NiCl_2$	300–500	NiF_2	
RuO_2	300	RuF_5	Hepworth *et al.* (1954); Ruff and and Vidic (1925)
$RhCl_3$	300–500	RhF_3, RhF_4, RhF_5	
RhI_3	400	RhF_3	Aynsley *et al.* (1952); Hepworth *et al.* (1957)
$PdCl_2$, PdI_2	400–500	PdF_3	Nyholm and Sharpe (1952); Hepworth *et al.* (1957)
ReO_2	—	ReO_2F_3	Aynsley *et al.* (1950)
$PtCl_2$	$(3F_2/1N_2)$	PtF_5	Bartlett and Lohmann (1960, 1964)
CuO, CuCl	400–500	CuF_2	Billy and Haendler (1957)
AgCl	150–200	AgF_2	Gruner and Klemm (1937)
$AuCl_3$	300–600	AuF_3	Jack (1959)
Hg_2Cl_2		HgF_2	Ebert and Woitinek (1933)

with small amounts of gas; furthermore, the fluorine can be diluted with nitrogen and it is possible to control the relative amounts of the two gases. A sodium fluoride column traps traces of hydrogen fluoride which may be present; the NaF can be regenerated simply by heating. The oven is heated by a resistance wound around the nickel tube. The temperature is controlled by a thermocouple inside a nickel sheath placed next to the reaction boat; the fluorine that has not reacted is trapped in an anhydrous alumina column. After each operation, it is necessary to sweep out residual fluorine using a stream of nitrogen.

Fluorine reacts with most transition metals at normal pressure and gives rise to compounds having generally a high oxidation state (Table II);

TABLE IV

ACTION OF FLUORINE ON VARIOUS COMPOUNDS

Initial product	Experimental conditions (°C)	Compound formed	References
$AF + NiCl_2$ (A = alkali metal)	500–600	$ANiF_3$	
$ACl + HgCl_2, 2NH_3$ (A = alkali metal)	300–350	$AHgF_3$	Hoppe and Homan (1969)
$ACl + AgNO_3$ (A = alkali metal)	—	$AAgF_4$	Hoppe (1958)
$2KF + M_2O_3$ [M(III) = In, Tl]	400–500	KMF_4	Grannec (1970)
$AFeF_3$ (A = alkali metal)	500	$AFeF_4$	
$KMnF_3$	400–500	$KMnF_5$	Hoppe *et al.* (1959)
$BaFeO_3$	—	$BaFeF_5$	Peacock (1960)
$KCl + RuCl_3$	300	$KRuF_6$	Weise and Klemm (1955)
$(NH_4)_2PbCl_6 + MO$ [M(II) = Mg, Zn, Cd, Hg]	400–500	$MPbF_6$	Homann and Hoppe (1969)
$(NH_4)_2PbCl_6 + [Ni(NH_3)_6]Cl_2$	500	$NiPbF_6$	Homann and Hoppe (1969)
$2(NH_4)_2PbCl_6 + Ag_2SO_4$	450	$AgPbF_6$	Hoppe and Müller (1969)
$2KF + TiO_2$	350	K_2TiF_6	
A_2MCl_4 [A = Rn, Cs; M(II) = Co, Pd]	300–400	A_2MF_6	Hoppe and Klemm (1952), Hoppe (1956)
$2KCl + CrCl_3$	<300	K_2CrF_6	Huss and Klemm (1950)
$4KF + 2NaF + M_2O_3$ [M(III) = In, Tl]	400–500	K_2NaMF_6	Grannec (1970)
$3LiF + 3NaF + 2CoO$	500	$Li_3Na_3Co_2G_{12}$	de Pape *et al.* (1967)
$3LiF + 3NaF + M_2O_3$ [M(III) = In, Tl, Rh]	350–400	$Li_3Na_3M_2F_{12}$	de Pape *et al.* (1969)

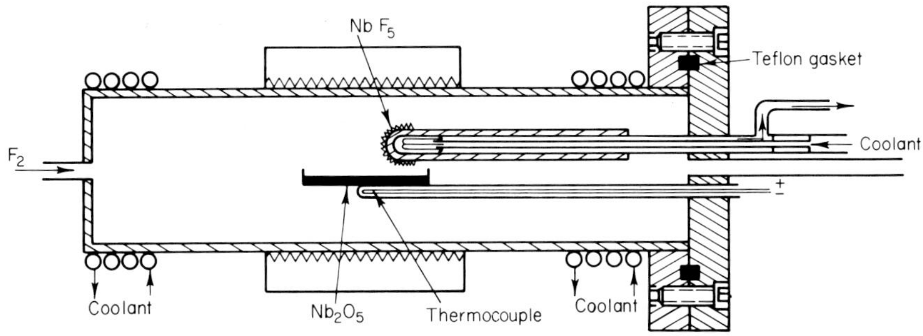

FIG. 3. Apparatus for condensation of volatile fluorides.

often, however, a mixture of fluorides is obtained. Furthermore, if the metal is not in powdered form, the reaction does not proceed homogeneously and layers of products having different compositions can be formed (Gillardeau, 1970). If small iron samples are fluorinated at 300°C, for example, a surface layer of ferric fluoride is formed, while underneath and next to the metallic iron, ferrous fluoride FeF_2 is produced (Claverie *et al.*, 1969). The passivation of the metal prevents further attack of the sample.

Because of these difficulties, it is often better to use halides or oxides as starting materials even though the latter can sometimes yield oxyfluorides (Table III). Double fluorides can also be obtained with fluorine if mixtures of oxides or salts are used (Table IV).

If the fluoride formed is volatile at the reaction temperature, the device shown in Fig. 3 can be added to the conventional apparatus; it is designed to condense the vapors formed (Grannec and Lozano, 1969). The condenser is made of a double nickel tube in which a strong current of coolant circulates. It is placed above the reaction boat and when atmospheric contact is reestablished after the reaction, it can be rapidly transferred to a sealed enclosure. In this way, it is possible to collect very simply, for example, the pentafluorides MF_5 (M = Nb, Ta, or Ru) after fluorination of the corresponding oxides.

Mention may be made of the preparation of osmium oxyfluoride OsO_3F_2 by the action of a mixture of oxygen and fluorine on the metal (Wiechert, 1950; Cady and Hargreaves, 1961).

2. *Reaction using fluorine under pressure*

Recent developments in fluorine handling technology have made it possible to carry out fluorinations under pressure to obtain, among other products, fluorine compounds in high oxidation states.

This technique, which might at first appear to be dangerous, can be applied satisfactorily if the experimental procedure, which is determined essentially by safety considerations, is carefully respected. Under these conditions, fluorinations under pressure can become as simple as conventional chemical manipulations. In this chapter, we will only briefly mention aspects connected with the use of fluorine under pressure, and we will end by giving some examples of this technique.

a. Reasons for using fluorine under pressure. The first question which may be posed is, of course, why and when should fluorine be used under pressure and what advantages does it possess compared to a simple fluorine gas flow at a pressure close to atmospheric? The answers are best given by a consideration of the reaction from a thermodynamic point of view. It is well known that up to pressures of about 1000 atm, the greatest effect of pressure is on chemical reactions that involve gaseous products. In effect, the latter are very compressible and the pressure will produce a large increase in their concentrations in the reaction volume. In most cases, the law of mass action makes it possible to explain quantitatively the effects produced in the system.

For higher pressures, of the order of 10,000 atm, the compressibility of gases decreases rapidly and the difference between gases and liquids is not noticeable in the dense reaction media obtained. It is then not the increases in concentration which are of importance, but rather effects due to the intermolecular forces produced, in particular between the transient states created by the pressure.

Here, we are only concerned with the study of reactions involving fluorine under pressure in media of average density and we will only consider the first pressure zone.

First of all, we note the beneficial effect of pressure in the case of reactions occurring with a decrease in volume, as predicted by the Le Châtelier principle. However, for a quantitative appreciation of this phenomenon, it is necessary to make use of value of the fugacity of the gases under the conditions of the synthesis. Reaction under pressure can also be considered as a means of obtaining an activation of the fluorine; such an activation is usually brought about only by methods such as heating, ultraviolet irradiation, or the use of an electrical discharge. Finally, the pressure technique has the advantage over gas-flow methods of allowing the reaction to be followed by the pressure changes in the reaction volume.

In order to predict the behavior of a reaction mixture, it is necessary to calculate as accurately as possible the thermodynamic data such as enthalpy, free energy, entropy, etc. under the conditions of the experiment.

For calculating these values, it is necessary to make use of the equation of state relating the pressure P, the volume V, and the temperature T.

b. Choice of an equation of state. Various equations exist and can be used according to the domain being studied. The simplest is the ideal gas law

$$PV = nRT$$

where n is the number of moles and T the absolute temperature. This equation is sufficiently precise for low pressures, and also for high-pressure regions for which the temperature is much higher than the critical temperature. This law, however, is not at all satisfactory for explaining the behavior of real gases in certain pressure and temperature regions. In these cases, one has to take into consideration the compressibility coefficient C, $C = PV/nRT$. This is equal to unity for perfect gases.

The application of this modified equation ($PV = CnRT$) is made easy by the use of the law of corresponding states; this law states that if perfect gases are brought to a single temperature and a single pressure, qualified as "reduced" and defined by

$$T_R = T/T_c \qquad \text{and} \qquad P_R = P/P_c$$

(T_c is the critical temperature and P_c the critical pressure), then one observes the same compressibility coefficient for all gases. The gases are then said to be in corresponding states. It should be noted that this is not an exact natural law; it explains satisfactorily, however, to within a few percent, the behavior of all real gases.

It is therefore possible in the case of a pressure fluorination reaction to predict with a fairly high accuracy the behavior of fluorine under pressure using tables giving the compressibility of gases at several temperatures as a function of the reduced pressure.

Nelson and Obert (1954) have given the diagrams corresponding to three pressure zones, P_R varying from 0 to 1, 0 to 10, and 0 to 40 atm. By taking the following critical constants for fluorine (Horvitz, 1950)

$$T_c = 144°\text{K} \qquad \text{and} \qquad P_c = 55 \quad \text{atm.}$$

one obtains from the Nelson–Obert diagrams the values shown in Table V for compressibility coefficients for $T = 15°\text{C}$ and $T = 447°\text{C}$. From these values of the compressibility coefficients, it can be seen that in the second case, one always has a value greater than 1, whereas in the first case, the coefficient initially has a value lower than 1, which then increases. It should be recalled that the maximum temperature for which the compressibility coefficient is less than 1, for a given reduced pressure, is the so-called Boyle

TABLE V

COMPRESSIBILITY COEFFICIENTS FOR F_2

	P (atm)			
	10	100	1000	2200
C for $T = 288°K$ ($T_R = 2$)	0.995	0.97	1.58	2.63
C for $T = 720°K$ ($T_R = 5$)	1	1.03	1.33	1.75

point (413.8°K) in the case of fluorine, i.e., $T_R = 2.88$ (Weinstock *et al.*, 1966). It is observed that at the lowest temperatures and highest pressures, one has considerable deviations from the ideal gas law and the pressure is higher than that given by this law.

Use can be made of another equation of state, in which the compressibility factor is expressed in series form:

$$PV/RT = C = 1 + (\beta/V) + (\gamma/V^2) + \cdots$$

β, γ, etc., being respectively the second, third, etc. virial coefficients, which can be expressed as a function of the temperature and of molecular parameters for various types of intermolecular forces.

The virial equation has been used with only the second coefficient by Weinstock *et al.* (1966) in a systematic study of the xenon–fluorine system under pressure, which necessitated a high accuracy as far as the equilibrium measurements were concerned. Using the force constants determined by White *et al.* (1953), they obtained the values for β as a function of the temperature. These are given in Table VI.

The van der Waals equation

$$[P + (a/V^2)](V - b) = RT$$

describes satisfactorily the behavior of real gases over a large temperature and pressure range, but it is not very precise around the critical point of the gases.

TABLE VI

VALUES OF THE SECOND VIRIAL COEFFICIENT FOR FLUORINE AT DIFFERENT TEMPERATURES

T (°K)	300	523.15	573.15	623.15	673.15	774.15
β (cm³)	−19.11	9.75	12.80	15.28	17.40	20.57

The constant b, or covolume, represents approximately the volume occupied by the molecules themselves; it is related to the attractive forces between the molecules.

The values of these constants can be determined from the relationships

$$a = 27R^2T_c^2/64P_c, \qquad b = RT_c/8P_c$$

which are obtained by equating to zero the first- and second-order differentials of the equation of state at the critical point,

$$(\partial P/\partial V)_{T_c} = 0; \qquad (\partial P^2/\partial^2 V)_{T_c} = 0$$

In the case of fluorine, one obtains, with the critical constants already given,

$$a = 1.075 \quad \text{liter}^2 \text{ atm mole}^{-2}$$

$$b = 2.68 \times 10^{-2} \quad \text{liter mole}^{-1}$$

Another equation with two constants has been proposed by Kwong and Redlich (1949); it makes it possible to represent the behavior of the gas over a wide range of pressure and temperature, though with a limited accuracy.

In the cases where, in addition to the fluorine, there are other gases present, either as the reacting species or the reaction products, it is possible to use a compressibility coefficient calculated with a fair degree of accuracy with the help of the law due to Amagat; this states that

$$C_m = C_A x_A + C_B x_B + \cdots$$

In this expression, C_m is a mean compressibility factor for the mixture and C_A, C_B, etc. are the compressibility factors for the components A, B, etc. calculated for the conditions of temperature and of total pressure applicable to the mixture. x_A, x_B, etc. are the molar fractions of the components A, B, etc.

Other methods can also be used. For example, that of Kay (1936), which uses the idea of the pseudocritical point of the mixture.

c. Displacement of the chemical equilibria. One can consider the case of fluorination reactions that involve a chemical equilibrium. Here, if one wants to know the influence of the pressure on the amounts of product formed, it has to be remembered that the equilibrium constant at a given temperature is independent of the total pressure only if the law of mass action is expressed in terms of the activities of the components of the system.

If the activity a_i of one component of the system is defined as the ratio of its fugacity f_i under the conditions of the experiment to the fugacity

f_i° under the standard conditions of the same state, then

$$a_i = f_i/f_i^\circ$$

For gases, $f_i^\circ = 1$ atm:

$$a_i = f_i$$

The fugacity of a gas is defined by the expression

$$F_i - F_i^\circ = RT \log_e f_i$$

in which F_i and F_i° are the free enthalpies under the conditions considered and in the standard state, respectively. The fugacity is thus a modified value of the partial pressure, which can be used for real gases; at low pressure, the fugacity tends to the partial pressure as the latter tends to zero, whereas at high pressures, the fugacity can be markedly different from the partial pressure. In this case, use is made of the fugacity coefficient

$$\phi_i = f_i/P_i$$

determined either from measurements made on each gas, or using the law of corresponding states; in effect, this fugacity coefficient can be obtained from the previously mentioned compressibility factor C, and it can be shown [see, for example, Comings (1956, p. 267)] that

$$\log_e \phi = \int_0^P (C - 1) \frac{P_c}{P} \frac{dP}{P_c} = \int_0^{P_R} \frac{C - 1}{P_R} dP_R$$

where the reduced pressure $P_R = P/P_c$, and P_c is the critical pressure.

According to the principle of corresponding states, the integral calculated over the isotherm at a reduced temperature T_R, up to some reduced pressure P_R, will have the same value for all gases. Newton (1935) carried out graphical integrations of experimental data and gave tables of the fugacity coefficient ϕ as a function of the reduced pressure at different temperatures. The values of the fugacity coefficient for fluorine deduced from this work are given in Table VII.

Using this coefficient for each of the gaseous components of the system (for the other states, it is possible to neglect the effect of the pressure on the activity), the equilibrium constant can be determined on condition that use is made of the approximation due to Lewis and Randall for ideal mixtures. According to this, the fugacity of each component of an ideal mixture of real gases is equal to the product of its molar fraction and the fugacity that it would have in the pure state at the temperature and total pressure of the mixture. If

$$f_i = f_i^* x_i$$

TABLE VII

FUGACITY COEFFICIENT FOR F_2

	P (atm)			
	10	100	1000	2200
ϕ for $T = 288°K$ ($T_R = 2$)	~1	0.95	0.12	Out of diagram; >2.5
ϕ for $T = 720°K$ ($T_R = 5$)	~1	1.02	1.37	2.03

where x_i is the molar fraction of the component i ($x_i = n_i/n_T$, namely the ratio of the number of moles of the component n_i to the total number of moles n_T), and f_i^* is the fugacity of the pure component at the temperature and total pressure of the mixture, then for a gas,

$$a_i = f_i = f_i^* x_i$$

and since $\phi_i = f_i^*/P$, where ϕ_i is the fugacity coefficient found for the gas i at the total pressure P of the mixture at the temperature considered, then

$$a_i = \phi_i P x_i$$

If only the gaseous components are considered in an equilibrium,

$$\alpha A + \beta B + \cdots \rightleftarrows \gamma C + \delta D + \cdots$$

then

$$K = \frac{(a_C)^\gamma (a_D)^\delta \cdots}{(a_A)^\alpha (a_B)^\beta \cdots} = \frac{(\phi_C)^\gamma (\phi_D)^\delta \cdots}{(\phi_A)^\alpha (\phi_B)^\beta \cdots} \frac{(n_C)^\gamma (n_D)^\delta \cdots}{(n_A)^\alpha (n_B)^\beta \cdots} \left(\frac{P}{n_T}\right)^{\gamma+\delta+\cdots-(\alpha+\beta+\cdots)}$$

or

$$K = K_P K_\phi$$

with

$$K_P = \frac{(P_C)^\gamma (P_D)^\delta \cdots}{(P_A)^\alpha (P_B)^\beta \cdots} = \frac{(n_C)^\gamma (n_D)^\delta}{(n_A)^\alpha (n_B)^\beta} \left(\frac{P}{n_T}\right)^{\gamma+\delta+\cdots-(\alpha+\beta+\cdots)}$$

and

$$K_\phi = \frac{(\phi_C)^\gamma (\phi_D)^\delta \cdots}{(\phi_A)^\alpha (\phi_B)^\beta \cdots}$$

For a given temperature, K is independent of the pressure, but K_ϕ varies both with the pressure and with the composition of the mixture; it can deviate quite appreciably from unity.

d. Technological and safety aspects. Quite apart from the specific technological considerations arising from the use of fluorine, certain safeguards must be taken because of the high pressures employed in this work.

The safety criteria depend on the size of the reactor and the strength of the materials employed for the reactor and accessories. Great care must always be taken when handling fluorine under pressure, the main danger arising from the possibility that traces of oil or organic products can initiate a combustion reaction with the fluorine. Under these conditions, the protective fluorine layer can be destroyed and the combustion may spread rapidly with the risk of an explosion due to the increase in pressure.

It should be remembered that a compressed gas represents, at the same pressure and volume, a much greater risk than a liquid, since the energy contained represents the work necessary to compress the fluid and this is of course greater in the case of a gas than in the case of a liquid. Furthermore, an apparatus will have some parts weaker than others, e.g., valves and manometers. Therefore, it is necessary, as far as possible, to protect the operators from metallic parts which might be projected in the event of an explosion of some part of the apparatus. Shielding screens made from mobile metallic plates are suitable for laboratory work. Other systems include a concrete wall fitted with passages through which manipulations may be made; this, however, is a less flexible solution to the problem.

e. Reactor technology. Nickel and Monel have the reputation of being suitable for fluorine pressure work from the corrosion point of view; using these materials, a reactor can be designed to stand up to the maximum pressures produced during a reaction.

The dimensions of the reactor required for volume and temperature conditions can be calculated from the mechanical characteristics of the metal in the elastic zone. It should be noted in this connection that if it is intended to carry out reactions with relatively small quantities of product (a few grams, say), it can be advantageous to use reactors made from metal tubing, from which it is easy to obtain fairly thick wall thicknesses for low internal diameters. This type of reactor also has the advantage of being easy to weigh on precision balances. As an example, we cite the work of Kelly *et al.* (1963), who used a tube made of nickel for carrying out high-pressure syntheses.

In the case of a cylindrical reactor with a relatively low wall thickness, for which the ratio R_0 of the external (R_e) to internal (R_i) radii is about 1.2, the relationship between the internal pressure P and the tangential tensile stress σ_t is

$$P = \sigma_t e/r$$

where e is the wall thickness and r the cylinder radius. Since the wall thickness is never negligible, the mean radius must be taken as

$$r = (R_e + R_i)/2$$

whence

$$P = 2\sigma_t(R_0 - 1)/(R_0 + 1)$$

It has been shown that the longitudinal stress is half as great as the tangential tensile stress and therefore is not the most important factor.

If, for a given mean radius r, the wall thickness is increased, the pressure can be increased proportionately, at least initially; for higher thicknesses, it is no longer valid to assume that the tangential tensile stress is the same at the inside and outside surfaces. The wall thickness must be increased more than proportionately to the pressure. In this case, use must be made of the formulas due to Lamé which give the stress distribution in the *elastic* zone for the metal. It is found, for example, that the tangential stress is a maximum on the internal surface and is given by the following expression (for a negligible external pressure):

$$\sigma_t = P(R_0^2 + 1)/(R_0^2 - 1)$$

Under these same conditions, the radial (σ_r) and longitudinal (σ_l) stresses have the respective values

$$\sigma_r = P \qquad \text{and} \qquad \sigma_l = P/(R_0^2 - 1)$$

The maximum calculated pressure, depending on the metal chosen and the ratio R_0, thus corresponds in fact only to the upper limit of the elastic threshold for the metal, considered at the internal surface. If the pressure is increased, the inelastic deformation zone spreads toward the outside, and it is often necessary to reach much higher pressures before the whole of the wall thickness is in the plastic zone and before risk of explosion occurs.

In addition to the effect of the temperature on the elastic limit of the metal, the temperature gradient across the wall must also be taken into account; if the wall is hotter on the outside, the tensile stress thus produced in the wall is added to that due to the internal pressure, whereas if the temperature gradient is in the opposite direction (the case of cooling of the external wall or internal heating), the stress produced by the internal pressure is partially compensated.

Several factors must be considered in choosing the dimensions of an appropriate reactor. Help may be obtained in this connection from the theories concerning the maximum principal stress, the maximum deformation, the maximum shear stress, etc.

For example, with the maximum principal stress theory, the maximum permissible pressure without a temperature gradient, $P = \sigma_t(R_0^2 - 1)/(R_0^2 + 1)$, will depend on the maximum stress chosen, i.e., the elastic limit of the metal. For Monel (International Nickel Co.), for example, the values at room temperature for the hardened metal are as follows after various treatments: cold-drawn, 6650 kg/cm^2; hot-rolled, 4550 kg/cm^2; cast, 2100 kg/cm^2 (Comings, 1956, p. 523).

If one wishes to calculate the design characteristics of a pressure reactor with higher strength characteristics and the same corrosion resistance, it will be necessary to use special techniques. For example, it is possible to make composite wall structures in which the outside wall, less corrosion resistant, has a higher strength.

Since the elastic limit depends on the composition, on the prior treatment, on the form of the metal used, and on the temperature, it is advisable to consult tables of metallurgical constants for help with the design of pressure reactors.

f. Technology of accessory equipment: *Valves.* At the present time, several commercial products exist which are quite suitable for working with fluorine under pressure, for example, those made by Autoclave Engineer, Inc. These are made of Monel and some of them may be used for higher temperatures, since the Teflon packing is separated from the main body of the valve.

Manometers. Bourdon-type gauges made of Monel or nickel are the most suitable; they can be used up to 5000 atm. A wide pressure range and good sensitivity are available in this type of equipment. Some of these manometers make it possible to record the pressure, e.g., those of the Heise Bourdon Tube Company, Inc. As far as connecting tubes, joints, and types of welding are concerned, Dodge (1950) may be consulted for further details.

From the point of view of safety, mention may be made of pressure safety disks designed to yield at a given gas pressure, which can be varied in each case according to requirements. In the case of fluorine, which is very corrosive, and depending on the amount of gas used and the nature of the site, a large volume can be placed on the other side of the disk to capture any jet of fluorine which might be produced on rupture of the disk.

g. Examples of fluorinations under pressure. We will consider the case of the preparation of hexafluoronickelates using fluorine under pressure. It is possible with this technique to prepare sodium hexafluoronickelate (Bougon, 1968). The overall reaction may be written

$$2NaF + NiF_2 + F_2 \rightarrow Na_2NiF_6$$

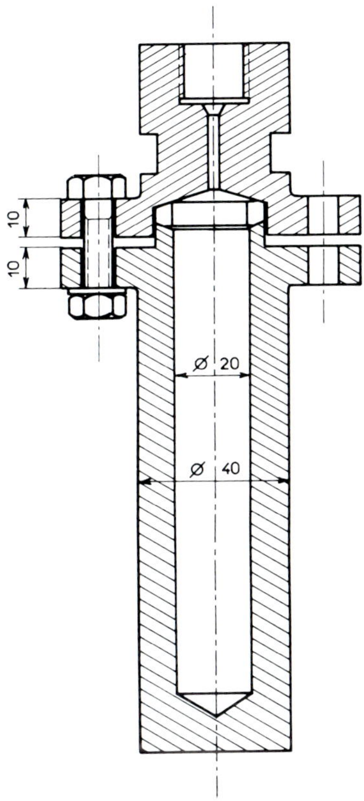

FIG. 4. High pressure reactor. Scale 1/1.

It occurs with a reduction in the mass of gas and is therefore favored by the high pressure.

The reactor used in the work cited is of the simplest possible type (Fig. 4). The joint is the lens-ring type made of copper (Dodge, 1950). This type of joint is increasingly effective as the pressure rises. The head of the valve is made up of a threaded portion into which is adapted a high-pressure Engineer Co. autoclave valve. It is possible in certain cases to fit a connection next to the valve so that a manometer may be added for following the course of the reaction. The reactor is pretested in a hydraulic press at 600 bar; it is then outgassed *in vacuo* at 350°C and given a pretreatment with a few atmospheres of ClF_3 at 350°C. After this operation the reactor is dismantled in a dry glove box and NaF–NiF_2 mixture is introduced. A prefluorination of the solid mixture using ClF_3 is carried out,

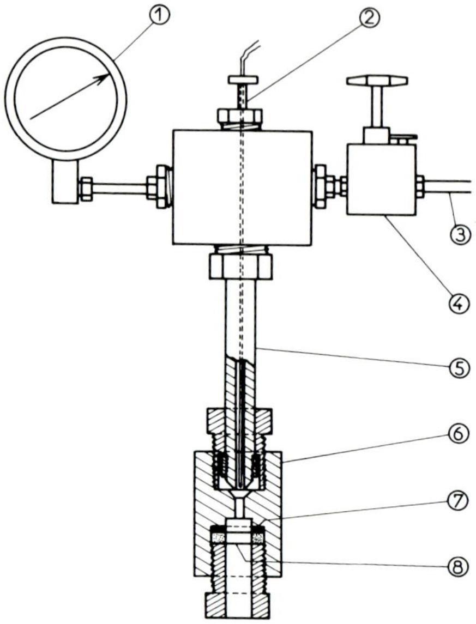

FIG. 5. High-pressure reactor of Kelly *et al.* (1963); (1) pressure gauge; (2) iron–constantan thermocouple (nickel clad); (3) fluorine inlet; (4) high-pressure valve; (5) nickel reactor; (6) rupture assembly; (7) copper gasket; (8) rupture disk (nickel). Adapted from *J. Amer. Chem. Soc.* **85,** 307 (1963). Copyright by the American Chemical Society. Used by permission of the copyright owner.

followed by another outgassing. The amount of fluorine required to give an initial pressure of 350 bar at 350°C is then introduced into the reactor. This quantity is evaluated using the compressibility coefficient calculated for these conditions; the fluorine is measured volumetrically and is introduced by condensation into the reactor cooled to −195°C by liquid nitrogen. When the saturated vapor pressure of fluorine is attained, the reactor is isolated and heated with a ring oven. A shield is placed in front of the apparatus and the pressure is made to rise gradually.

When the reaction is considered to be over, and before evacuation of the fluorine, the reactor is cooled with liquid nitrogen in order to avoid a sudden pressure drop which might send the solid product into other parts of the apparatus when the valve is opened. In the present case, the fluorine is allowed to leak slowly into the introductory enclosure; the pressure in this section, as read on the manometer, gives an idea of the amount of reaction which has occurred. The method has been applied to very many syntheses. It has the advantage of being simple to carry out and of not requiring a too complicated apparatus.

Other examples of fluorination under pressure may be mentioned. Kelly *et al.* (1963) studied the fluorination of alkali metal chlorides at pressures of up to 4000 atm and at temperatures of about 300°C. By fluorinating NaCl, LiCl, KCl, and RbCl, they obtained, respectively, NaF, LiF. $KClF_4$, and $RbCl_4$; the two latter products were contaminated by the binary fluoride. The fluorination of CsCl did not lead to a well-defined crystallized product under these conditions, although Asprey *et al.* (1961) have synthesized $CsClF_4$. In this work, the reactor was made from a nickel tube having a volume of about 9 ml on which was fixed a manometer and a nickel guard disk designed to yield at 5000 atm (Fig. 5). The authors state that, as far as fluorine corrosion is concerned, Monel is quite satisfactory at low temperatures, but that at higher temperatures, it is better to use nickel. In the example here, the fluorine is introduced as before by condensation into the reactor which is cooled to liquid nitrogen temperature.

J. Slivnik *et al.* (1965) synthesized the perfluorides of rhenium, osmium, indium, and platinum with a yield of 90–100% by reacting together the metal and fluorine at 50–60 atm. The reaction was carried out at 125–500°C for 1–2 hr in a nickel autoclave. Glemser *et al.* (1963, 1966) have described the construction and use of nickel reactors for the preparation under pressure of the compounds CrF_6 and OsF_7.

3. *Reactions using gaseous hydrogen fluoride*

At low pressure, gaseous hydrogen fluoride reacts with metals only at high temperature. The reactivity of the metals is considerably increased by a prior treatment with hydrogen. As an example, mention may be made

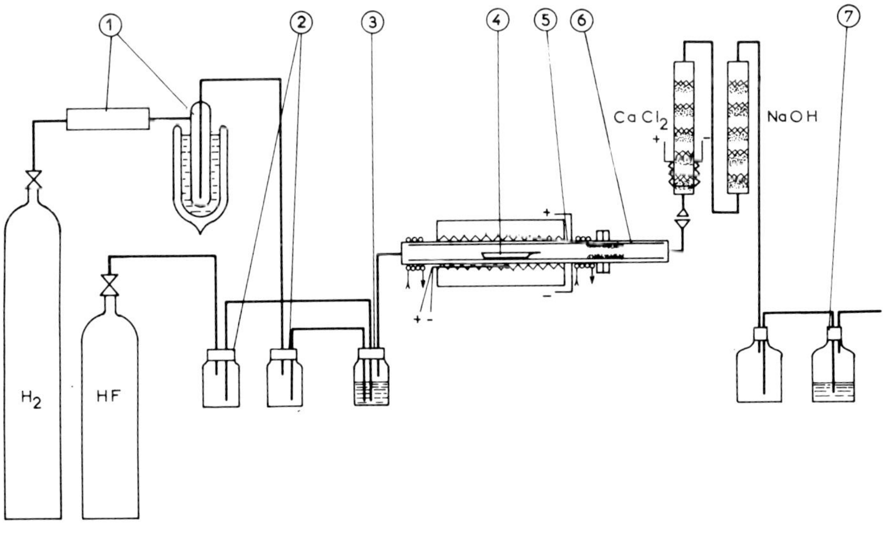

FIG. 6. Hydrogen fluoride apparatus: (1) deoxygenation and dehydration of the hydrogen; (2) gas mixers; (3) gas flow control; (4) gold crucible; (5) nickel tube; (6) mobile nickel tube; (7) flow control for end of reaction.

of the preparation of divalent vanadium fluoride VF_2 by a reaction of HF with powdered vanadium at 1200°C (Shafer, 1969).

The most commonly used technique for obtaining fluorides and oxyfluorides of transition elements is the fluorination of the oxide, the halide, or the oxychloride (Table VIII). Figure 6 shows an apparatus for carrying out fluorination with HF at low pressures up to 700°C (Grannec, 1970).

TABLE VIII

ACTION OF HYDROGEN FLUORIDE ON VARIOUS OXIDE AND HALIDE COMPOUNDS

Initial compound	Experimental conditions (°C)	Compound formed	References
$LaCl_3$	500	LaF_3	
$TiH_{0.7}$, $TiCl_3$	700	TiF_3	Ehrlich and Pietzka (1954), Siegel (1956)
$TiCl_4$	—	TiF_4	Haendler *et al.* (1954)
$HgCl_4$	300	HgF_4	Sense *et al.* (1954)
VCl_3	300–500	VF_3	Jack and Gutmann (1951), Nyholm and Sharpe (1952)
VCl_4	0–20	VF_4	Ruff and Lickeftt (1911)
$VOBr_2$	—	VOF_2	Ruff and Lickfett (1911)
Nb_2O_5	500	NbO_2F	
Ta_2O_5	250	TaO_2F	
	>300	TaF_5	
$CrCl_3$	—	CrF_3	Hepworth *et al.* (1957), Nyholm and Sharpe (1952)
CrO_3	—	CrO_2F_2	Wartenberg (1941), Englebrecht and Grosse (1952)
$MoCl_3$, $MoBr_3$	—	MoF_3	Emeleus and Gutmann (1949, 1950)
$MOCl_4$ (M = Mo, W)	—	MOF_4	Ruff and Eisner (1907), Ruff *et al.* (1907)
WO_2	300–800	WOF_2	Priest and Schumb (1948)
ReO_3Cl	—	ReO_3F	Englebrecht and Grosse (1954)
$FeCl_2$	500	FeF_2	Baur (1958)
$FeCl_3$	200	FeF_2	
$CoCl_2, 6H_2O$	550	CoF_2	Baur (1958), Henkel and Klemm (1935)
$NiCl_2$	500	NiF_2	Baur (1958), Henkel and Klemm (1935)
ZrH	HF/H_2; 750	ZrF_4	Ehrlich *et al.* (1964)
VF_3	HF/H_2; 1250	VF_2	Stout and Boo (1966)
NbH	HF/H_2; 570	NbF_3, NbF_5	Ehrlich *et al.* (1964)
Ta, TaH	HF/H_2; 300	TaF_3, TaF_5	Emeleus and Gutmann (1949)

The sample is placed in a gold crucible placed in an Inconel tube. In order to obtain a controlled fluorination, use is made of a mixture of hydrogen fluoride and either nitrogen or hydrogen which has been previously dried and freed of its oxygen. The flow rate is controlled throughout the duration of the fluorination reaction. The volatile products formed can be collected on an internal mobile nickel tube. If an HF/H_2 mixture is passed over NbH at 570°C, a blue-black solid NbF_3 is formed in the crucible, whereas NbF_5 crystals are condensed on the cooled walls of the tube.

Double fluorides can also be obtained by the action of HF on mixtures of halides or oxyhalides:

$$K_2ReBr_6 \xrightarrow{HF/200^\circ C} K_2ReF_6$$ (Weise, 1956)

$$MF + FeCl_2 \xrightarrow{HF/500^\circ C} MFeF_3 \quad (M = \text{alkali metal})$$

(Portier *et al.*, 1969)

Hydrogen fluoride is also used to treat fluorides prepared in aqueous solution; the role of the HF, which passes over the heated product, is merely to prevent hydrolysis during the dehydration process (YF_3, CrF_3, VF_3) (Zalkin and Templeton, 1953; Hepworth *et al.*, 1957).

4. *Reactions using other fluorinating gases*

a. Chlorine trifluoride. This is a very strong fluorinating agent; even in the gas phase, it is often more strongly fluorinating than fluorine itself. Because of this, it can be used for carrying out reactions at lower temperatures. Chlorine trifluoride is now widely produced commercially and can be transported in the liquid form at room temperature, this being an additional advantage with respect to fluorine. It has the disadvantage, however, of giving with oxides or water adsorbed on the product to be treated certain oxides and oxyfluorides of chlorine; reactions are therefore often less simple than in the case of fluorine.

An example of such a fluorination is the preparation of AgF_2 from AgF and ClF_3 (Charpin *et al.*, 1966). This reaction occurs at 100°C, whereas in the case of fluorine, a temperature of 160°C is required. According to the authors, complete transformation of the initial AgF to AgF_2 only occurs if ClF_3 is used, and then in a much shorter time: 24 hr.

In these types of fluorination, it is very often necessary to use a flow of gas over the solid reactant unless sufficient ClF_3 can be introduced in one single operation. The gas required is drawn off from the ClF_3 source either with the use of a dry nitrogen stream or by distillation of the liquid;

in the latter case, it is necessary to compensate for the cooling effect due to evaporation, by heating the ClF_3 source.

If a nitrogen stream is used, the fluorination will be much less vigorous, because of the dilution effect. An example is the action of ClF_3 on NH_4F to give the difluoramine NF_2Cl. This is a violent, if not an explosive reaction, in the condensed phase. Gardner *et al.* (1963) operated as follows. The NH_4F is put in a copper boat which is placed in a Kel F reactor; in this way, the changes in the solid NH_4F can be followed visually. The ClF_3 gas, either pure or diluted with nitrogen at a total flow rate of 85–160 ml min^{-1}, is sent over the sample. For a ClF_3 concentration greater than 30% by volume in the nitrogen, combustion of the NH_4F occurs, sufficient heat being evolved for the reactor chamber to melt. With a concentration of under 30% and a flow rate of 140–160 ml min^{-1}, the reaction leads to volatile NF_2Cl, which is condensed downstream from the reactor and then purified by successive distillations.

These same authors have carried out the reaction using a suspension of NH_4F in an organic fluorinated oil which ensures a better distribution of the heat produced. Since the reactor was here again made up of a 60-mm outside diameter Kel F tube, stirring was effected magnetically; the tube was placed in a bath thermostated to $\pm 1°$. The reactor was filled with amounts of up to 5 g of NH_4F in 40 ml of oil; the best yields (about 22%) were obtained with nitrogen containing 11.7% ClF_3 at a total flow rate of 178 ml min^{-1} and with an initial temperature in the suspension of 60–73°C.

b. Sulfur tetrafluoride SF_4. This compound is a much more moderate fluorinating agent than ClF_3, so much so that it can be used in organic chemistry.

In inorganic chemistry, many fluorides have been prepared by the action of SF_4 on oxides or sulfides. For uranium oxides, for example, Johnson *et al.* (1961) give the following reactions:

$$UO_3 + 3SF_4 \xrightarrow{300°C} UF_6 + 3SOF_2$$

$$UO_2F_2 + 2SF_4 \xrightarrow{300°C} UF_6 + 2SOF_2$$

$$U_3O_8 + 8SF_4 \xrightarrow{400°C} 2UF_6 + UF_4 + 8SOF_2$$

$$UO_2 + 2SF_4 \xrightarrow{500°C} UF_4 + 2SOF_2$$

In addition to these two fluorinating agents, ClF_3 and SF_4, others having intermediate fluorinating powers can be used for gas-phase reactions. They are, for example, BrF_3 (on condition that a fairly high temperature is used, to compensate for its low volatility), BrF_5, ClF_5, and XeF_4.

B. Action of Nonfluorinating Gases

In the field of gas–solid reactions, mention must also be made of the use of gases that have no fluorinating properties. These involve essentially reduction and oxidation processes.

1. *Reduction*

The most commonly used reduction method for preparing a fluoride is to pass hydrogen over the corresponding fluoride in a higher oxidation state:

$$MF_n + mH_2 \rightarrow MF_{n-2m} + 2mHF$$

The hydrogen must be carefully dried in order to avoid any hydrolysis of the fluorides at high temperatures. It should be also freed of its oxygen to prevent formation of oxyfluorides. Examples of this type of reaction are the preparation of CrF_2 and of EuF_2 by reduction of the corresponding trivalent fluorides (Sturm, 1962; Klemm and Doll, 1939) and of the phases $AMnF_4$ (A = alkali metal) starting from the tetravalent compounds $AMnF_5$ using a temperature of 250°C (Hoppe *et al.*, 1959).

The action of the volatile hexafluorides MF_6 (M = Mo, Os, Re) on tungsten or molybdenum hexacarbonyl at 25°C leads to the formation of the pentavalent compounds MF_5 (Hargreaves and Peacock, 1960a; Beveridge and Clark, 1967):

$$5MoF_6 + Mo(CO)_6 \xrightarrow{25°C} 6MoF_5 + 6CO$$

Ruthenium pentafluoride RuF_5 is reduced to the trifluoride by action of carbon monoxide at 149°C (Peacock, 1960) or of iodine at 500°C (Hepworth *et al.*, 1957).

2. *Oxidations*

The action of oxygen can lead in certain cases to the formation of oxyfluorides: the oxyfluoride TlOF is formed by the action of oxygen at 4.5. kbar on TlF at 500°C, according to the following equations (Demazeau *et al.*, 1969):

$$4TlF + O_2 \rightarrow 2Tl_2OF_2$$

$$2Tl_2OF_2 + O_2 \rightarrow 4TlOF$$

Chlorine gas can also be used to oxidize monovalent mercury compounds (Ruff and Bahlau, 1918):

$$Hg_2F_2 + Cl_2 \rightarrow HgCl_2 + HgF_2$$

3. *Action of water vapor*

A general method of preparing oxyfluorides is to hydrolyze the corresponding fluorides. For example, water vapor (steam) can hydrolyze the fluorides of the lanthanides and of yttrium, leading to the formation of the oxyfluoride LnOF (Zalkin and Templeton, 1953; Zachariasen, 1949, 1951; Klemm and Klein, 1941; Finkelburg and Stein, 1950). The oxyfluorides InOF and TlOF can be obtained in a similar way (Grannec *et al.*, 1967, 1970). The oxyfluorides MOF_4 (M = W, Mo) hydrolyze gradually, giving the compounds MO_2F_2 (Hultgren and Brewer, 1956; Zado, 1963; Colton *et al.*, 1964). Mention may also be made of the preparation of $XeOF_4$ by controlled hydrolysis of XeF_6 (Smith, 1963).

4. *Action of nitrogen*

An interesting method of preparation has been proposed for uranium fluoronitride UNF. This is the simultaneous action of silicon and nitrogen on the fluoride UF_4 (Yoshihara *et al.*, 1969):

$$4UF_4 + 3Si + 2N_2 \rightarrow 4UNF + 3SiF_4$$

V. Solid-Phase Reactions

A. Low Pressure

The double fluorides of the transition elements are usually prepared from the binary fluorides in a direct solid-phase synthesis. There is no need to go into the details of this method, which hardly differs from conventional chemical syntheses. It can be noted, however, that certain precautions should be taken. In most cases, the mixtures must be ground in a closed volume because they are hygroscopic. The reaction mixture is heated to the required temperature either in a crucible or in a sealed platinum or gold tube which has been argon-sealed; the latter technique is surer from the point of view of preventing hydrolysis of the products. When it cannot be used, the reaction may be carried out in an open crucible, but the reaction chamber must be carefully evacuated and outgassed using a secondary pumping unit. The reactions are usually fairly rapid. The method can be applied to complexes of nontransition metals. For example, NH_4MgF_3 and $(NH_4)_2MgF_4$ can be prepared by simply heating together ammonium bifluoride and magnesium carbonate in varying proportions in an open crucible (Roux *et al.*, 1968). The same process can be used in the case of oxyfluorides, but the syntheses are more difficult to

TABLE IX

REACTIONS IN THE SOLID STATE

Initial product	Experimental conditions (°C)	Compound formed	References
Fluoride–fluoride	300–1100	—	Babel (1967), Tressaud (1969)
$2KF + NaF + MF_3$ [M(III) = Cr, Fe, Ga]	500	K_2NaMF_6	Knox and Mitchell (1961)
$2KF + 2NaF + TiO_2 + K_2TiF_6$	400–900	$K_2NaTiOF_5$	Pausewang and Rüdorff (1969)
$2KF + NaF + NbO_2F$	400–900	$K_2NaNbO_2F_4$	Pausewang and Rüdorff (1969)
$2KF + NaF + MoO_3$	400–900	$K_2NaMoO_3F_3$	Pausewang and Rüdorff (1969)
$FeO_3 + FeF_3$	15 min, 950	FeOF	Hagenmuller *et al.* (1965)
$NbO_2F + Nb_2O_5$ system	500–1100	Nb_3O_7F, $Nb_5O_{12}F$	Andersson and Åström (1964) Andersson (1964)
		$Nb_{17}O_{42}$, $Nb_{31}O_{17}F$	Andersson (1965)
$xMF + xFeF_2 + (1-x)FeF_3$ (M = alkali metal)	300–700	M_xFeF_3	de Pape (1965), Tressaud *et al.* (1970)
$xKNbO_3 + (1-x)NbO_2F$	500	$K_xNbO_{2+x}F_{1-x}$	de Pape *et al.* (1968)

carry out and the temperatures of reaction are higher, generally speaking. Some examples of such reactions are given in Table IX.

Certain fluorides can be made by a solid-phase reduction:

$$2CrF_3 + Cr \xrightarrow{1100^\circ C} 3CrF_2$$

(Sturm, 1962; Fergusson, 1967)

$$3MoF_5 + 2Mo \longrightarrow 5MoF_3$$

(La Valle *et al.*, 1960)

Mention may also be made of the preparation of titanium garnet $Li_3Na_3Ti_2F_{12}$ by reduction with silicon of a LiF–NaF–TiF_4 mixture (de Pape *et al.*, 1967).

B. High Pressure

High-pressure techniques are of course not used only in fluorine chemistry; they are, however, of the greatest use in this field.

The equipment used is not very different from that used, for example, for work on oxides. Pressure can be applied to the sample either through the intermediate of a fluid (see Section III.A) or by using presses (belt apparatus, tetrahedral anvil press) (Bradley, 1966; Hall, 1960). It is important that the reactants be placed in gold or platinum receptacles to avoid corrosion of the apparatus.

Although they are not strictly speaking syntheses, allotropic transformations can be effected by this method (Longo and Kafalas, 1969).

The pressure plays a double role in these synthesis reactions. First, it favors the reactions by ensuring an intimate contact between the solids, and second, it prevents the decomposition of unstable phases or the sublimation of volatile compounds. Chamberland and Sleight (1967) have prepared in this way a series of oxyfluorides with the rutile structure (TiOF, VOF, FeOF, $Ti_2O_{4-x}F_x$, $V_2O_{4-x}F_x$, etc.) by heating the corresponding oxides and fluorides under pressure 900–1300°C; 0–1 to 65 kbar). Crockett and Grossman (1964) have prepared at 20°C, but at a pressure of 7 kbar, perovskites of the type $NH_4M(II)F_3$ (M(II) = Co, Cu, Mn, Cd, Ni, Zn). This interaction is not due merely to the relatively high volatility of NH_4F; similar reactions can be observed with, for example, KBr.

VI. Gas-Phase Reactions

Mention may be made of the reactions of nitric oxide NO with the hexafluorides MoF_6 and UF_6. Geichman *et al.* (1962) have carried out these reactions either in a glass apparatus made up of two one-liter flasks connected by a tap, or in a metallic apparatus composed of two volumes separated by a wide-bore tap. Other experiments were carried out in nickel containers connected with Teflon or aluminum joints, and in Kel F tubes. All parts of the apparatus were exposed to fluorine or ClF_3 before being used for the reactions. In this way, the compounds $NOUF_6$ and $NOMoF_6$ were obtained almost instantaneously by mixing the two gases at temperatures between 25 and 60°C; the authors point out that the reactions are much more rapid than if liquid MoF_6 and solid UF_6 are used.

In the new field of noble gas chemistry, reactions have been obtained between xenon and fluorine which lead to the formation of XeF_2, XeF_4, and XeF_6 according to the conditions of temperature, pressure, and fluorine to xenon ratio (Weinstock *et al.*, 1966).

A further example is the preparation of the compounds $Xe(PtF_6)_n$ carried out in either a glass, silica, or nickel apparatus (Bartlett and Jha, 1963).

TABLE X

DECOMPOSITION OF SOME FLUORINE COMPOUNDS

Initial compound	Experimental conditions	Compound formed	References
$(NH_4)_3AlF_6 \xrightarrow{170^\circ C} NH_4AlF_4 + 2NH_4F$			Shinn *et al.* (1966)
$\xrightarrow{300^\circ C} \gamma\text{-}AlF_3 + NH_4F \xrightarrow{720^\circ C} \alpha\text{-}AlF_3$			
VF_4	600°C	$VF_3 + VF_5$	Ruff and Lickfett (1911)
NH_4CrF_3	350°C	CrF_2	Dumora (1969)
$(NH_4)_3CrF_6$	5 hr, 1100°C	CrF_2	Fergusson (1967)
K_2CrF_6	*a*	CrF_5	Huss and Klemm (1950)
$(NH_4)_2MF_6$ (M = Zr, Hf)		ZrF_4, HfF_4	Sense *et al.* (1954)
Mo_2F_9	170°C	$MoF_4 + MoF_5$	Peacock (1957)
ReF_5	*a*	ReF_4	Hargreaves and Peacock (1960a)
OsF_6	UV, 20°C	OsF_5	Hargreaves and Peacock (1960b)

[a] Temperature is unknown in these cases.

VII. Other Methods

Finally, we note preparative methods based on the thermal degradation of fluorides. The starting products are usually ammonium salts, which decompose at relatively low temperatures, or compounds in an unstable oxidation state. These reactions must be monitored very closely by gravimetric methods to obtain the required product (Table X).

Ultraviolet radiation has been used for the preparation of OsF_5 by photochemical decomposition of OsF_6 (Hargreaves and Peacock, 1960b). Electrolytic reductions have also been used.

REFERENCES

Andersson, S. (1964). *Acta Chem. Scand.* **18,** 2339.

Andersson, S. (1965). *Acta Chem. Scand.* **19,** 1401.

Andersson, S., and Åström, A. (1964). *Acta Chem. Scand.* **18,** 2233.

Asprey, L. B., Margrave, J. L., and Silverthorn, M. E. (1961). *J. Amer. Chem. Soc.* **83,** 2955.

Aurivillius, B. (1955). *Acta Chem. Scand.* **9,** 1206.

Aynsley, E. E., Peacock, R. D., and Robinson, P. L. (1950). *J. Chem. Soc. (London)* **1950,** 1622.

Aynsley, E. E., Peacock, R. D., and Robinson, P. L. (1952). *Chem. Ind.* **1951,** 1002.

Babel, D. (1967). *In* "Structure and Bonding" (C. K. Jorgensen *et al.*, eds.), Vol. 3, pp. 1–87. Springer-Verlag, Berlin.

Barber, E. J., and Cady, G. H. (1956). *J. Phys. Chem.* **60,** 505.

Bartlett, N., and Jha, N. J. (1963). *In* "Noble Gas Compounds" (H. H. Hyman, ed.), pp. 23–30. Univ. of Chicago Press, Chicago, Illinois.

Bartlett, N., and Lohmann, D. H. (1960). *Proc. Chem. Soc. (London)* **1960,** 14.

Bartlett, N., and Lohmann, D. H. (1964). *J. Chem. Soc. (London)* **1964,** 619.

Bartlett, N., and Rao, P. R. (1965). *Chem. Commun.* **1965,** 252.

Baur, W. H. (1958). *Acta Crystallogr.* **11,** 488.

Bayard, M. L. F. (1969). Thesis, Brown Univ., Providence, Rhode Island.

Benedict, M., Webb, G. B., and Rubin, L. C. (1940). *J. Chem. Phys.* **8,** 334.

Beveridge, A. D., and Clark, H. C. (1967). *In* "Halogen Chemistry" (V. Gutmann, ed.), Vol. 3, pp. 179–226. Academic Press, New York.

Billy, C., and Haendler, H. M. (1957). *J. Amer. Chem. Soc.* **79,** 1049.

Bougon, R. (1968). *C. R. Acad. Sci. Paris* **267,** 681.

Bradley, S. (1966). "Advances in High Pressure Research." Academic Press, New York.

Cady, G. H., and Hargreaves, G. B. (1961). *J. Chem. Soc. (London)* **1961,** 1568.

Canterford, J. H., and O'Donnell, T. A. (1968). *In* "Technique of Inorganic Chemistry" (H. B. Jonassen and A. Weissberger, eds.), Vol. 7, pp. 273–306. Wiley (Interscience), New York.

Canterford, J. H., Colton, R., and O'Donnell, T. A. (1967). *Rev. Pure Appl. Chem.* **17,** 123.

Chamberland, B., and Sleight, A. W. (1967). *Solid State Commun.* **5,** 765.

Charpin, P., Dianoux, A. J., Marquet-Ellis, H., and Nguyen-Nghi, H. (1966). *C. R. Acad. Sci. Paris* **263,** 1359.

Christe, K. O., and Pilipovich, D. (1969). *Inorgan. Chem.* **8,** 391.

Christe, K. O., and Sawodny, W. (1969). *Inorgan. Chem.* **8,** 212.

Clark, H. C., and Emeleus, H. J. (1957). *J. Chem. Soc. (London)* **1957,** 119.

Clark, H. C., and Emeleus, H. J. (1958). *J. Chem. Soc. (London)* **1958,** 190.

Clark, H. C., and Sadana, Y. N. (1964). *Can. J. Chem.* **42,** 50.

Claverie, J., Lozano, L., and Portier, J. (1969). Rapports I.R.S.I.D., 305.68 and 355.69.

Colton, R., Tomkins, I. B., and Wilson, P. (1964). *Aust. J. Chem.* **17,** 496.

Comings, E. W. (1956). *In* "High Pressure Technology." McGraw-Hill, New York.

Crockett, D. S., and Grossman, R. A. (1964). *Inorgan. Chem.* **3,** 644.

Demazeau, G., Grannec, J., Marbeuf, A., Portier, J., and Hagenmuller, P. (1969). *C. R. Acad. Sci. Paris* **269,** 987.

Dodge, B. F. (1950). *In* "Chemical Engineers' Handbook" (J. H. Perry, ed.), pp. 1233–1262. McGraw-Hill, New York.

Dumora, D. (1969). Thesis, Univ. of Bordeaux, France.

Ebert, F., and Woitinek, H. (1933). *Z. Anorg. Allg. Chem.* **210,** 269.

Edwards, A. J., Peacock, R. D., and Small, R. W. H. (1962). *J. Chem. Soc.* **1962,** 4486.

Edwards, A. J., Hugill, D., and Peacock, R. D. (1963). *Nature London* **200,** 672.

Ehrlich, P., and Pietzka, G. (1954). *Z. Anorg. Allg. Chem.* **275,** 121.

Ehrlich, P., Ploeger, F., and Koch, E. (1964). *Z. Anorg. Allg. Chem.* **333,** 209.

Emeleus, H. J., and Gutmann, V. (1949). *J. Chem. Soc. (London)* **1949,** 2979.

Emeleus, H. J., and Gutmann, V. (1950). *J. Chem. Soc. (London)* **1950,** 2115.

Engelbrecht, A., and Grosse, A. V. (1952). *J. Amer. Chem. Soc.* **74,** 5262.

Englebrecht, A., and Grosse, A. V. (1954). *J. Amer. Chem. Soc.* **76,** 2042.
Fairbrother, F. (1967). *In* "Halogen Chemistry" (V. Gutmann, ed.), Vol. 3, pp. 123–178. Academic Press, New York.
Fairbrother, F., and Frith, W. C. (1951). *J. Chem. Soc. (London)* **1951,** 3051.
Fairbrother, F., Frith, W. C., and Woolf, A. A. (1954). *J. Chem. Soc. (London)* **1954,** 1031.
Fergusson, J. E. (1967). *In* "Halogen Chemistry" (V. Gutmann, ed.), Vol. 3, pp. 227–303. Academic Press, New York.
Finkelburg, W., and Stein, A. (1950). *J. Chem. Phys.* **18,** 1296.
Forsberg, H. E. (1957). *Acta Chem. Scand.* **11,** 676.
Gardner, D. M., Knipe, W. W., and Mackley, C. J. (1963). *Inorgan. Chem.* **2,** 413.
Garton, G., and Wanklyn, B. M. (1967). *J. Crystal Growth* **1,** 49.
Gaunt, J. (1953). *Trans Faraday Soc.* **49,** 1122.
Geichman, J. R., Smith, E. A., Trond, S. S., and Ogle, P. R. (1962). *Inorgan. Chem.* **1,** 661.
Geichman, J. R., Swaney, L. R., and Ogle, P. R. (1963). U.S. Rep. GAT-T-808.
Gillardeau, J. (1970). *In* "Oxidation of Metals" (K. Hauffe, ed.), Vol. 2, No. 3. Plenum, New York.
Gladney, H. M., and Street, G. B. (1968). *J. Inorgan. Nucl. Chem.* **30,** 2949.
Glemser, O., Roesky, H., and Hellberg, K. H. (1963). *Angew. Chem.* **75,** 346.
Glemser, O., Roesky, H. W., Hellberg, K. H., and Werther, H. U. (1966). *Chem. Ber.* **99,** 2652.
Grannec, J. (1970). Thesis, Univ. of Bordeaux, France.
Grannec, J., and Lozano, L. (1969). Private communication.
Grannec, J., Portier, J., de Pape, R., and Hagenmuller, P. (1967). *Bull. Soc. Chim. Fr.* **1967,** 4281.
Grannec, J., Portier, J., von der Muhl, R., Demazeau, G., and Hagenmuller, P. (1970). *Mater. Res. Bull.* **5,** 185.
Gruner, E., and Klemm, W. (1937). *Naturwissenschaften* **25,** 59.
Haendler, H. M., Bartram, S. F., Becker, R. S., Bernard, W. J., and Bukata, S. W. (1954). *J. Am. Chem. Soc.* **76,** 2177.
Hagenmuller, P., Portier, J., Cadiou, J., and de Pape, R. (1965). *C. R. Acad. Sci. Paris* **260,** 4768.
Hall, T. (1960). *Rev. Sci. Instrum.* **31,** 125.
Hargreaves, G. B., and Peacock, R. D. (1960a). *J. Chem. Soc. (London)* **1960,** 1099.
Hargreaves, G. B., and Peacock, R. D. (1960b). *J. Chem. Soc. (London)* **1960,** 2618.
Hassel, O., and Nilssen, S. (1929). *Z. Anorg. Allg. Chem.* **181,** 172.
Henkel, P., and Klemm, W. (1935). *Z. Anorg. Allg. Chem.* **222,** 73.
Hepworth, M. A., and Jack, K. H. (1957). *Acta Crystallogr.* **10,** 345.
Hepworth, M. A., and Robinson, P. L. (1957). *J. Inorgan. Nucl. Chem.* **4,** 274.
Hepworth, M. A., Robinson, P. L., and Peacock, R. D. (1954). *J. Chem. Soc. (London)* **1954,** 1197.
Hepworth, M. A., Jack, K. H., Peacock, R. D., and Westland, G. J. (1957). *Acta Crystallogr.* **10,** 63.
Holloway, J. H., and Peacock, R. D. (1963). *J. Chem. Soc.* **1963,** 527.
Holloway, J. H., Rao, P. R., and Bartlett, N. (1965). *Chem. Commun.* **1965,** 306.
Homann, R., and Hoppe, R. (1969). *Z. Anorg. Allg. Chem.* **368,** 271.
Hoppe, R. (1956). *Rec. Trav. Chim.* **75,** 569.
Hoppe, R. (1958). *Z. Anorg. Allg. Chem.* **294,** 135.
Hoppe, R., and Homann, R. (1969). *Z. Anorg. Allg. Chem.* **369,** 212.

Hoppe, R., and Klemm, W. (1952). *Z. Anorg. Allg. Chem.* **268,** 364.

Hoppe, R., and Müller, B. (1969). *Naturwissenschaften* **1,** 56.

Hoppe, R., Blinne, K., and Liebe, W. (1959). *Int. Conf. Coordination Chem., London* Chem. Soc. Special Publ., No. 13, p. 132.

Horvitz, D. (1950). Theoretical, Laboratory, and Experimental Investigations of High Energy Propellants: Review of the Physical and Chemical Properties of Fluorine and Certain of Its Compounds, Rep. No. RM1-293-55, p. 26. Reaction Motors Inc., Rockaway, New Jersey.

Hultgren, N., and Brewer, L. (1956). *J. Phys. Chem.* **60,** 947.

Huss, E., and Klemm, W. (1950). *Z. Anorg. Allg. Chem.* **262,** 25.

Hyman, H. H., and Katz, J. J. (1965). *In* "Non-Aqueous Solvent Systems" (T. C. Waddington, ed.), p. 47. Academic Press, New York.

Iwasaki, H., Yahata, T., Suzuki, K., and Oshina, K. (1962). *Kogyo Kagaku Zasshi* **65,** 1165.

Jack, K. H. (1959). Communication to the *Int. Symp. Fluorine Chem., Birmingham.*

Jack, K. H., and Gutmann, V. (1951). *Acta Crystallogr.* **4,** 246.

Johnson, C. E., Fischer, J., and Steindler, M. (1961). *J. Amer. Chem. Soc.* **83,** 1620.

Kay, W. B. (1936). *Ind. Eng. Chem.* **28,** 1014.

Kelly, D. H., Post, B., and Mason, R. W. (1963). *J. Amer. Chem. Soc.* **85,** 307.

Kirk, R. E., and Othmer, D. F. (eds.). (1966). "Encyclopedia of Chemical Technology," Vol. 9, pp. 506–847. Wiley, New York.

Klemm, W., and Doll, W. (1939). *Z. Anorg. Allg. Chem.* **241,** 234.

Klemm, W., and Klein, H. A. (1941). *Z. Anorg. Allg. Chem.* **248,** 167.

Knox, K., and Mitchell, D. W. (1961). *J. Inorgan. Nucl. Chem.* **21,** 253.

Kwong, J. N. S., and Redlich, O. (1949). *Chem. Rev.* **44,** 233.

Laudise, R. A. (1962). *In* "Progress in Inorganic Chemistry" (A. Cotton, ed.), Vol. 3, p. 1. Wiley (Interscience), New York.

La Valle, D. L., Steele, R. M., Wilkinson, M. K., and Yakel, H. L. (1960). *J. Amer. Chem. Soc.* **82,** 2433.

Longo, J. M., and Kafalas, J. A. (1969). *J. Solid State Chem.* **1,** 103.

Nelson, L. C., and Obert, E. F. (1954). *Mech. Eng.* **76,** 280.

Newton, R. H. (1935). *Ind. Eng. Chem.* **27,** 302.

Nyholm, R. S., and Sharpe, A. G. (1952). *J. Chem. Soc.* **1952,** 3579.

O'Donnell, T. A. (1956). *J. Chem. Soc. (London)* **1956,** 4681.

Palmer, W. G. (1954). "Experimental Inorganic Chemistry." Cambridge Univ. Press, London and New York.

Pape, R. de (1965). *C. R. Acad. Sci. Paris* **260,** 4527.

Pape, R. de, Portier, J., Gauthier, G., and Hagenmuller, P. (1967). *C. R. Acad. Sci. Paris* **265,** 1244.

Pape, R. de, Gauthier, G., and Hagenmuller, P. (1968). *C. R. Acad. Sci. Paris* **266,** 803.

Pape, R. de, Portier, J., Grannec, J., Gauthier, G., and Hagenmuller, P. (1969). *C. R. Acad. Sci. Paris* **269,** 1121.

Pausewang, G., and Rüdorff, W. (1969). *Z. Anorg. Allg. Chem.* **364,** 69.

Peacock, R. D. (1957). *Proc. Chem. Soc.* **1957,** 59.

Peacock, R. D. (1960). *In* "Progress in Inorganic Chemistry" (F. A. Cotton, ed.), Vol. 2, pp. 193, 249. Wiley (Interscience), New York.

Portier, J., Tressaud, A., Dupin, J. L., and de Pape, R. (1969). *Mater. Res. Bull.* **4,** 45.

Portier, J., Tanguy, B., Morell, A., and Pouchard, M. (1970). *In* "Colloque International C.N.R.S. sur les Elements des Terres Rares." CNRS, Paris.

Priest, H. F., and Schumb, W. C. (1948). *J. Amer. Chem. Soc.* **70,** 3378.
Rault, M., Demazeau, G., Portier, J., and Grannec, J. (1970). *Bull. Soc. Chim. Fr.* **1970,** 74.
Roesky, H. W., Glemser, O., and Hellberg, K. H. (1965). *Chem. Ber.* **98,** 2046.
Roux, N., Charpin, P., and Ehretsmann, J. (1968). *C. R. Acad. Sci. Paris* **267,** 484–486.
Ruff, O., and Bahlau, G. (1918). *Ber.* **51,** 1752.
Ruff, O., and Eisner, F. (1907). *Ber.* **40,** 2926.
Ruff, O., and Lickfett, H. (1911). *Ber.* **44,** 2539.
Ruff, O., and Vidic, E. (1925). *Z. Anorg. Chem.* **143,** 163.
Ruff, O., Eisner, F., and Heller, W. (1907). *Z. Anorg. Chem.* **52,** 256.
Sense, K. A., Snyder, M. J., and Filbert, R. B. (1954). *J. Phys. Chem.* **58,** 995.
Shafer, M. W. (1969). *Mater. Res. Bull.* **4,** 905.
Sharpe, A. G. (1959). *In* "Advances in Fluorine Chemistry." Butterworth, London and Washington, D.C.
Shinn, D. B., Crocket, D. S., and Haendler, H. M. (1966). *Inorgan. Chem.* **5,** 1927.
Siegel, S. (1956). *Acta Crystallogr.* **9,** 684.
Sleight, A. W. (1969). *Inorgan. Chem.* **8,** 1764.
Slivnik, J., Smalc, A., and Zemljic, A. (1965). Nukl. Inst. Josef Stefan, NIJS Rept. No. R-472.
Smith, D. F. (1963). *Science* **140,** 899.
Stout, J. W., and Boo, W. O. J. (1966). *J. Appl. Phys.* **37,** 966.
Sturgeon, G. D., Penneman, R. A., Kruse, F. H., and Asprey, L. B. (1965). *Inorgan. Chem.* **4,** 748.
Sturm, B. J. (1962). *Inorgan. Chem.* **1,** 665.
Tressaud, A. (1969). Thesis, Univ. of Bordeaux, France.
Tressaud, A., Galy, J., and Portier, J. (1969). *Bull. Soc. Fr. Mineral. Cristallogr.* **92,** 335.
Tressaud, A., Pape, R. de, Portier, J., and Hagenmuller, P. (1970). *Bull. Soc. Chim. Fr.* **1970,** 3411.
Trevorrow, L. E. (1958a). *J. Chem. Soc.* **1958,** 362.
Trevorrow, L. E. (1958b). *J. Phys. Chem.* **62,** 362.
Trevorrow, L. E., Fischer, J., and Steunenberg, R. K. (1957). *J. Amer. Chem. Soc.* **79,** 5167.
von Wartenberg, H. (1941). *Z. Anorg. Chem.* **247,** 135.
von Wartenberg, H. (1942). *Z. Anorg. Chem.* **249,** 100.
Weinstock, B., Weaver, E. E., and Knop, C. P. (1966). *Inorgan. Chem.* **5,** 2189.
Weise, E. (1956). *Z. Anorg. Allg. Chem.* **283,** 377.
Weise, E., and Klemm, W. (1955). *Z. Anorg. Allg. Chem.* **279,** 74.
White, D., Hu, J., and Johnston, H. L. (1953). *J. Chem. Phys.* **21,** 1149.
Whitney, E. D., Mac Laren, R. D., Fogle, C. E., and Hurley, T. J. (1964). *J. Amer. Chem. Soc.* **86,** 2583.
Wiechert, K. (1950). *Z. Anorg. Allg. Chem.* **261,** 310.
Wold, A. (1969). Private communication.
Yoshihara, K., Kanno, M., and Mukaibo, T. (1969). *J. Inorgan. Nucl. Chem.* **31,** 985.
Zachariasen, W. H. (1949). *Acta Crystallogr.* **2,** 388.
Zachariasen, W. H. (1951). *Acta Crystallogr.* **4,** 231.
Zado, F. (1963). *J. Inorg. Nucl. Chem.* **25,** 1115.
Zalkin, A., and Templeton, D. H. (1953). *J. Amer. Chem. Soc.* **75,** 2453.

Boron Syntheses

R. NASLAIN

Service de Chimie Minérale Structurale associé au CNRS, Université de Bordeaux, Talence, France

Introduction

Boron is one of the light elements which have been the subject of much research in the last 20 years in view of potential applications in the fields of electronic components, refractory materials, and composites. It has a low density, a high melting point, a hardness close to that of diamond, a semiconducting behavior, and remarkable mechanical properties. From a critical study of the literature, it appears that the results for both fundamental and applied research have not been as good as expected mainly, because boron of insufficient purity was used in most of the experiments.

The problem of obtaining boron of high purity is dominated by the strong affinity of boron for oxygen, carbon, and many common metals. The chemical behavior of the element and the necessary requirement of high temperature make the use of standard techniques for preparing pure elements quite difficult in the case of boron. The synthesis of pure boron is also complicated by analytical difficulties. The best method for the analysis of boron gives a standard error of $\pm 0.3\%$. Lacking an accurate analytical determination of the element, test of high purity implies the careful search for and analysis of all the impurities, the boron content being equal to the remainder.

Another feature of the preparation of boron comes from the fact that boron may exist either in an amorphous glassy state or in an unusually large number of crystalline modifications. The polymorphism of boron has been reviewed in detail by Hoard and Newkirk (1960), Hoard (1961), Hoard and Hughes (1967). From studies on single crystals, it now appears that boron may form two rhombohedral modifications and two tetragonal ones. There seems to be no doubt about the rhombohedral modifications, which are true boron varieties. On the contrary, it is not yet clearly established whether or not the tetragonal structures require small amounts of foreign atoms for the stabilization of their boron framework. It now appears more and more evident that the β-rhombohedral form is probably the only thermodynamically stable variety of boron from room temperature to the melting point. The other modifications could be either metastable boron frameworks grown under favorable kinetic conditions, or nonstoichiometric, boron-rich borides containing a small amount of foreign atoms. Apart from these four well-characterized forms, one can find, in the literature, many so-called "new boron modifications" which, from all evidence, are either mixtures of previously reported varieties or unknown borides.

Two types of boron samples are required for research. For an accurate study of the physical properties of boron, the most suitable samples are large single crystals (or at least, bulk specimens with a low density of grain boundaries) with a wide range of thermal stability and as low an impurity content as possible (of the order of a few ppm). On the contrary, for the synthesis of boron compounds with a view toward chemical or crystallographic studies, the best samples are finely divided powders, more reactive than bulk materials, purity of 99.5 being sufficient in general.

After recent progress in the manufacture of boron halides and in their reduction with hydrogen, rods of boron claimed to be free of any spectrographically detectable impurities are now available. However, the price of such material is very high and the content of carbon still significant. If the situation is quite good for the samples used by physicists for the study of elemental boron, it is rather poor for the powdered samples used for the synthesis of most of the boron compounds on which, unfortunately, more and more physical measurements are being reported. It has often been pointed out that most of the so-called "99% boron" commercial samples are in fact of poor grade (95%). In as much as powdered samples obtained by grinding high-purity boron rods are contaminated by the grinding equipment, and also because they are not always sufficiently reactive, any serious work on boron compounds, e.g., borides, has to start with the preparation of the element.

All the various methods for boron synthesis have been reviewed recently in detail by Newkirk (1961, 1964) in two very complete articles with numerous references. Our purpose here is to give only selected examples of methods which have been checked by many authors. In Section II, several methods involving the reduction of boron compounds by metals are briefly studied. The resulting boron is not of high purity, but is so easy to prepare cheaply that it may be used as starting material for upgrading or purification techniques. In Section III, the electrolytic processes are presented, mainly because most of the commercial boron samples are prepared by electrolysis of fused salt mixtures; it is also of interest to point out what impurities have to be expected in these materials, which are utilized daily by chemists. In Sections IV, the synthesis of high-purity boron by hydrogen reduction of boron halides is described in detail, and purification by zone melting is discussed in Section V.

I. The Various Forms of Boron in the Solid State

From the structural point of view, boron presents an uncommon complexity among the elements; it may exist either in a large number of crystalline modifications or in an amorphous form.

A. Crystalline Modifications of Boron

From the results of single-crystal studies, it now appears that boron may form two rhombohedral modifications and two tetragonal ones. Except for one tetragonal form whose structure is not yet fully known, the three other varieties have structures based on B_{12} boron clusters in which 12 boron atoms occupy the vertices of a more or less regular icosahedron. Table I gives the cell data for the four boron polymorphs.

The α-rhombohedral variety (α-rh) has the simplest structure: it consists of nearly regular icosahedra in a slightly deformed cubic close-packing. One icosahedron is centered at each of the eight vertices of the elementary cell. In the three-dimensional boron framework, 50% of the B—B bonds between icosahedra are typical two-electron covalent bonds, the remaining bonds being three-center bonds. These weak and electron-poor δ-bonds give a low thermal stability to the boron skeleton: at about 1200°C, the α-rh phase transforms irreversibly into the second rhombohedral variety (Decker and Kasper, 1959).

The boron skeleton of the β-rhombohedral form (β-rh) is also based upon B_{12} icosahedra which may be associated to form complex B_{84} units with the structural formula $B_{12}(B_6)_{12}$. In the structure of the β-rh variety, the B_{84} units play the same role as the B_{12} units in the structure of the α-rh variety; they are centered at the vertices of the elementary rhombo-

TABLE I

Cell Data for Crystalline Boron Polymorphs

Polymorph	Cell parameters	Space group	Atoms per unit cell	Primitive structure elements	Density	References
α-Rhombohedral	$a = 5.057$ Å, $\alpha = 58°06'$	$R\bar{3}m$	12	B_{12}	2.46	Decker and Kasper (1959)
β-Rhombohedral	$a = 10.145$ Å, $\alpha = 65°17'$	$R\bar{3}m$	105	$B_{84} + 2B_{10} + B$	2.35	Hoard *et al.* (1958)
α-Tetragonal	$a = 8.75$ Å, $c = 5.06$ Å	$P4_2/nnm$	50	$4B_{12} + 2B$	2.31	Hoard *et al.* (1960)
β-Tetragonal	$a = 10.12$ Å, $c = 14.14$ Å	$P4_122$	192	Unknown	2.36	Amberger and Ploog (1968), Talley *et al.* (1960b)

hedral cell. The packing of the B_{84} units leaves an important void in the center of the cell, which is filled by additional boron atoms in the form of a small chain of structural formula B_{10}–B–B_{10} (Hoard *et al.*, 1970).

The structure of the α-tetragonal variety (α-tetra) also contains boron icosahedra: 48 of the 50 atoms in the unit cell belong to four B_{12} icosahedra. Two isolated boron atoms, located in tetrahedral positions, complete the bonding between neighboring icosahedra and are fundamental for the stabilization of the boron framework (Hoard *et al.*, 1958). From recent studies, it now appears that the formation of this structure is strongly favored by impurities, e.g., carbon, capable of replacing boron in the tetrahedral positions. However, it is not yet definitely established whether the structure may exist in the absence of any traces of such impurities (Amberger and Ploog, 1968).

The structure of the β-tetragonal modification (β-tetr) is not fully known. In addition, the thermal stabilities of the boron polymorphs have not been studied in detail: Some varieties are obtained only in narrow temperature ranges and all transform irreversibily into the β-rh form. It seems more and more accepted that β-rh variety is the only thermodynamically stable modification from room temperature to the melting point (2200–2300°C).

B. Amorphous Boron

Amorphous boron is generally obtained by vapor-phase deposition; it has the properties of a glass and its x-ray diffraction pattern shows only a small number of broad and diffuse rings. The nature of this amorphous state of boron is the subject of controversy: According to Otte and Lipsitt (1966), it might be a disordered packing of sheets of boron icosahedra, while according to Lindquist *et al.* (1966), amorphous boron is no more than an assemblage of small grains of the α-rh modification, the size of the crystals (30 Å) being mainly responsible for the broadening of the diffraction rings. Amorphous boron exhibits remarkable mechanical properties, which are used in an important class of composite materials, the boron fiber composites (Wavner, 1967).

II. Reduction of Boron Compounds by Metals

Boron compounds, such as boric oxide, metal fluoroborates, or boron halides, are easily reduced to boron by the most electropositive metals:

alkali metals, alkaline-earth metals, or even aluminum. Reduction occurs with a considerable evolution of heat, which considerably raises the temperature of the reaction products. The extraction of boron from the reaction products is the main problem of these methods of boron synthesis.

A. Reduction of Boric Oxide by Metals

Moissan (1892b) and later Mikheeva *et al.* (1957) established that magnesium is the best reducing agent for B_2O_3. The reaction product is not pure boron, as thought by Moissan, but probably a mixture of amorphous boron and of a boron suboxide (Weintraub, 1909; Kahlenberg, 1925).

Moissan recommended the use of three parts of B_2O_3 for one part of magnesium (by weight) as the most favorable ratio for the starting materials. According to a recent work by Yannacakis and Nies (1960), it seems more judicious to choose a mixture poorer in B_2O_3; in these conditions, the product has a higher content of magnesium, but this impurity is more easily eliminated than oxygen in the washing treatment.

The application of the method is very simple. The mixture of boric oxide and magnesium is put into a refractory or metallic crucible and placed into a furnace having preferably an inert atmosphere and which has been preheated to about 1000°C. The reaction is strongly exothermic and takes place after a few minutes. After cooling, the excess of boric oxide, the magnesium borate, and eventually magnesium borides are removed by leaching with a solution of HCl. The residue is a brownish powder which is known as "Moissan boron" and has a boron content of 80–90%. If the reduction is carried out in an inert atmosphere from high-purity materials, the main impurities are oxygen and magnesium.

Moissan boron is so easy to prepare that considerable effort has been made in attempts to purify this product with chemical or physical processes. Figure 1 gives three methods for upgrading Moissan boron to purity claimed to be as high as 98.3–99.8%.

From our own experience, amorphous boron seems to be very sensitive to oxidation when in a finely divided state and any chemical treatment with aqueous solutions may introduce oxygen in the samples. This suggests that the purity claimed by Moissan (98.3%) is very optimistic. Weintraub (1909) and later Kahlenberg (1925) did not succeed in obtaining purity higher than 83–92% when following Moissan's original procedure.

Heating under vacuum appears to be a more interesting upgrading process, since most of the impurities sublime at high temperature (Mikheeva *et al.*, 1957; Yannacakis and Nies, 1960). To be effective, the process requires temperatures close to the melting point. In these conditions, the product, obtained as a bulk material in the β-rh form, has a boron content higher than 99%, but has a poor chemical reactivity.

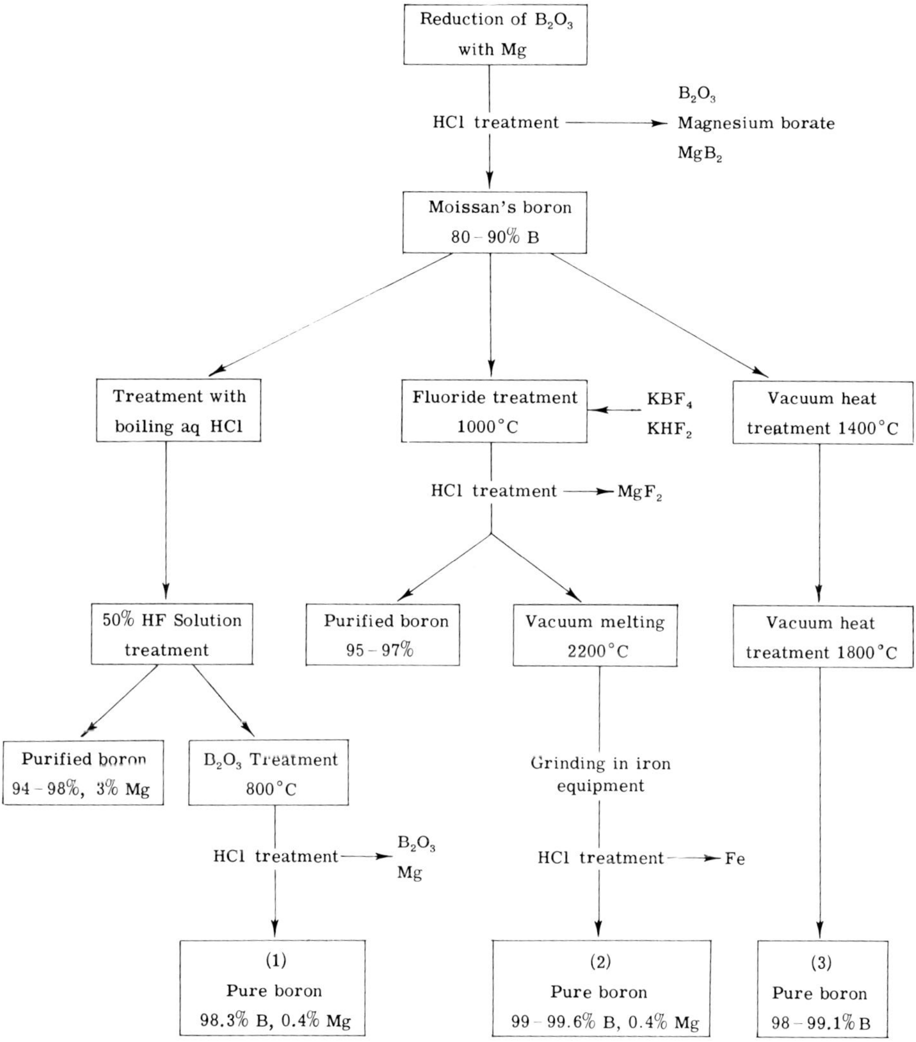

FIG. 1. Purification of so-called "Moissan's boron" according to (1) Moissan (1892b), (2) Yannacakis and Nies (1960), and (3) Mikheeva *et al.* (1957).

B. Reduction of Boron Halides and Fluoroborates by Alkali Metals

The possibility of obtaining boron with the reduction of boron halides BX_3 (X = F, Cl, Br) or metal fluoroborates by alkali metals has been known for a long time (Moissan, 1892a). More recently, these reactions have been the subject of much study (Sowa, 1949; Spevack, 1954; Haag, 1957; Finn *et al.*, 1960; Schott and Hansley, 1963).

The reduction of a boron halide BX_3 (X = F, Cl, Br) by an alkali metal, generally sodium, proceeds according to the reaction

$$BX_3 + 3Na \rightarrow B + 3NaX \tag{1}$$

The reaction is strongly exothermic; it may be controlled by regulation of the boron halide flow rate or by introduction of an inert gas for constant temperature operation. For maximizing the interface between the reagents to obtain a higher yield, the liquid alkali metal is sprayed into an atmosphere of boron halide which may or may not contain an inert gas diluent. The resulting boron is amorphous or crystalline depending on the temperature. It is separated from the alkali metal halide by leaching with water.

By the use of metal fluoroborates MBF_4 (M = Na, K), the problems common to the handling of boron halides are avoided. The generally preferred method is the reduction, by sodium, of potassium fluoroborate, which has a higher thermal stability, according to the reaction

$$KBF_4 + 3Na \rightarrow B + KF + 3NaF \tag{2}$$

The reaction, which is highly exothermic, starts at low temperature (250°C), but it is commonly carried out between 400 and 700°C. Boron is extracted from the mixtures of alkali metal fluorides by leaching the product with water.

When reactions (1) and (2) are carried out in an atmosphere free of air, one would expect that they would lead to boron of high purity, i.e., free of oxygen. On the contrary, chemical analysis shows that this is not the case, the boron content of the product being about 90% after washing with water.

Following the hypothesis formulated by Moissan and Andrieux on the existence of unstable alkali borides, most authors have supposed that a by-product of the reduction was a sodium boride which might be destroyed by hydrolysis during a leaching with an aqueous solution of HCl (Moissan, 1892a, Andrieux and Barbetti, 1932). It has been shown in this laboratory that the affinity of boron for alkali metals is weak. It has also been established that sodium and potassium react with boron only at high temperature (900°C for sodium and 1150°C for potassium) and under a pressure of the

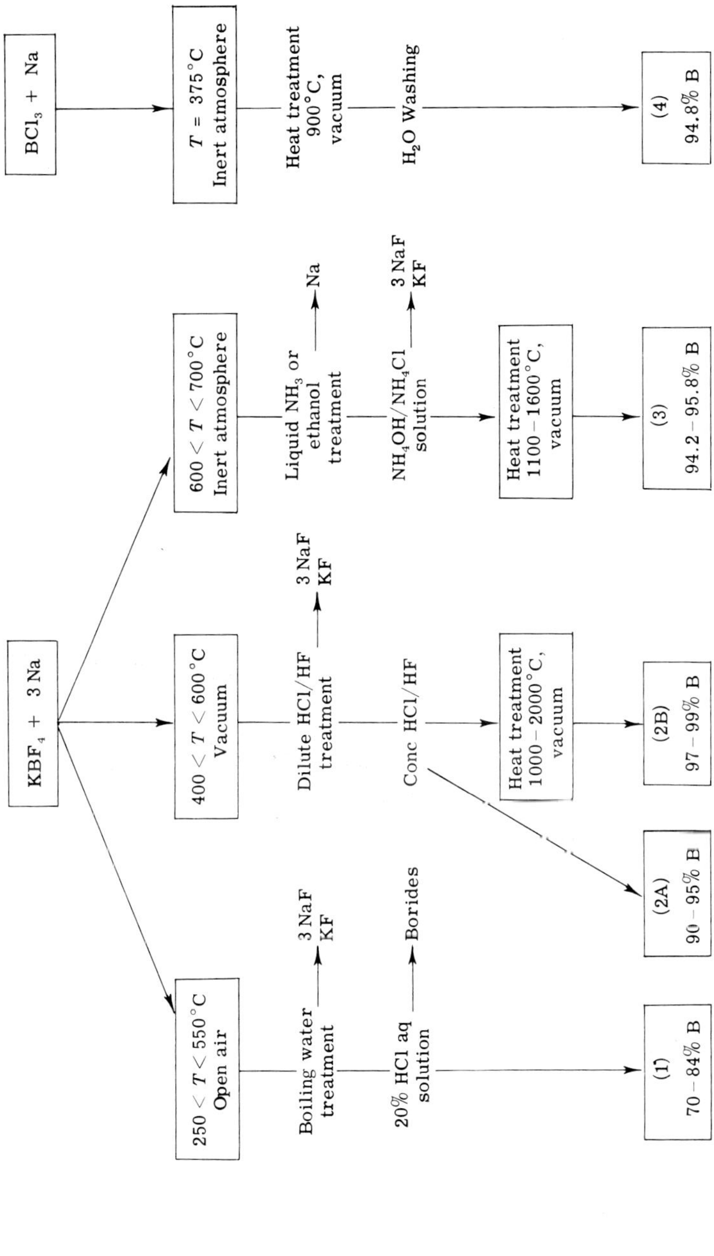

Fig. 2. Processes for obtaining boron with alkali-metal reduction of KBF_4 or BCl_3 according to (1) Sowa (1949); (2A, B) Haag (1957); (3) Finn *et al.* (1960); and (4) Schott and Hansley (1963).

alkali metal. Furthermore, the only alkali borides known up to now—NaB_6, NaB_{15}, and KB_6—do not dissolve in aqueous solution of HCl. Consequently it appears improbable to us that a sodium boride might be formed in the temperature range (400–700°C) used for the application of reactions (1) and (2). There is also evidence that leaching the product with aqueous solution does not dissolve a hypothetical sodium boride but merely produces a serious oxidation of the amorphous boron which has been prepared at very low temperature, i.e, in a form chemically very reactive (Hagenmuller and Naslain, 1963; Naslain and Etourneau, 1966; Naslain *et al.*, 1967; Naslain, 1967; Naslain and Kasper, 1970).

Figure 2 clearly shows that heating under vacuum and at high temperature is again the best upgrading process; it is effective for the volatile impurities—boron suboxide, for example—which might be introduced during the washing.

Spevack (1954) has proposed the utilization of the evolution of heat during the reduction for raising the temperature of the reaction products and obtaining boron in the β-rh variety, which is less sensitive to oxidation at room temperature than the amorphous form. Boron prepared under these conditions is probably not oxidized during the elimination of NaF in aqueous solution, but unfortunately, the author does not report any indication of the purity of his samples.

III. Reduction of Boron Compounds by Electrolysis

In the molten state, boric oxide B_2O_3 does not exhibit sufficient electrical conductivity for direct electrolysis. Electrolytic synthesis of boron is possible, however, starting from complex baths containing, as a source of boron, either a borate, a fluoroborate, or boric oxide itself. In most of the processes, an alkali-metal halide or an alkaline-earth halide, is added to increase the electrical conductivity and the fluidity of the bath while also participating in the mechanism of the electrolysis. It is generally accepted that boron, which appears in the neighborhood of the cathode, results from a secondary reaction between the components of the bath and alkali metal (or alkaline-earth metal) which is produced by a primary reaction at the cathode.

The earliest information on the electrolytic synthesis of boron appears in the work of Hampe (1888) and Kahlenberg (1925). More recently, the work of Andrieux (1929) and Cooper (1959) has been particularly important for the elaboration of modern methods of obtaining boron by molten salt electrolysis.

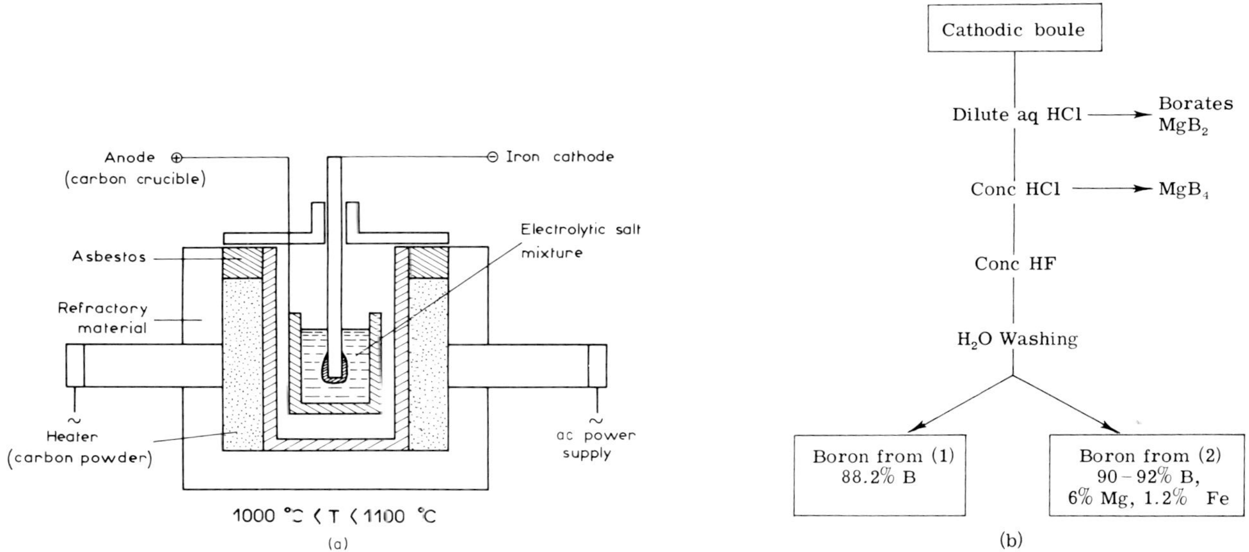

FIG. 3. Andrieux's process for obtaining elemental boron by electrolysis of (1) $MgO–B_2O_3$ or (2) $MgO–B_2O_3–MgF_2$ fused mixtures. (a) Electrolytic cell. (b) Chemical treatments of the cathodic boule. (c) Proposed mechanism for the electrolysis:

Anodic reaction: $6O^{2-} \rightarrow 3O_2 + 12e^-$

Cathodic reactions:
$$\begin{cases} 6Mg^{2+} + 12e^- \rightarrow 6Mg^0 \\ 5B_2O_3 + 6Mg^0 \rightarrow 3Mg_2B_2O_5 + 4B \end{cases}$$

A. Reduction of Molten Magnesium Borate by Electrolysis: Andrieux's Process

In 1929, Andrieux showed that the electrolytic reduction of molten borates between 800 and 1000°C leads to an evolution of oxygen at the anode and to the formation of the corresponding metal at the cathode. The active metal may reduce boric oxide in the neighborhood of the cathode to give boron, a mixture of boron and borides, or a mixture of borides alone according to the nature of the metal. A cathodic boule forms around the electrode; it consists of boron, borides, borate, and a variable quantity of molten salt mixture in which boron is partly protected from the oxidizing atmosphere of the cell. At the end of the electrolysis, the cathodic boule, which adheres strongly to the electrode, is slowly withdrawn from the bath in order that it may be covered with a liquid layer protecting it from atmospheric oxidation. After grinding, the cathodic boule is purified chemically following the method of Moissan (Fig. 3).

The chemical analysis reported by Andrieux shows that the electrolytic reduction of borates generally gives rather impure boron samples except in the case of magnesium borate, for which the boron content, after chemical purification, may reach 88% or even 92% if magnesium fluoride has been added to the bath. The main impurities are magnesium, which could be present in the form of borides insoluble in HCl (MgB_4 and MgB_6), iron and carbon coming from the electrodes of the cell, and oxygen.

B. Electrolytic Reduction of Molten Potassium Fluoroborate: Cooper's Process

The electrolytic reduction of molten borates gives boron samples which are strongly oxidized (electrolysis of alkali-metal borates) or which contain a high percentage of metal (electrolysis of magnesium borate). Cooper, following upon some observations of Kahlenberg, has proposed a process for obtaining pure boron by the electrolysis of a bath free of oxygen compounds of boron (Cooper, 1951a, 1959).

In its original version, Cooper's process is based on the electrolysis at 800–900°C of a molten bath having the composition $KBF_4 + 3KCl$. Potassium chloride is used for dissolving and stabilizing KBF_4. According to the mechanism shown in Fig. 4, the primary reaction of the electrolysis leads to an evolution of chlorine at the anode and to the formation of potassium at the cathode. As in Andrieux's process, it is the active metal which reduces potassium fluoroborate in the neighborhood of the cathode

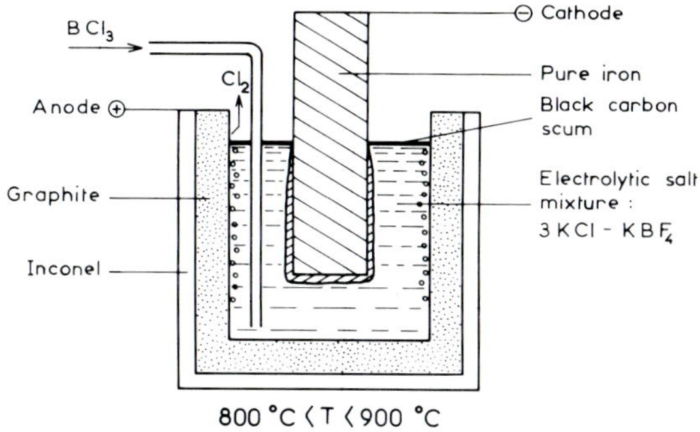

FIG. 4. Cooper's process for producing pure elemental boron by electrolysis of oxygen-free fused salt mixtures. Proposed mechanism for the electrolysis:

Anodic reaction: $6Cl^- \rightarrow 3Cl_2^0 + 6e^-$

Cathodic reactions:
$$\begin{cases} 6K^+ + 6e^- \rightarrow 6K^0 \\ 2KBF_4 + 6K^0 \rightarrow 2B + 8KF \end{cases}$$

Regeneration of the electrolytic salt mixture with BCl_3:

$$8KF + 2BCl_3 \rightarrow 6KCl + 2KBF_4$$

and leads to the formation of a cathodic boule. Boron formed by this secondary reaction is mixed with potassium fluoride and liquid in the cathodic deposit, from which it is extracted by chemical treatment.

The formation of potassium fluoride during the reduction of KBF_4 considerably decreases the electrical conductivity of the bath and leads to the appearance of an anode effect which must be avoided in order to keep constant the electrical parameters of the cell. In order to regenerate the bath and to avoid the accumulation of KF, a boron halide, generally BCl_3, is introduced into the bath. The flow rate of boron halide is adjusted so that potassium fluoride is transformed at just the same rate as it is formed, according to the reaction

$$BCl_3 + 4KF \rightarrow KBF_4 + 3KCl \tag{3}$$

Not only does the reaction of BCl_3 with KF result in the formation of the two starting materials, but in addition these are reformed in the exact ratio required by the electrolytic reaction. The use of BF_3 is less interesting, since it requires the extra addition of KCl for regeneration of the starting mixture.

The electrolysis is carried out in the cell shown in Fig. 4. The graphite crucible, protected on its outside by a sheet of Inconel, is the anode. The cathode is copper or molybdenum for the laboratory scale apparatus and low-carbon iron such as Armco iron for large cells. The electrolysis is carried out at 800–900°C with a voltage of about 6–8 V and a current density at the cathode of a few amperes per square centimeter.

The cathodic deposit is leached with hot water and concentrated HCl solution. The resulting boron is in the form of microcrystalline aggregates. According to Cooper, the boron content may be as high as 99.5%, the main impurities being carbon, silicon, and iron. Carbon is presumably produced by the erosive action of the hot bath on the graphite crucible; it accumulates at the surface as a black scum which is skimmed off every 10–15 min in order to keep the carbon content of the resulting boron below 0.5%.

The process is suitable for the preparation of ^{10}B and ^{11}B isotopes. Separation of the two isotopes of boron is generally accomplished by first converting ordinary boron trifluoride into a volatile dimethylether–boron trifluoride complex. The complex is then subjected to a series of multistage distillations whose final fractions are enriched with ^{10}B or ^{11}B. For example, $^{10}BF_3$ is obtained by simple thermal dissociation of the corresponding $(CH_3)_2O$–$^{10}BF_3$ complex; it may be converted into $^{10}BCl_3$ by reaction with $AlCl_3$. Elementary ^{10}B isotope is easily prepared by Cooper's process: $^{10}BCl_3$ is dissolved in molten potassium fluoride up to the composition $K^{10}BF_4$–3KCl and the bath electrolyzed.

C. Electrolysis of Complex Baths

Whereas the electrolysis of molten borates generally yields boron of poor grade, Cooper has established that the electrolysis of B_2O_3 dissolved in KBF_4–KCl complex baths gives a product having a boron content as high as that obtained with the electrolysis of oxide-free KBF_4–3KCl bath (Cooper, 1951). The composition of the bath is close to 60KCl–12KBF_4–7B_2O_3; KCl is a solvent for KBF_4 and increases the fluidity of the bath; KBF_4 acts as a solvent for B_2O_3. Although the exact mechanism is not fully known, it seems that during the electrolysis, only B_2O_3 is decomposed since oxygen, and not chlorine, is evolved at the anode and the boron is formed at the cathode.

The electrical conductivity of the complex bath containing dissolved B_2O_3 is higher than that of the oxide-free bath, with the result that the cell can be operated with higher current. The purity of the product is generally 95–98% and may even be better than 99%.

D. Conclusions

Although both electrolytic reduction of boron compounds and direct chemical reduction by metals are in fact based on the same reactions, the former is definitely superior.

Many authors have used Cooper's processes, but all of them do not agree with the purity claimed by Cooper for the products obtained from KBF_4–3KCl bath or from complex bath with dissolved B_2O_3. One suggests that the disagreement between the reported purities—from 91 to 99.5% B—may be due to differences in the techniques themselves and also to the difficulties in the chemical analysis of boron (Cooper, 1951a; Andrieux and Deiss, 1955; Samsonov *et al.*, 1960; Newkirk, 1964).

Boron obtained by electrolysis is very often amorphous or poorly crystallized because it is prepared at low temperature. It is very interesting to make a connection between the high carbon impurity content of Cooper's boron and the fact that the only diffraction pattern given by Cooper (1961) is that of the α-tetragonal modification.

Although Cooper's boron is of sufficient purity for many uses, it is not pure enough for the synthesis of boron compounds on which physical measurements have to be done. This is especially true for carbon, the main impurity of Cooper's boron, which, even in the range of a few ppm, greatly modifies the semiconducting properties of boron.

IV. Reduction of Boron Halides by Hydrogen

Weintraub was the first to clearly point out, in 1909, the potential of hydrogen reduction of boron halides for obtaining boron of high purity. He succeeded in preparing the very first samples of crystalline boron by reducing BCl_3 with hydrogen in an electrical arc. Since this first experiment, much work has been devoted to this method, the main variations being with regard to the nature of the hot zone in which the reduction is carried out.

A. Principle and Features of the Method

The method is based on the vapor-phase reduction of a boron halide BX_3 by hydrogen according to the reaction

$$2BX_3 + 3H_2 \rightarrow 2B + 6HX \qquad (X = F, Cl, Br) \tag{4}$$

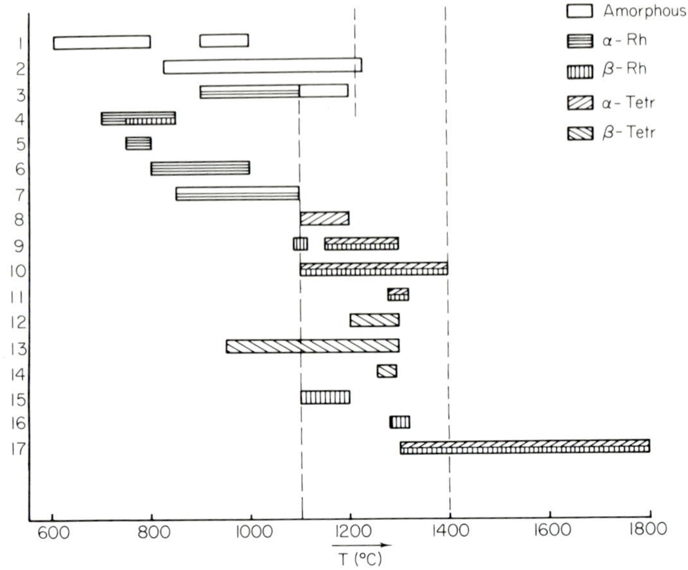

FIG. 5. Crystalline and amorphous varieties of boron obtained by hydrogen reduction of boron trihalides. References: (1) Laubengayer *et al.* (1943). (2) Talley *et al.* (1960a). (3) Boiret (1968). (4) Naslain *et al.* (1968). (5) Robertson (1943). (6) McCarty and Carpenter (1960). (7) Amberger and Dietze (1965). (8) Boiret (1968). (9) Uno (1958). (10) Becher and Schäfer (1960). (11) Laubengayer *et al.* (1943). (12) Boiret (1968). (13) Amberger and Ploog (1968). (14) Talley (1960a). (15) Stern and Lynds (1958). (16) Bean and Medcalf (1960). (17) Olempska *et al.* (1966).

Stern and Lynds (1958) have studied the thermodynamics of this reaction. The calculation of the free enthalpy of reaction versus temperature shows that the reduction of BF_3 requires temperatures too high to be suitable for the technical preparation of boron. On the other hand, the reduction of both BCl_3 and BBr_3 is thermodynamically possible in a more convenient range of temperature. The reduction of BBr_3 starts at about 650°C, while that of BCl_3 is technically possible only above about 1000°C. The use of BBr_3 allows the study of the reaction in a wider range of temperature. For BI_3, simple thermal decomposition under vacuum is generally preferred to hydrogen reduction.

The reduction occurs on a hot substrate on which boron deposits as a solid; at the reduction temperature, all the chemical species involved in the reaction are gaseous except boron. Consequently, the first important advantage of the method is the fact that the resulting boron is extracted very easily from the reaction medium.

Another feature of the method is that it gives directly the various amorphous or crystalline modifications, depending upon the experimental parameters, among which the temperature of the substrate and the rate of deposition of boron are the most important. Figure 5 shows that in first approximation, the α-rhombohedral form and the amorphous modification are commonly obtained at low temperature ($T < 1100°C$) and that the β-rhombohedral form is preferably synthesized at high temperature ($T > 1400°C$). Both tetragonal modifications appear at intermediate temperatures.

The reduction can be performed by different procedures, depending on the nature of the substrate and the method of heating the reduction zone. The first experiments were carried out with an arc and water-cooled electrodes (Weintraub, 1909; Kahlenberg, 1925; Hackspill *et al.*, 1931; Cueilleron, 1944; Hackspill and Cueilleron, 1944). The use of a quartz tube externally heated by an electrical furnace is satisfactory if the reduction is done at low temperature (Kiessling, 1948; Becher and Schäfer, 1960; Hutchinson and Eick, 1962; Niemyski and Olempska, 1962; Hagenmuller and Naslain, 1965). However, the method which is most commonly used and which gives the product of highest purity is the reduction on a filament heated by the Joule effect (Warth, 1923; Dupuy and Hackspill, 1923; Freymann and Stieber, 1934; Laubengayer *et al.*, 1943; Stern and Lynds, 1958; Powell *et al.*, 1960; Talley *et al.*, 1960a; Hutchinson and Eick, 1962; Hagenmuller and Naslain, 1965). Since boron has a strong affinity for most of the common metals, diffusion phenomena might occur at the boron–substrate interface. Consequently, delicate filaments made with boron are used to avoid any contamination of the boron deposit by the substrate (Bean and Medcalf, 1960; Starks and Buford, 1963).

The hydrogen reduction of boron halides is undoubtedly the best method for boron synthesis, and purity better than 99% is commonly obtained. The impurities come from the reagents themselves and from the substrate on which the deposition of boron occurs.

If one neglects the case of hydrogen—a gas easy to purify—the main source of impurities is evidently the boron halide. BCl_3, BBr_3, and BI_3 are boron compounds which can be purified to a very high degree by classical fractional distillation, or by zone melting in the case of BI_3. This is one of the main reasons for the success of this method of boron synthesis. Boron bromide boils at 91.3°C under atmospheric pressure and melts at −46°C. Thus it is a liquid in a very broad range of temperature, including room temperature: This makes BBr_3 easy to handle and to purify by distillation. On the contrary, this no longer obtains for BI_3 (mp 48–49°C; bp 208–210°C).

The substrate is another source of impurities for the boron deposit. The contamination is either minimized by the use of a metallic filament of very small diameter, or reduced by removing the metallic core from the boron deposit by chemical or physical treatment, or, still better, eliminated by the use of a pure boron substrate.

We shall first study the methods for the synthesis of boron halides BX_3 (X = Cl, Br, I) and shall discuss the impurities they may contain. Then, a detailed description will be given for the reduction techniques in a hot tube, on a metallic filament, and on a boron substrate.

B. Synthesis and Purification of BCl_3, BBr_3, and BI_3

Three methods are commonly used for the synthesis of BCl_3 and BBr_3: chlorination of mixtures of boric oxide and carbon, direct synthesis from the elements, and reaction of BF_3 with $AlCl_3$ or $AlBr_3$. The first method is the basis of the industrial process. The second is often used in pilot plants and has the advantage of limiting the risk of contamination. The last is useful for obtaining small quantities of either BCl_3 or BBr_3 in the laboratory.

Although direct synthesis of BI_3 from the elements occurs with a much lower yield than does that of BCl_3 or BBr_3, it nevertheless constitutes the best method of synthesis. The reaction of iodine on sodium or lithium borohydride also produces BI_3.

1. *Synthesis of BCl_3 and BBr_3 from boric oxide B_2O_3*

The method is based on the reaction

$$B_2O_3 + 3C + 3X_2 \rightarrow 2BX_3 + 3CO \qquad (X = Cl, Br) \tag{5}$$

The preparation is the same for both BCl_3 and BBr_3. The following details, given as illustration for BCl_3, are according to the procedure published by Cooper (1945a,b).

An intimate mixture of boric oxide and an excess of carbon (15%) is sintered at high temperatures (1200–1300°C) to eliminate moisture and adsorbed gases, to obtain a partial reduction of B_2O_3 to boron suboxides (which are less sensitive to atmospheric alteration than B_2O_3), and to transform the mixture in a porous aggregate which may be crushed into sized fragments for chlorination. The chlorination reaction proceeds smoothly at 1200°C. A condensation at −10°C separates gaseous carbon monoxide from a liquid phase containing mainly BCl_3 (bp 12.5°C) with dissolved chlorine and phosgene $COCl_2$ (bp 8.3°C). Since the phosgene results from the direct reaction between free chlorine and carbon monoxide, the flow rate of chlorine may be adjusted in order to give an effluent gas which is as poor as possible in chlorine.

Since BCl_3 and $COCl_2$ have closely similar boiling points, it is not possible in practice to separate $COCl_2$ by fractional distillation. Cooper has suggested a chemical conversion of $COCl_2$ into carbon monoxide and chlorine by passing the vapors over a boron carbide column heated to about 800–1000°C. The resulting chlorine reacts with boron carbide to give BCl_3 according to the reactions

$$6COCl_2 \rightarrow 6CO + 6Cl_2 \tag{6}$$

$$B_4C + 6Cl_2 \rightarrow 4BCl_3 + C \tag{7}$$

A second condensation of the effluent gases at −10°C gives gaseous carbon monoxide and a liquid boron chloride which is almost free of phosgene and chlorine.

In the case of BBr_3, $COBr_2$ may also be formed if the effluent gases from the reactor contain free bromine. However, $COBr_2$ may be extracted from the product by distillation since its boiling point (64.5°C) is far enough from the boiling point of BBr_3 (91.3°C). Since carbon is a very important and tenacious impurity in boron, especially as far as application in electronics is concerned, one understands why this easy removal of carbon by distillation gives an important reason for the choice of BBr_3 as an intermediate for obtaining samples of the highest purity.

2. *Synthesis of boron halides from the elements*

The preparation is based on the reaction of gaseous halogen on boron, in pellets or in a fluidized bed, heated to moderate temperature in a vertical reactor. The reaction may proceed either under reduced pressure of halogen or with an inert gas carrier (argon or helium) (Stieber, 1933; Cueilleron, 1944; Ivanov-Emin *et al.*, 1961; Starks and Buford, 1963; Armington *et al.*, 1965; Gaulé, 1965; Hagenmuller and Naslain, 1965).

With chlorine and bromine, the reaction may be quantitative at 400°C, but it is generally carried out at higher temperature (700°C) so as to increase the rate of production per hour. In contrast, the yield is temperature-dependent in the case of BI_3 and goes through a maximum at 850°C; even in optimal conditions, it does not exceed 60% (McCarty and Carpenter, 1960; Armington *et al.*, 1964).

This method has been used by many investigators. The following details have been published recently by Hagenmuller and Naslain for the preparation of BBr_3 on a pilot plant scale (Hagenmuller and Naslain, 1965; Naslain, 1967; Naslain *et al.*, 1968).

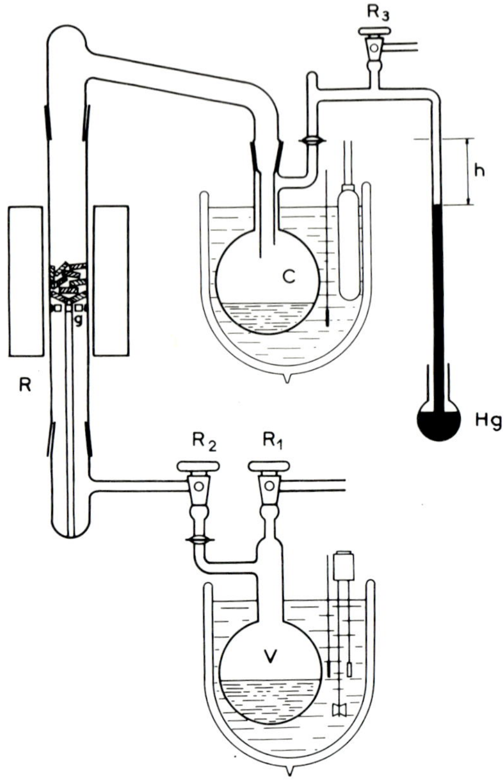

FIG. 6. Apparatus for the synthesis of boron tribromide from commercial boron and pure bromine.

Boron tribromide is obtained by reacting dry bromine under reduced pressure, on boron pellets, at 750°C:

$$2B + 3Br_2 \rightarrow 2BBr_3 \quad (8)$$

Boron is used in excess—two or three times the stoichiometric quantity—in order to give a total transformation of bromine into boron tribromide. The starting boron need not to be of high grade (85–90%). The products obtained by Moissan's method or by electrolysis of fused salt mixtures, which are chemically quite reactive, are very good starting materials for the synthesis. The boron that has not been consumed can be reused after washing and drying, thereby giving a very high yield.

The synthesis is carried out in the apparatus shown in Fig. 6. The reaction vessel R is a silica tube heated by an external electrical furnace.

Boron is introduced as pellets onto a silica grating g supported on a silica rod or tube. Since boron halides are very sensitive to moisture, the apparatus and the starting materials are carefully dried.

Dry bromine is distilled under reduced pressure (100 Torr) from the evaporator V to the reaction vessel R heated to 750°C. The resulting boron bromide is collected in a condenser C maintained at −40°C. The rate of bromine distillation should be adjusted so that the product condensing in C is colorless, i.e., free of uncombined bromine.

The crude boron bromide contains solids (boron, boron suboxides) carried by the vapors as well as impurities dissolved in the liquid, such as volatile bromides or even free bromine. The solids are easily separated from the liquid phase by a simple distillation under reduced pressure in an argon atmosphere. The distillate is generally colorless, but it may be slightly yellowed by traces of free bromine resulting from imperfect control of the flow rate of bromine during the synthesis.

3. *Preparation of BCl_3 or BBr_3 from BF_3 (or from a metal fluoroborate)*

For the synthesis of a small quantity of BCl_3 or BBr_3, it is very convenient to use one of the two following reactions (X = Cl, Br):

$$BF_3 + AlX_3 \rightarrow AlF_3 + BX_3 \qquad (9)$$

$$KBF_4 + AlX_3 \rightarrow AlF_3 + KF + BX_3 \qquad (10)$$

These methods eliminate the problems of handling halogens; they are simple, but the yield of boron halide (BCl_3 80%, BBr_3 70%) is smaller than in the case of direct synthesis from the elements. The boron halide, which is formed between 150 and 250°C, is more volatile than the corresponding aluminum halide. It is easily separated by fractional distillation and condensed in a trap maintained at low temperature (Gamble *et al.*, 1940).

4. *Preparation of BI_3 from alkali-metal borohydride*

Boron triiodide may be prepared by reacting iodine on an alkali-metal borohydride: The yield is 45% from $NaBH_4$ and 65% from $LiBH_4$. The mechanism for the formation of BI_3 is complex, the fundamental reaction being

$$3MBH_4 + 8I_2 \rightarrow 3MI + 3BI_3 + 4H_2 + 4HI \qquad (M = Li, Na) \qquad (11)$$

The reaction takes place at low temperature, occurring at 200°C for $NaBH_4$ and 125°C for $LiBH_4$. The BI_3 is extracted from the reaction vessel by sublimation and condensation on a cool wall; it is always mixed with a high percentage of free iodine (Schumb *et al.*, 1949).

5. *Purification of the boron halides BCl_3, BBr_3, and BI_3*

The purification of the boron halides is the fundamental step in the process of obtaining pure boron by the reduction or the thermal decomposition of these halides. The purity of boron is closely related to the efficacity of the methods used for the purification of the halide intermediates.

Table II gives some of the possible impurities which may be found in crude boron halides. Except for certain compounds, e.g., $COCl_2$, most of the impurities have boiling points very far from those of the boron halides or are decomposed by heating. It is thus possible to purify the boron halides by fractional distillation, the degree of purification being monitored by spectrographic analysis. On the other hand, the melting point of BI_3 is high enough to permit its purification either by sublimation or by zone melting.

a. Chemical purification. Chemical purification is necessary for those impurities that are not eliminated by distillation (phosgene in boron chloride) or whose content is too high, such as results from an imperfect control of the synthesis (BCl_3, BBr_3) or from a poorly equilibrated reaction (BI_3).

In Cooper's process, $COCl_2$ is eliminated from crude boron chloride by thermal dissociation and transformation of the resulting chlorine into BCl_3 over boron carbide (Cooper, 1945b). Liquid zinc is also used for absorbing chlorine (Gould and Gardner, 1962).

Free chlorine and bromine are easily removed from crude BCl_3 and BBr_3 by agitating with mercury for a few minutes at room temperature (Hagenmuller and Naslain, 1965). The extraction of iodine dissolved in

TABLE II

Process	Boron halide BX_3	Impurity						
		Free halogen	AlX_3	SiX_4	CX_4	COX_2	MgX_2	CaX_2
Chloride intermediate	12.5	−34.6	178[b]	57.57	76.8	8.3	1412	>1600
Bromide intermediate	91.3	58.78	263.3 (747)	154	189.5	64.5	(*d*)	810
Iodide intermediate	208–210	184.35	360	287.5	170[c]	—	700[c]	—

[a] Weast (1968). Values given are for a pressure of 760 Torr, except where a figure is given in parentheses.
[b] Sublimation.

BI_3 is more difficult since both methods of synthesis give a crude material which contains a high percentage of free halogen. Schumb *et al.* (1949) proceed by dissolving the crude boron iodide in carbon disulfide and reducing the free iodine by either mercury or zinc. McCarty and Carpenter (1960) prefer a direct elimination by multistage distillation, while Armington *et al.* (1965) used both a reduction by tin and a distillation.

b. Purification by fractional distillation. Fractional distillation is used for obtaining boron halides of very high purity. In order to limit the number and the content of the impurities, one generally starts from a raw material prepared by direct synthesis from the elements. The impurities in such a material are halogen or oxyhalogen compounds of the impurities of the initial boron and similar compounds resulting from an attack on the reaction vessel, e.g., $SiBr_4$. Figure 7 shows that commercial boron (90–92%) used for boron halide synthesis contains a large number of impurities (Fe, Cr, Mn, Ni, Cu, Co, Ti, Ca). To these impurities one has to add light elements which are not detected in trace amounts by x-ray fluorescence (C, O, Al, Mg) (Hagenmuller and Naslain, 1965; Naslain, 1967).

Table II clearly indicates that, among the impurities which might be present in crude boron halides, halogen or oxyhalogen compounds of carbon, silicon, iron, sulfur, tin, titanium, and aluminum would be the most difficult to extract by distillation.

To minimize the risk of contamination, fractional distillation of boron halides must be carried out in dried apparatus, without ground glass joints, and under reduced pressure so as to avoid overheating the raw material. Various types of columns are utilized, generally of large size

Boiling Points (°C) of Possible Impurities in Boron Halides[a]

Impurity								
FeX_3	CrX_3	NiX_2	TiX_4	SnX_4	HgX_2	ZnX_2	SOX_2	CuX
310[b]	1300[b]	973[b]	136.4	114.1	302	732	78.8 (746)	1490
(*b*)	(*d*)	(*d*)	230	202	322	650	138 (773)	1345
(*c*)	350[b]	(*d*)	377.1	364.5	354	624	—	1290

[c] Decomposition.
[d] Sublimation at high temperature but the true sublimation temperature is unknown.

(150 cm height) (Ivanov-Emin *et al.*, 1961; Armington *et al.*, 1964; Hagenmuller and Naslain, 1965).

For BBr_3, which is liquid at room temperature, conventional apparatus as shown in Fig. 8 is satisfactory. The raw material is transferred to the previously dried distillation apparatus in such a way as to avoid any hydrolysis of BBr_3. The transfer takes place through a high-porosity frit P under a small positive pressure of argon. Following transfer, the distillation unit is sealed off at XX and the liquid BBr_3 distilled at about 60°C under reduced pressure of argon. The purified BBr_3 is collected into ampoules which are connected to a collecting adapter D. The ampoules are sealed off as soon as they are filled (Hagenmuller and Naslain, 1965; Naslain, 1967).

For BCl_3, which boils at 12.5°C, the distillation requires either cooled apparatus or a refrigerated laboratory. On the contrary, for BI_3, which boils at 208–210°C and melts at 49.9°C, the whole distillation unit must be heated above the latter temperature.

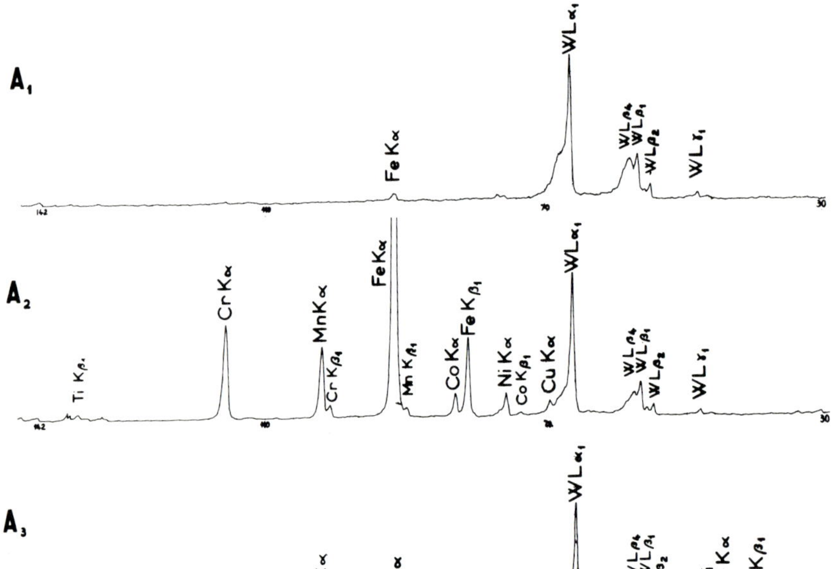

FIG. 7. X-ray fluorescence analysis (tungsten tube) of (A_2), commercial boron used for the synthesis of BBr_3; pure boron resulting from the hydrogen reduction of BBr_3 at 1300°C (A_1), on a tantalum filament; and at 750°C (A_3), on the inner wall of a silica tube.

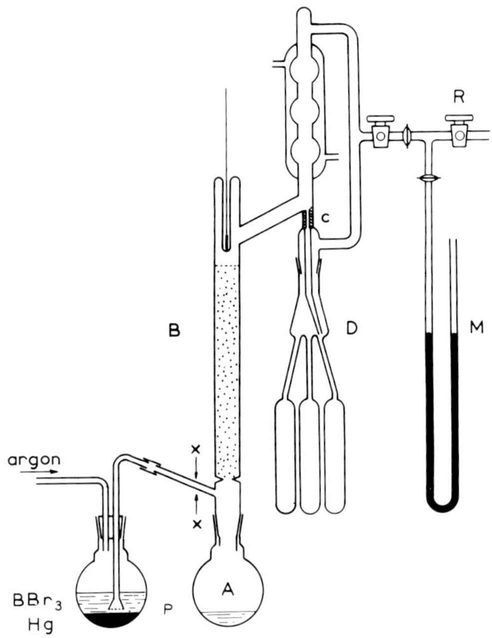

FIG. 8. Apparatus for the distillation of boron tribromide under a reduced pressure of argon.

Generally, two or three fractional distillations with a 10:1 reflux give a purified boron halide spectrographically free of any detectable impurities except a few ppm of silicon and carbon.

c. *Other purification methods for BI_3*. The high melting point of BI_3 (48.4–49.9°C) permits two other purification techniques: sublimation and zone melting.

Armington *et al.* (1965) have studied the purification of BI_3 by sublimation; they operated at 60°C under vacuum and condensed the sublimated BI_3 on a cooled wall. Table III indicates that sublimation lowers the content of the main impurities, but that it is rather less effective for silicon.

Zone refining of BI_3 has been described in detail by Amberger and Dietze (1965) and Armington *et al.* (1965). From the results of Amberger and Dietze, it appears that purification by zone melting is the most effective for SiI_4 ($K = 0.3$), less effective for AlI_3 ($K = 0.5$), and the least effective for MgI_2 ($K = 0.7$). Zone-melted samples are claimed by the authors to be free of carbon tetraiodide.

TABLE III

PURITY ACHIEVED BY SUBLIMATION OF BI_3 [a]

	Elements present (ppm)					
	Mg	Si	Ca	Ti	Al	Fe
Before sublimation	155	>5000	275	310	>5000	86
After sublimation	<3	2400	<50	<6	<68	<10

[a] Armington *et al.* (1965).

C. Hydrogen Reduction of Boron Tribromide

The last chemical step in the process for obtaining pure boron through the intermediate of boron halides is a hydrogen reduction according to reaction (4). Boron, which is formed from the vapor phase, deposits on a hot substrate kept at constant temperature. As shown by Kiessling in 1948, one of the simplest procedures to carry out this reaction is to use a quartz tube heated by an electrical furnace. When the gaseous hydrogen–boron halide mixture enters the hot reduction zone, boron deposits on the inner wall of the tube, which acts as a substrate. Since boron reduces silica at high temperature, the process can give pure boron only if the reduction zone is heated to moderate temperatures, thus excluding the use of boron chloride BCl_3.

The hydrogen reduction of BBr_3 in a silica tube has been studied in detail by Hagenmuller and Naslain (1965), Naslain (1967), and Naslain *et al.* (1968). Under the experimental conditions stated, it can yield pure α-rhombohedral boron.

1. *Principle and apparatus*

The reaction starts at 600°C, but at this temperature, the boron is poorly crystallized and the yield low. Above 850°C, the rate of boron deposition becomes significant, but the boron begins to reduce silica. The temperature range 700–850°C has been chosen as a compromise which yields an appreciable rate of deposition while avoiding extensive contamination by silicon.

In this range of temperature, the reaction is slow and therefore only a small fraction of the boron bromide in the vapor phase is reduced in the hot zone. To achieve as high a yield as possible versus the starting amount of boron halide, the boron bromide that has not been reduced is condensed at the exit of the silica tube and recycled.

At the beginning of a run, boron deposits only on the inner wall of the silica tube as porous granules of very high specific area. After a few hours, the substrate on which boron deposits is no longer the silica wall, but rather the first deposited boron granules themselves, acting as seeds. Their very high specific area gives the reaction a significant rate despite the use of low temperature. The hot zone becomes progressively filled with boron. In order to continue the reduction, the hot zone must be shifted along the tube axis as a function of time by displacing the electrical furnace. The time τ spent by the boron deposit in the reduction zone is an important parameter. There is a critical time τ_c above which the boron deposit consists of a mixture of α- and β-rh modifications. τ_c is determined by the geometry of the reaction vessel and by the temperature. When $\tau < \tau_c$, the deposit is in the form of α-rh boron granules which have few contacts either among themselves or with the wall of the reaction tube. When $\tau > \tau_c$, the product is a compact powder which is a mixture of both rhombohedral varieties. In as much as the transformation temperature for α-rh $\rightarrow$ β-rh is 1200°C, it appears quite improbable that this same transformation could occur in the much lower temperature range (700–850°C) used here.

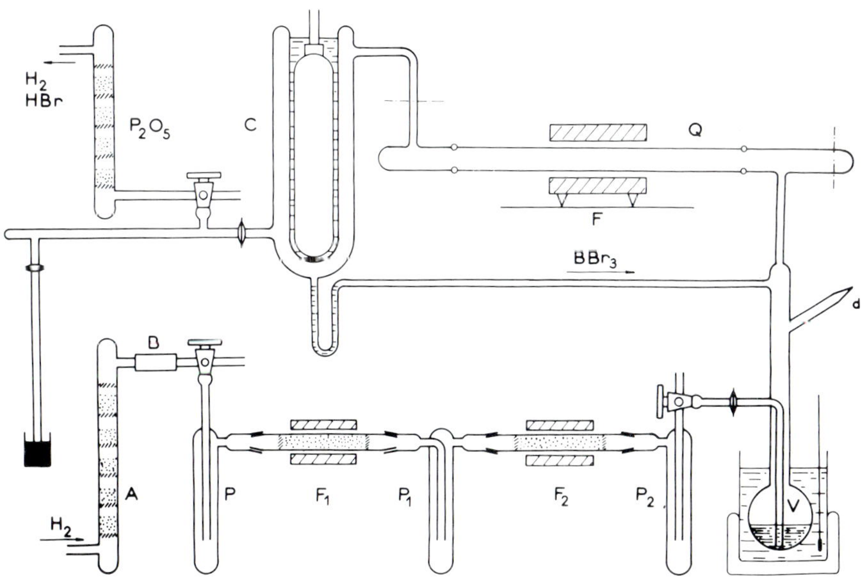

FIG. 9. Apparatus for the hydrogen reduction of boron tribromide on the inner wall of a silica tube.

It seems more plausible to us that β-rh variety forms directly from the vapor phase and that kinetic factors are responsible for its appearance above τ_c in the presence of porous granules of the α-rh type.

Our experimental arrangement is shown schematically in Fig. 9. Commercial hydrogen (99%) is dried over CaH_2 (A); oxygen and nitrogen are removed by passing through Deoxo catalyst (B), liquid nitrogen traps (P, P_1, and P_2), and magnesium furnaces (F_1, F_2). Purified hydrogen bubbles in the BBr_3 tank (V); the boron bromide and hydrogen gaseous mixture is admitted into the silica reaction tube (Q), where boron deposits. The unreacted BBr_3 is condensed at $-40°C$ (C) and recycled. The HBr is absorbed by passing the exit gases from the P_2O_5 column through several bubblers filled with an aqueous solution of sodium hydroxide. Since the vapor pressure of BBr_3 is low at room temperature (70 Torr at 25°C), all the glass tubing is electrically heated so that any desired BBr_3 partial pressure can be maintained in the vapor phase. To obtain a gaseous mixture

FIG. 10. Automatic unit for the synthesis of finely divided α-rh boron by hydrogen reduction of BBr_3 at low temperature. V: BBr_3 tank. F: Reduction zone. Q: Silica tube. C: BBr_3 condenser. I: Cryostat for condenser C. H: Recycling of unreacted BBr_3. G: HBr absorber.

sufficiently rich in BBr_3, it is necessary to heat the vaporizer V at about 60°C (280 Torr BBr_3), to have tubing without ground glass joints, and to heat stopcocks between V and C at about 80°C.

The unit which actually operates in this laboratory is shown in Fig. 10; it gives 50–100 g of boron per run with a yield of 90% and a deposition rate of about 250 mg hr^{-1}.

2. *Analysis of the product*

Boron prepared at low temperature according to this process is in the form either of a brown powder or as coarse porous granules. The latter are easily pulverized without any contamination: This is a definite advantage over the filament techniques, which yield very hard bulk samples.

The product is mainly α-rh boron: The observed broadening of its diffraction rings is a proof of its finely divided state (Naslain *et al.*, 1968). The particle size and the metastability of the α-rh form give the product a high chemical reactivity: Boron prepared at low temperature must be kept in sealed ampoules filled with argon, otherwise it is slowly oxidized by air. Samples of boron are easily dissolved in either nitric acid or mixtures of nitric and sulfuric acids, with a large evolution of heat. The alkalimetric titration of the resulting boric acid in the presence of mannitol gives a purity of 99%.

It is often claimed that hydrogen reduction of boron halides in silica tubes must *a priori* lead to a strong contamination of the product by silicon. Analytical control tests have been carried out in this laboratory. The x-ray fluorescence patterns given in Fig. 7 clearly show that the only impurity detected is bromine. Optical emission spectrography reveals that the main trace impurities are silicon (250 ppm), copper (50 ppm), and aluminum (25 ppm). Light elements (H, C, N, O) were not analyzed. These results demonstrate that the reduction of BBr_3 may be carried out in a silica tube if its temperature is kept low and if a tube of large diameter is utilized; in these conditions, the boron granules themselves act as a substrate and their contact with silica is minimized.

Boron prepared according to this process, which is of high grade and chemically very reactive, is an ideal material in the synthesis of boron compounds. For example, the synthesis of KB_6 was achieved in this laboratory for the first time, despite the weak affinity of potassium for boron; the same synthesis would have been impossible or at least very difficult starting from the bulk β-rh variety prepared on a hot filament (Naslain, 1967; Naslain *et al.*, 1967).

3. *Upgrading treatment*

Volatile impurities—bromine and probably hydrogen—still remain in the boron prepared at low temperature despite the outgassing treatment *in situ* at 700°C under vacuum.

In order to upgrade the product, the boron may be pressed into pellets and submitted to a new outgassing treatment at elevated temperature and under high vacuum (Etourneau, 1970). The purification becomes effective when the boron is heated over the transformation temperature of α-rh $\rightarrow$ β-rh (1200°C) (Naslain *et al.*, 1968). If temperature is limited to about 1600–1700°C, the treatment causes only a mild sintering and yields samples of β-rh boron which may still be easily pulverized. There is a definite improvement in purity (99.5–99.9%), but a decrease in chemical reactivity. This upgraded boron, however, is very satisfactory for most of the syntheses of boron compounds involving elements having a strong affinity for boron.

D. Hydrogen Reduction of BCl_3 and BBr_3 on Hot Metal Filaments

Reduction or thermal decomposition of volatile chemical species on metallic, resistance-heated wires has been used for the preparation or purification of numerous metals or metal compounds (Campbell *et al.*, 1949; Powell *et al.*, 1965). Warth appears to have been the first, in 1923, to prepare pure boron from BCl_3 according to this general method. Limited for a long time to the synthesis of small amounts of boron, the method has been recently developed for the preparation of larger quantities of pure boron intended for fundamental studies of the electronic properties of the element (Dupuy and Hackspill, 1923; Freymann and Stieber, 1934; Laubengayer *et al.*, 1943). Several experimental setups for preparing pure boron on a 100-g scale are described in the proceedings of the first International Conference on Boron (Kohn *et al.*, 1960). Nevertheless, the preparation of large amounts of boron on heated filaments remains difficult and requires considerable experience.

1. *Principle and apparatus*

In the hot-wire technique, reduction of boron halides (BCl_3 or BBr_3) occurs on a resistively heated refractory metal filament. Deposit of a large amount of boron requires that the temperature of the substrate be kept constant as a function of time during the experiment. In fact, it is only at the very beginning that boron deposits directly on metal; soon thereafter, the boron initially deposited acts as the substrate. The electrical

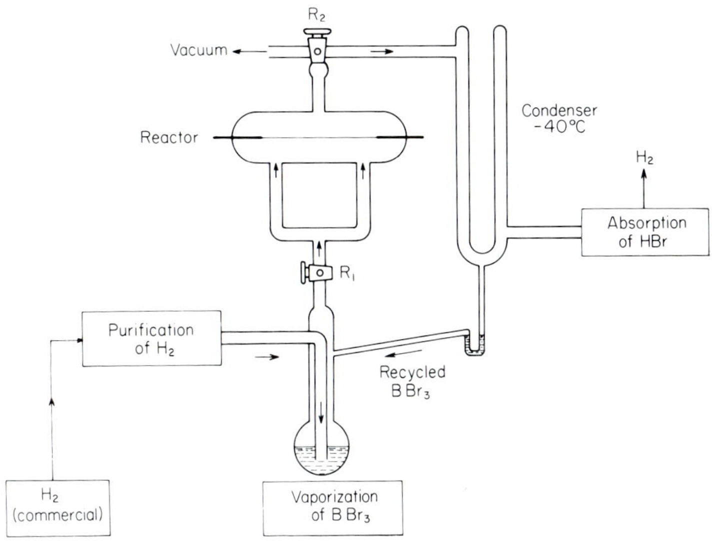

FIG. 11. Unit for the hydrogen reduction of BBr_3 on a hot filament.

conductivity of boron, which is low at room temperature (10^{-6}–10^{-7} ohm^{-1} cm^{-1}), rises rapidly with increasing temperature to reach about 10 ohm^{-1} cm^{-1} at 900°C. As a result of this semiconducting behavior, a growing boron deposit can be maintained at constant temperature only by progressively increasing the applied electrical power, but it cannot be cooled and then restarted.

In order to avoid extensive contamination of boron by the metal substrate, the filaments have very small diameters (80 200 μm). Consequently the deposition area is small and the substrate must be operated at high temperature (1200–1400°C), where the reaction is very fast. In this temperature range, both BCl_3 and BBr_3 may be used as starting material, and in either case, boron deposits as a homogeneous bulk solid, which in turn can act as a heating resistance.

Powell *et al.* (1960) have described the technique of this process; it was later taken up by Hagenmuller and Naslain (1965). With boron bromide, the reduction may be performed in the apparatus shown schematically in Fig. 11. It is generally similar to the one shown in Fig. 7, except that the silica tube Q is replaced here by a hot-filament reactor. The reactor can be isolated by two valves R_1 and R_2 so that several deposits may be obtained from the same BBr_3 charge. The reactor is shown in Fig. 12; it is made of Pyrex glass and cooled with a water spray. The

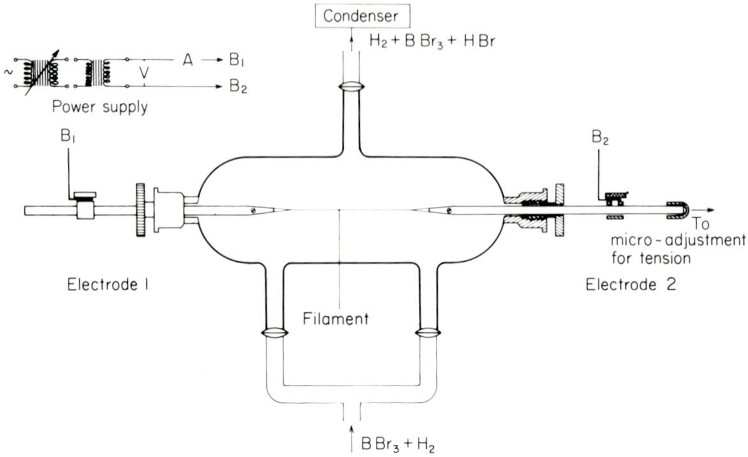

FIG. 12. Hot-filament ampoule for the hydrogen reduction of BBr_3 at high temperature.

filament is held in the axis of the ampoule between two molybdenum electrodes and subjected to an adjustable tension to prevent sagging as the filament grows. Teflon O-rings (or mercury wells) permit the electrodes to slide, while isolating the inside of the reactor from the atmosphere. The filament is fed with a low-voltage alternating current through B_1 and B_2.

The most critical period in the growth of boron deposits occurs in the first minutes of reaction. Vibrations set up in the growing filament can easily snap the small boron deposit unless there is rapid control of the tension on the filament. The filament temperature is measured with an optical pyrometer and controlled by variation of the input voltage. The rate of deposition is a function of filament temperature, hydrogen flow rate, and vaporizer temperature. While it is easy to control the first two parameters, it is difficult to increase the temperature of the vaporizer without condensing BBr_3 on cold sections of the apparatus.

Powell *et al.* (1960) report a rate of deposition ranging from 8 to 10 g hr^{-1} on a 9-in. filament and a total mass of 100 g per run. As in the case of the quartz tube unit shown in Fig. 7, condensing and recycling unreacted BBr_3 increase the yield ($\sim$90%).

2. *Diffusion phenomena at the interface between boron and metal core*

The choice of the substrate metal is important. The metal must keep a sufficient strength at the reaction temperature and must not diffuse into the boron deposit. An extensive study of diffusion phenomena at the interface between boron and metal has been done in this laboratory (Naslain, to be published).

Starting at temperatures as low as 1000°C, most common metals, even the refractory ones, react with boron; the most interesting are transition metals from groups IV, V, and VI of the periodic table.

During the first minutes of deposition, boron atoms diffuse in the substrate and form a layer of borides around the pure metal filament whose thickness increases with time. For titanium, zirconium, niobium, and tantalum, a diboride MB_2 single phase is observed. For molybdenum and tungsten, however, diffusion leads to a two-phase layer (M_2B_5 and MB_4) between 1000 and 1400°C. Substrate boridization is rapid; for example, a 200-μm tantalum filament is completely transformed into TaB_2 after 2 hr of deposition at 1100–1200°C, as shown in Fig. 13.

Substrate boridization has several consequences. The increase in volume of the substrate (see Fig. 13) which occurs inside a rigid boron cylinder

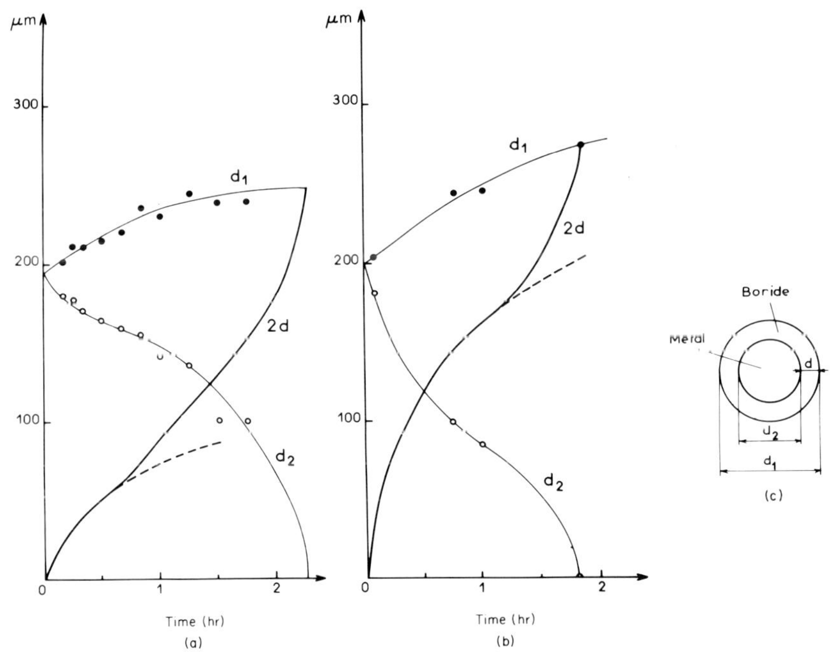

Fig. 13. Growth of a TaB_2 layer by diffusion of boron atoms into a tantalum filament, during hydrogen reduction of BBr_3 at high temperature: (a) at 1100°C, (b) at 1200°C, (c) cross section of filament.

gives rise to a stress, which could explain the existence of radial cracks in the deposit. Diffusion of boron atoms in transition metals is mainly a one-way phenomenon; consequently, to the flux of boron atoms diffusing in the metal there is associated a flux of vacancies diffusing in the opposite direction. Vacancies gather in the neighborhood of the interface and form macroscopic pores which decrease the cohesion between boron deposit and metal core. The existence of this porosity decreases the strength of the deposit, but often makes the extraction of the metal core easier. The effects of diffusion are minimized for filaments of very small diameter. Tantalum, molybdenum, and tungsten seem to be the best metal substrates.

3. *Purity of boron prepared on a metal substrate*

The main impurities come from the starting material and also from the substrate; this latter contamination remains the chief disadvantage of the process. However, for 100-g deposits obtained on small filaments (80 μm diameter), it is not necessary to separate the metal core, which represents only an extremely weak relative mass. Powell *et al.* report a 0.02% tantalum content for boron prepared on tantalum without any separation of the metal core (Table IV) (Powell *et al.*, 1960).

Very often, the weight of the boron deposit is smaller and the metal core can no longer be neglected. In this case, the substrate must be extracted by either physical or chemical methods.

Sometimes, the metal core breaks off easily in tiny cylinders when one tries to reduce the boron deposits into fragments before the grinding

TABLE IV

Typical Analysis of Boron Produced by the Hot-Wire Reduction of Boron Bromide[a]

Impurity	Per cent
Carbon	0.02–0.08
Sulfur	<0.01
Silicon	<0.03
Tantalum	0.02
Other metallics	<0.01
Nitrogen	<0.003
Oxygen	<0.01
Hydrogen	0.02
Boron (by difference)	99.9+

[a] Powell *et al.* (1960).

TABLE V

DENSITIES OF SOME METALLIC PHASES IN BORIDIZED METAL SUBSTRATES

	Tantalum		Molybdenum			Tungsten		
	Metal	TaB_2	Metal	Mo_2B_5	MoB_4	Metal	W_2B_5	WB_4
Density ($g\ cm^{-3}$)	16.6	12	10.2	7.5	~ 5	19.3	13	~ 10

operation; in this case, the substrate may be separated under a microscope. Generally, however, separation of the metal core is more difficult; Hagenmuller and Naslain (1965) have proposed a method which is based on the differences between the densities of boron (2.35 g cm^{-3}) and metal compounds of the substrate. Table V shows that, even in the unfavorable case of molybdenum, the difference in the densities is sufficient for an easy separation of substrate by levigation in bromoform ($d = 2.9$ g cm^{-3}) (Naslain, 1967).

Chemical methods have also been proposed; they are in general poorly specific because metals in boron deposits are combined as compounds having properties close to that of boron itself. Stern and Lynds (1958) removed titanium from their deposits on titanium wire by a chloridization at 300–350°C. Formstecher and Ryskevic (1945) lowered the molybdenum content of their boron to 0.0045% by a treatment with a mixture of dilute nitric acid and hydrogen peroxide followed by a washing with a solution of sodium hydroxide. For tantalum and tungsten, which are the best metal substrates, no chemical method has been reported for the extraction of the metal core.

4. *Crystallinity*

The crystalline modifications of boron prepared on metal filaments depend upon a large number of experimental parameters. Among the most important factors are substrate temperature, rate of deposition, and nature of the substrate. According to the choice of the experimental parameters, either amorphous or crystalline boron may be obtained.

a. Amorphous boron. Deposition of amorphous boron on metal filament has been much studied in view of its application in the field of boron fibers in composite materials, but for mechanical and technical reasons, the diameter of these special amorphous boron deposits is limited to 100–150 μm

(Wavner, 1967). On the other hand, outside of the fundamental work of Talley *et al.* (1960a), there have been very few publications bearing on the synthesis of large-diameter amorphous boron deposits.

The experimental conditions for the growth of amorphous boron have been studied in this laboratory (Boiret, 1968). Two parameters are dominant: substrate temperature and rate of deposition. Boron is amorphous

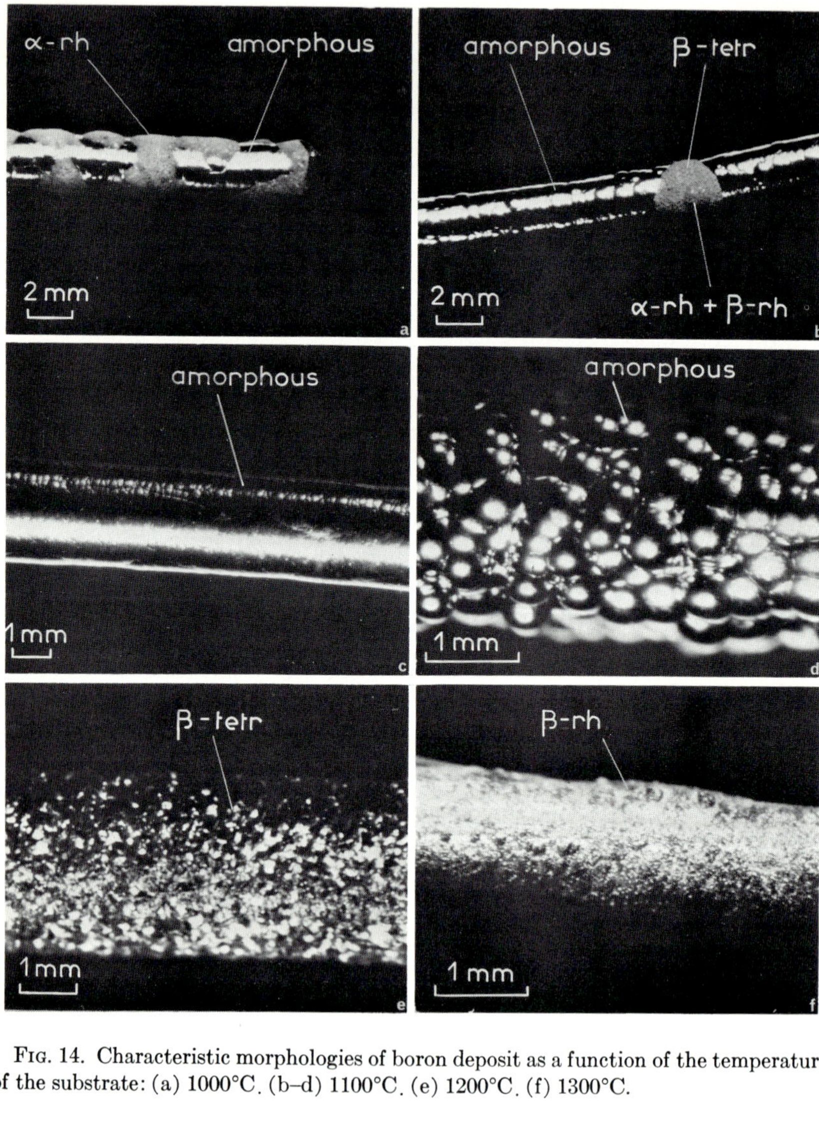

FIG. 14. Characteristic morphologies of boron deposit as a function of the temperature of the substrate: (a) 1000°C. (b–d) 1100°C. (e) 1200°C. (f) 1300°C.

if the rate of deposition is high and if the substrate temperature is sufficiently low to avoid the appearance and growth of crystalline nuclei. The first condition is achieved by operating with a high hydrogen flow rate (500–1000 cm^3 min^{-1}) and a vapor phase rich in boron halide. As shown in Fig. 5, the temperature must not exceed 1200°C, a limit above which the boron deposit is crystalline. Even if the preceding conditions are maintained, it seems difficult to prepare amorphous boron rods with a diameter larger than 2–3 mm. Very often, crystal growth nuclei appear on the surface of the rod at the expense of the amorphous boron. The existence of surface defects favors the appearance of crystal growth nuclei (Fig. 14a,b).

Amorphous boron deposits exhibit a typical appearance: the surface of the rod is composed of nodules separated by well-defined boundaries (Fig. 14c,d). This characteristic results from a radial growth in cones analogous to that observed in pyrolytic carbons. Most of the applications of amorphous boron stem from its remarkable mechanical properties (Wavner, 1967).

b. Crystalline modifications. Hydrogen reduction of boron halides on a metal substrate may lead to a surprisingly large number of crystalline modifications (see Fig. 5). At low temperature ($T \leq 1100°C$), the only crystalline variety in significant amount is the α-rh form. At high temperature ($T > 1400°C$), homogeneous boron deposits are generally constituted of the β-rh form. Between 1100 and 1400°C, deposition phenomena are more complex and several crystalline modifications are observed. The experimental conditions for obtaining each of these forms alone are still not well known.

The experimental conditions for obtaining the tetragonal forms have been studied recently. According to Amberger and Ploog (1968), hydrogen reduction of very pure boron bromide between 1000 and 1300°C should lead to a single variety, β-tetr boron (Fig. 14e); the higher the temperature, the better the crystallinity. On the other hand, this same reaction in the same range of temperature leads only to α-tetr boron when the vapor phase is doped with methane. This influence of a carbon impurity on the growth of α-tetr boron was confirmed in this laboratory (Boiret, 1968).

In conclusion, to obtain homogeneous boron rods of large diameter, it is recommended to maintain the substrate temperature at about 1300°C; the resulting boron is generally the β-rh variety (Fig. 14f).

E. Hydrogen Reduction of Boron Halides on Boron Substrates

In the hydrogen reduction of boron halides on metal substrates, there remains a risk of contamination of boron by the substrate. In order to

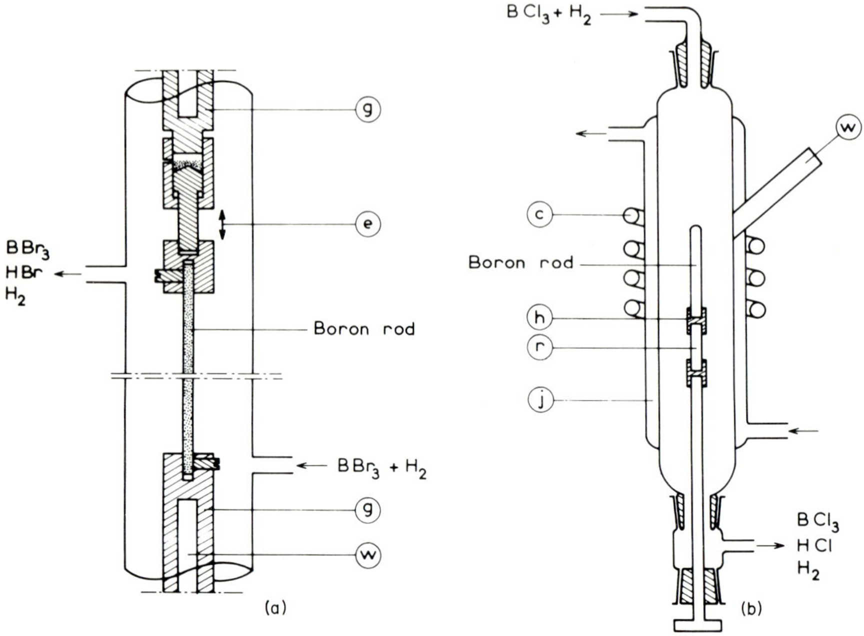

FIG. 15. Apparatus used for the hydrogen reduction of BBr_3 on a boron substrate. (a) Piston-type electrode tube (according to Bean and Medcalf, 1960): g, graphite electrode; e, expansion zone. w, Water-cooling channel. (b) Substrate heated with hf inductive current (according to Olempska *et al.*, 1966): c, induction coil; h, molybdenum holder; r, ceramic rod; j, water-cooling jacket; w, quartz window for temperature control.

avoid this contamination completely, boron itself may be used as the substrate (Bean and Medcalf, 1960; Olempska *et al.*, 1966). The method first requires the preparation of boron "filaments," which are then utilized as substrates.

1. *Preparation of the boron substrates*

Bean and Medcalf (1960) have found that boron deposits can be slipped intact, under certain special conditions, from straight tantalum filaments in such a manner that they are free of metal contamination; these hollow boron cylinders ($\phi_o = 4$ mm, $\phi_i = 1.5$ mm, length $= 150$ mm) are mounted directly in a floating-zone unit and are zone-melted to provide strength and current-carrying capacity. The possibility of slipping boron from the metal core is connected to diffusion phenomena at the boron–substrate interface, i.e., to the growth of the TaB_2 layer, which is sensitive to temperature and time. If the deposition occurs at moderate temperature (1100°C) for a short time on an outgassed tantalum filament, then the thin TaB_2 layer does not adhere strongly to boron; after quenching, it can be removed along with the pure metal core without breaking the boron deposit.

2. *Heating the boron substrate by the Joule effect*

The boron substrate may be heated by the Joule effect (i.e., resistively) if first it has been preheated externally to a moderate temperature. The experimental equipment described by Bean and Medcalf (1960) is schematically shown in Fig. 15a. The boron substrate is mounted between two water-cooled graphite electrodes along the axis of a vertical quartz tube. The upper, piston-type electrode permits expansion of the boron filament during the deposition. This type of vertical reactor may be inserted in a unit such as the one shown in Fig. 11; it yields homogeneous boron bars ($\phi = 2$ cm, length $= 15$ cm) in which it is usually impossible to observe a demarcation line between the deposit and the substrate.

3. *Heating the boron substrate with hf current*

The electrical conductivity of boron at high temperature also allows the substrate to be heated inductively with an hf coil, after preheating with a metallic susceptor, for example.

This very simple technique has been extensively used by Olempska *et al.* (1966). Their deposition vessel is shown schematically in Fig. 15b. The boron rod ($\phi = 1$ cm, length $= 10$ cm) is set vertically on the axis of the ampoule on a molybdenum holder.

TABLE VI

SPECTROGRAPHIC ANALYTICAL DATA FOR BORON DEPOSITED FROM THE VAPOR PHASE ON BORON SUBSTRATE[a]

	Impurity (ppm)					
	Si	Mg	Fe	Cu	Ca	Al
Spectrographic detection limit	1	0.5	1	0.5	0.5	2
Boron deposited on boron filament	5–10	ND[b]	ND	ND	ND	ND
Selected samples on boron	ND	ND	ND	ND	ND	ND

[a] Bean and Medcalf (1960).
[b] Not detected.

4. *Purity of boron prepared on boron substrates*

Having eliminated the risks of contamination, this method leads to boron which is free of any detectable impurities, provided the starting materials—hydrogen and boron halides—are carefully purified. Boron obtained by Bean and Medcalf contains, at the very most, a few ppm of silicon, in the least favorable cases (Table VI).

F. Thermal Decomposition of BI_3 on Hot Filaments

The thermal dissociation of boron halides proceeds according to the reaction

$$2BX_3 \rightarrow 2B + 3X_2 \qquad (X = Br, I) \tag{12}$$

Although some preparations have been carried out with BBr_3, it is the iodide BI_3 that is commonly used (Naray Szabo and Tobias, 1949; McCarty and Carpenter, 1960; Armington *et al.*, 1964; Amberger and Dietze, 1964, 1965).

Thermal dissociation of boron halides is carried out, *in vacuo*, in a hot-wire reactor similar to the ones previously described for the hydrogen reduction technique; since it does not require a gas carrier, the apparatus is very much simplified. The substrate is either tantalum wire or boron filament. The unit is operated under high vacuum, which allows boron halide to be transported from the vaporizer to the filament by distillation. Since only a small fraction of boron halide is dissociated on the hot substrate, a mixture of unreacted boron halide and free halogen is condensed in a cooled trap at the reactor exit (Amberger and Dietze, 1964).

The purity of the resulting boron is very high. In particular, according to Armington *et al.* (1965), it seems that the choice of BI_3 as halide intermediate considerably lowers the carbon content of the boron. This important advantage of BI_3 over BBr_3 comes from the fact that no carbon–iodine compounds are known which are stable at 200°C, and hence any carbon-containing compound should decompose during purification of BI_3 by distillation.

The crystallinity of the boron depends on the experimental parameters, as previously indicated. Thermal dissociation of BI_3 is recommended for the preparation of single crystals of the α-rh variety (McCarty and Carpenter, 1960; Amberger and Dietze, 1965).

G. Conclusions

The boron halides (BCl_3, BBr_3, and BI_3) appear to be excellent intermediates mainly because they are easy to purify by fractional distillation up to a very high degree of purity. Of these, BBr_3 and BI_3 give the highest-purity boron and the former, which is liquid at room temperature, is easier to handle than the latter.

Hydrogen reduction or thermal dissociation of boron halides results in boron which is either crystalline or amorphous and which can be obtained as powders, bulk samples, or even single crystals according to the choice of the experimental parameters. It is thus possible with these methods to prepare products that fit most of the potential applications of boron.

V. Purification of Boron by Zone Melting

Zone melting is a classical method for the ultimate purification of many semiconductor materials. In the case of boron, this method is difficult because of the high melting point of the element (~2200°C) and the fact that boron attacks most of the refractory materials at high temperature.

A. Technique

The melting of boron is achieved either with an electron gun in high vacuum or with hf current after preheating the sample with a metal susceptor. Horn (1960) utilized horizontal boron nitride boats which had been previously fired for a long time at 1500°C in pure hydrogen. Niemyski *et al.* (1965) preferred silver boats internally cooled with a rapid flow of water. In both methods, the fact that the boron sample lies on a substrate presents a common disadvantage. In the former method, liquid boron may be contaminated by volatile impurities emitted by the BN boat at high temperature, while in the latter, the boron may form alloys with silver if the boat is insufficiently cooled.

On the other hand, the risks of contamination of liquid boron from

TABLE VII

Distribution Coefficients K of Impurities between Solid and Liquid Boron

Si	Cu	Fe	V	Ca	Ni	Cr	Mg	Al	Ti	C	References
<1	<1	<1	<1	<1	—	—	>1	—	—	—	Starks and Medcalf (1960)
<1	<1	≪1	—	—	≪1	≪1	—	≫1	>1	≫1	Horn (1960)
—	—	—	—	—	—	—	—	—	—	4	Hinz and Wirth (1965)

boats or crucibles are completely eliminated in the floating-zone technique. The high surface tension of liquid boron stabilizes the molten zone and facilitates the application of this technique (Greiner, 1960; Starks and Medcalf, 1960; King *et al.*, 1965).

B. Results

Zone melting has two effects on a boron rod: The impurities are redistributed along the rod and the crystal defects are partly eliminated.

Only a few papers have been published on measurements of the distribution coefficients K of impurities between solid and liquid boron. As

TABLE VIII

Method	Boron content (%)	Impurities	State
Moissan's process	Very low	O_2, Mg	Powder
Reduction of BF_3 or KBF_4 by sodium	Low	O_2, Na	Powder
Andrieux's process	Low	O_2, C, metallics	Powder
Cooper's process	95–99	C and many metallics	Powder or small aggregates
Hydrogen reduction or thermal dissociation of BCl_3, BBr_3, and BI_3:			
BBr_3 in silica tube at low T	98–99	Few hundred ppm Si, O_2, probably H_2	Powder
on metal substrate (Ta, W, Mo)	>99.9	Contamination by substrate	Bulk
on B substrate	Highest purity	No impurities detectably by spectrographic analysis	Bulk

shown in Table VII, most of the impurities concentrate in the liquid phase and are therefore transported at the end of the rod, notable exceptions being aluminum, magnesium, titanium, and carbon, which seem to be concentrated in the front (Horn, 1960; Starks and Medcalf, 1960; Hinz and Wirth, 1965). From the more accurate results given by Horn, one can deduce that zone melting of boron is effective for iron, nickel, chromium, aluminum, and carbon. This result concerning carbon is important if one remembers that carbon is sometimes difficult to eliminate from boron halides by fractional distillation.

Vertical zone melting also results in a noticeable decrease of the crystal defect density: Cracks and grain boundaries are partly removed and

SOME IMPORTANT METHODS FOR THE SYNTHESIS OF BORON

Crystallinity	Disadvantages	Advantages	Uses
Amorphous	Very impure	Very cheap, easy to prepare	Often unsiutable for chemical uses
Amorphous	Handling of sodium	—	May be utilized for boron halide synthesis or upgrading techniques
Amorphous	—	—	—
Amorphous or α-tetr	High C and metallic content	Medium purity	Can be used for some chemical synthesis, but often unsuitable for physical use
α-rh form, α-rh + β-rh mixture	Slow rate of deposition; sensitive to atm. oxidation; Si content	Chemically very reactive; low metallic content; no grinding problem; easily upgraded by sintering	Suitable for most chemical syntheses; physical measurements on resulting compounds possible
Amorphous or crystalline	Difficult; grinding problems	—	Synthesis of compounds corresponding to elements having a strong affinity for boron; physical studies of the element and of resulting compounds
In general, β-rh form	Expensive; difficult; grinding problems	Highest-grade B	

monocrystalline zones of a length of several centimeters may be obtained after several passes without the use of a seed (Niemyski *et al.*, 1965). crystallization from the liquid zone always yields β-rh crystals if the purity of boron is sufficiently high. Furthermore, during zone melting, the growth of the single crystals occurs in such a manner that the hexagonal *c* axis of the crystals is roughly along the length of the rod, varying from 5° to 15° from the cylinder axis (King *et al.*, 1965).

Summary

The methods for the synthesis of boron which have been reviewed here are briefly summarized in Table VIII. They can be divided into three classes: (1) processes leading to amorphous powders which are of poor grade but are cheap and very suitable for the synthesis of boron halides; (2) Cooper's process, yielding boron which is a commercial product of medium purity; and (3) methods based on hydrogen reduction or thermal dissociation of boron halides, which yield the samples of the highest purity.

The highest purity is achieved through the boron halide intermediates. For laboratory research, two methods are of interest: the hydrogen reduction of boron bromide in a silica tube at low temperature, and the reduction of boron halides on boron substrate at high temperature. The former, which gives highly reactive boron powders with 99% purity and well-defined crystallinity, is ideal for inorganic synthesis. The latter results in the highest-purity samples and is strongly recommended for preparing the bulk specimens required for the physical study of the element, while its uses in chemistry are limited by very serious contamination-free grinding problems.

Two upgrading processes can be applied with success: sintering in vacuum at high temperature and the floating-zone technique. The former can be applied to products prepared according to one of the processes of class 1 with a significant decrease of the impurity content if temperature is close to the melting point of boron. It can also be carried out at moderate temperature (1600–1700°C) on products obtained by hydrogen reduction of BBr_3 at low temperature in a silica tube; in this case, the α-rh $\rightarrow$ β-rh transformation yields 99.9% boron samples, easily pulverized, which are recommended for the synthesis of compounds on which physical measurements have to be done. The floating-zone technique results in a significant increase of purity and in the growth of large, monocrystalline zones, both of which are required for the careful study of the physical properties of the element.

REFERENCES

Amberger, E., and Dietze, W. (1964). *Z. Anorg. Allg. Chem.* **332,** 131.

Amberger, E., and Dietze, W. (1965). *In* "Boron II Preparation, Properties, and Applications" (G. K. Gaulé, ed.), p. 1. Plenum Press, New York.

Amberger, E., and Ploog, K. (1968). *J. Less Common Metals* **15,** 240.

Andrieux, L. (1929). *Ann. Chim.* (*Paris*) **10**(12), 423.

Andrieux, J. L., and Deiss, W. J. (1955). *Bull. Soc. Chim.* **1955,** 838.

Andrieux, L., and Barbetti, D. (1932). *C. R. Acad. Sci. Paris* **194,** 1573.

Armington, A. F., Dillon, G. F., and Mitchell, R. F. (1964). *Trans. Met. Soc. AIME* **230,** 350.

Armington, A. F., Buford, J. T., and Starks, R. J. (1965). *In* "Boron II Preparation, Properties, and Applications" (G. K. Gaulé, ed.), p. 61. Plenum Press, New York.

Bean, K. E., and Medcalf, W. E. (1960). *In* "Boron Synthesis, Structure, and Properties" (J. A. Kohn, W. F. Nye, and G. K. Gaulé, eds.), p. 48. Plenum Press, New York.

Becher, H. J., and Schäfer, A. (1960). *Z. anorg. Allg. Chem.* **306,** 260.

Boiret, M. (1968). Private communication.

Campbell, I. E., Powell, C. F., Nowicki, D. H., and Gonzer, B. W. (1949). *J. Electrochem. Soc.* **96,** 318.

Cooper, H. S. (Feb. 13, 1945a). U.S. Patent 2,369,212.

Cooper, H. S. (Feb. 13, 1945b). U.S. Patent 2,369,214.

Cooper, H. S. (Oct. 23, 1951a). U.S. Patent 2,572,248.

Cooper, H. S. (Oct. 23, 1951b). U.S. Patent 2,572,249.

Cooper, H. S. (Dec. 22, 1959). U.S. Patent 2,918,417.

Cooper, H. S. (1961). *In* "Rare Metal Handbook" (C. A. Hampe, ed.), Chapter 5. Van Nostrand-Reinhold, Princeton, New Jersey.

Cueilleron, J. (1944). *Ann. Chim.* (*Paris*) **19,** 459.

Decker, B. F., and Kasper, J. S. (1959). *Acta Crystallogr.* **12,** 503.

Dupuy, E., and Hackspill, L. (1923). *C. R. Acad. Sci. Paris* **197,** 229.

Etourneau, J. (1970). Thesis no. 285. Univ. of Bordeaux, Centre de Doc. C.N.R.S. No. A0.4312.

Finn, J. M., Jr., Nelson, E. N., and Mercuri, R. A. (Dec. 29, 1960). Brit. Patent 857,060.

Formstecher, M., and Ryskevic, E. (1945). *C. R. Acad. Sci. Paris* **221,** 747.

Freymann, R., and Stieber, A. (1934). *C. R. Acad. Sci. Paris* **199,** 1109.

Gamble, E. C., Gimlmont, P., and Stiff, J. (1940). *J. Amer Chem. Soc.* **62,** 1257.

Gaulé, G. K., ed. (1965). "Boron II-Preparation, Properties and Applications." Plenum Press, New York.

Gould, J. R., and Gardner, D. M. (July, 10, 1962). U.S. Patent 3,043,665.

Greiner, E. S. (1960). *In* "Boron Synthesis, Structure, and Properties" (J. A. Kohn, W. F. Nye, and G. K. Gaulé, eds.), p. 70. Plenum Press, New York.

Haag, H. (June 4, 1957). U.S. Patent 2,794,708.

Hackspill, L., and Cueilleron, J. (1944). *C. R. Acad. Sci. Paris* **218,** 458.

Hackspill, L., Stieber, A., and Hocart, R. (1931). *C. R. Acad. Sci. Paris* **193,** 776.

Hagenmuller, P., and Naslain, R. (1963). *C. R. Acad. Sci. Paris* **257,** 1294.

Hagenmuller, P., and Naslain, R. (1965). *Rev. Hautes Temp. Refract.* **2,** 225.

Hampe, W. (1888). *Chem. Z.* **12,** 84.

Hinz, I., and Wirth, H., (1965). *In* "Boron II Preparation, Properties, and Applications" (G. K. Gaulé, ed.), p. 9. Plenum Press, New York.

Hoard, J. L. (1961). *In* "Borax to Boranes" (R. F. Gould, ed.), p. 42. Amer. Chem. Soc., Washington, D.C.

Hoard, J. L., and Hughes, R. E. (1967). *In* "The Chemistry of Boron and Its Compounds" (E. L. Muetterties, ed.), Vol. 2, p. 25. Wiley, New York.

Hoard, J. L., and Newkirk, A. E. (1960). *J. Amer. Chem. Soc.* **82,** 70.

Hoard, J. L., Hughes, R. E., and Sands, D. E. (1958). *J. Amer. Chem. Soc.* **80,** 4507.

Hoard, J. L., Sullinger, D. B., Kennard, C. H. L., and Hughes, R. E. (1970). *J. Solid State Chem.* **1,** 268.

Horn, F. H. (1960). *In* "Boron Synthesis, Structure, and Properties" (J. A. Kohn, W. F. Nye, and G. K. Gaulé, eds.), p. 70. Plenum Press, New York.

Hutchinson, W. E., and Eick, H. A. (1962). *Inorg. Chem.* **1,** 434.

Ivanov-Emin, B. N., Nisselson, L. A., and Petrussevitch, I. V. (1961). *J. Prikl. Khim.* **34,** 2378.

Kahlenberg, H. H. (1925). *Trans. Amer. Electrochem. Soc.* **47,** 23.

Kiessling, R. (1948). *Acta Chem. Scand.* **2,** 707.

King, I. R., Wavner, F. E., Jr., Taylor, G. R., and Talley, C. P. (1965). *In* "Boron II Preparation, Properties, and Applications" (G. K. Gaulé, ed.), p. 45. Plenum Press, New York.

Kohn, J. A., Nye, W. F., and Gaulé, G. K. eds. (1960). "Boron Synthesis, Structure and Properties." Plenum Press, New York.

Laubengayer, A. W., Hurd, D. T., Newkirk, A. E., and Hoard, J. L. (1943). *J. Amer. Chem. Soc.* **65,** 1924.

Lindquist, P. F., Hammond, M. L., and Bragg, R. H. (1966). *Phys. Status Solidi* **17,** 25.

McCarty, L. V., and Carpenter, D. R. (1960). *J. Electrochem. Soc.* **107,** 38.

Mikheeva, V. I., Shamrai, F. I., and Krylova, E. Ya. (1957). *Russ. J. Inorg. Chem.* **2**(6), 12, 27, 42, 53.

Moissan, H. (1892a). *C. R. Acad. Sci. Paris* **114,** 319.

Moissan, H. (1892b). *C. R. Acad. Sci. Paris* **114,** 392.

Naray Szabo, S. V., and Tobias, C. W. (1949). *J. Amer. Chem. Soc.* **72**(2), 1882.

Naslain, R. (1967). Thesis no. 188, Univ. of Bordeaux, Centre de Doc. C.N.R.S., No. AO 1332.

Naslain, R., and Etourneau, J. (1966). *C. R. Acad. Sci. Paris* **263,** 484.

Naslain, R., and Kasper, J. S. (1970). *J. Solid State Chem.* **1,** 150.

Naslain, R., Etourneau, J., and Hagenmuller, P. (1967). *Bull. Soc. Chem. Fr.* **1967,** 2529.

Naslain, R., Etourneau, J., and Hagenmuller, P. (1968). *Proc. 3rd Int. Symp. Boron, Warsaw, Elect. Technol.* **3** (1).

Naslain, R., Boiret, M., and Becanne, B. (to be published).

Newkirk, A. E. (1961). *In* "Borax to Boranes" (A. F. Gould, ed.), p. 27. Amer. Chem. Soc., Washington, D.C.

Newkirk, A. E. (1964). *In* "Boron, Metallo-Boron Compounds and Boranes" (R. M. Adams, ed.), Vol. 4, p. 233. Wiley, (Interscience), New York.

Niemyski, T., and Olempska, Z. (1962). *J. Less Common Metals* **4,** 235.

Niemyski, T., Pracka, I., Szczerbinski, R., and Frukacz, Z. (1965). *In* "Boron II Preparation, Properties, and Applications" (G. K. Gaulé, ed.), p. 35. Plenum Press, New York.

Olempska, Z., Badzian, A., Pietrzak, K., and Niemyski, T. (1966). *J. Less Common Metals* **11,** 351.

Otte, H., and Lipsitt, H. A. (1966). *Phys. Status Solidi* **13,** 439.

Powell, C. F., Ish, C. J., and Blocher, J. M. (1960). *In* "Boron Synthesis, Structure, and Properties" (J. A. Kohn, W. F. Nye, and G. K. Gaulé, eds.), p. 7. Plenum Press, New York.

Powell, C. F.. Campbell, I. E., and Gonzer, B. W. (1965). *In* "Vapor Plating," p. 103. Wiley, New York.

Robertson, J. A. (1943). Thesis, Cornell Univ., Ithaca, New York.

Samsonov, G. V., Obolonchik, V. A., and Koulichina, G. N. (1960). *Zh. Prikl. Khim.* **33,** 1365.

Schott, S., and Hansley, V. L. (Apr. 23, 1963). U.S. Patent 3,086,847.

Schumb, W. C., Gamble, E. L., and Banus, M. D. (1949). *J. Amer. Chem. Soc.* **71,** 3225.

Sowa, F. J. (Apr. 5, 1949). U.S. Patent 2,465,989.

Spevack, J. S. (Aug. 3, 1954). U.S. Patent 2,685,501.

Starks, R. J., and Buford, J. T. (1963). *Electrochem. Technol.* **1,** 108.

Starks, R. J., and Medcalf, W. E. (1960). *In* "Boron Synthesis, Structure, and Properties" (J. A. Kohn, W. F. Nye, and G. K. Gaulé, eds.), p. 59. Plenum Press, New York.

Stern, D. R., and Lynds, L. (1958). *J. Electrochem. Soc.* **105,** 676.

Stieber, A. (1933). Thesis 298/A274, Univ. of Paris, Paris, France.

Talley, C. P., Line, L. E., and Doverman, Q., Jr. (1960a). *In* "Boron Synthesis, Structure, and Properties" (J. A. Kohn, W. F. Nye, and G. K. Gaulé, eds.), p. 94. Plenum Press, New York.

Talley, C. P., Post, B., and La Placa, S. (1960b). *In* "Boron Synthesis, Structure, and Properties" (J. A. Kohn, W. F. Nye, and G. K. Gaulé, eds.), p. 83. Plenum Press, New York.

Uno, R. J. (1958). *J. Phys. Soc. Japan* **13,** 667.

Warth, A. H. (1923). *Maryland Acad. Sci. Bull.* **3**(3), 8.

Wavner, F. E., Jr. (1967). *In* "Modern Composite Materials" (L. J. Broutman and R. H. Krock, eds.), Vol. 10. p. 245. Addison-Wesley, Reading, Massachusetts.

Weast, R. C., ed. (1968). "Handbook of Chemistry and Physics" 49th ed. The Chem. Rubber Publ. Co., Cleveland, Ohio.

Weintraub, E. (1909). *Trans. Amer. Electrochem. Soc.* **16,** 165.

Yannacakis, J., and Nies, N. P. (1960). *In* "Boron Synthesis, Structure, and Properties" (J. A. Kohn, W. F. Nye, and G. K. Gaulé, eds.), p. 38. Plenum Press, New York.

Preparation Conditions of the Ferrites

M. PAULUS

Groupe d'Étude et de Synthèse des Microstructures
Centre National de la Recherche Scientifique,
Meudon-Bellevue, France

I. Introduction

Ferrites play an important part in the field of telecommunications. The required properties may be classified as: intrinsic, i.e., those defined only by chemical composition and crystal structure (saturation magnetization, magnetic anisotropies, dc electrical conductivity, etc.), and extrinsic, i.e., those also affected by the microstructures (permeability, coercive force, losses, etc.). The relationship of intrinsic properties to chemical composition and crystal structure is quite well understood, but the situation with respect to extrinsic properties is not yet clear. It is necessary first to know the various relations between one property and the different microstructures, and second to optimize each microstructure taking into account the desired properties, which may require contradictory microstructure characteristics. Thus, ferrites must be tailor-made with a view to each specific application. Since a large number of these applications depend on the extrinsic properties, we have to control not only the composition, but also the microstructures (chemical, crystalline, granular, porous, mechanical, electrical, magnetic, dielectric, and optical properties). Moreover, each microstructure is very sensitive to each stage of the thermochemical treatment from the raw material to the end product.

Our purpose here is to show with some examples the main parameters and techniques for each stage of preparation. All these parameters and techniques interact with each other and we must always consider the preceding and the following stages. The choice of a preparation scheme

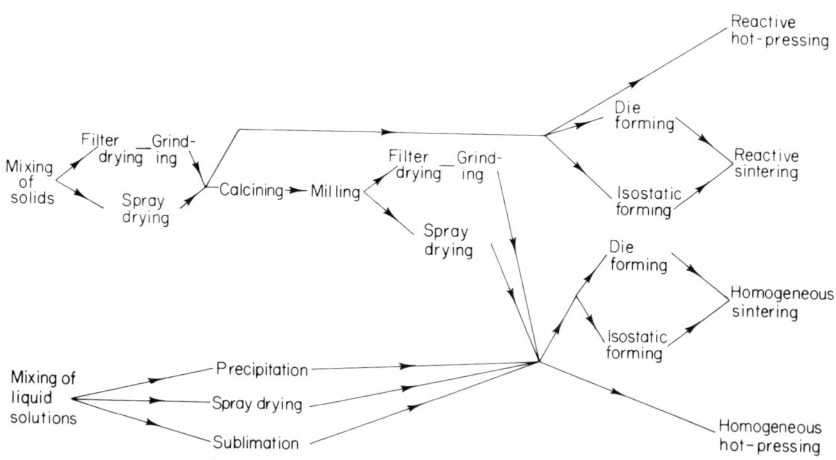

FIG. 1. Main preparation routes for sintered ferrites.

(Fig. 1) depends upon available starting materials, desired properties, scale of preparation, and available technical means. Nevertheless, it seems that, although the high-temperature solid state reaction process is used on an industrial scale, starting with a pure solid solution with very fine particle size and the right amount of cations and anions leads to the most uniform and well-defined microstructures.

Special attention is paid to thermochemical equilibrium and to the technical problems related to it, since it is one of the most important preparation parameters, and research in this field has not yet solved the many problems involved. We must always keep in mind during the course of preparation that the oxygen content, which is an important determinant of the properties, depends simultaneously upon temperature, the partial pressure of oxygen, diffusion coefficients, and diffusion paths in the sample.

II. Thermodynamic Equilibria

A. Phase Diagrams

The study of ferrite phase diagrams is quite complex, since their composition is determined not only by the cations, but also by their oxygen content, which is a function of other parameters: temperature and oxygen partial pressure. Very often, it is not possible to represent the limit of solid solution exactly, but only through several diagrams. The three most frequent types of diagram are the isothermal section of the ternary diagram, the pseudobinary section, and the equilibrium diagram with isocomposition lines.

1. *Isothermal section*

This kind of diagram gives the phases present (Fig. 2) for each composition at a certain temperature. To allow the processing of ferrites, the diagram must show the isobars of oxygen partial pressure, the ferrite oxygen content, and the phases formed as a function of oxygen partial pressure. These will be straight lines in regions where two phases are in equilibrium and their extremities will denote the composition of the two phases in equilibrium.

However, this approach hardly allows a study of ferrite preparation in relation to temperature, even when displaying isothermal sections at very short intervals, showing the isobar displacement and phase limits. It is necessary to know the function $P_{O_2} = f(T)$ for each composition.

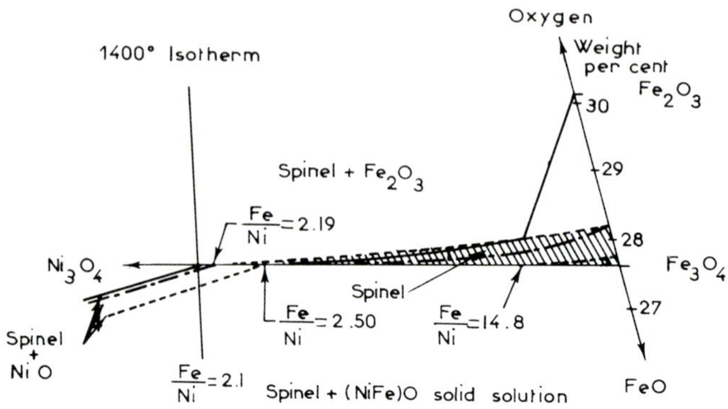

FIG. 2. Phase equilibria in the ferrite region of the Ni–Fe–O system. Isotherm at 1400°C. The solid, dashed-dotted, and dashed curves are the oxygen isobars 10^0, $10^{-0.7}$, and 10^{-2} atm, respectively. These will be straight lines in regions where two condensed phases are in equilibrium and their extremities will denote the compositions of the two phases in equilibrium. (Adapted from Shafer, M. W., *J. Phys. Chem.* **65**, 2055 (1961). Copyright 1961 by the American Chemical Society. Used by permission of the copyright owner.)

2. *Pseudobinary section*

This gives the phase diagram with respect to temperature and composition in a form similar to the one used for metallic alloys (Fig. 3). These diagrams must be used carefully, since, quite frequently, a diagram is determined at constant oxygen partial pressure (pure oxygen, air, etc.). Every point of the diagram represents a different oxygen content of the ferrite, though the variation of oxygen content is not plotted on the diagram. In fact, P_{O_2} is not only a function of anionic and cationic compositions, but also of temperature. This diagram cannot be considered as a pseudobinary section of a ternary diagram. It may be used for the preparation of ferrites, but only as an approximation, except in the very rare cases where the ferrite composition is not very sensitive to the oxygen pressure. To be valid, the diagram should be established at a constant oxygen concentration in the ferrite. It should show the isobars allowing the determination of the partial equilibrium pressure of the ferrite in relation to the temperature and the cationic composition. Such diagrams do not seem to have been prepared, except for iron ferrite, but in this particular case, oxygen is a variable of the composition (Fig. 4). According to this diagram, it is clear that the region of existence of magnetite from 400 to 1500°C (Fig. 3) cannot be in equilibrium with only one oxygen pressure. For $P_{O_2} = 10^{-4}$ atm, $Fe_3O_{4\pm\gamma}$ exists only between approximately 1530 and 1090°C, and in this temperature interval the stoichiometry parameter γ varies.

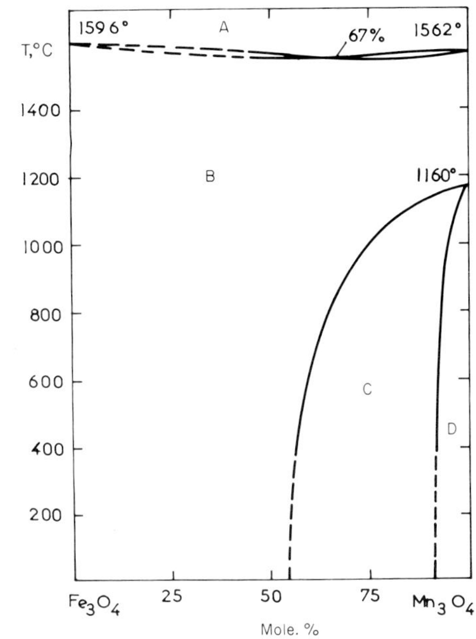

FIG. 3. The system Fe_3O_4–Mn_3O_4 (Van Hook, 1958). A, Liquid; B, magnetite–high hausmanite SS; C, magnetite–high hausmanite SS + low hausmanite SS; D, low hausmanite SS.

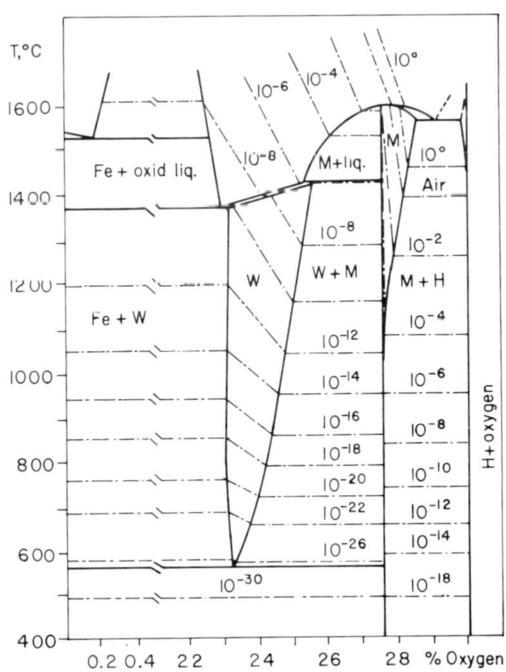

FIG. 4. System Fe–O-Dashed-dotted lines are the oxygen isobars; M, magnetite; W, wustite; H, hematite (Muan and Osborn, 1964).

3. *Equilibrium diagram with isocomposition lines*

Thermal treatments with constant composition are better shown on the plot of $\log P_{O_2} = f(1/T[°K])$ of Fig. 5. It is valid only for one cationic composition. Blank (1961) has established that when cationic and anionic compositions are constant and in the zone of solid solution, $\log P_{O_2}$ is nearly a linear function of $1/T$. Every straight line of the diagram corresponds to a degree of ferrite oxidation. Limits of solubility of the ferrite for every degree of oxidation are plotted, and seem to cut some of the straight isocomposition lines. Blank (1961) presented this diagram as a universal one. A ferrite of any composition and in equilibrium at a given temperature and oxygen partial pressure will keep the equilibrium if the variation of P_{O_2} with respect to temperature follows the straight line going through the figurative point (P_{O_2}, T) of the initial equilibrium. Under these conditions, the degree of oxidation associated with a straight line and the solubility limits vary with the cationic composition. However, generalizations must be made with discretion, for the measurements are not yet very accurate or numerous. Furthermore, Macklen (1967) showed, for lines representing a lower degree of oxidation, a change in the slope above 1225°C (Fig. 6). This behavior would be consistent with either a structural rearrangement or a change in cation distribution at this temperature.

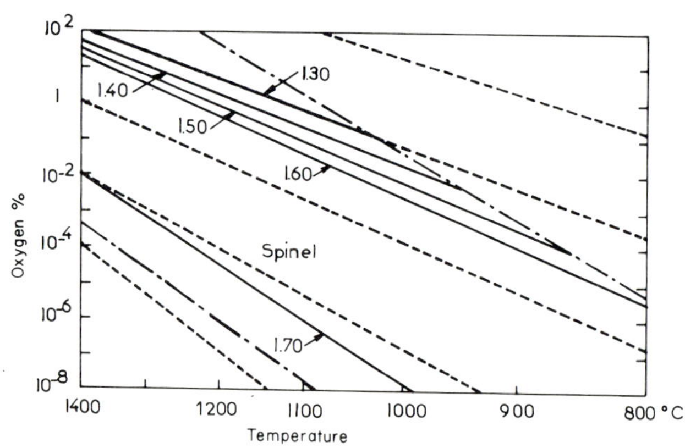

FIG. 5. Isocomposition phase diagram. Each dashed or solid line is an isocomposition one. Dashed lines, universal diagram of Blank (1961); solid lines, isocomposition lines for a starting composition 54.6 mole% Fe_2O_3, 29.3 mole% MnO, and 16.1 mole% ZnO; dotted-dashed lines, phase limits. The oxidation level is expressed in wt% Fe^{2+}.

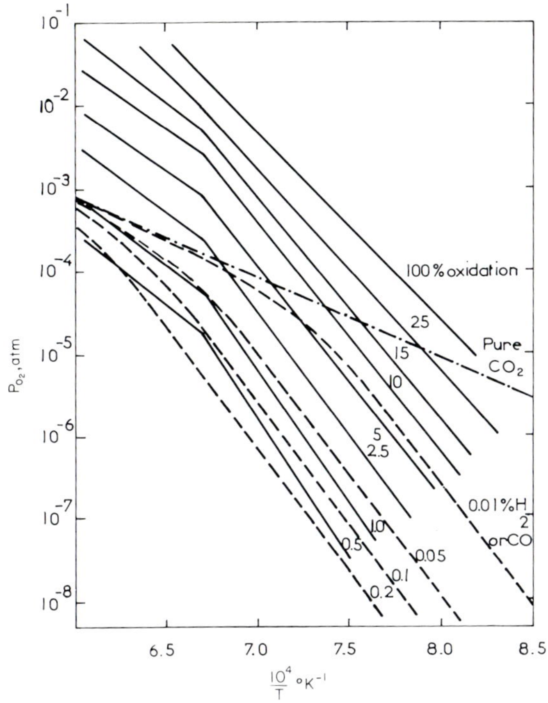

FIG. 6. Isocomposition diagram for $Co_{0.165}Fe^{2+}{}_{0.835}Fe_2O_4$. The oxidation level is expressed as the percentage of total Fe^{2+} ions that have been oxidized to Fe^{3+}. Dashed lines represent the oxygen content of mixtures of CO_2/CO and CO_2/H_2 calculated from the existing thermodynamic data (Macklen, 1967).

It is obvious that to obtain a homogeneous ferrite, the straight isocomposition line must not cut the phase limit up to the working temperature for the ferrite. To follow the variation of P_{O_2} down to low temperatures, it may be convenient to use the diagram for the variation of the free energy of oxide and nitride formation, as elaborated by Olette and Ancey Moret (1963), which has been applied by Paulus and Morineau (1967) to follow the nickel ferrite equilibrium at low temperature. In fact, the isocomposition curves are straight lines or segments of straight lines, standard enthalpy and entropy varying only slightly with temperature. Moreover, this diagram can give directly the mixtures CO/CO_2 and H_2/H_2O in equilibrium with the ferrite.

The isocomposition equilibrium diagrams are coming into general use and are the easiest to use.

B. Establishment of Phase Diagrams

It is only for the elementary ferrites Fe_3O_4, $Mn_xFe_{3-x}O_4$, and $Ni_xFe_{3-x}O_4$ and for some compositions quite interesting for their magnetic properties that we begin to find valid equilibrium diagrams. Thus, the worker in the field is often compelled to establish for himself the diagram corresponding to the ferrite that he wants to prepare. Three techniques are generally used for this purpose.

1. *Quenching of a sample in equilibrium*

The samples are set in equilibrium with different oxygen partial pressures at different temperatures, and then quenched. The oxygen content of the samples is titrated, and the phase limits are determined by x-ray diffraction and metallography. The method is precise if the atmosphere of the furnace is well-defined and if the sample is homogeneous (see Section II.C.1). It is time-consuming, since numerous thermal treatments are necessary to establish the phase and the curves $P_{O_2} = f(T)$. Nevertheless, it is the easiest process for establishing diagrams for an isolated study.

2. *Thermogravimetric equilibrium*

The curves $P_{O_2} = f(T)$ are determined, for constant mass, by means of a thermobalance. This method may be more rapid, but requires discretion and corrections because the dynamic gas pressure varies according to the flow and with the static pressure change with temperature, and also because of the unequal dilatation of the balance beams during the thermal variations. This method is thus difficult and requires further development for its perfection. Some of these problems may be solved by means of a thermobalance with double beams or a torsion balance (Reijnen and Roksnoer, 1969). Furthermore, we note that there is a problem when there is a cation with high vapor pressure in the ferrite, such as zinc or lithium in a ferrite with an oxygen content $4 \pm \gamma < 4$. However, this is one of the methods best adapted to the establishment of curves of $P_{O_2} = f(T)$.

Paulus and Guillaud (1966) have shown it is possible to use a thermobalance coupled with the preparation furnace to regulate the oxygen partial pressure. In this process, a pilot sample of a previously prepared ferrite with known degree of oxidation, but not always the desired one, is continually weighed in the treatment furnace during the heat cycle. Monitoring the weight of the pilot sample acts as a means to adjust the oxygen partial pressure of the furnace, whereby the weight of the pilot sample and hence also the charge is maintained substantially constant during the heat cycle at the value corresponding to the optimum desired oxygen content.

3. *Reversible solid galvanic cell*

Measurements of the reversible emf's of galvanic cells of the type Fe, $Fe_{0.95}O/0.15CaO–0.85ZrO_2$/Sp Wp (Sp designates a spinel ferrite or solid solution, and Wp a wüstite phase) have made it possible to determine $\Delta\bar{G}_{O_2}$, the partial molar free energy, and therefore $\log P_{O_2} = \Delta\bar{G}_{O_2}/RT$, the equilibrium dissociation pressure of the ferrite or solid solution. This work has been performed in large part by the Russians (Tret'yakov *et al.*, 1965; Schmalzried and Tret'yakov, 1966).

4. *Control of the oxygen partial pressure*

Two important operations are carried out during the thermochemical treatment of ferrites: production of a specific partial pressure of oxygen and its measurement.

a. Measurement of the partial pressure of oxygen. The measurement of the partial pressure of oxygen must be done between 1 and 10^{-6} atm, at the extreme, 10^{-8} atm.

The paramagnetic oxygen analyzers give valid results with oxygen content over 1%. Marion (1967) has described an analysis technique for active oxygen that consists in measuring the resistance of cobalt or nobium oxide wire in equilibrium at 1000°C with the atmosphere that is to be analyzed. The measurement interval is 1–10^{-12} atm. This device is made by Hermann-Moritz (France).

A high-temperature (850°C) galvanic cell, consisting of a calcium-stabilized zirconium oxide electrolyte with porous platinum electrodes, is also used to measure pressure. The measurement range is 1–10^{-6} atm (Westinghouse analyzer).

b. Production of a specified oxygen partial pressure. To change the partial pressure of oxygen from 10^{-8} to 1 atm, several methods can be applied:

1. A mixture of two gases with O_2 content measured by means of a flow meter is the easiest to employ, but the variation intervals of P_{O_2} are rather limited.

For low O_2 content, the following methods are available.

2. Oxygen is produced by electrolysis and carried by a neutral vector gas. The regulation of oxygen flow is easy to check, but the polarization voltage limits the flow to a certain minimum, and, moreover, water and acid vapors are carried away with the vector gas and must be trapped completely (Marion, 1967).

3. Small quantities of oxygen are introduced into a vector gas through capillaries, the openings of which are operated by rapid rate electrovalves. Flow is regulated by changing the duration and frequency of opening of the valves. For small volume gas circuits, with quick response, the fluctuations of gas composition are substantial (produced by Herrmann-Moritz, France).

4. The volumetric pumps of "Wosthoff" type from Beckmann allow the preparation of gaseous mixtures from 0.02 ppm to 100%. They are easy to use, but the variation of oxygen partial pressure is not continuous.

5. Diffusion of oxygen into or from the vector gas may be obtained at 850°C through a tube of calcium-stabilized zirconium oxide electrolyte metallized on the inside and outside walls, similar to the galvanic cell of Section II.B.3 (Baukal, 1969). If a potential difference is applied between the two electrodes, oxygen is carried through the solid electrolyte. The device makes it possible to increase or decrease the partial pressure of oxygen by a factor of 100. The change of composition is quite progressive and easy to regulate and automate.

However, while the other processes described are available commercially, this one still seems to be a laboratory device.

6. With suitable selection of the gas mixture CO_2/CO or CO_2/H_2, it may be possible to match the oxygen content of the furnace to that required by the ferrite. The required ratio of CO_2/CO or CO_2/H_2 is easier to achieve for the same partial pressure of oxygen than with a mixture N_2/O_2. Furthermore, for some ferrite compositions, the same ratio of CO_2/CO or CO_2/H_2 yields the same equilibrium values of P_{O_2} over an extended temperature range (Fig. 6).

C. Deviation from Thermochemical Equilibrium

Even when the curve $P_{O_2} = f(T)$ and phase limits are quite well known, the thermal treatment of a ferrite in thermodynamic equilibrium is hampered by different factors related to the treatment chamber and the ferrite itself.

1. *Causes for deviation from equilibrium*

The use of a gas-tight furnace is not sufficient to achieve a correct thermochemical treatment. We must consider the important variations in the oxygen content of the oven atmosphere during the heat treatment. A ferrite in equilibrium at 1400°C with $P_{O_2} = 1$ atm, remains in equilibrium at 700°C only at $P_{O_2} = 10^{-6}$–10^{-8} atm. This requires a perfect draining of the furnace, which is quite difficult to achieve if we consider the oxygen adsorption by the refractories that desorb it at low temperature (low oxygen partial pressure). Even in a perfectly drained and desorbed furnace, the use of vector gas with low oxygen content ($N_2 + O_2 < 5$ ppm) does not permit the creation of the desired oxygen partial pressure without certain precautions. The rubber pipes or gaskets of the gas distribution system may play the part of an oxidizing or reducing agent: A gas containing 10 ppm of O_2 may contain from 1 to 150–200 ppm after having passed through one meter of tube. The same remarks are valid for piston reducing valves which pump oxygen, and those with porous diaphragms requiring a long time for degassing.

The sample itself may be a very important oxygen source which changes the composition of the atmosphere when a mixture of oxides is treated. Thus the formation of $MnFe_2O_4$ (i.e., ~230.6 g of ferrite) from Fe_2O_3 and Mn_2O_3 releases about six liters of oxygen at room temperature and ~32 liters at 1300°C. Even when the ferrite $NiFe_2O_4$ is formed from Fe_2O_3 and NiO, which does not involve a loss or a gain of oxygen, we observe (Section III) first the release and then the absorption of oxygen by the sample.

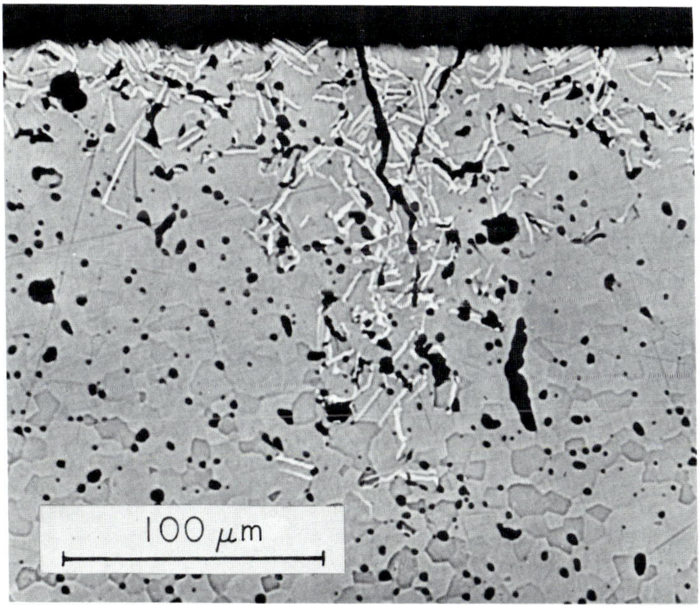

FIG. 7. Mn–Zn ferrite slightly oxidized during cooling. Fe_2O_3 precipitates around a crack.

2. *Consequences of deviation from equilibrium*

For the study of their physical properties, ferrites are mostly prepared as polycrystalline sintered materials. Therefore, the reaction with the oven atmosphere occurs at the surface of the sample. Even if the ferrite is fired as a loose powder, it agglomerates to a gas-tight compact. Every deviation

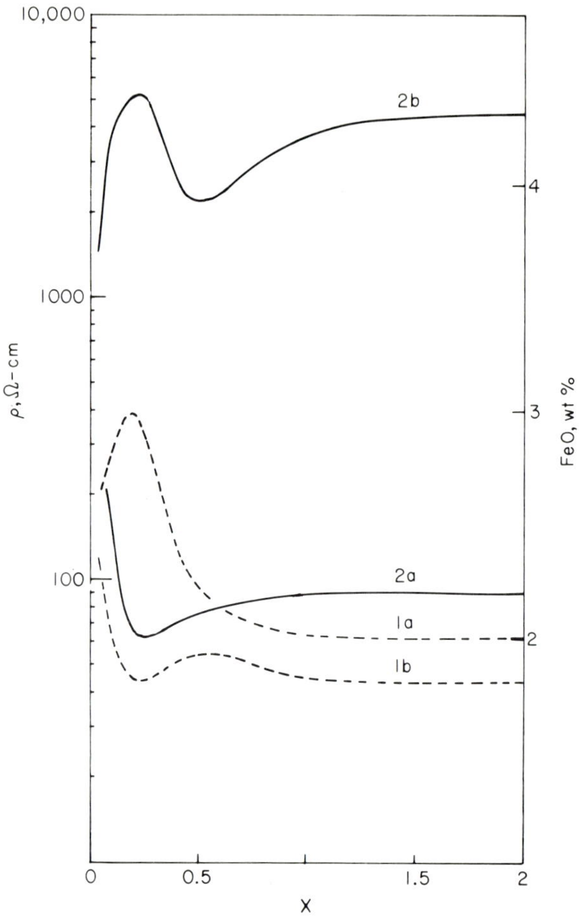

FIG. 8. Change in FeO content (1) (dashed lines) and resistivity (2) (solid lines) versus the distance x from the surface of the sample. Mn–Zn ferrite (a) heated to 1260°C in N_2, 5 hr isothermal treatment at 1260°C in 5% $O_2 + N_2$, and then quenched; (b) heated to 1260°C in O_2, 5 hr isothermal treatment at 1260°C in 5% $O_2 + N_2$, and then quenched.

from the $P_{O_2} = f(T)$ law leads to a heterogeneity. Depending on the extent of deviation, a superficial oxidation or reduction gives rise either to a parameter decrease or increase, sometimes with cracks in the sample, or to a precipitate of Fe_2O_3 or wüstite (Fig. 7). The mechanical strains induced in the sample drastically modify some extrinsic properties, such as permeability (Paulus and Guillaud, 1964). The oxidation heterogeneities which occur at the beginning of the heat treatment take very long to eliminate. Figure 8 shows the variation of resistivity, lattice parameter, and FeO content with distance from the surface for two samples which were heated in two different atmospheres but homogenized at 1260°C for 5 hr.

Deviation from equilibrium with respect to the oxygen partial pressure not only changes the superficial oxygen content, but also: (a) induces cation migration (see Section II.D.2); (b) changes the evaporation rate of such cations as Li (Ridgley *et al.*, 1970) and Zn [the loss of Zn decreases when P_{O_2} increases; this is interpreted by Veith (1970) as the effect of an oxygen zone of enhanced density at the surface of the ferrite as a result of the adsorption potential, which brakes the rate of Zn evaporation]; (c) promotes discontinuous crystal growth (Michalowsky and Vogler, 1963); (d) modifies the disorder $\rightarrow$ order transition rate in Li ferrite; Amemiya (1970) shows that this transition rate is affected by the oxidizing rate, but not by the oxygen content of the ferrite.

D. Equilibrium Segregation

1. *Lattice deformation segregation*

It is not sufficient to cool a ferrite in constant equilibrium with the atmosphere, i.e., without release or absorption of oxygen, to obtain a single-phase ferrite. The cooling path on the diagram may cross a phase limit and a very fine precipitate can be observed in the core of the grain (Fig. 9). Even if a phase boundary is not crossed, some large cations like Ca and Y (Paulus, 1966), Ba (Paulus and Hamelin, 1968), and Mn (Hamelin and Paulus, 1970), which introduce a distortion energy in the lattice, migrate to the grain boundary and modify the composition near the boundary.

2. *Oxidation–reduction segregation*

The oxidation or reduction of a ferrite through its surface may induce a cation migration from or to the surface according to its valence. If we

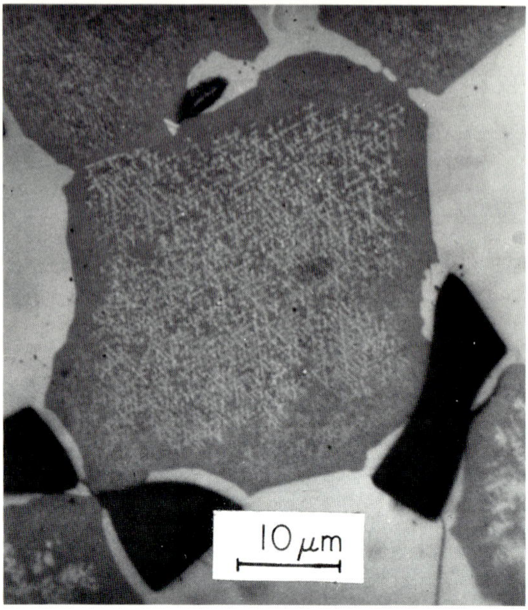

FIG. 9. Precipitates in the spinel phase due to the crossing of the spinel phase boundary during rapid cooling. 37.8 mole% Fe_2O_3, 62.2 mole% NiO. Heat treatment: 48 hr at 1350°C in nitrogen and then quenched in water (Eveno, unpublished results.)

consider a stoichiometric ferrite

$$2Fe^{3+},\quad (1-x)Fe^{2+},\quad xMe^{2+},\quad 4O^{2-} \tag{1}$$

the partial pressure of oxygen can be given by the equation

$$P_{O_2} = \{[2Fe^{3+}/(1-x)Fe^{2+}K_2]K_1{}^{3/16}]\}^4 \tag{2}$$

It is clear from Eq. (2) that during oxidation, equilibrium may be restored by oxidation, with an increase of the ratio of vacancies to interstitial cations, but also by migration of divalent cations to the ferrite–atmosphere interface or of trivalent cations to the core of the sample (variation of x). This effect has been observed by Paulus and Guillaud (1964) for Ca, Y, and Fe and by Hamelin and Paulus (1970) for Mn in spinel ferrites.

III. Solid State Formation

Knowledge of the formation mechanism for ferrites is necessary to understand the different behaviors of oxide mixtures during calcination or sintering. For Ni ferrite and Mg ferrite, the formation occurs by counter diffusion of divalent ions through the solid, but it is also accompanied by a release or an absorption of oxygen at the interface between oxide and ferrite and by a Kirkendall effect (Fig. 10).

Figure 11 shows the different mechanisms proposed for Mg ferrite and Ni ferrite formation. From the discussion of the results, it appears that the mechanism is controlled by the extent of diffusion of Fe into the divalent oxide (Paulus and Eveno, 1968). For Ni ferrite formation, the release of oxygen alone occurs at the beginning, and the absorption takes place only in the last step of formation in order to oxidize the mixed wüstite FeO,NiO into $NiFe_2O_4$ when the Fe content in NiO is sufficient. This is achieved in limited samples or particles and is always the case for a mixture of oxides. This explains the results of Elwell *et al.* (1967), who show that the Ni ferrite forms first with a low oxygen content and oxidizes afterward.

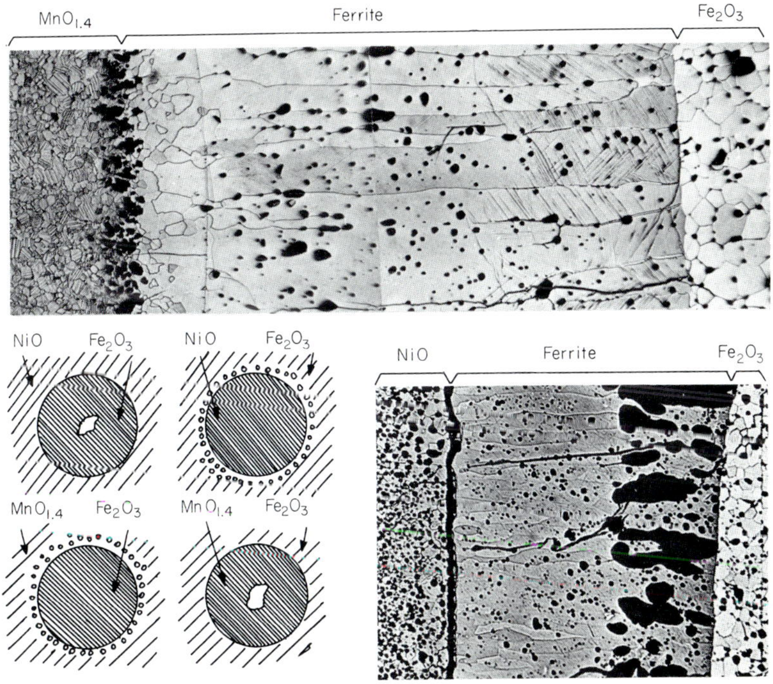

FIG. 10. Diffusion couples for the formation of manganese ferrite and nickel ferrite showing the gathering of pores due to a Kirkendall effect, and schematic representation of the consequences of the porosity location, depending on the kind of large particles embedded in fine particles (Paulus, 1970).

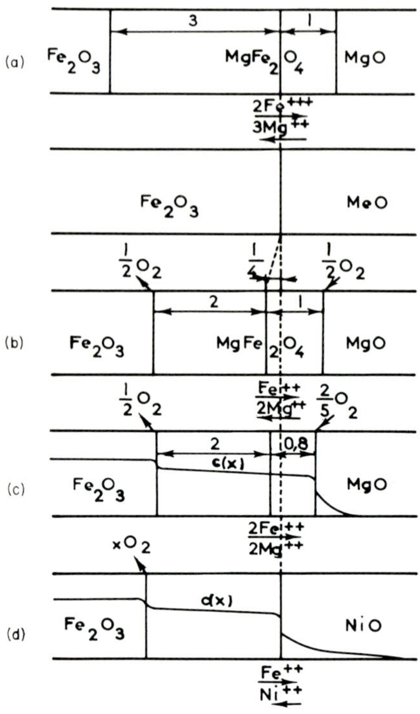

FIG. 11. Proposed mechanisms for the formation of nickel ferrite and magnesium ferrite. The $c(x)$ curve represents the Fe content in the diffusion couples. (a) Carter (1961), (b) Kooy (1964), (c) Paulus and Eveno (1968), (d) Paulus and Eveno (1968).

These different effects must be kept in mind in the study of reactive sintering. They control the kind of porosity and oxygen content of the ferrite, depending on the particle size and type of gathering and on the thermochemical cycle (see Section VI.A).

IV. Preparation of Ferrites

We must distinguish between the chemical and crystalline preparation of the ferrite and the sintering of the final compact, which must be characterized not only by its crystalline structure and its chemical composition, but also by its different microstructures (chemical, crystalline, granular, mechanical, magnetic, electrical, dielectric), though these two stages are sometimes achieved during the same thermochemical treatment. In this section, we deal only with the chemical preparation. This preparation should lead to a ferrite with well-defined, homogeneous cationic and anionic compositions and formed with fine, dense particles and, in some cases, with narrow diameter dispersion, since it is quite difficult to homogenize the ferrite during the thermochemical treatment of the compact and the above-mentioned parameters control the different microstructures in the final material (see Sections II.C, III, VI, and VII).

A great number of methods has been developed to meet these requirements, and, in fact, this excess of methods shows the difficulties encountered in obtaining a ferrite with homogeneous composition and defined particle size.

The formation of ferrite by reaction between the salt particles occurs at high temperature and yields ferrites with relatively large particle size. Advances in knowledge concerning ferrite formation have emphasized the need for very homogeneous powders with a fine particle size and high purity. To this end, numerous methods of preparing salt solid solutions and mixed organometallic compounds and of studying their decomposition have been considered. We give only one example for each method and will not make judgments on the different methods, since, in fact, the results cannot be separated from the subsequent thermochemical treatment.

A. High-Temperature Reaction of Oxides or Salts

Of the three following methods, the calcination of mixed salt particles is the only one that is now being used to any great extent. Nevertheless, the new fused salt process must not be neglected, since it seems that it should be very easy to scale up this process to produce large quantities of ferrite.

1. *Calcination of mixed oxides, oxalates, sulfates, nitrates, carbonates*

With this conventional technique, starting oxides (Fe_2O_3, NiO, ZnO, MnO_x, Y_2O_3, etc.) or other compounds (oxalates, sulfates, nitrates, or carbonates) are first mixed as intimately as possible. The apparent particle size is mostly between 0.01 and 10 μm. Many commercial systems of mixing or mixing and grinding have been used, such as mortar, ball-mill-grinder, vibrating or barrel systems and their numerous variants. This operation is generally carried out in a wet medium (water or alcohol). There are still few theoretical and experimental studies on mixing for ferrite preparation (Chol *et al.*, 1968; Chol and Aubaile, 1970). Generally, every improvement of homogeneity has, as corollary, a contamination by the mixed material. On mixing for 24 hr in a ball mill, the iron content of the mixture may increase by 0.5%, for which due allowance must be made in initial composition. In spite of this contamination and an imperfect mixing, this method is one of the most used. Results are strongly dependent on the features of the raw material (Paulus and Mercier, 1966) and (Paulus, 1970) (see Section VI). The hardness and shape of the particles change the abrasion rate of the grinder, and the dispersion of particle size affects the segregation by sedimentation when the slurry is decanted or filtered to reduce the water content.

After drying, the powder mixture, either unpressed or pressed at $\sim$0.5 ton cm^{-2}, is calcinated at a temperature depending on the ferrite that is to be formed and the partial pressure of oxygen. Although the spinel phase appears after reacting the oxides at 700°C, the last traces of ZnO, NiO, and Fe_2O_3 finally disappear at about 800, 900, and 1100°C, respectively. These results may vary to a considerable extent with the raw material and the thermochemical heat treatment (Paulus and Mercier, 1966; Paulus, 1970). The solid state reaction takes place by mutual diffusion of cations through the oxygen lattice by way of contact areas between the particles of the different oxides. During this chemical reaction, exchange of oxygen with the atmosphere occurs as long as the porosity remains open. Thus, account must be taken of the sintering accompanying the solid state reaction, which reduces the exchange of oxygen with the atmosphere. The reaction temperature and time should be chosen to obtain a good compromise between homogeneity and keeping an open porosity. This treatment has usually been carried out in air. However, it should take place at the equilibrium partial pressure of oxygen, at least at the end of the treatment (release of the oxide oxygen) (Section II.C.1). In most cases, the calcinated or prefired powder must be milled to reduce the particle size.

Guillaud (1957) has emphasized the importance of pure reactive oxides and found that carefully calcinated oxalates provide excellent raw materials for preparing high-quality Mn–Zn ferrites by high-temperature solid state reaction.

Recent work of Reijnen *et al.* (1970) has shown that mixing sulfates in a ball mill is conducive to a better homogeneity and higher density after sintering than preparations from an oxide mixture.

2. *Steam process*

Buckley and Owen (1953) developed a process which has been successfully used for producing high-quality Ni–Zn ferrite. This process involves the blending of the carbonyl Fe powder and oxides of Ni and Zn in a cone mixer, heating these for 6–10 hr in steam at 500–600°C in steel trays, so that at least 98% of the Fe is converted to magnetite, and ball-milling in a steel mill as an aqueous slurry. The water content is reduced below 20% by vacuum filtering. The final drying occurs in large dough mixers. The temperature is slowly raised to 275°C to convert most of the Fe_3O_4 to

Fe_2O_3. The resulting mixture of oxides may be used directly for sintering after dry grinding, but complex shapes require calcination at 925°C following further milling.

3. *Fused salts*

In the two previous processes, lengthy milling and sometimes several milling–prefiring operations are often necessary in order to produce a completely reacted, homogeneous ferrite. Wickham (1970) accelerates the reaction between the oxides or accomplishes it at relatively low temperature by adding to the reaction mixture a substance which melts and furnishes a liquid phase through which the reacting material can be transported more rapidly than it can through solid phases.

Many substances may be used as a flux for this purpose. We give here examples of the most promising compound mixture. Lithium ferrite $LiFe_5O_8$ has been successfully prepared from Fe_2O_3 and Li_2CO_3 with Li_2SO_4 and Na_2SO_4 as a flux, mixed together in a mortar. The molar ratio of Li_2SO_4 to Na_2SO_4 corresponds to the composition that has the minimum melting point, 600°C. The mixture is ignited for 1 hr at 800°C in a platinium or Al_2O_3 crucible. The Li_2CO_3 dissolves completely in the molten salt and rapidly attacks the iron oxide. The cold product is treated with water to wash away the inert salts and finally the ferrite is collected on a filter. The optimum quantity of flux depends upon the nature of the iron oxide used and is determined by experiment. The amount required must be large enough to permit the uniform distribution of Li_2CO_3 throughout the mass of the iron oxide powder, but an excess will cause a liquid layer at the bottom of the crucible, and the lithium carbonate is then not uniformly distributed with respect to the iron oxide. The solubilities of iron oxide and ferrite in the molten salt are apparently very small and a finely divided ferrite is obtained.

Many other fluxes dissolve both iron oxide and ferrite and cause the growth of small crystals. At 800–1000°C, the cubic ferrites $MgFe_2O_4$, $NiFe_2O_4$, $CoFe_2O_4$, and $ZnFe_2O_4$ can be prepared by the fused salt technique in the presence of molten Li_2SO_4–Na_2SO_4. These ferrites are more stable than $LiFe_5O_8$, which is in turn more stable than $BaFe_{12}O_{19}$, $SrFe_{12}O_{19}$, and $CaFe_2O_4$. As a consequence, it is not possible to prepare these latter alkaline-earth ferrites with fluxes containing lithium. In all cases, the reaction forms $LiFe_5O_8$, as shown by the equation

$$BaCO_3 + Li_2SO_4 + 5Fe_2O_3 \rightarrow 2LiFe_5O_8 + BaSO_4 + CO_2 \quad (3)$$

The compounds $BaFe_{12}O_{19}$, $SrFe_{12}O_{19}$, and $CaFe_2O_4$ can be prepared with the flux $[Na_2SO_4]$–$[K_2SO_4]$ at 940°C.

The cubic spinel ferrites (Ni, Mg, Co, Zn ferrites) can also be prepared with aid of Na_2SO_4–K_2SO_4 mixtures, but higher temperatures are necessary.

B. Thermal Decomposition of Coprecipitated Salt Solid Solutions

1. *Carbonate, oxalate, formate, acetate, acetate–pyridine complex*

Ferrites can be prepared from a starting solution containing cations in the right ratio, mixed with a solution of sodium carbonate. After filtration, the precipitate is washed. The carbonate solid solution decomposes at relatively low temperature. This old technique was used for different ferrites, such as $NiFe_2O_4$.

To avoid sodium ions, Krieger *et al.* (1970) have prepared carbonate powder by coprecipitation of a mixture of chlorides by ammonium carbonate. When heated at a rate of 10°C min^{-1}, the carbonate solid solution decomposes into spinel ferrite (Mn–Mg ferrite). At 900°C, the spinel phase content reaches 90%.

Precipitated salts may also be oxalates, formates, or acetates, since the decomposition products during subsequent calcination are gaseous. However, the oxalate is usually preferred, due to its low solubility and low decomposition temperature. Very fine particles (100–2000 Å) may be obtained. Nickel ferrite powder prepared at 300°C from coprecipitated oxalate with ammonium oxalate has particles smaller than 100 Å (Stuijts, 1969). Kosinskaya and Tret'yakov (1969) have shown that for the same thermal prehistory, ferrite powders prepared by decomposing precipitated oxalate solid solution or acetate–pyridine complex $Ni_3Fe_6(CH_3CO_2)_{17}O_3OH \cdot 12C_6H_5N$ are quite poorly sintered in comparison with the ferrite powders prepared by decomposing schoenite, which sinters intensively (Section IV.D.6).

2. *Hydroxides*

Usually, the procedure for obtaining coprecipitated hydroxide is the following: An aqueous solution of the salt of each cation is prepared. These solutions are mixed together to give the right ratio of cations and the mixture is added to a vigorously stirred solution of a base held at a defined temperature. The *p*H is monitored and more base is added, as necessary, to keep the *p*H at a predetermined value. After addition of the cation solution is completed, stirring is continued and the suspension is boiled with a refluxing condenser for several hours. Upon cooling, the *p*H is readjusted, because during boiling, the *p*H is lowered as hydroxide ions replace salt anions which have been absorbed, occluded, or chemically

TABLE I

CHARACTERIZATION OF PRECIPITATES AND MICROSTRUCTURE

Source salts	Precipitating base	After sintering[a]: Theoretical density (%)	After sintering[a]: Grain size (μm)	Before drying at 500°C: Phase[b]	Before drying at 500°C: Size (μm)	After drying at 500°C: Phase[b]	After drying at 500°C: Size (μm)
$FeCl_2 + Ni(NO_3)_2$	K	76–85	3				
	TMAH	86	3	Sp	0.05	Sp	0.05
	Li	96	30				
$Fe(NO_3)_3$	K	71	5				
$+ Ni(NO_3)_2$	Li	79	DG	Sp	0.02	Sp	0.04
$FeSO_4 + NiSO_4$	K	98	4				
	Na	96	3				
	TMAH	98	DG	Sp	0.02	Sp	0.035
	Li	98	DG				
$Fe_2(SO_4)_3 + NiSO_4$	K	96	7			Fe_2O_3	
	Na	94	17	G	0.005	+ Ni	0.02
	TMAH	96	25			+ Sp	
	Li	96	27				
Mixed oxide	—	94	7	—	—	—	—

[a] After sintering at 1300°C for 4 hr in 1% $O_2 + N_2$ (O'Bryan *et al.*, 1969).
[b] Sp = Spinel; G = Goethite; DG = discontinuous grain growth.

bound to the precipitate. The precipitate is then filtered, washed repeatedly with dilute ammonium hydroxide, and dried in air at 110°C.

O'Bryan *et al.* (1969) used this technique for $Ni_{0.6}Fe_{2.4}O_4$. They show that the anions present in the salts used as sources of ions determine the porosity of the fired microstructure of coprecipitated $Ni_{0.6}Fe_{2.4}O_4$. Table I summarizes the results for different salts and bases.

a. Influence of the salt anions on sinter density. Ferrites prepared from $FeCl_2 + Ni(NO_3)_2$ solution and KOH show a retrograde sintering (Fig. 12). This is the result of high internal gas pressure, which expands the closed pores. The gaseous species responsible here has been identified as ferrous chloride, which appears to be impossible to eliminate. The mixture $Fe(NO_3)_3 + Ni(NO_3)_2$ gives low-density compacts. The mixture $FeSO_4 + NiSO_4$ coprecipitated with KOH gives the highest density with a small grain size (Table I).

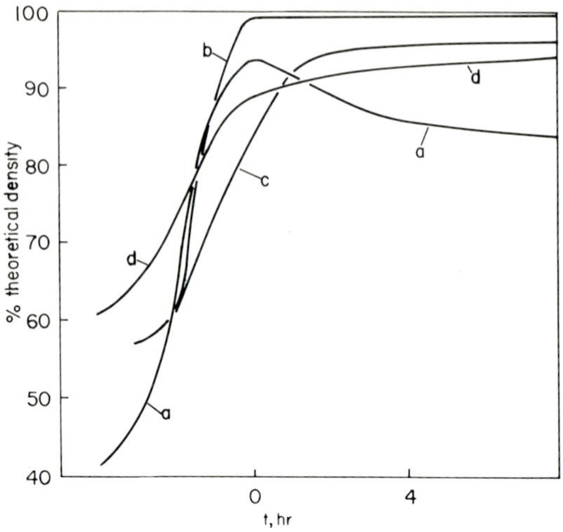

FIG. 12. Densification curves for coprecipitated ferrites prepared from: (a) $FeCl_2$ + $Ni(NO_3)_2$ solution with KOH, atmosphere 1% $O_2 + N_2$, heating rate 100°C hr^{-1}. (b) $FeSO_4$–$NiSO_4$ solution, atmosphere N_2, heating rate 400°C hr^{-1}. (c) $Fe_2(SO_4)_3$–$NiSO_4$ solution, atmosphere N_2, heating rate 400°C hr^{-1}. (d) Mixed oxide ferrite, atmosphere 1% $O_2 + N_2$, heating rate 100°C hr^{-1}. (O'Bryan, 1969.)

b. Influence of the cation valence on sinter density. The effect of iron valence on microstructure is observed when $Fe_2(SO_4)_3$ replaces $FeSO_4$. The grain size increases and the density decreases slightly. Changes with iron valence are related to the physical characteristics of coprecipitated powders before sintering. Prior to sintering, the divalent iron material is single phase (spinel) with an average particle size of 0.035 μm. Material from Fe^{3+} solutions, on the other hand, contains three phases (hematite, nickel oxide, and spinel) with an average particle size of 0.02 μm (Table I). Together, the three phases present contain more oxygen than is required to form a single-phase ferrite. This extra oxygen is released in the temperature range where rapid densification occurs. The released oxygen becomes trapped in the closed pores and its removal can only occur by solid state diffusion. The high oxygen pressure in the closed pores retards their shrinkage and therefore their densification. Another effect of the physical character of the Fe^{3+} coprecipitated material concerns the tendency of very small grains to undergo discontinuous grain growth (Burke, 1959).

c. Influence of the base cation on discontinuous grain growth. The cation present in the base used to precipitate ferrite powders determines whether the fired microstructure is uniformly fine-grained or exhibits discontinuous grain growth. The microstructure variations are caused by trace amounts of alkali metal ions. For ferrite precipitated with tetramethyl ammonium hydroxide which contains no added alkali ions, the discontinuous grain growth is believed to be typical of pure, submicron ferrite during sintering (Burke, 1959). According to the work of O'Bryan *et al.* (1959), Li < 0.001 wt% has no effect on grain growth and therefore also promotes discontinuous grain growth, but on the other hand, $K = 0.007$ wt% and Na $<$ 0.001 wt% retard grain growth and suppress discontinuous growth (Table I).

3. *Emulsion coprecipitation*

To avoid composition heterogeneities and segregation observed in coprecipitates obtained by the previously described conventional methods, Reijnen *et al.* (1970) have employed an emulsion coprecipitation process.

The aqueous solution of sulfates of the ferrite cations and the aqueous solution of the corresponding sodium carbonate are separately emulsified in kerosene. Both emulsions are mixed together and energetically stirred to obtain coprecipitation. This technique has been applied to $Ni_{0.8}$, $Mg_{0.15}$, $Co_{0.15}$, $Mn_{0.04}$, $Fe_{1.98}$, O_4 for comparison with other methods, such as ball-mill mixing of oxides and of sulfates, spray drying, emulsion drying, and conventional coprecipitation, to determine the influence of the method of preparation on ferrite homogeneity. The sharpness of the second maximum on a permeability versus temperature and density curve is chosen as the main criterion. The emulsion coprecipitation process yields a better density after sintering and considerably improves homogeneity in comparison with the ball mixing of oxides and the conventional coprecipitation technique.

C. Thermal Decomposition of Complex Compounds

The hydroxide or basic carbonate deposits of most metals are almost always gelatinous, very difficult to wash throughly, and contain large, variable quantities of adsorbed ions. Moreover, precipitation of the hydroxides or basic carbonates is often more or less complete and thus the desired compositions for solid solutions are no longer maintained.

In addition, the coprecipitates are dehydrated by gradual heating, and to expel the last water molecules, in the case of certain very stable hydroxides, one must heat at such temperatures that a complete crystallization of the phase cannot be avoided. This detracts somewhat from the main advantage of the method.

1. *Complex combination*

To alleviate these difficulties, Pâris (1963) has elaborated a process of preparation of solid metallic oxide solutions which involves essentially the pyrolysis of complex combinations at temperatures between 200 and 500°C in an appropriate inert oxidizing or reducing atmosphere, depending on the material. These complex combinations include cations in the proper ratio in their crystalline structure, together with ionic and molecular species containing the necessary oxygen for the formation of solid solution, the remainder being volatile or decomposable into volatile elements. For instance, the preparation of $MgFe_2O_4$ begins with the complex salt:

$$2Fe(OH)_3 + 6H_2C_2O_4 + MgO \rightarrow H_4Mg[Fe(C_2O_4)_3]_2 + 7H_2O \tag{4}$$

Evaporation of the solution gives complex salt crystals. Decomposition of these crystals between 200 and 500°C in air yields $MgFe_2O_4$. The average grain size is smaller than 1 μm (Pâris *et al.*, 1962).

2. *Organic resin*

To avoid segregation and to obtain decomposition at low temperature, Pechini (1968) prepared manganese ferrite solution, YIG, and hexaferrite by forming organic resin intermediates which are ignited to remove the organic portion and leave chemically combined, mixed oxides which are then pressed and sintered. The organic resins are obtained by dissolving hydrated ferric oxide in an aqueous solution containing citric acid. Then, the right amount of carbonate is added to the solution. The solution is then heated to increase solubility and expel CO_2, and 15% of glycol is added. The solution is further heated in an open container to remove excess solvent until a solid resin intermediate is formed. The organic intermediate is calcinated at 450°C. The mixed oxide is single phase. The lactate may also be used.

D. Thermal Decomposition of Dehydrated Salts

Other approaches have been developed to avoid contamination by the ball-mill material in conventional processes, or by the coprecipitating ions in the preparation of coprecipitated salt solid solutions. These other approaches include evaporation of the salt liquid solution, sublimation of the quickly frozen solution, and chemical dehydration.

1. *Spray drying*

Ceramic powders of ferrite may be prepared from mixed sulfate solutions. The liquid is divided into droplets of about 10–20 μm by a centrifugal disk atomizer in a chamber in which the droplets come in direct contact with a stream of hot gas. The drying is very rapid and suppression of the segregation of the different components is expected. After evaporation, solid particles form in the shape of hollow spheres with approximately the same diameters as those of the original droplets (De Laue, 1969). The sulfate solid solution is then decomposed into ferrite between 800 and 1000°C. Particles are about 0.2 μm, but they are sintered to an agglomerate. They can easily be broken up into particles smaller than 1 μm by a treatment for a few minutes in a turbine mixer.

A wide range of compounds has been made from spray-dried solutions of mixed sulfates (Stuijts, 1969). They show great purity and homogeneity (Reijnen *et al.*, 1970). However, due to the voluminous network structure and more or less elongated shape of the particles, the press densities are rather low, 45% of the theoretical densities even after hydrostatic pressing at 1000 kg cm^{-2}. Nevertheless, nickel ferrites could be hot-pressed to practically the theoretical density.

2. *Spray roasting*

Spray roasting provides a single-step process for the production of ferrite powders from the salt solution. Akashi *et al.* (1965) prepared various kinds of powders, such as $(Mn_xZn_{1-x}O)_y(Fe_2O_3)_{1-y}$, $ZnFe_2O_x$, $MnFe_2O_x$, and $NiFe_2O_x$, by the spray roasting technique. The desired metallic components are dissolved in a small excess of HNO_3 aqueous solution. By use of a centrifugal atomizer, the solution is sprayed into a roasting chamber similar to that used for spray drying, but hot air is introduced into the chamber at 350–400°C instead of 100–150°C. Each spherical particle has a diameter ranging from 5 μm to 50 μm, and is actually an aggregate of very fine particles, the mean size being estimated to be ~100 Å. The powder is not in the final spinel form, though the crystal structure is still unknown. Treating the powder above 400°C yields spinel. The spray-roasted powder exhibits a remarkable reactivity with high purity. The various steps from ball-mill mixing to presintering become unnecessary. Ruthner *et al.* (1970) used this process for production of very reactive Fe_2O_3 from chloride solution. This Fe_2O_3 was used for the preparation of ferrites by the conventional solid state reaction method, but at lower temperature.

3. *Flame-spraying*

Some years ago, Wenckus *et al.* (1956) developed the atomizing burner technique. A solution of the nitrates in alcohol, with the right ratio of ions, is sprayed in a burner to decompose the salt. This process was used by Malinofsky *et al.* (1961) to prepare nickel ferrite densified by hot pressing to nearly 100% of the theoretical density, with a very small grain size, 0.08 μm. Zneimer *et al.* (1964) prepared nickel–aluminum–gallium spinel ferrite by putting the nitrate salts of the desired metallic ions in solution with methanol, atomizing the solution with oxygen, and burning the atomized spray. This process yields a powder with an average particle size less than 0.1 μm.

4. *Emulsion drying*

A new technique was developed by Reijnen *et al.* (1970) to avoid demixing during evaporation. The solution of sulfate in water, with the desired ion ratio, is mixed and stirred vigorously in kerosene. The sulfate emulsion in kerosene, is made up of small particles (5–10 μm). The emulsion is kept at 0°C under vacuum. On subsequent heating, the mixed sulfate particles precipitate. The mixed sulfate is then converted to the spinel phase by heating. The sintered ferrite shows good homogeneity and high density.

5. *Cryogenic process*

Schnettler *et al.* (1968) developed a new cryochemical method for preparing ceramic materials and recently applied this technique to a NiCrAlMn ferrite (Schnettler and Johnson, 1970).

An aqueous salt solution is first prepared with the exact requirement of ions. This solution is then subjected to quick freezing by spraying it into a bath of hexane chilled with dry ice and acetone. This solidifies the solution as fine beads of salts and ice. The frozen material is transferred to a freeze-dryer and the pressure dropped below that of the cryohydric point, and then heat is supplied to sublime off the ice. The resulting product of anhydrous salt is homogeneous on a molecular scale and free of contaminants from the starting material. The salt is converted to the oxide by decomposition at an elevated temperature. This material was used to prepare hot-pressed ferrites with small grains and theoretical density.

6. *Salting out in acetone*

Kosinskaya and Tret'yakov (1969) have made a comparison of the products of the thermal decomposition of oxalates, hydroxides, and acetate–pyridine with isomorphous mixtures of salts of the schoenite type.

The schoenite-type solid solutions are prepared by the salting out of double sulfates from aqueous solutions with acetone. Decomposition of the schoenite occurs between 200 and 900°C (Tret'yakov, 1965; Tret'yakov *et al.*, 1969a). This extended study of $MnFe_2O_4$, $MgFe_2O_4$, $NiFe_2O_4$, and $ZnFe_2O_4$ shows that for the same thermal prehistory, ferrite powders prepared by decomposing schoenite are the best sintered, while hydroxide materials are the most poorly sintered, and the decomposition products of oxalate and of acetate–pyridine complex are quite poorly sintered.

An analysis of the microstructure shows that the most perfect microstructure is possessed by samples produced by sintering the decomposition products of a solid solution of schoenite. These ferrites consist of crystals of uniform size; they contain no flaws, and the pores are uniformely distributed throughout the volume of the sample.

The remarkable results obtained from schoenite material are attributed to the sharp increase of their surface catalytic activity (i.e., the activity

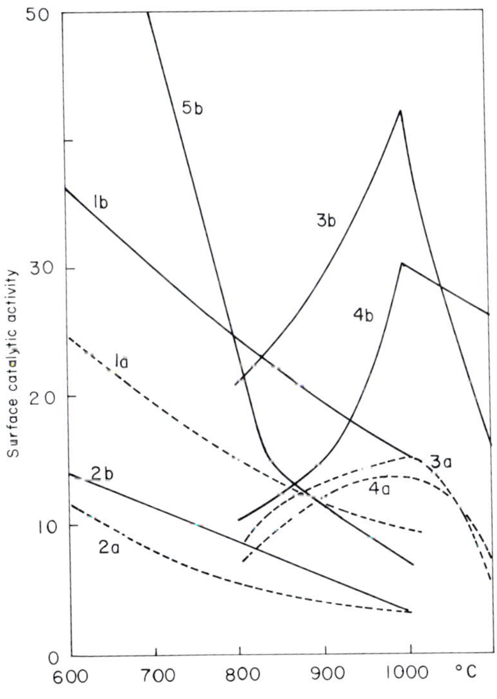

FIG. 13. "Surface" catalytic activity of ferrite powder as a function of chemical prehistory: (a) manganese ferrite (dashed line); (b) nickel ferrite (solid line); (1) hydroxide, (2) oxalate, (3) schoenite salting out, (4) schoenite equilibrium crystallization, (5) acetate–pyridine complex (Tret'yakov *et al.*, 1969).

of a unit surface) at 1000°C (Fig. 13) (Tret'yakov *et al.*, 1969b) and its uniform distribution among the particles. On the other hand, other decomposition products from oxalate, hydroxide, and acetate–pyridine complex show a decrease of surface catalytic activity versus temperature (Fig. 13). Furthermore, their surface catalytic activity is nonuniformly distributed in the individual particles. This promotes aggregation of the most active particles. Naturally, the sintering of such dispersed material, after pressing, leads to the formation of samples with crystallites differing sharply in dimensions, and with greater or lower porosity, due to the fact that these materials lose their activity before the intensive process of sintering begins..

E. Hydrothermal Oxidation

Some attempts have been made to prepare the ferrite directly from the solution, to avoid the thermal decomposition of the salts into spinel ferrite. The wet methods permit the recrystallization of the powder and the monitoring of the grain size and shape. The grains are mostly monocrystalline.

1. *Oxidation of coprecipitated hydroxide solid solutions*

a. Thermal process. Stuijts (1969) reported the conversion of a coprecipitated mixture of hydroxides into oxides or oxygen compounds by heating in water at high temperature. Powders of nickel ferrite consisting of cubes with a particle size of 250 Å have been obtained, as well as hexagonal plates of $BaFe_{12}O_{19}$ (basal plane size 1000–10,000 Å) formed by reaction of an iron oxide with barium ion in an alkaline solution at about 300°C.

b. Bubbling process. Takada and Kiyama (1970) formed the ferrite by bubbling the oxidizing gas into the suspension of hydroxides obtained by coprecipitation from a solution containing only ferrous ions and other divalent metal ions. The suspension is heated at temperatures between 60 and 90°C and the oxidizing gas, such as air, is uniformly bubbled at a constant rate until the precipitate of hydroxides disappears completely.

In solutions with *p*H value above 10, the ferrite precipitate is directly formed from the solution containing the primary precipitates (ferrous precipitates) by oxidation. The ferrite particles thus formed are cubic and their size depends strongly on the reaction conditions, such as the content of primary precipitate, the excess of alkali in the solution, the reaction temperature, and the oxidation rate.

With a *p*H between 6 and 9, the ferrite precipitate is formed by oxidation from a suspension containing the intermediate precipitate, with 40% or less of trivalent metal ion making up the total metal ions. The particles are isotropic in shape and the sizes ($\sim$0.1 μm) do not depend strongly on the parameters mentioned for pH > 10.

$BaO–6Fe_2O_3$, $2BaO–9Fe_2O_3$ and $BaO–2Fe_2O_3$ were also formed with pH > 11 and at a temperature between 150 and 300°C.

2. *Oxidation of electrolytic coprecipitates.*

Glaister *et al.* (1964) used a method due to Beer and Planer (1958) which is based on the bubbling of air through an electrolyte solution containing the metallic ions in the required concentrations.

The metallic ions are produced by anodic dissolution; the concentration ratio of each ion is adjusted by the currents passing through the anodes of the metals needed for the particular ferrite. This electrolyte is a solution such as sodium sulfate or chloride. The particle size varies with the electrolyte temperature and the magnitude of the currents.

High-purity ferrites could be obtained by using anodes of metals purified by the zone-refining technique, the chief difficulty being the removal of the last traces of electrolyte. Thermogravimetric analysis indicates a weight loss of 17% in the temperature range 200–400°C, increasing to 18% by 700°C, with a further loss between 750 and 800°C. The loss of 18.6% actually corresponds to three molecules of water per ferrite molecule.

3. *Direct wet process*

Sato *et al.* (1970) have succeeded in producing Fe_3O_4, $MnFe_2O_4$, $CoFe_2O_4$, and $Mn_{0.5}Co_{0.5}Fe_2O_4$ directly by coprecipitation in the higher alkaline aqueous solution in few minutes at temperatures below 100°C.

A solution of divalent ions Mn, Co, Ni, etc. and ferric ions is mixed with a 6N NaOH aqueous solution to adjust to the right pH value, stirred, and kept at a temperature of less than 100°C. Various ferrites are formed under the conditions pH > 11.5, 100°C for 1 hr. $CoFe_2O_4$ is magnetic at pH = 10.5, but $NiFe_2O_4$ is nonmagnetic and amorphous under the formation conditions mentioned. The main parameters for the preparation of ferrite by this technique are the pH value of the solution, the kind of alkali, the reaction temperature and time, and the type of metallic ion and its concentration. These parameters influence the lattice constants, the particle size and shape, and the magnetic properties.

Schule and Livingston (1961) reported previously on the direct formation of $NiFe_2O_4$ and $CoFe_2O_4$ from solution, but the process took from 10 to 100 hr.

V. Forming of Compacts

Before sintering, the ferrite powder must be compacted, in order to shape the end product and to densify the loose particulate matter into a coherent body and increase the strength of the sample. Many techniques have been used: *Slip casting*—a highly concentrated suspension of ferrites or oxides is poured into a mold of plaster of Paris, and water is drained from the suspension by capillarity. *Pastic forming*—this technique is used for extrusion, rolling, injection-moulding, and tape punching (memory cores) (Koppens, 1970). A ferrite paste must be prepared with the ferrite powder, water, plasticizers, and wetting agents like polyvinylalcohol. *Die-pressing*—this technique is the most commonly applied. The process is rapid. Only very small quantities of binder and lubricant have to be added and the firing shrinkage is lower than with the preceding methods. The main difficulty is in obtaining a homogeneous compact. First, the powder must fill the die uniformly. This requirement is achieved by spray-drying the ferrite slurry, which produces a narrow particle-size distribution of free-flowing spheres (50–250 μm). Second, the pressure (1–10 ton cm^{-2}) is not uniformly distributed in the compact, due to a frictional resistance on the die wall (Train, 1957). Some improvements in this respect are obtained by *double-effect die-pressing*. *Isostatic pressing*—this gives the most homogeneous compact. The powder is encapsulated in a rubber envelope which is closed up under vacuum and pressurized in a fluid. However, the shape of the sample is not easy to define.

Whatever techniques are used, care must be taken upon: segregation during drying (decantation, filtering, or spray-drying) and granulation, which modify to a great extent the porosity distribution (Stuijts and Oudemans 1965), and elimination of binders and lubricants at low temperatures (500–700°C).

VI. Sintering

"Sintering" has various meanings, depending on the usage. Here, we consider the term in its restricted meaning, i.e., the coalescence of particulate matter into massive form. However, we shall see that sintering is not always associated with apparent shrinkage and densification of the compact, especially in the case of multicomponent sintering or reactive sintering. For instance, the "sintering" of ferrites direct from a mixture of starting oxides involves a spinel formation at the common interface between Fe_2O_3 and MeO particles (Me = Ni, Zn, Mn, . . .). This reaction generally entails a volume increase, $\Delta V > 0$. In other parts of the material, due to imperfect mixing, Fe_2O_3 or MeO particles are in contact and may sinter without ferrite formation, $\Delta V < 0$. These two processes may occur at the same time and temperature, or separately.

The sintering process, which can not be dissociated from grain growth and oxygen exchange with the atmosphere, is a function of the thermochemical prehistory of the raw material. Each sintering procedure is a specific case. We have to take into account the mean grain size, the grain-size dispersion, the degree of homogeneity of the presintered ferrite, the degree of mixing for oxide mixtures, specific surface reactivity of the powder, thermal and mechanical defects, deviation from stoichiometry, homogeneity of compactness, existence of a liquid phase at high temperature, etc.

We do not list here the various sintering procedures, which are mainly industrial procedures that can be modified for specific raw materials to obtain well-defined properties. We only emphasize the influence of the main parameters on sintering. Knowledge of them permits us to choose the most suitable thermochemical treatment for sintering, at least from a theoretical point of view, since there are always interactions among the various parameters.

The sintering processes may be divided into the following classes: reactive sintering, homogeneous sintering, and hot-pressing.

A. Reactive Sintering

Reactive sintering or multicomponent sintering is simple from the experimental point of view, but the most complicated one concerning the mechanism. Although it is not commonly used in industry, it is of interest to study it, since the results may be extrapolated to homogeneous sintering, which, as a matter of fact, is really nonhomogeneous.

Figure 14 shows a schematic representation of density versus time and temperature for different kinds of atmospheres and various types of gathering for raw material particles (oxides) (Paulus, 1970). In the cases E and D, the interstitial particles are very small and of the same kind, and their sintering is very rapid even at low temperature, the sintering rate being inversely proportional to the diameter of the particle. For a volume diffusion process (Kingery and Berg, 1955; Paulus and Lacour, 1966), for the early stage of sintering (neck formation),

$$\Delta l/l = -(K'D/\bar{d}^3RT)^{2/5}t^{2/5} \tag{5}$$

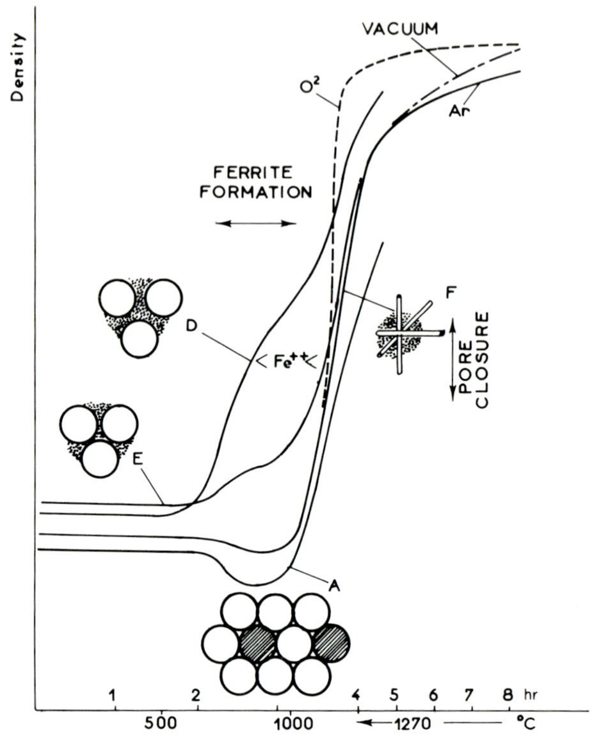

FIG. 14. Schematic representation of densification curves of manganese zinc ferrites for various particle arrangements, shapes, and sizes and different atmospheres. For the same constant atmosphere, the Fe^{2+} content increases with the temperature of pore closure (Paulus, 1970).

and for the last stage of sintering (reduction of pore diameter),

$$P = P_0 - (K''D/\bar{d}^3RT)t \tag{6}$$

where l is the sample length, $\bar{d}$ is the mean diameter of the particles or grains, D is the diffusion coefficient, t is the sintering time, P is the porosity, and P_0 is the porosity for $t = 0$. Turk (1970) confirms, from densification kinetics, that the mechanism for final-stage densification of NiZn ferrite is predominantly lattice diffusion.

The sintering of the small particles occurs before the ferrite formation. This formation begins at higher temperature, between the large and the partly sintered small particles. This last reaction is accompanied by a local volume increase, and therefore the densification rate slows down. For the two models E and D, the large particles are embedded in a tight matrix which hinders the direct gas exchange with the atmosphere during the ferrite formation.

On the other hand, with a particle gathering of type A, the gas exchange with the atmosphere is easier, since the sintering occurs only after the ferrite formation, provided that the particles have the same diameter and are perfectly mixed together. Evidence for this was given by Paulus and Mercier (1966). Fe_2O_3 and Mn_2O_3 with the same particle size, $\sim 0.04\ \mu m$, were mixed together in order to prepare the ferrite. At 700°C, the volume shrinkage was $\Delta V = -3\%$. However, for the same thermal treatment, shrinkage was $\Delta V = -40\%$ for Fe_2O_3 and -10% for Mn_2O_3 alone. The decrease of the shrinkage for the mixed powders is due to the fact that ferrite formation occurs only at higher temperature. Afterward, the sample behaves almost as in the case for homogeneous sintering.

Needlelike particles (model F) hinder the shrinkage of the whole sample as long as they are not destroyed by ferrite formation.

We have shown (Section III) that the ferrite formation gives rise to a Kirkendall effect, with coalescence of the pores in one part of the sample. As a consequence, the results change if we interchange the nature of the oxides. Depending on the oxide in interstitial position and on the kind of ferrite, either there is pore formation in the center of the large particles or the porosity is rejected in the very-fine-grain matrix. The rate of elimination of porosity and to some extent the release or absorption of oxygen during the ferrite formation are affected (Fig. 10).

Until now, we have considered homogeneous mixtures. If we are not dealing with a homogeneous mixture, a Kirkendall effect appears during homogenization, which gives rise to a clustering of pores which is difficult to eliminate. Figure 15 shows an example of pore segregation by the Kirkendall effect in an iron-rich heterogeneity. This effect is observed on a smaller scale in comparison with the porosity of ferrite prepared from imperfectly mixed oxides (Fig. 16) and from a very homogeneous mixture (Fig. 17).

All these observations explain the great deal of work (Section IV) being performed in order to obtain a very homogeneous ferrite as starting powder for sintering.

In addition, we have to consider the time required to homogenize a sample. Figure 18 shows a transmission electron micrograph of a MnFeCo ferrite after heat treatment at 1150°C for 4 hr. Although this micrograph is for a single-crystal grain and a single phase, the composition varies in it, and we can see the shape and size of the starting oxides. The crystal is not yet completely homogenized.

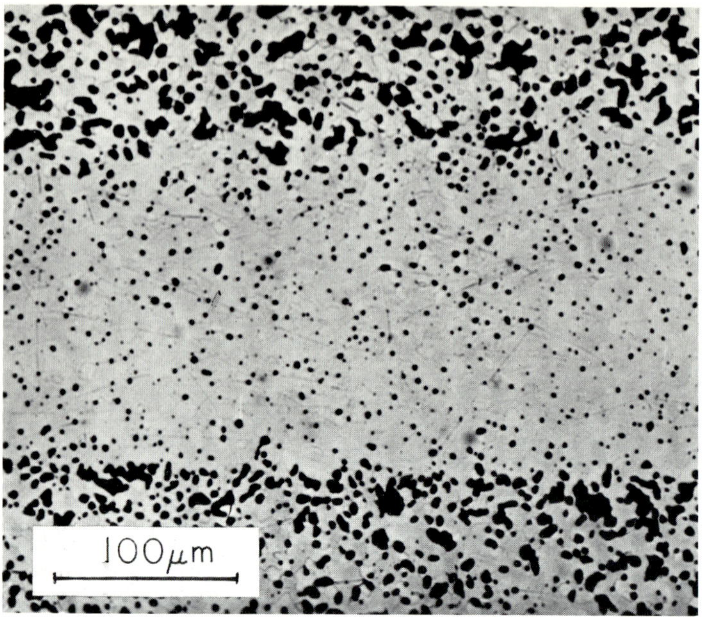

FIG. 15. Micrograph of a manganese–zinc ferrite with a large iron-rich heterogeneity (Paulus, 1970).

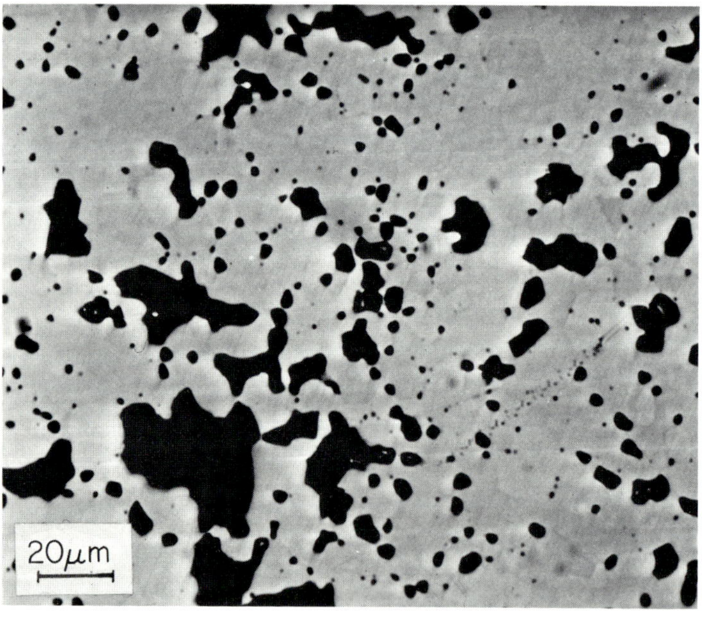

FIG. 16. Micrograph of a manganese–zinc ferrite prepared from an insufficiently homogenized oxide mixture (Paulus, 1970).

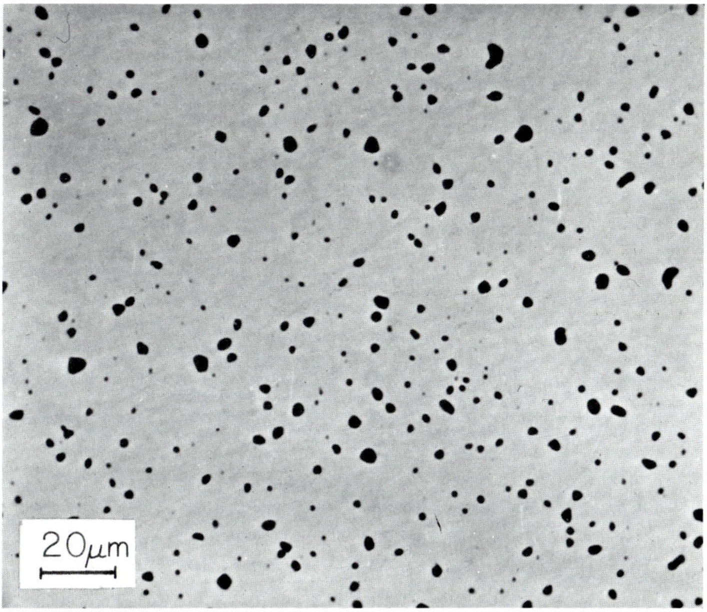

FIG. 17. Micrograph of a manganese–zinc ferrite prepared from a very homogeneous oxide mixture (Paulus, 1970).

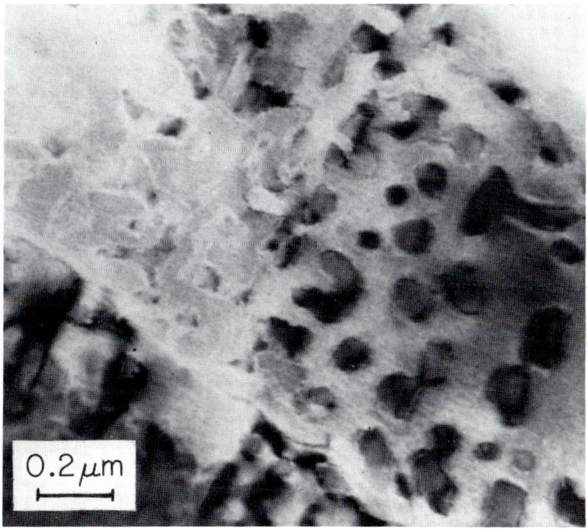

FIG. 18. Transmission electron micrograph of a manganese ferrite sintered 4 hr at 1150°C. The sample is thinned by ionic bombardment under grazing incidence (Paulus and Reverchon, 1961).

The sintering parameters that act almost in the same way for reactive and homogeneous sintering are dealt with in the next section on homogeneous sintering.

B. Homogeneous Sintering

We consider here the sintering of particles of ferrites formed with homogeneous composition; therefore, the main sintering mechanism is a volume-diffusion process in which the vacancies sink out at the grain boundaries. Hence, Eqs. (5) and (6) are valid, and it is clear from these equations that it is necessary to use small particles to increase the sintering rate. This is confirmed by all the work on the sintering of ferrites. Among the most recent work, Akashi *et al.* (1970) show that manganese–zinc ferrite with a very small particle size, 0.067 μm, can be sintered to a density of 5 g cm^{-3} at 1000°C.

However, it is not sufficient to start with very fine ferrite particles. We have to take into account the crystal growth that is activated by temperature in the sintering process and interferes with the latter. The sintering rate decreases with crystal growth since the whole boundary area decreases (vacancies sink). In the case of discontinuous grain growth, the effect is more drastic, for the pores are trapped in the core of the grain far from a grain boundary. The shrinkage stops at a limiting value which depends on temperature and on the time for the appearance of discontinuous grain growth (Paulus, 1962).† Stuijts and Kooy (1965) show that, due to the temperature gradient within the specimen, discontinuous grain growth appears at the periphery of the specimen. A rigid shell is formed, hindering further densification. The parameters affecting the rate of crystal growth are given in Section VII.

The other main parameter of Eqs. (5) and (6) is the diffusion coefficient. It increases with the number of defects. In order to increase the sintering rate, defects can be introduced by an imperfect formation of the ferrite. In this respect, it may be interesting to prepare a single-phase ferrite with a crystalline structure different from that of the final sintered ferrite, but with the same composition (Section IV.D.2) (Akashi *et al.*, 1965). Defects may also be introduced by grinding of the ferrite (Paulus and Lacour, 1968) or by irradiation.

The diffusion coefficient changes with the number of oxygen vacancies, i.e., with the composition both of cations and anions. Paulus (1962)‡ shows the increase of sintering rate for manganese–zinc ferrite when the partial pressure of oxygen is lowered below the equilibrium partial pressure for the stoichiometric ferrite, $\gamma = 0$ (Fig. 19). Reijnen (1969) points out the

† Paulus (1962a,b). ‡ Paulus (1962a,c).

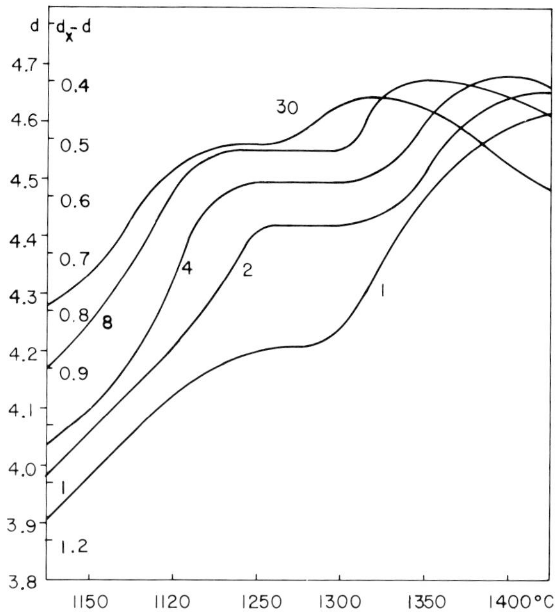

FIG. 19. Densification curves of manganese–zinc ferrite versus temperature for different sintering times (in hours). Atmosphere 1% O_2 + N_2 (Paulus, 1962a).

effect of addition of Me^+ or Me^{3+} on the number of anion vacancies and therefore on the sintering of ionic compounds. Addition of Me^+ increases their number and the sintering rate, while Me^{3+} ions give the opposite result. For nickel ferrites sintered in oxygen, it is observed that samples with excess Fe_2O_3 have lower density than samples with excess NiO, but the difference is small and the changeover is very smooth. When the samples are sintered in nitrogen, there is no change in density at all. The dissolution of Fe_2O_3 in the spinel lattice is compensated mainly by formation of Fe^{2+} ions at low oxygen partial pressure (Fig. 20).

The presence of inert gas in the pores lowers and finally stops the sintering process. When the volume of the pores decreases, the pressure of the gas increases and counteracts the shrinkage. Furthermore, such gases as N_2 and Ar diffuse only short distances from small pores (high equilibrium gas pressure) to larger pores (lower pressure). This gives rise to a coalescence of the pores (Paulus and Mercier, 1966). These large pores expand under the increasing internal gas pressure and we observe a retrograde sintering. On the other hand, if there is vacuum or pure oxygen with the right pressure in the pores, the shrinkage may proceed to higher density

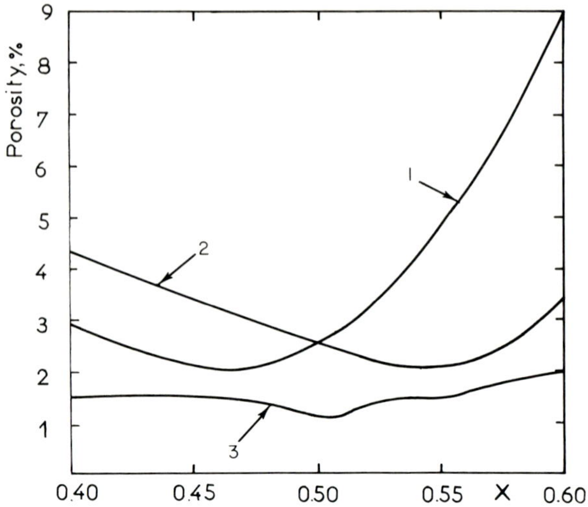

FIG. 20. Porosity of nickel ferrites sintered 24 hr at 1350°C in (1) oxygen, (2) nitrogen + 1% oxygen, (3) nitrogen, as a function of the composition parameter x: $(1 - x)NiO + xFe_2O_3$ (Reijnen, 1968).

since the internal oxygen pressure may be lowered by a slight oxidation of the ferrite. Figure 14 shows a comparison of the last stage of sintering in argon, vacuum, and oxygen. The rapid density increase for pure oxygen is due to the presence of small inclusions of Fe_2O_3 which remain at higher temperature than for low oxygen pressure, and stop the grain growth. However, when the inclusions of Fe_2O_3 dissolve at high temperature, a discontinuous grain growth appears and the shrinkage stops.

Paulus (1962a) observed that there is a correlation between density and final grain size. This suggests an increase of vacancy diffusion by grain boundary migration, which may be explained by a sliding of adjacent grains and by a change of the radius of curvature of the drained pores.

C. Hot-Pressing

With normal reactive or homogeneous sintering, it is not always possible to obtain microstructures that result in the required physical properties. The ability to apply pressure to a material, in addition to the firing parameters, increases the possibility of obtaining a wider range of microstructures and physical properties. Hot-pressing is mainly used for obtaining a small grain size with low porosity in cases that require a composition difficult to sinter by the normal processes, e.g., materials with an excess of cation vacancies. It can also be used to prepare ferrites that tend to volatize at elevated temperatures.

Paladino *et al.* (1966) prepared various ferrites, such as $NiFeO_4$, $Ni_{1-x}Mn_xFe_2O_4$, and $Ni_{1-x}Co_xFe_2O_4$, by hot-pressing in alumina dies at temperatures ranging from 1150 to 1250°C and at pressures between 6400 and 7200 psi. To prevent the material from sticking to the die, and to ensure the safe removal of the sample, the powder was pressed into a cylinder and packed in 325 mesh ZrO_2 sand. Grains with sizes ranging from 2 to 5 μm with a porosity between 2 and 0.1% was obtained.

West and Blankenship (1967) obtained lithium and lithium–zinc ferrite by the same technique. The hot-pressed ferrite is composed of equiaxed grains with an average diameter of 3–5 μm and a porosity of 0.6%.

In the same manner, Nicolas *et al.* (1969) hot-pressed manganese–magnesium ferrites, YIG, and gadolinium–aluminum-substituted yttrium garnets. The porosity was not over a few thousandths and the grain size varied from one to a few microns.

Sawaoka and Saito (1970) prepared manganese–zinc ferrites with a grain size 1–2 μm and 2% porosity in a very short time (100 sec) using a uniaxial pressure of 100–300 Kg cm^{-2} and alternating current of 2 kHz through the sample. The die is made of pyrophyllite. The same authors obtained ferrites with a density over 99.5% of the theoretical density by using isostatic hot-pressing. The sample was compressed by means of fluid pressure through a rubber wall and a thermal insulating powder.

Schnettler and Johnson (1970) reached the theoretical density with a grain size under 2 μm for a Ni–Cr–Al–Mn ferrite by isostatic hot-pressing. The sample was fired for 20 min at 960°C with an applied pressure of 1.7 kbar. The ferrite powder was first encapsulated in a thin, gas-tight platinum envelope and pressurized with helium in a cold-wall steel autoclave.

Oudemans (1968) developed a continuous hot-pressing technique. It permits the use of pressures over 1000 Kg cm^{-2} at temperatures above 1000°C with a brittle material die like alumina, by means of six pressure cylinders which counteract the internal die pressure. The wear of the die is avoided by continuous moving of the ferrite rod through the die. The die is fed periodically with the powder. Li–Zn ferrites, Ni–Zn ferrites, Mg–Cu–Zn ferrites, and Mg–Mn ferrites have been hot-pressed to a porosity ranging from 0.2 to 0.9% with a grain size of 0.5 μm. The rate of the process is a few centimeters per hour, but can be much greater.

Kimura (1963) has shown that it is possible to obtain Zn ferrite by explosive compression for an equimolecular mixture of ZnO and Fe_2O_3.

VII. Crystal Growth

The grain size and its distribution is a very important parameter for the magnetic and electrical properties of ferrites. Magnetization and loss mechanisms depend strongly upon crystalline microstructure (Guillaud and Paulus, 1956; Guillaud *et al.*, 1956). Therefore, intensive studies on crystal size and porosity have been made in connection with these properties. Nevertheless, very few studies appear in the literature on the mechanism of crystal growth in ferrites. The main reason for this is the number of parameters which hinder or promote crystal growth.

A. Normal Crystal Growth

From studies of the first stage of crystal growth of Mn–Zn ferrites and Ni–Zn ferrites, Paulus (1962)† has shown that all ferrites with the spinel structure should have approximately the same value of activation energy for grain boundary migration, irrespective of their composition. The crystal growth rate within the polycrystalline aggregates of ferrites is represented by the following relations:

$$\bar{d}_{3(t_1)}{}^2 - \bar{d}_{3(t_0)}{}^2 = (22\nu\sigma_3\phi^4\lambda/KT \times 10^{0.6\sigma_3})(t_1 - t_0)e^{-Q/RT} \tag{7}$$

$$\bar{d}_{3(t_1)}{}^2 - \bar{d}_{3(t_0)}{}^2 = (K_0/T)(t_1 - t_0)e^{-Q/RT} \tag{8}$$

where Q is the activation energy for the ionic jumps; T is the temperature of sintering; t is the time of sintering; $\bar{d}_3$ is the median of the crystal diameters at the indicated times (t_1 and t_0); σ_3 is the standard deviation of the crystal diameters; ϕ is the ionic diameter (oxygen); ν is the frequency of ionic vibration; λ is the specific grain boundary energy; $K_0/T = 6.5$ cm sec^{-1} when $T > T_{tr}$, and $K_0/T = 15.0$ cm sec^{-1} when $T < T_{tr}$; Q is taken as 83,000 cal mole^{-1}. This relation is valid as long as the migration of grain boundaries is not hindered by inclusions or pores. Furthermore, the mean free energy λ of the boundaries (as calculated from the kinetics of crystal growth) doubles when the reducing atmosphere is replaced by an oxidizing one. The transition temperature T_{tr} between the two values of λ varies with oxygen partial pressure. Figure 21 shows that the rate of crystal growth doubles when the reducing atmosphere is replaced by an oxidizing one.

The crystal growth rate may be modified by another parameter, the standard deviation σ_3 of the grain size of the raw material [Eq. (7)]. Paulus and Lacour (1968) demonstrate the influence of this factor on crystal growth rate and discontinuous grain growth.

† Paulus (1962a,c).

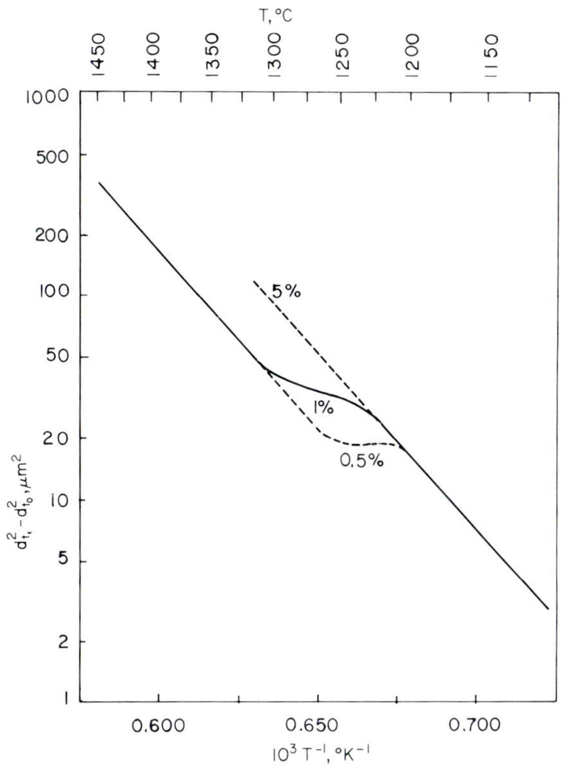

FIG. 21. Rate of grain growth as a function of temperature for different partial pressures of oxygen. d_{t_1} is the median grain size at time t_1; $t_1 - t_0 = 4$ hr (Paulus, 1962a, c).

B. Abnormal Crystal Growth

Pores hinder the migration of the grain boundaries. The limiting value of the average crystal diameter is a function of the porosity and pore size. When the average crystal size tends toward the calculated limit, the standard deviation of the distribution decreases. The same effect is found with inclusions like SiO_2, Cr_2O_3, or Fe_2O_3 (maintained in an oxidizing atmosphere), but they give rise to a discontinuous grain growth due to their dissolution at higher temperature (Paulus, 1962b; Paulus and Mercier, 1966). The effect of other cations has been presented in Section IV.B.2c (O'Bryan *et al.*, 1969).

Discontinuous grain growth is also promoted by the formation of a thin liquid phase at the grain boundary. This liquid phase can be obtained with some additions, such as Ba in $MnFe_2O_4$ (Paulus and Hamelin, 1968) or $2SiO_2$ and CaO in $MnZnFe_2O_4$ (Bando *et al.*, 1970).

Change in the oxygen partial pressure is sometimes sufficient to induce discontinuous grain growth in Ni–Mg ferrite (Michalowsky and Vogler, 1963).

C. Texture

Ferrites with good texture can be made by first orienting the ferrite particles of a slurry in a magnetic field. This is followed by pressing the slurry while simultaneously removing the liquid. The compact is then fired normally. High-anisotropy ferrites such as barium hexaferrite (Stuijts *et al.*, 1955), cobalt ferrite, and even nickel ferrite (Rodrigue and Crouch, 1966) can be obtained in the grain-oriented state.

Cubic ferrites can be prepared by topotactic reaction. A powder mixture of nonmagnetic metal oxides and a ferromagnetic hexagonal ferrite are oriented in a magnetic field and then normally fired. The following reaction take place at high temperature:

$$PbFe_{12}O_{19} + 6NiO \rightarrow 6NiFe_2O_4 + PbO(liq.) \tag{9}$$

After heating above 1100°C, the liquid PbO can be evaporated completely. Broese van Groenou *et al.* (1969) obtained a nickel ferrite with a strong [111] texture (±4°). $NiFe_2O_4$ proved to be a most successful compound, because substitution of Mn, Mg, Co, or Zn for Ni led to a decreased orientation factor.

Oriented hexagonal ferrites can also be obtained by uniaxial pressing and then sintering at high temperature (Stablein and Willbrand, 1966; Paulus and Lacour, 1968).

REFERENCES

Akashi, T., Tsuji, T., and Onoda, Y. (1965). *Conf. Sintering Related Phenomena 2nd, June 21–23, Univ. of Notre Dame.*

Akashi, T., Sugano, I., Okuda, T., and Tsuji, T. (1970). *In* "Ferrites" (*Proc. Int. Conf. Ferrites, July 6–10, Kyoto, Japan*) (Y. Hoshino, S. Iida, and M. Sugimoto, eds.), pp. 96–98. Univ. Park Press, Baltimore, Maryland. (1971).

Amemiya, M. (1970). *In* "Ferrites" (*Proc. Int. Conf. Ferrites, July 6–10, Kyoto, Japan*) (Y. Hoshino, S. Iida, and M. Sugimoto, eds.), pp. 154–157. Univ. Park Press, Baltimore, Maryland. (1971).

Bando, Y., Ikeda, Y., Akashi, T., and Takada, T. (1970). Inst. for Chem. Res., Kyoto Univ., Kyoto, Japan.

Baukal, W. (1969). *Chem. Ing. Tech.* **41**, 791–798.

Beer, H. B., and Planer, G. V. (1958). *Brit. Commun. Electron.* **5**, 939.

Blank, J. M. (1961). *J. Appl. Phys.* **S 32,** 378S.

Broese van Groenou, A., Beljers, H. G., and Driessens, F. C. M. (1969). *J. Appl. Phys.* **40,** 1424–1426.

Buckley, S. E., and Owen, D. H. (1953). Standard Telephones and Cables. Brit. Patent 780 975.

Burke, J. E. (1959). *In* "Kinetics of High Temperature Processes" (W. D. Kingery, ed.), pp. 109–116. MIT Press, Cambridge, Massachusetts.

Carter, R. E. (1961). *J. Amer. Ceram. Soc.* **44,** 116.

Chol, G., Auradon, J. P., and Damay, F. (1968). *Int. Congr. X-Ray, Opt. Microanal., 5th Tubingen, Germany. Sept. 9–14* (G. Mollenstedt and K. H. Gaukler, eds.). Springer-Verlag, New York.

Chol, G., and Aubaille, J. P. (1970). *In* "Ferrites" (*Proc. Int. Conf. Ferrites, July 6–10, Kyoto, Japan*) (Y. Hoshino, S. Iida, and M. Sugimoto, eds.), pp. 243–248. Univ. Park Press, Baltimore, Maryland. (1971).

DeLau, J. G. M. (1969). *Ann. Meeting Amer. Ceram. Soc., May 7, Washington, D.C.* (1970). *Amer. Ceram. Soc. Bull.* **49,** 572–574.

Elwell, D., Parker, R., and Tinsley, C. J. (1967). *Czech. J. Phys.* **B–17,** 382.

Glaister, R. M., Allen, N. A., and Hellicar, N. J. (1964). *Meeting Basic Sci. Sect. Oct. 12–14 Brit. Ceram. Soc., Brighton, England.*

Guillaud, Ch. (1957). *Proc. Inst. Elect. Eng. Suppl.* **104B,** 5, 165.

Guillaud, Ch., and Paulus, M. (1956). *C. R. Acad. Sci. Paris* **242,** 2525–2528.

Guillaud, Ch., Paulus, M., and Vautier, R. (1956). *C. R. Acad. Sci. Paris* **242,** 2712–2715.

Hamelin, A., and Paulus, M. (1970). *In* "Ferrites" (*Proc. Int. Conf. Ferrites, July 6–10, Kyoto, Japan*) (Y. Hoshino, S. Iida, and M. Sugimoto, eds.), pp. 134–136. Univ. Park Press, Baltimore, Maryland. (1971).

Kimura, Y. (1963). *Jap. J. Appl. Phys.* **2,** 312.

Kingery, W. D., and Berg, M. (1955). *J. Appl. Phys.* **26,** 1205.

Kooy, C. (1964). *Int. Symp. Reactivity Solids, Aug. 2–8, Munich, Germany,* pp. 21–28. Elsevier, Amsterdam. (1965).

Koppens, L. J. (1970). Joint Meeting on Electrical and Magnetic Ceramics, April 13–14, Eindhoven, Netherlands.

Kosinskaya, I. Ya., and Tret'yakov, Yu. D. (1969). *Inorg. Mater.* **55,** 1482–1486.

Krieger, E. M., Nazarova, E. A., and Yakushovskaya, F. T. (1970). *Sov. Powder Met. Metal Ceram.* **89,** 418–421.

Lacour, C., and Paulus, M. (1968). *Int. Conf. Crystal Growth, July 15–19, Birmingham, England*; *J. Crystal Growth* **3, 4,** 814–817.

Macklen, E. D. (1967). *Czech. J. Phys.* **B17,** 376.

Malinofsky, W. W., and Babbitt, R. W. (1961). *J. Appl. Phys.* **S32,** 2375.

Marion, F. (1967). *Chim. Anal.* **49,** 224.

Michalowsky, L., and Vogler, G. (1963). *Phys. Status Solidi* **3,** K312–316.

Muan, A., and Osborn, E. F. (1964). "Phase Equilibria among Oxides in Steel Making," p. 28. Addison-Wesley, Reading, Massachusetts.

Nicolas, J., Lagrange, A., Sroussi, R., and Hilderbrandt, H. (1969). *Rev. Tech.* **1,** 504–522.

O'Bryan, H. M., Gallagher, P. K., Monforte, F. R., and Schrey, F. (1969). *Amer. Ceram. Soc. Bull.* **48,** 203–208.

Olette, M., and Ancey Moret, M. F. (1963). *Rev. Met.* **60,** 569–582.

Oudemans, G. J. (1968). Philips Tech. Rundschau No. 1/2, 25–33.

Paladino, A. E., Waugh, J. S., and Green, J. J. (1966). *J. Appl. Phys.* **37,** 3371–3377.

Pâris, J. M. (1963). Thesis, Faculté des Sci. de l'Univ. de Lyon, July 12.

Pâris, J. M., Merlin, J. C., and Paris, R. A. (1962). French Patent 898 516; 898 517.

Paulus, M. (1962a). Thesis, Faculté des Sci. de l'Univ. de Paris, March 29; (1963). *J. Rech. C.N.R.S., France* **63,** 141–183.

Paulus, M. (1962b). *Phys. Status Solidi* **2,** 1325–1341.

Paulus, M. (1962c). *Phys. Status Solidi* **2,** 1181–1194.

Paulus, M. (1966). *Conf. Role Grain Boundaries Surfaces Ceram., Nov. 16–18, Raleigh, North Carolina; Mater. Sci. Res.* **3,** 31–47.

Paulus, M. (1970). *In* "Ferrites" (*Proc. Int. Conf. Ferrites, July 6–10, Kyoto, Japan*) (Y. Hoshino, S. Iida, and M. Sugimoto, eds.), pp. 114–120. Univ. Park Press, Baltimore, Maryland. (1971).

Paulus, M., and Eveno, P. (1968). *Int. Symp. Reactivity Solids, Aug. 25–30, Schenectady, New York.* (1969). *In* "Reactivity of Solids" (J. W. Mitchell, R. C. DeVries, R. W. Roberts, and P. Cannon, eds.), pp. 585–594, Wiley (Interscience), New York.

Paulus, M., and Guillaud, Ch. (1964). *Int. Colloq. "Les Composès Oxygénés des Eléments de Transition à l'Etat Solide," C.N.R.S., Sept. 24–27, Bordeaux, France; Bull. Soc. Chim. Fr.* **1965,** 1175–1185.

Paulus, M., and Guillaud, Ch. (1966). French Patent 1459 875.

Paulus, M., and Hamelin, A. (1968). *Int. Conf. Crystal Growth, July 15–19, Birmingham, England; J. Crystal Growth* **3, 4,** 500–503.

Paulus, M., and Lacour, C. (1966). *Z. Angew. Phys.* **21,** 539–545.

Paulus, M., and Lacour, C. (1968). *Europaisches Symp. uber Pulver Met., May 8–10, Stuttgart, Germany; Conf. Proc.* Vol. III, pp. 1–11, 143–145.

Paulus, M., and Mercier, D. (1966). *Int. Conf. Powder Met., Sept. 27–30, Stary Smokovec, Czecho-Slovakia.*

Paulus, M., and Morineau, R. (1967). *Le Vide* **130,** 203–212.

Paulus, M., and Reverchon, F. (1961). *Colloq. Int. Bombardement Ionique C.N.R.S., Dec. 4–8.* Published by C.N.R.S. (1962) No. 113, 223–232.

Pechini, M. P. (1968). Canadian Patent 832-365.

Reijnen, P. J. L. (1968). *Conf. Reactivity Solids, 6th, Aug. 25–30, Schenectady, New York,* (1969). *In* "Reactivity of Solids" (J. W. Mitchell, R. C. DeVries, R. W. Roberts, and P. Cannon, eds.), pp. 99–114. Wiley (Interscience), New York.

Reijnen, P. J. L., and Roksnoer, P. (1969). *In* "Thermal Analysis," Vol. 1 (R. F. Schwenker and P. D. Garn, eds.). Academic Press, New York.

Reijnen, P. J. L., Aarts, G. P. Th. A., van de Heuvel, R. M., and Stuijts, A. L. (1970). *Joint Meeting Elec. Magn. Ceram., April 13–14, Eindhoven, Netherlands.*

Ridgley, D. H., Lessoff, H., and Childress, J. D. (1970). *J. Amer. Ceram. Soc.* **53,** 304–311.

Rodrigue, G. P., and Crouch, L. A. (1966). *J. Appl. Phys.* **37,** 923.

Ruthner, M. J., Richter, H. G., and Steiner, I. L. (1970). *In* "Ferrites" (*Proc. Int. Conf. Ferrites, July 6–10, Kyoto, Japan*) (Y. Hoshino, S. Iida, and M. Sugimoto, eds.), pp. 75–78. Univ. Park Press, Baltimore, Maryland. (1971).

Sato, T., Kuroda, C., Saito, M., and Sugihara, M. (1970). *In* "Ferrites" (*Proc. Int. Conf. Ferrites, July 6–10, Kyoto, Japan*) (Y. Hoshino, S. Iida, and M. Sugimoto, eds.), pp. 72–74. Univ. Park Press, Baltimore, Maryland. (1971).

Sawaoka, A., and Saito, S. (1970). *In* "Ferrites" (*Proc. Int. Conf. Ferrites, July 6–10, Kyoto, Japan*) (Y. Hoshino, S. Iida, and M. Sugimoto, eds.), pp. 102–104. Univ. Park Press, Baltimore, Maryland. (1971).

Schule, W. J., and Livingston, V. D. (1961). *J. Appl. Phys.* **32,** S235.

Schmalzried, H., and Tret'yakov, Yu. D. (1966). *Ber. Bunseng. Phys. Chem.* **70,** 180–189.

Schnettler, F. J., and Johnson, D. N. (1970). *In* "Ferrites" (*Proc. Int. Conf. Ferrites, July 6–10, Kyoto, Japan*) (Y. Hoshino, S. Iida, and M. Sugimoto, eds.), pp. 121–124. Univ. Park Press, Baltimore, Maryland. (1971).

Schnettler, F. J., Monforte, F. R., and Rhodes, W. W. (1968). *Sci. Ceram.* **4**, 79–90.

Shafer, M. W. (1961). *J. Phys. Chem.* **65**, 2055.

Stablein, H., and Willbrandt, J. (1966). *IEEE Trans. Magn.* **2**, 459–463.

Stuijts, A. L. (1964). *Proc. Brit. Ceram. Soc.* **2**, 73–81.

Stuijts, A. L. (1969). *Proc. Int. Conf. Sci. Ceram., 5th, April 20–23, Ronneby Brunn, Sweden.*

Stuijts, A. L., and Kooy, C. (1965). *Sci. Ceram.* **2**, 231–242.

Stuijts, A. L., and Oudemans, G. J. (1965). *Proc. Brit. Ceram. Soc.* **3**, 81–99.

Stuijts, A. L., Ratheneau, G. W., and Weber, G. H. (1955). *Rev. Tech. Philips* **16**, 205–232.

Takada, T., and Kiyama, M. (1970). *In* "Ferrites" (*Proc. Int. Conf. Ferrites, July 6–10, Kyoto, Japan*) (Y. Hoshino, S. Iida, and M. Sugimoto, eds.), pp. 69–71. Univ. Park Press, Baltimore, Maryland. (1971).

Train, D. J. (1957). *Trans. Inst. Chem. Eng.* **35**, 258.

Tret'yakov, Yu. D. (1965). Doctoral Dissertation, Moscow.

Tret'yakov, Yu. D., Saksonov, Yu. G., and Gordeev, I. V. (1965). *Inorg. Mater.* **1**, 382.

Tret'yakov, Yu. D., Kosinskaya, I. Ya., Oleinikov, N. N., and Saksonov, Yu. G. (1969a). *Inorg. Mater.* **5**, 1067–1070.

Tret'Yakov, Yu. D., Kosinskaya, I. Ya., and Petrova, A. A. (1969b). *Inorg. Mater.* **5**, 1487–1491.

Turk, H. L. (1970). *In* "Ferrites" (*Proc. Int. Conf. Ferrites, July 6–10, Kyoto, Japan*) (Y. Hoshino, S. Iida, and M. Sugimoto, eds.), pp. 99–101. Univ. Park Press, Baltimore, Maryland. (1971).

Van Hook, H. J., and Keith, M. L. (1958). *Amer. Mineral.* **43**, 80.

Veith, H. (1970). *Z. Angew. Phys.* **30**, 277–282.

Wenckus, J. F., and Leavitt, W. Z. (1956). *Proc. Conf. Magn. Magn. Mater. AIEE Special Publ.* (1957) **T-91**, 526.

West, R. G., and Blankenship, A. C. (1967). *J. Amer. Ceram. Soc.* **50**, 343–349.

Wickham, R. G. (1970). *In* "Ferrites" (*Proc. Int. Conf. Ferrites, July 6–10, Kyoto, Japan*) (Y. Hoshino, S. Iida, and M. Sugimoto, eds.), pp. 105–107. Univ. Park Press, Baltimore, Maryland. (1971).

Zneimer, J. E., Kaplan, B., Lehman, K., and Lepore, D. A. (1964). *J. Appl. Phys.* **35**, 1020–1022.

Preparation of Transition-Metal Carbides and Related Compounds

S. WINDISCH and H. NOWOTNY

University of Vienna, Austria

I. Introduction

In 1865 Troost had already prepared zirconium carbide by the carbothermic reaction of zirconia with carbon. Later, in a series of creative work, Moissan (1893, 1895, 1896, 1897a,b, 1904) set the foundation for today's increasing interest in refractories, particularly in transition element carbides. Following an idea of Just and Hanamann (1903), van Arkel and de Boer (1925) were able to demonstrate the formation of high-melting carbides by the reaction of two gases, metal chlorides and hydrocarbons.

Generally speaking, carbides can be prepared by reaction of a metal or a metal compound with carbon or carbonaceous compounds. The reaction can be performed in the solid, liquid, or gaseous state, depending on the starting material and on the equipment used. The specific mode of reaction depends on the nature and the composition of the carbide to be prepared. Combined methods, e.g., carbidizing of solid metals by means of gaseous hydrocarbons, are pertinent, too. The aim of obtaining pure and homogeneous material of a definite composition usually demands considerable effort because of the high melting temperature and the readiness of carbides to dissolve other elements, particularly oxygen and nitrogen.

Numerous publications cover a large area concerned with the preparation of carbides (e.g., Kieffer and Benesovsky, 1963; Eisenkolb, 1963; Russell *et al.*, 1964; Storms, 1967; Stanford Research Institute, 1969). This review discusses approved preparation methods and new information on metallic carbides, mainly of group 3A to 6A elements.

One can divide the means of preparation as follows: powder metallurgical procedures, melting methods (including fusing, casting, and the menstruum process), formation by means of volatile components, and special operations.

II. Methodology

A. Powder Metallurgical Methods

The bulk of all metallic carbides are produced for industrial purposes and on a laboratory scale by powder metallurgical reaction. Such powder metallurgical reactions between the metal or metal oxide and carbon (graphite) are considered to occur mainly in the solid state. Therefore the powders ought to be as fine-grained as practicable and mixed as intimately as possible. Wet mixing is strongly recommended, since the dry powders are inclined to separate during operation. The liquid to be used ought to be a solution of an agent which later acts as a pressing aid. The solvent for the agent to be used may be water if oxides are the starting material, but an organic solvent is preferred with elemental powders. It is advisable to combine pulverizing and blending of the constituents. The duration of blending depends very much on the available apparatus; it should be remembered that thriftiness in this inexpensive operation might become quite costly. Coarse powders require extended reaction time, and inadequate mixing yields diverse phases. All solvent should be removed before the powder is compacted into pellets or bricks. Frequently, the pressure applied for compacting is unnecessarily high; the pellets should be just strong enough to be handled. The pressing aid is removed by heating the compacts to 200°C in vacuum or to 400°C under purging inert gas. This can be done either in a separate kiln or in the reaction furnace. In the latter case, a condenser trap is recommended to prevent contamination of the furnace and/or the pumping system.

The conditions for the carbide preparation will be given under the individual compounds. These values are to be considered as an approximation, since they also depend on, besides the starting material, the size and type of the furnace, the amount of the batch, and the given specifications.

The reaction is generally performed in an induction or resistance furnace capable of heating up to high temperatures (2500°C and higher). Generally, it is useful to adapt the furnace not only for vacuum, but also for static and flushing atmospheres of hydrogen, which promote the formation of carbides.

A furnace for industrial purposes was constructed by Krall.† It contains a vertical graphite heater, and operates without protective gas, since the carbon monoxide formed during the reaction provides sufficient protection. Carbidization is usually done from 2300 to 2700°C, although the furnace design permits operation at still higher temperatures.

Recently, furnaces have been proposed by DuPont (1970) for laboratory-scale (vertical version) and for continuous production (horizontal arrangement). In both cases, the container is made of graphite and heated by induction. The carbothermal reaction can be controlled by measuring the CO evolution in the exit gas as a function of time and temperature.

Some of the carbides form solid solutions of high stability with oxygen and nitrogen. Thus, the formation of Ti(C, N, O) is well-known. Since the efflux of these elements from the crystals is governed by a diffusion process, the particles ought to be very small, but one cannot prevent grain growth at the high temperature which is necessary for the purification procedure. Furthermore, the vaporization of the metal becomes significant; for instance, Williams (1964) observed on TiC that titanium evaporates 10 times faster than carbon at about 1950°C. If oxides are reacted, the pellets ought to be small in order to promote the expulsion of oxygen, and they should not be heaped up too high. Removing all detectable oxygen causes

† See Kieffer and Benesovsky (1963), p. 88.

some severe problems, especially when oxides and carbon are reacted. The need for cleanliness of the furnace is self-evident. Small amounts of dirt can ruin a whole batch even in cases where such dirt does not evaporate during firing, because it usually contains water or other volatile compounds.

The carbon monoxide generated during the reactions ought to be removed as fast as possible by an arrangement which keeps the carbon monoxide from accumulating in the furnace, e.g., by frequent evacuation. One also has to keep in mind that in the presence of H_2, water is formed, which may condense on cooled components, such as furnace walls or power leads. The same is true for water that is brought in by hydrogen or an inert gas.

In powder metallurgy, a good vacuum is considered the most convenient "protective atmosphere" if the furnace is actually leakproof at the operation temperature; however, it is not the most effective one for removing the last residues of volatile impurities. The vacuum treatment approaches an equilibrium state, whereas a streaming gas carries away the undesired products. In the latter case, the reaction can be followed by gas chromatographic analysis of the effluent gas stream (Fletcher *et al.*, 1964; Harrison and Isaacs, 1964). Of course, the carrier gas has to be purified before use. If large quantities of the inert gas are necessary, it is advisable to reuse the gas in a cycle. This can be done by chemical getters and/or by freezing out the impurities. Finally, it seems reasonable that by repeating the annealing and regrinding procedure, an oxygen-free carbide will result, especially if some getter material like zirconium sponge is present in the chamber.

Another way to remove oxygen and nitrogen is to arc-melt the reacted carbide, although a few shortcomings are connected with this method (solutions to these will be discussed in the section on arc-melting). Consequently, in several cases, the ultimate choice is the combination of three steps: powder metallurgical carbothermal reaction, followed by arc-melting the carbide, and completed by annealing the corrected composition.

B. Melting Methods

1. *Fusing*

Melting in general serves many purposes. Mostly, it is used to obtain a castable material, or to convert powders into a coherent ingot, or to perform a reaction. Direct fusing, which yields in a continuous operation a workpiece of the final shape, still has a very limited range of application. Most of the difficulties are engineering problems because of the high temperature of the melt involved. The understanding of solidification and the ability to control the various types of eutectics will very likely improve the practicability of the fusing method. Presently, it is mainly used for the technical preparation of tungsten carbide and tungsten carbide alloys.

2. *Arc-melting*

The electric arc is a gaseous plasma at a temperature between 5000 and 10,000°K, in exceptional cases approaching 50,000°K in the center (Hoyaux, 1968, 1970). Therefore it is frequently used as a heat source for high-temperature operations. The material to be melted should be well-mixed and compacted; however, under certain circumstances, coarse powders or even pastilles can be used because of the stirring effect of the arc. On the other hand, inhomogeneities can be caused by various reasons: rapid reaction of the starting material, wide differences in the melting temperatures of the components, and the high temperature gradient of the plasma. In such cases, it is advisable to arc-melt prereacted carbides instead of the elements, and if necessary, to homogenize the powdered and corrected material by annealing at high temperatures (1500–2000°C).

Sometimes, arc-melting is performed to produce a brittle material which is easy to pulverize, resulting in irregularly shaped particles. While this technique serves diverse purposes in the laboratory, it has gained only modest importance for the technical production of a few carbides, e.g., Mn_3C and chromium carbide (a mixture of Cr_3C_2 and Cr_7C_3). This method is especially effective for refining and for the preparation of solid solutions, particularly when carbides of groups 4A and 5A are involved. A further advantage of most arc-melted pieces is their smooth surface, which is less susceptible to contamination.

A short report (Jones, 1961) on electron-beam melting of transition element carbides disclosed relatively high volatilization of the components. Thus the preparation of carbides by such a method appears to be unsatisfactory.

C. Menstruum Process

Similar to the hot extraction method (Lebeau process), the menstruum process is used for the preparation of carbides which are distinguished by good wettability and a very low content of impurities such as oxygen and nitrogen. Kieffer (1969) found as the most suitable auxiliary metal bath for the reaction between transition metals and carbon, an alloy of 70 wt % iron and 30 wt % nickel. The amount of auxiliary metal bath (menstruum) to be used is about four times the volume of the transition metal. The latter may be applied as powder, chips, or sponge, but graphite only as powder, in order to attain a saturated melt easily. The ingredients are put into a graphite crucible; therefore it is not necessary to use an excess of carbon over the calculated value. The batch is heated up to 2000°C in a purified static atmosphere of noble gas. Some titanium or zirconium sponge in a separate boat on top of the crucible acts as a getter. The duration of the reaction depends largely on the original form of the transition metal. If the starting material was powdered, the reaction is completed on a laboratory scale after 30 min. The reaction product is crushed and put in a warm solution of about 3–6 *N* hydrochloric acid to dissolve the menstruum alloy.

D. Formation by Means of Volatile Compounds

A number of halides, especially chlorides, display a high volatility in a temperature range which is convenient to work with. They can be produced from oxides or scrap metal and chlorine, and may be separated in one single distillation. In the gaseous state, they are easy to mix. At elevated temperatures, they are highly reactive and therefore are utilized for the preparation of various compounds. The mode of preparation can be directed to form powders, as well as coatings or coherent solids.

For carbide preparation, quite diverse carbonaceous compounds, such as methane, toluene, carbon monoxide, and others, have been tried. However, the starting carbonaceous compounds do not influence the nature of the carbides obtained. The reaction is usually carried out in the presence of hydrogen; this indicates that the chloride seems to be first reduced to the element, which subsequently reacts to form the carbide. However, as was found by Pearce and Marek (1968) under the conditions commonly applied, no elemental titanium is formed.

Some halides are easily reduced without hydrogen to the elements by the carbonaceous compounds. In these cases, the formation of carbides becomes significant only at high temperatures, which was already observed by Pring and Fielding (1909).

By the deposition method, a layer (crust) can be formed on a heated substrate, the latter being a solid piece or just a powder of a high-melting metal or graphite. The nature of the final deposit depends on the surface of the substrate and on other experimental conditions. The various possibilities of this method for coating purposes are demonstrated in a comprehensive investigation by Campbell *et al.* (1949).

When a gaseous mixture of $MeCl_x + C_nH_m$ is heated, it is evident that these compounds dissociate considerably into fragments which will subsequently either conglomerate, forming particles of Me and C and molecules of Cl_2 and H_2, or react to carbides. In order to favor the carbide formation and to suppress the conglomeration, one has to adapt the ratio of $MeCl_x$ to C_nH_m to the reaction temperature. The nature of the final product also depends on the concentration, thus on the pressure, of the participating components; the lower it is, the better developed are the crystals. According to Bourdeau (1968), a powder is formed when the pressure exceeds 100 Torr, while a dense coating can be obtained below 100 Torr and superior hardness can be achieved when the pressure is kept below 20 Torr.

Ultrafine powders of carbides, nitrides, and elements were prepared by Neuenschwander (1966) by means of a plasma gun: A stream of hydrogen is blown through an arc between a tungsten cathode and a copper anode (117 V dc, 200 A). The vapors of refractory metal chlorides and methane are fed into the hot zone of hydrogen, using a ratio of one mole chloride to 1.2 moles methane to 12–15 moles hydrogen. The particle size of the resulting product is in the range 100–1000 Å; it coordinates with an extensive surface area. The latter accounts for the relatively high oxygen content; nevertheless, it has been observed that the powders are not pyrophoric at room temperature.

Besides the practical aspects, this method was also applied on basically scientific problems: Meerson *et al.* (1965) determined the solubility of boron in silicon carbide, and Ambarcumjan and Babich (1969) measured the growth rate of TiC and ZrC on graphite as a function of temperature.

III. Special Operations

A. Hot Pressing

As a mode of preparation, hot pressing is widely used in laboratories as well on a technical scale. The hot press shown in Fig. 1 can be used for preparation of samples of diverse sizes, as well as for quenching experiments (see below). The theory and procedure of hot pressing for the preparation of carbides are extensively discussed by Ramqvist (1966).

B. Quenching

Various carbides are stable only within a certain temperature range. Some of them change in crystal structure at a definite temperature, others decompose eutectoidly. For the preparation of high-temperature carbide modifications, quenching, occasionally by dropping into a tin bath, has been

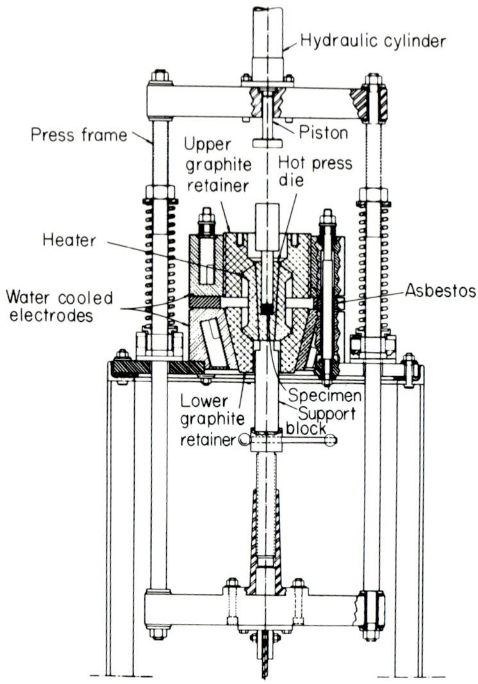

FIG. 1. Scheme of a hot press (after E. Rudy and G. Progulski, unpublished).

successfully applied for retaining these structures. Thus, e.g., the cubic modifications of MoC_{1-x} and WC_{1-x} were prepared in this way: a flat piece of the alloy was inserted into a shortened sample container of the hot press (Fig. 2) and equilibrated at a specific temperature. With the aid of a hydraulic piston mounted on top of the sample holder, the sample was then ejected into a preheated tin bath (300°C) by breaking the thin diaphragm beneath the specimen.

C. SINGLE CRYSTALS

For a complete understanding of those physical phenomena that are orientation-dependent, single-crystal measurements are essential. The technique of crystal growth can be classified as follows: (a) liquid (melt or solution) → solid; (b) vapor → solid; (c) solid → solid.

(a) The crystal growth from a solution can be performed either by a slow temperature change or by a modification of the menstruum process.

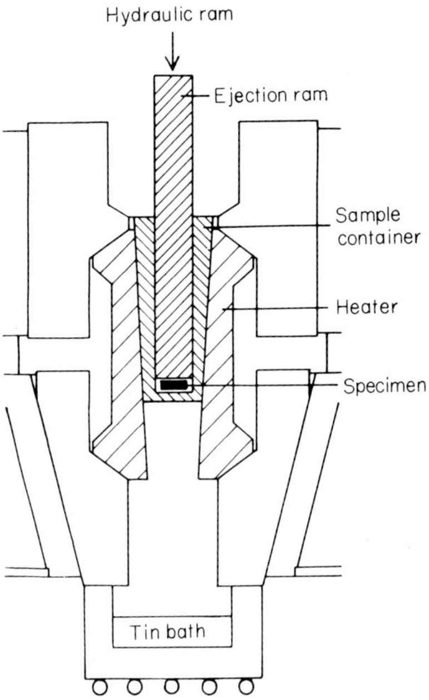

FIG. 2. Experimental setup for quenching studies on carbides.

This mode of preparation is most frequently used in inorganic chemistry, and it has also been applied to the preparation of carbide crystals. Because of the high melting temperature of the carbides, new problems arise when such crystals are pulled out of a melt of their own (Czochralski method). Crystals of TiC were prepared by this method in an arc-melting furnace equipped with three arcs, which melted only the upper part of the charge (Reed, 1969). Partial melting also occurs when single crystals are formed by the floating-zone technique, where a molten zone travels through a solid rod. Precht and Hollox (1968) reported the formation of $VC_{0.85}$ single crystals when they heated a sintered rod with a radiofrequency induction coil.

Most of the carbide crystals have been prepared by the Verneuil method, modified insofar as the hydrogen–oxygen burner is replaced by another nonoxydizing heat source (e.g., susceptor ring heaters, dc plasma, induction heating, etc.). Finely divided powder is blown together with a carrier gas through the hot zone. A powder cone builds up when it reaches the hot zone, and then the top of the cone melts and starts to solidify as the cone is lowered. The crystals obtained are usually termed a "boule."

(b) Quite frequently, single crystals of metals have been prepared by pyrolytic decomposition of halides on a wire (Koref, 1922; Brenner, 1963).

The same method is used to produce carbide crystals, but these are usually very small. For instance, TiC single crystals of 0.5 mm length grew from the reaction between $TiCl_4$ + toluene on a heated tungsten wire (Chermant and Deschanvres, 1968).

(c) Sintering at a high temperature, usually close to the melting temperature, can produce large crystals (see, for instance, Pons, 1967). The formation can be enhanced by superimposing an external strain or by slight deforming (drawing, rolling, bending). The deformation causes a "nonequilibrium strain energy distribution" which is relieved during subsequent annealing by grain growth. This method was utilized by Fleischer and Tobin (1969) to produce single crystals of ZrC, HfC, VC, and NbC. First, the molten metal in a graphite tube forms a layer of carbide at the interface; then, increasing the temperature (in small steps) produces strain on the carbide formed. This causes microstructural transformation, leading to discontinuous grain growth. After secondary recrystallization of the primary structure, large single crystals can be obtained.

IV. Binary Carbides

A large number of phase diagrams of transition-element–carbon systems are now available in a fairly completed form: Ti–C, and Zr–C (Storms, 1967; Rudy *et al.*, 1965a); Hf–C (Storms, 1967); V–C and Nb–C (Rudy *et al.*,1968); Ta–C (Storms,1967; Rudy and Harmon,1966); Cr–C (Storms, 1967; Rudy,1969); Mo–C (Rudy *et al.*, 1967); and W–C (Storms, 1967; Rudy *et al.*, 1966).

From these diagrams, one can derive some typical features among the carbides to be found. Thus, e.g., most of the cubic monocarbides exhibit an extended region of homogeneity which depends on the temperature. In view of the preparation of these carbides, an interesting point concerns the maximum carbon content, which definitely turns out to be substoichiometric for TiC, ZrC, VC, etc. This means such a compound should be formulated as TiC_{1-x}, etc. In contrast, UC displays a wide homogeneity domain at high temperatures, including compositions UC_{1-x} and UC_2 as well.

Among other peculiarities, the so-called subcarbides of formula Me_2C also exist within a homogeneous range and split according to composition and temperature treatment, sometimes into diverse phases, frequently forming an ordered crystal structure at low temperature. On the other hand, several carbides without a detectable homogeneous region occur as "line compounds," e.g., Cr carbides and WC. Beyond that, there are high-pressure carbides also existing, such as Th_2C_3 (Krupka, 1970). The preparation of these or other, obviously metastable carbides, like the zeta carbides of formulas V_4C_3, Nb_4C_3, and Ta_4C_3 (Yvon and Parthé, 1970), require elaborate methods.

The influence of small quantities of nonmetal elements (H, N, O, . . .) on the stability of the various monocarbide and subcarbide phases is not yet fully understood as far as various transition element carbides are concerned. Some of these problems will be discussed in Section VI.

A. Carbides of Group 3A

Powder metallurgy as well as arc-melting are the predominant techniques used to prepare carbides of group 3A. Scandium carbides, for instance, have been synthesized from the elements above 1200°C in high vacuum (Rassaerts *et al.*, 1967; Jedlicka *et al.*, 1971). The carbothermal reduction is frequently applied for the preparation of carbon-rich carbides, such as $Sc_{15}C_{19}$ (Jedlicka *et al.*, 1971), YC_2, LnC_2 (Achard, 1966), ThC_2, and UC_2. Matyushenko and Svinarenko (1968) reported the formation of EuC_2 by heating $Eu_2O_3 + C$ for 2–4 hr at 1800°K in an atmosphere of Ar.

1. *Thorium carbides*

The following carbides have been found at normal pressure: a cubic monocarbide (ThC) and three dicarbides (ThC_2) which are, respectively, monoclinic, tetragonal, and cubic when the temperature is raised.

ThO_2 can be converted to carbides between 1800 and 2150°C, depending on the desired composition. First, the firing can be done in an atmosphere of technical hydrogen; in the final stage of the heating, a purified gas or vacuum is applied. A second heat treatment of the pulverized material improves purity and permits correction of the composition.

All subsequent operations ought to be done in an inert atmosphere; samples are preferentially immersed in dry paraffin or CCl_4.

Lofgren and Krikorian (1964) prepared coatings of ThC_2 in the following manner: the inside wall of a graphite tube was swabbed with an organic glue; a layer of Th metal powder was applied on this surface. The tube was heated to 2000°C in vacuum and subsequently above 2500°C under 50 Torr of argon.

Spherical particles of ThC_2, wrapped by a dense coating, are used as fuel in high-temperature reactors. These particles can be prepared by forming, sintering, and subsequent coating. Knotik *et al.* (1965) developed a procedure which combines carbidization and coating in one operation. A suspension of ThO_2 or Th metal in paraffin is sprayed into liquid nitrogen, forming small droplets. These particles are mixed with an excess of lampblack and fired in streaming argon. After the carbidization is completed, the temperature is further raised to 2500–2600°C; ThC_2 melts and forms spheres with an adherent graphite skin.

Thorium carbides of highest purity are preferably prepared by arc melting, using the elements as starting material (Storms, 1967).

The high-pressure phase Th_2C_3 is formed when an equimolar mixture of ThC + ThC_2 is subjected to annealing above 1200°C under a minimum pressure of 30 kbar (Krupka, 1970).

2. *Uranium carbides*

Three compounds, UC, U_2C_3, and UC_2, have been determined in the U–C system, the latter stable only above 1500°C and occurring in two different modifications, tetragonal below 1765°C and cubic above this temperature. The cubic phases of UC and UC_2 form a solid solution at high temperature (Storms, 1966, 1967). The equilibrium is strongly influenced by oxygen and nitrogen.

a. Powder metallurgical procedures. Numerous publications deal with the preparation of uranium carbides. Although the experimental conditions seem to be identical in many cases, the properties of the reaction products differ seriously. It is likely that contamination by oxygen accounts for this phenomenon, because it is exceptionally severe when uranium oxide is reacted with carbon. In addition, the heating rate is frequently neglected: The formation of the carbides starts at 1250°C (Atoda *et al.*, 1964), and therefore one cannot ignore the amount of carbon dioxide which is generated besides monoxide, thus leaving more carbon for the carbide. Likewise, Bečvař (1969) found less oxygen and nitrogen after slow heating to 1700°C than after rapid firing.

b. Melting methods. UC and UC_2 can be easily prepared to a high degree of purity by arc-melting the elements. Storms (1967) noticed that the use of helium instead of argon allows shorter melting times.

Single crystals of UC have been prepared by melting in an electron-beam floating-zone apparatus, described by Nadler (1963) and Calverly *et al.* (1957).

Dense UC_2 kernels of spherical shape coated by an adherent graphite layer can be produced by the method previously described (see discussion of Th–C), and by a modified procedure which starts from aqueous solutions; it is distinguished by the following advantages: reclaimed solutions can be used directly, particle formation by spraying is simple, and no dust is generated (Bildstein and Knotik, 1967).

B. Carbides of Group 4A

1. *Titanium carbide*

a. Powder metallurgical procedures

1. Carbothermal reduction of TiO_2. The compacted powder mixture is slowly heated in vacuum to 800°C, then further heated with purging hydrogen to 1900°C, followed by a vacuum annealing for 1–2 hr, and finally is cooled under vacuum.

Meerson *et al.* (1970) obtained nearly stoichiometric TiC by reacting TiO + 2C at 1700 (1 Torr) or 1750°C (Ar atmosphere).

2. Reaction of the elements. Ogawa and Bando (1959) obtained TiC with 19.2 wt % C (48.6 atom % C) by reacting the powder mixture for 1 hr at 1200°C under vacuum. Kudaka and Nakase (1968) used TiH_2 powder of grain sizes 1.2 and 5.5 μm. Dehydrogenation of the hydride was ascertained between 200 and 700°C. The formation of TiC was established as a solid state reaction, with graphite from 700°C up, but with CO just above 1200°C. Grain growth as a function of annealing temperature reveals that the grain size after a reaction at 1600°C is twice as large as after 1200°C.

b. Melting methods. Because of the high melting point of TiC (~3100°C), the fusing method has achieved only limited significance. However, the addition of metals that form a eutectic with TiC may make the fusing method adaptable in preparing plain carbides and in casting work pieces.

Spherical particles up to 1 mm in diameter, consisting of a dense TiC kernel coated by a thin layer (less than 10 μm) of graphite, have been prepared by Bildstein and Knotik (1970) in a further development of their work on ThC_2 and UC_2 (Knotik *et al.*, 1965). The principle is outlined in the discussion of thorium carbides.

c. Menstruum process. Taken as menstruum: 100 g of titanium sponge + 25 g of graphite powder + 400 g of iron–30 nickel alloy. This batch was reacted for 30 min at 2000°C in a vacuum, yielding TiC with 48.6 atom % C (Kieffer, 1969).

d. Volatile components as starting material. Mixtures of $TiCl_4 + CH_4 + H_2$ have been used to prepare ultrafine TiC powder (Neuenschwander,

1966), to deposit TiC on graphite at 1200–1300°C (Pearce and Marek, 1968), to grow TiC single crystals on a W wire (Chermant and Deschanvres, 1968), and to synthesize specimens in the composition range TiC–C (Kilin *et al.*, 1969).

e. Technical production. Generally, the carbothermal reduction of TiO_2 is applied. Since a continuous operation is favored, the carbidization is not performed in a vacuum, but in an inert atmosphere such as hydrogen or noble gases.

It is assumed that all TiC produced in this way actually consists of a solid solution of TiO in TiC. However, views differ about the possibility of determining the correct oxygen content and about the influence of small oxygen impurities on the properties. For this reason, TiC is also frequently prepared by carbidizing elemental titanium. In this case, the metal is precarbidized, and thus embrittled for easier grinding.

2. *Zirconium carbide*

a. Powder metallurgical procedures.

1. Carbothermal reduction of ZrO_2. The compacted powder mixture is slowly heated under purging hydrogen to 2000°C, and subsequently annealed at this temperature for at least 2 hr in a high vacuum. The high stability of solid solutions Zr(C, O) makes it extremely laborious to remove all oxygen immediately after the reaction. The most promising method is to arc-melt the lumps in a purified atmosphere, followed by pulverizing and homogenization annealing at 1800°C.

2. Reaction of the hydride. Norton and Lewis (1963) fired a mixture of zirconium hydride + graphite for 3 hr at 2200°C.

b. Melting methods. Fleischer and Tobin (1969) melted zirconium metal in a graphite tube, forming ZrC, which was transformed into single crystals. Spherical particles of ZrC were prepared by Bildstein and Knotik (1970) (see ThC_2).

Arc-melting of the elements yields a pure carbide, but composition homogeneity is not satisfactory. Therefore, this process is frequently used as a mode of purification.

c. Menstruum process. Kieffer (1969) obtained ZrC with 49 atom % C under the following conditions: 100 g of zirconium sponge, 13.2 g of lampblack and 300 g of iron–30 nickel were heated for 30 min at 2000°C in a vacuum.

d. Formation by means of volatile compounds. Sugiyama and Takahashi (1970) succeeded in preparing ZrC fibers (diameter 12 μm) by the reaction of $ZrCl_4$ + *i*-butane + H_2 at 350–700°C in an argon atmosphere under the influence of electrical discharge (0.5–0.7 mA).

e. Technical production. Kroll and co-workers (1947) suggested an arc-melting process for the preparation of ZrC on a technical scale. In the first instance, the ZrC prepared in this way is used as an intermediate compound for the production of $ZrCl_4$.

3. *Hafnium carbide*

a. Powder metallurgical procedure.

1. Carbothermal reduction of HfO_2. The reaction is performed in the same manner as described for ZrC. In case of carbon deficiency and/or oxygen content, the lumps are powdered, their composition corrected by adding the proper amount of carbon, and finally heated for several hours at 2200°C in a high vacuum.

2. Reaction of the elements. Hafnium powder of grain size <44 μm and lampblack were heated for 2 hr at 1950°C in a vacuum of 10^{-5} Torr, yielding HfC with 49.3 atom % carbon (Rudy, 1965).

Fleischer and Tobin (1969) prepared HfC single crystals by the annealing procedure described in Section III.C.

b. Melting methods. Spherical particles consisting of HfC kernels and coated by graphite were produced by Bildstein and Knotik (1970) in the same way as discussed for thorium carbide.

Hafnium carbide was prepared by Kieffer (1969) in a menstruum, using a weight ratio of 100Hf : 7C : 300Co.

c. Formation by means of volatile compounds. Most of the publications in this field deal with carbide coating on a substrate like graphite or tungsten. Wilcox *et al.* (1968) studied pyrolytic deposition at 1800°C on graphite. They used various gas mixtures ($HfCl_4$, toluene) and dopants (WCl_6) and obtained whiskers, needles, dendrites, and faceted crystals. It is very likely that with an apparatus larger than the one described, the reproducibility will improve.

d. Technical production of HfC. If HfC achieves technical importance, it very likely will be produced by the powder metallurgical reduction of HfO_2.

C. Carbides of Group 5A

1. *Vanadium carbides*

V_2C, or a mixture containing this carbide, is used as a master alloy in metallurgical processes. While the carbides of formula MeC are easily produced by the carbothermal reduction of the oxides, this mode is not advisable for making the subcarbide of formula Me_2C. V_2C was produced (Rudy *et al.*, 1968) by reacting the cold-compacted mixture of vanadium powder and lampblack in a graphite element furnace. The batch was heated up to 1300°C under vacuum, and the reaction then completed at 1750°C under a high-purity helium atmosphere.

VC

a. Powder metallurgical procedures.

1. Carbothermal reduction of ammonium vanadate or vanadium oxide. The well-blended mixture is heated at 1800°C under hydrogen for several hours. Usually, the oxygen cannot be completely removed in a single operation. Therefore the lumps are pulverized, loosely piled up in a graphite container, and finally heated preferentially in a high vacuum.

A patent was granted DuPont (1970) for the preparation of VC in one step on a laboratory scale and on an industrial basis. The mixture of vanadium oxide and graphite is piled up with sufficient interstices. The reaction is performed at 1400–1700°C under a stream of inert gas which carries away carbon monoxide.

Based upon thermodynamic considerations, Paris and Clar (1969) reacted vanadium oxide with a mixture of methane and hydrogen at 700–900°C. The resulting VC (carbon content 46.3–47.6 atom % C) was, according to the authors, well-crystallized and more reactive than vanadium carbide prepared at higher temperatures.

2. Reaction of the elements. Vanadium powder must not be exposed to air for extended periods of time. During dry mixing, the blender should be filled with inert gas. Wet mixing should be done with waterless liquid. The maximum carbon content of VC is 47 atom %. However, since most of the finely powdered vanadium is slightly oxidized—perceptible by the greenish tint—a small excess of carbon is suitable. The heating conditions are the same as given for V_2C.

Single crystals of VC were prepared by "annealing under stress" by Fleischer and Tobin (1969) (see Section III.C).

b. Melting methods. Frequently, coarse powders or granules of vanadium are available as starting material. In these cases, melting, especially arc-melting, becomes superior to the powder metallurgical preparation.

c. Menstruum process. In a menstruum bath of iron–30 nickel, high-purity VC was prepared at 2000°C (30 min) using the elements as starting material (Kieffer, 1969).

d. Formation by means of volatile compounds. When a mixture of vanadium halide + hydrocarbon + hydrogen is heated, not only VC, but also elemental vanadium is formed. The latter alloys easily with a metallic substrate, e.g., with tungsten. Therefore the chemical composition of the deposit is not uniform, but the method might be useful on graphite as substrate. The preparation of VC in a plasma arc has not yet been reported.

e. Technical production of VC. Usually, V_2O_3 and graphite powders are reacted at 1800°C (Kieffer and Benesovsky, 1963). The quality improves on a subsequent heat treatment at 1600–1700°C in a high vacuum.

2. *Niobium carbides*

a. Powder metallurgical procedures. Both Nb_2C and NbC were prepared from a mixture of niobium powder (grain size $<74\ \mu m$) and lampblack by firing for 1 hr at 1900°C in a high vacuum (Rudy *et al.*, 1968). On the other hand, Storms and Krikorian (1960) reported that 38 hours were necessary to complete the formation of $NbC_{0.74}$ at 1800°C.

Frequently, Nb_2O_5 is used as starting material and is reacted with graphite powder at 1300–1400°C (Kieffer and Kölbl, 1949). An additional heat treatment in a high vacuum above 1900°C is recommended to eliminate oxygen and nitrogen below a detectable level (Storms and Krikorian, 1960).

b. Melting methods. Long-duration firing can be avoided by either arc-melting or the menstruum process. The reaction in the auxiliary bath is carried out at 1800°C (Kieffer, 1969) in order to advance the formation of NbC, although the participating metals niobium and iron–30 nickel already form a liquid solution at 1600°C.

c. Formation by means of volatile compounds. When a substrate like tungsten is heated to >900°C in a gas containing $NbCl_5$ + hydrocarbon + hydrogen, a deposit is formed, consisting of niobium carbides and elemental niobium. The metal can be converted to NbC by heating to 1300°C with an excess of hydrocarbon (Campbell *et al.*, 1949).

Ultrafine NbC containing 47.6 atom % carbon and accompanied by a rather high amount of free carbon was produced by Neuenschwander (1966) in the plasma arc chamber.

d. Technical production of NbC. Oxygen-free NbC is produced without difficulties by powder metallurgical reaction of either niobium oxide or niobium metal with carbon at 1300–1400°C.

3. *Tantalum carbides*

a. Powder metallurgical procedures. A significant reaction occurs between tantalum powder (grain size $<44\ \mu m$) and lampblack at 1900°C (E. Rudy and S. Windisch, unpublished). Therefore Ta_2C and TaC were synthesized at 1900–2000°C in an evacuated furnace. The formation of these carbides is completed in less than 1 hr when such fine-grained powders are applied. Purity and homogeneity can be improved by a subsequent annealing of the pulverized material at 2000–2200°C (Lesser and Brauer, 1958).

The prevailing mode of technical production is the powder metallurgical reduction of tantalum oxide with carbon at 1500–1600°C under hydrogen. A consecutive heat treatment at 1500–1800°C in a high vacuum (better than 10^{-5} Torr) removes all detectable oxygen.

Kolchin and Berlin (1964) described the continuous production of tantalum carbide (carbon content 5.1–6.2 wt % C_{tot}) in a Tammann furnace at 1750–1800°C, using Ta_2O_5 as starting material.

b. Melting. Arc-melting of TaC tends to produce a carbon deficiency, apparently when the samples are too large and thus partly overheated by the center of the plasma. Storey and Landise (1970) were able to grow single crystals of Ta_2C in a dc plasma. McKenna (1939) synthesized TaC in an aluminum bath heated to 2000°C. Kieffer (1969) obtained TaC with 49.4 atom % carbon by heating the constituents in an equal amount of iron–30 nickel for 2 hr at 1800°C.

c. Formation by means of volatile compounds. When a mixture of $TaCl_5$ + hydrocarbon + hydrogen is heated, the pyrolytic formation of elemental tantalum becomes significant at 900°C (Moers, 1931), but raising the temperature causes augmented carbidization. Neuenschwander (1966) produced ultrafine TaC and observed that the high content of free carbon is considerably reduced after blowing the reacted powder again through the arc.

D. Carbides of Group 6A

1. *Chromium carbides*

a. Powder metallurgical procedures. The proper mixtures of chromium and carbon can be reacted without difficulty to the carbides $Cr_{23}C_6$, Cr_7C_3, and Cr_3C_2 at 1500, 1700, and 1750°C, respectively. The formation is completed within 1 hr if lampblack and chromium powder of grain size $<44\ \mu m$ are used.

The carbothermal reduction of chromium oxide yields mainly Cr_3C_2 if the reaction is performed below 1300°C; however, Cr_7C_3 is the chief product at higher temperature (Kosolapova and Samsonov, 1959; Grigoreva and Klimenko, 1959).

b. Melting methods. Single crystals of Cr_3C_2 were obtained directly from the melt in an arc furnace (Rundqvist and Runnsjö, 1969).

Only a ternary phase of $(Cr,Me)_7C_3$ is formed in the menstruum bath of cobalt or iron–30 nickel (Kieffer, 1969); however, in an overheated copper bath, the carbon-rich Cr_3C_2 can be prepared at 1800°C (Kieffer and Benesovsky, 1963).

c. Technical production. Chromium carbide of formula Cr_3C_2 can be produced from pure chromium oxide Cr_2O_3; the conditions, especially temperature, duration, and heating rate, have to be carefully determined in order to avoid the formation of metal-rich carbides.

Chromium carbides of heterogeneous composition are among the few that are produced industrially by arc-melting a mixture of chromium oxide + carbon.

2. *Molybdenum carbides*

It is not advisable to prepare any molybdenum carbides from the oxide or oxygen-containing compounds because it is easier to reduce molybdenum oxide than to purify a contaminated carbide.

Mo_2C

a. Powder metallurgical procedures. Lampblack and molybdenum powder (grain size $<44\ \mu m$) react at 1600–1900°C in a vacuum within 2 hr to Mo_2C (Rudy *et al.*, 1967).

The technical production starts from the oxide, which is first reduced by means of hydrogen at 900°C; the resulting metal powder is carbidized, usually in an atmosphere of hydrogen at 1400–1500°C.

b. Melting. Single crystals of Mo_2C up to a length of 40 mm were prepared by the Verneuil method (Vahldiek *et al.*, 1966). Kieffer (1969) obtained Mo_2C by heating molybdenum + graphite powder in a cobalt menstruum for 3 hr at 1800°C.

c. Deposition. Deposition from the gas phase can be accomplished by heating a substrate, e.g., tungsten wire or graphite rod, in an atmosphere of $MoCl_5$ + methane. Pring and Fielding (1909) had already observed that below 1300°C, mainly molybdenum is formed, but above 1300°C, molybdenum carbides are obtained.

Weiss (1946) and Andrieux and Weiss (1948) deposited Mo_2C by electrolysis of MoO_3 dissolved in a molten mixture of alkali–carbonate–borate–fluoride. The carbon content of the carbide depends on the ratio of MoO_3 to carbonate.

MoC_{1-x}

Two high-temperature phases appear adjacently in the vicinity of 40 atom % C. The hexagonal MoC_{1-x} contains slightly less carbon, while the cubic MoC_{1-x} consists of more than 40 atom % C. Nowotny and Kieffer (1952) detected in a quenched Mo_2C–C alloy the cubic phase, which was later stabilized by small additions of hafnium or boron (Nowotny, 1962). Claugherty *et al.* (1961) obtained the cubic modification when they applied a pressure of about 20 kbar. Rudy *et al.* (1962) observed the formation of the cubic carbide on samples quenched from the molten state. Rudy *et al.* (1965) prepared homogeneous pieces of about 1 cm^3 of both the hexagonal and the cubic form MoC_{1-x} by quenching from 2400°C into a tin bath.

3. *Tungsten carbides*

Three different tungsten carbides have been established: W_2C, WC, and WC_{1-x}; the first compound additionally occurs in two high-temperature forms, the latter is stable only at high temperatures. It is common practice to carbidize elemental tungsten, because it is much more convenient to reduce tungsten oxide to metal than to purify a contaminated tungsten carbide. Nevertheless, numerous publications deal with the preparation by the carbothermic reduction of tungsten oxides (compilation in Kieffer and Benesovsky, 1963).

W_2C

With W powder of a grain size smaller than 50 μm and lampblack as starting materials, the subcarbide can be prepared at 2000°C under vacuum within 30 min (E. Rudy and S. Windisch, unpublished).

The high-temperature forms can be partially retained when disks of W_2C are dropped, after equilibration at 2200 or 2500°C, into a tin bath (Rudy and Windisch, 1967).

WC_{1-x}

Kirner (1957) detected a cubic phase when he splashed a W_2C–WC melt on a cold substrate. Rudy *et al.* (1966) tried to retain this cubic WC_{1-x} by quenching in the same way as described before; alloys with the highest content of the cubic phase contained 38 atom % C.

WC

a. Powder metallurgical procedures. The carbidization of tungsten powder (grain size $<50\ \mu m$) is completed within 2 hr by firing at 1700–2000°C in a hydrogen atmosphere (Rudy *et al.*, 1966).

b. Melting methods. Arc-melting is frequently employed to prepare specimens up to 42 atom % C. At higher carbon content, free graphite arises from the peritectic decomposition of WC.

Various auxiliary metal baths (iron, cobalt, nickel, iron–30 nickel) are equally suitable for the preparation of WC in a menstruum.

Single crystals of WC have been grown in a melt of either nickel or cobalt (Pfau and Rix, 1954; Corteville and Pons, 1965; Takahashi and Freise, 1965).

Fusing is applied in industrial manufacture (see below).

c. Formation by means of volatile compounds. Just and Hanamann (1903, 1905, 1906) developed the so-called "substitution method": graphite fiber is heated in an atmosphere of $WCl_6 + H_2$, causing a deposition of tungsten; the temperature is then raised to 2000°C to complete the formation of WC. The deposition of WC by electrolysis of dissolved tungstate was investigated by Weiss (1946) and Andrieux and Weiss (1948).

Hurd *et al.* (1952) were able to produce fine-grained WC powder, e.g., suitable for hard metal production without further grinding: $W(CO)_6$ is pyrolytically decomposed at 650–1100°C and subsequently heated in an atmosphere of $4CO + CO_2$.

d. Technical production of tungsten carbides. The quality, especially the grain size of the final product, depends very much on the properties of the original tungsten oxide. Fine-grained tungsten carbide is usually desired. Very fine powders result from precipitated tungstic acid. In order to avoid grain growth during the reduction, the most favorable conditions of temperature, duration, and hydrogen drift have to be established. The carbidization is performed under hydrogen between 1400 and 1600°C.

On the other hand, a one-step reaction between either tungstic acid or tungsten oxide and carbon yields coarse-grained WC. The amount of carbon to be used has to be determined experimentally because of varying humidity and oxygen content.

Besides the powder metallurgical carbidization, the fusing process has become a significant mode of production. Tungsten or tungsten-containing scrap alloys are reacted with carbon to produce diverse types of alloys, from W_2C to the eutectic W_2C–WC, and modified W_2C, which contains up to 10 wt % of other metals (e.g., Nb, Ta, Fe). The powder mixture is tamped into a graphite crucible and heated in a hydrogen atmosphere. Carbidization starts at 1200°C, melting occurs around 2700°C. After exceeding 3000°C, the batch is cast into a preheated graphite die.

The quality of such cast alloys, especially hardness and wear resistance, improves considerably when titanium carbide instead of carbon is added to the tungsten powder (16–20 g of TiC for 100 g of tungsten powder). The molten charge has to be kept for a short period at 2900–3000°C to eliminate the inhomogeneities. Otherwise, cracks appear on the final product. Similar flaws are found if the melt is poured into a cold die. Therefore, the graphite die has to be preheated to 500–800°C. On the other hand, the die should be large enough to take in the excessive heat without exceeding 1200°C.

V. Intercarbide Systems

The expression "intercarbide systems" is widely used in the book on interstitial alloys by Goldschmidt (1967). The most significant carbide combinations concern the so-called monocarbides of transition elements, such as TiC–TaC, TiC–TaC–WC, TiC–WC, etc. Within the ternary system Me(I)–Me(II)–C, a distinctive readiness to form solid solutions of the cubic monocarbides was recognized long ago (Nowotny and Kieffer, 1947). Later, the volume condition (Hume-Rothery 15% rule) was confirmed for systems containing HfC and UC (Nowotny *et al.*, 1958, 1959). The same condition holds for a large number of carbonitrides and related systems (Duwez and Odell, 1950; Nowotny *et al.*, 1960).

With respect to the preparation of solid solutions (Me(I)Me(II))C, attention has to be paid to the limited solubility due either to different crystal structures (e.g., TiC–WC or TiC–TaC–WC), or volume differences (e.g., ZrC–VC), or to miscibility gaps at low temperatures in some systems (e.g., TiC–ZrC, TiC–HfC, VC–NbC, VC–TaC) (Kieffer *et al.*, 1968).

Besides the phases mentioned above, complex carbide phases were found in ternary and multicomponent systems. The elucidation of double carbides, such as H-phases, perovskite carbides, eta and kappa carbides, etc., is of interest for an understanding of the chemistry of carbides, because little is known about the physical properties and the chemical behavior of these particular carbides. The main reason for this lack of knowledge lies in the difficulty in reproducing the preparation technique of such compounds, because only a few of these so-called complex carbides have been isolated in pure form.

A. Mixed Metal Carbides

Generally, mixed metal carbides can be prepared analogously to the binary carbides, but because some of the mixed carbides are stable only within a certain temperature range or a limited composition region, knowledge of the phase diagram is an essential prerequisite for determining the conditions of preparation. Frequently, the product is not homogeneous after a single heat treatment, especially when binary carbides are used as starting material, because of the slow diffusion process. However, one has to bear in mind that diffusion is governed not only by the familiar parameters such as temperature and specific experimental factors, but is also influenced by accompanying elements, which are sometimes present only as traces. Thus, the intercarbide solution process can be considerably enhanced by voluntary addition of small amounts of elements, such as 1% Co, which are finally vaporized in a vacuum (Nowotny and Glenk, 1946; Nowotny and Kieffer, 1947; Nowotny *et al.*, 1951; Hinnüber and Kinna, 1961). The vapor pressure and the rate of evaporation of cobalt as a sinter aid have been determined by Rüdiger and Röttger (1965).

According to Kieffer and Benesovsky (1964), solid solutions of transition element carbides are subjected to a type of autopurification. Thus, vacuum-annealed solid solutions of 4A and 5A group carbides display higher wettability than the synthetic binary carbides.

B. Complex Carbides T–M–C (T = transition metal, M = metametal)

A variety of carbides result from the reaction between a usually high-melting transition metal, rather low-melting elements such as Zn, Al, Ga, Sn, Pb, etc., and C. The preparation of homogeneous compounds with definite composition encounters some difficulties which arise from the high stability of transition element carbides and from the high vapor pressure of the metametals at the temperature needed just to perform the reaction within a reasonable duration. For these reasons, such compounds have been synthesized mainly by hot pressing and/or long-duration annealing (Nowotny *et al.*, 1965). Hot pressing is most successful if the components are heated rapidly under low pressure. To be safe, it is advisable to pulverize the specimen, and to anneal the compacted material under conditions which are better defined than hot pressing. Such homogenizing was done on a large number of complex carbides by either heating in sealed quartz tubes at 700–900°C for a period of up to 1500 hr (Jeitschko *et al.*, 1963a, 1964), or by annealing in a high vacuum at about 1200°C (Wolfsgruber *et al.*, 1967; Haschke *et al.*, 1967).

VI. Metal–Nonmetal Carbides

Long ago, it was recognized that solid solutions like Me(C, N, O) are easily formed, where Me is a transition element. Besides these carbonitrides and oxycarbides, other carbides containing H, B, Si, P, and S have been described in the literature. The preparation methods are not standardized, and are sometimes rather laborious. Only some examples are treated here.

A. Me–{H, N, O}–C

1. Me–H–C

Ternary metal–hydrogen carbides were first described by Rexer and Peterson (1964). These authors prepared Ti, Zr, Hf, V, Y, La, and Th hydride carbides, starting from the metal carbides, which were subsequently hydrogenated, e.g., Ti_2CH_x is formed at 940°C and 550 Torr after an equilibration of two months. Goretzki *et al.* (1964) applied higher hydrogen pressure (up to 75 atm) and obtained saturation within a few hours on originally substoichiometric carbides. The ternary compound depends on the initial ratio of metal to carbon, e.g., either the cubic zirconium hydrogen carbide or the hexagonal Zr_2HC can be obtained.

2. Me–N–C

Most of the carbides react with nitrogen above 1100°C (Zelikman *et al.* 1947; Kieffer *et al.*, 1970); only a few, e.g., TaC, WC, HfC–TaC, withstand the attack of elemental nitrogen. This reaction yields either a solid solution of carbide + nitride or, in a few instances, ternary compounds such as $Cr_3C(C,N)$ (Ettmayer, 1966; Ettmayer *et al.*, 1966). Correspondingly, these metal–nitrogen carbides, usually called carbonitrides, can be syn-

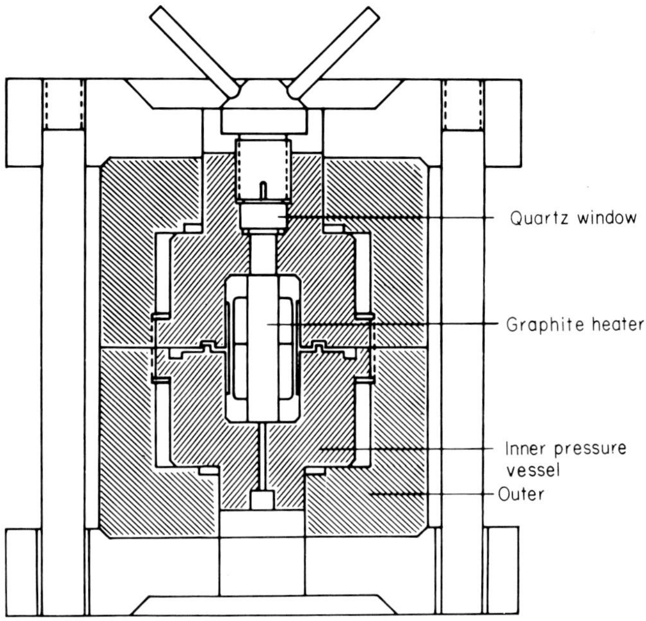

FIG. 3. A high-temperature autoclave (Ettmayer and Kieffer, 1970).

thetized by firing a mixture of carbide + nitride. Thus, Duwez and Odell (1950) prepared carbonitrides of Ti, V, Nb and solutions thereof at 2300–2600°C in an atmosphere of nitrogen (1 atm). Hot pressing has become the main mode of preparation (Bittner *et al.*, 1963). Some of these compounds exhibit a high stability, e.g., Ti(C, N), which is partly counterbalanced by the increasing dissociation pressure when the temperature is raised. For this reason, arc-melting is rejected as a preparation technique, and high-pressure furnaces have attracted increased interest. The progress in preparation by means of autoclaves was recently reviewed by Ettmayer *et al.* (1969) and Ettmayer and Kieffer (1970). Figure 3 shows the scheme of a furnace which is applicable for temperatures up to 2500°C and pressures up to 2000 atm.

3. Me–O–C

The refractory carbides, especially those of groups 4 and 5, are capable of dissolving oxygen, partly by saturating the carbon deficiency and partly by substituting the carbon atoms. To this effect is further superimposed the fact that some oxides form solid solutions with carbides (e.g., TiC–TiO,

VC–VO) with the oxygen atoms rather randomly distributed. Compounds like Nb(C, O) and Ta_2(C, O), whose stability at high temperatures is lower than that of the above-mentioned solid solutions, are prepared either by hot pressing or by annealing a mixture MeC + MeO in a sealed capsule to avoid loss of CO. On the other hand, rare-earth–oxygen carbides of the formula Ln_4O_3C (Ln = La, Nd, Gd, Ho, Er) have been prepared by arc-melting of mixtures of Ln_2O_3 + C in a helium atmosphere (Butherus and Eick, 1968).

4. Me–B–C

Ternary metal–boron–carbon compounds have been hitherto found with elements of groups 3A (especially lanthanides and actinides) and 6–8A. They were prepared either by melting—ScB_2C_2 (Smith *et al.*, 1965), LnB_2C_2 (Smith and Gilles, 1967), $Fe_{23}(B, C)_6$ (Stadelmaier and Gregg, 1963)—or by hot pressing—Th–B–C (Toth *et al.*, 1961b), UBC (Toth *et al.*, 1961a), Mo_2BC (Jeitschko *et al.*, 1963b; Rudy *et al.*, 1963).

5. Me–Si–C

Lanthanide–silicon carbides of formula Ln_5Si_3C have been prepared by Mayer and Shidlovsky (1969), reacting the elements at 1500–1600°C. Each step of the operation was carried out in an inert atmosphere.

Uranium–silicon carbides can be synthesized either by arc-melting of the components or by firing at 1500–2000°C of a mixture of finely divided UC and SiC (Smith *et al.*, 1964). A well-defined compound of the composition $U_3Si_2C_3$ was prepared by arc-melting and subsequent annealing at 1700°C (Blum *et al.*, 1965).

Another example worth mentioning is the compound Ti_3SiC_2, which is formed in a peritectic reaction: Pulverized TiC is heated with an excess of $TiSi_2$ in a closed graphite capsule to 2000°C and then cooled within 20 min to 1200°C; platelets of crystal fragments can be isolated from the matrix (Jeitschko and Nowotny, 1967).

B. Me–{P, As, S}–C

Complex carbides, containing either phosphorus, arsenic, or sulfur, are prepared most advantageously in a closed system. Sealed quartz tubes are very suitable for this purpose. Usually, a mixture of the elemental powders is reacted. Because of the relatively high vapor pressures of phosphorus, arsenic, and sulfur, the heating rate has to be very low. For instance, Nb_2PC, Nb_2SC, and V_3AsC were prepared by this mode; at the end of the reaction, the final (maximum) temperature was 1150°C (Boller and Nowotny, 1968). It is noteworthy that such compounds can be recrystallized at even higher temperatures.

The loss of the nonmetallic elements sometimes can be prevented by the reaction of master alloys instead of the elements. Thus, e.g., Kudielka (1965) prepared Hf_2FeSC_2 by melting two alloys at 1600°C, one consisting of iron and carbon, the other containing iron, hafnium, and sulfur.

REFERENCES

Achard, J. C. (1966). *Rev. Int. Hautes Temp. Refract.* **3,** 281–289.

Ambarcumjan, R. S., and Babich, B. N. (1969). *Neorg. Mater.* **5,** 301–304.

Andrieux, J. L., and Weiss, G. (1948). *Bull. Soc. Chim. Fr.* **1948,** 598.

Atoda, T., Kabayashi, M., Sasa, Y., Takahashi, Y., and Higashi, I. (1964). *In* "Carbides in Nuclear Energy" (*Proc. Symp., Harwell, England, 1963*) (L. E. Russell *et al.*, eds.), pp. 496–517. Macmillan, New York.

Bečvař, J. (1969). *J. Nucl. Mater.* **32,** 156–160.

Bildstein, H., and Knotik, K. (1967). *Kerntech. Isotopentech. Chem.* **9,** 57–62; Austrian Patents: 262 931, 267 006, 267 007, 267 008.

Bildstein, H., and Knotik, K. (1970). Private communication.

Bittner, H., Goretzki, H., Benesovsky, F., and Nowotny, H. (1963). *Mh. Chem.* **94,** 518–526.

Blum, P. L., Guinet, P., and Silvestre, G. (1965). *C. R. Acad. Sci. Paris* **260,** 1911.

Boller, H., and Nowotny, H. (1968). *Mh. Chem.* **99,** 721.

Bourdeau, R. G. (1968). US Patent 3,399,980; *Chem. Abstr.* **70,** 5579a.

Brenner, S. S. (1963). *In* "The Art and Science of Growing Crystals" (J. J. Gilman, ed.). Wiley, New York.

Butherus, A. D., and Eick, H. A. (1968). *J. Amer. Chem. Soc.* **90,** 1715.

Calverly, A., Davis, M., and Lever, R. F. (1957). *J. Sci. Instrum.* **34,** 142.

Campbell, I. E., Powell, C. F., Nowicki, D. H., and Gonser, B. W. (1949). *J. Electroch. Soc.* **96,** 318–333.

Chermant, J. L., and Deschanvres, A. (1968). *Bull. Soc. Chim. Fr.* **1968,** 2373–2374.

Claugherty, E. V., Lothrop, K. L., and Kafalas, J. H. (1961). *Nature* **191,** 1194.

Corteville, J., and Pons, L. (1965). *C. R. Aci. Sci. Paris* **260,** 4477–4480.

DuPont de Nemours, Wilmington, Delaware. (1970). Austr. Patent 271 383.

Duwez, P., and Odell, F. (1950). *J. Electrochem. Soc.* **97,** 299.

Eisenkolb, F. (1963). "Grundlagen der Pulvermetallurgie." Akad. Verlag, Berlin.

Ettmayer, P. (1966). *Mh. Chem.* **97,** 1248.

Ettmayer, P., Vinek, G., and Rassaerts, H. (1966). *Mh. Chem.* **97,** 1258–1262.

Ettmayer, P., Priemer, H., and Kieffer, R. (1969). *Metallurgia* **23,** 307–310.

Ettmayer, P., and Kieffer, R. (1970). *Radex Rdsch.* **70/3,** 191.

Fleischer, L. R., and Tobin, J. M. (1969). *In* "High Temperature Technology" (N. K. Hiester, ed.), pp. 665–672. Butterworth, London.

Fletcher, J. M., Hyde, K. R., and Landman, D. A. (1964). *In* "Carbides in Nuclear Energy (*Proc. Symp., Harwell, England, 1963*) (L. E. Russell *et al.*, eds.), pp. 550–555. Macmillan, New York.

Goldschmidt, H. J. (1967). "Interstitial Alloys." Butterworth, London.

Goretzki, H., Bittner, H., and Nowotny, H. (1964). *Mh. Chem.* **95,** 1521.

Grigoreva, V. V., and Klimenko, V. N. (1959). *Issl. Zharo. Splav. Akad. Nauk SSSR Inst. Met.* **4**, 317.

Harrison, I. D. L., and Isaacs, J. W. (1964). *In* "Carbides in Nuclear Energy" (*Proc. Symp., Harwell, England, 1963*) (L. E. Russell *et al.*, eds.), pp. 556–571. Macmillan, New York.

Haschke, H., Nowotny, H., and Benesovsky, F. (1966). *Mh. Chem.* **97**, 1469.

Hinnüber, J., and Kinna, W. (1961). *Tech. Mitt. Krupp.* **19**, 130–153.

Hoyaux, M. F. (1970). *High Temp. High Pressure* **2**, 17–30;

Hoyaux, M. F. (1968). "Arc Physics." Springer, New York.

Hurd, D. T., McEntee, H. R., and Brisbin, P. H. (1952). *Ind. Eng. Chem.* **44**, 2432–2435.

Jedlicka, H., Nowotny, H., and Benesovsky, F. (1971). *Mh. Chem.* **102**, 389–403.

Jeitschko, W., and Nowotny, H. (1967). *Mh. Chem.* **98**, 329.

Jeitschko, W., Nowotny, H., and Benesovsky, F. (1963a). *Mh. Chem.* **94**, 672, 844, 1198, 1201.

Jeitschko, W., Nowotny, H., and Benesovsky, F. (1963b). *Mh. Chem.* **94**, 65.

Jeitschko, W., Nowotny, H., and Benesovsky, F. (1964). *Mh. Chem.* **95**, 156, 178, 319, 431, 436; *J. Less Common Metals* **7**, 133–138.

Jones, F. O. (1961). Zone melting of carbides (A Report) Ass. El. Indust. Lim., Aldermaston, Berkshire, England.

Just, A., and Hanamann, F. (1903). D.R.P. 154 262; (1904). *Chem. Z.* **1904**(12), 927.

Kieffer, R. (1969). *J. Inst. Met.* **97**, 164–172.

Kieffer, R., and Benesovsky, F. (1963). "Hartstoffe." Springer, Wien.

Kieffer, R., and Benesovsky, F. (1964). *In* "Encyclopedia of Chemical Technology" (R. E. Kirk and D. F. Othmer, eds.), 2nd ed., Vol. 4, pp. 70–100. Wiley, New York.

Kieffer, R., and Kölbl, F. (1949). *Powder Met. Bull* **4**, 4–17.

Kieffer, R., Nowotny, H., Neckel, A., Ettmayer, P., and Usner, L. (1968). *Mh. Chem.* **99**, 1020.

Kieffer, R., Nowotny, H., Ettmayer, P., and Freudhofmeier, M. (1970). *Mh. Chem.* **101**, 65–82.

Kilin, V. S., Evstyukhin, A. I., and Dergunova, V. S. (1969). *Por. Met.* **12**, 32–38.

Kirner, K. (1957). Quoted as private communication in Kieffer and Benesovsky (1963).

Knotik, K., Bildstein, H., and Koss, P. (1965). *Kernt. Isotopent. Chem.* **7**, 361–365.

Kolchin, O. P., and Berlin, I. K. (1964). *Tsvet. Metall.* **37**(8), 67–71.

Koref, F. (1922). *Z. Elektrochem.* **28**, 511–517.

Kosolapova, T. Ya., and Samsonov, G. V. (1959). *Zh. Prikl. Khim.* **32**, 1505.

Kroll, W. J., Schlechten, A. W., Carmody, W. R., Yerkes, L. A., Holmes, H. P., and Gilbert, H. L. (1947). *Trans. Electrochem. Soc.* **92**, 187–201.

Krupka, M. C. (1970). *J. Less Common Metals* **20**, 135–140.

Kudaka, K., and Nakase, E. (1968). *J. Jap. Soc. Powd. Met.* **15**, 372–374.

Kudielka, H. (1965). *Z. Kristallogr.* **121**, 2–4.

Lesser, R., and Brauer, G. (1958). *Z. Metkde.* **49**, 622.

Lofgren, N. L., and Krikorian, O. H. (1964). *In* "Carbides in Nuclear Energy" (*Proc. Symp. Harwell, England, 1963*) (L. E. Russell *et al.*, eds.), pp. 315–329. Macmillan, New York.

Matyushenko, N. N., and Svinarenko, O. P. (1968). *Ukr. Fiz. Zh.* **13**(7), 1083–1085.

Mayer, I., and Shidolovsky, I. (1969). *Inorg. Chem.* **8**, 1240.

McKenna, P. M. (1939). *Metal Progr.* **36**, 152–155.

Meerson, G. A., Kiparisov, S. S., Gurevich, M. A., and Fen-Syan, Den. (1965). *Porosh. Met.* **2**(26), 15–21.

Meerson, G. A., Krejn, O. E., Kiparisov, S. S., and Narva, V. K. (1970). *Izv. Akad. Nauk SSSR, Ser. Neorgan Mater.* **6,** 10, 1749–1752.
Moers, K. (1931). *Z. Anorg. Allg. Chem.* **198,** 243.
Moissan, H. (1893). *C. R. Acad. Sci. Paris* **116,** 349, 1225.
Moissan, H. (1895). *C. R. Acad. Sci. Paris* **120,** 290, 1320.
Moissan, H. (1896). *C. R. Acad. Sci. Paris* **122,** 1297.
Moissan, H. (1897a). *Proc. Roy. Soc.* **60,** 156–160.
Moissan, H. (1897b). *C. R. Acad. Sci. Paris* **125,** 839.
Moissan, H., and Hoffmann, H. K. (1904). *C. R. Acad. Sci. Paris* **138,** 1558.
Nadler, H. (1963). Rep. NAA-SR-7465. Atomics International, Canoga Park, California.
Neuenschwander, K. (1966). *J. Less Common Metals* **11,** 365–375.
Norton, J. T., and Lewis, R. K. (1963). NASA-CR-321. National Aeronautics and Space Administration, Washington, D.C.
Nowotny, H. (1962). *In* "Advances in X-Ray Analysis" (W. M. Mueller, ed.), Vol. 5, pp. 13–32. Plenum, New York.
Nowotny, H., and Glenk, G. (1947). *Metallforschung* **2,** 265–269.
Nowotny, H., and Kieffer, R. (1947). *Metallforschung* **2,** 257–265.
Nowotny, H., and Kieffer, R. (1952). *Z. Anorg. Allg. Chem.* **267,** 261–264.
Nowotny, H., Kieffer, R., and Knotek, O. (1951). *Berg. Hüttenm. Mh.* **96,** 6–8.
Nowotny, H., Kieffer, R., and Benesovsky, F. (1958). *Rev. Metall.* **55,** 453.
Nowotny, H., Kieffer, R., Benesovsky, F., Brukl, Ch., and Rudy, E. (1959). *Mh. Chem.* **90,** 669.
Nowotny, H., Benesovsky, F., and Rudy, E. (1960). *Mh. Chem.* **91,** 348.
Nowotny, H., Jeitschko, W., and Benesovsky, F. (1965). *In* "Symposium sur la Metallurgie des Poudres," pp. 239–258. Ed. Metaux, Saint Germain-en-Laye.
Ogawa, K., and Bando, Y. (1959). *Funtai Oyobi Funmatsuyakin* **6,** 160.
Paris, R. A., and Clar, E. (1969). *Chim. Ind. Gen. Chim.* **102,** 57.
Pearce, M. L., and Marek, R. W. (1968). *J. Amer. Ceram. Soc.* **51,** 84–87; disc. p. 355.
Pfau, H., and Rix, W. (1954). *Z. Metkd.* **45,** 116–118.
Pons, L. (1967). *In* "Anisotropy in Single Crystal Refractory Compounds" (F. W. Vahldiek and S. A. Mersol, eds.). Plenum, New York.
Precht, W., and Hollox, G. E. (1968). *J. Crystal Growth* **3, 4,** 818–823.
Pring, J. N., and Fielding, W. (1909). *J. Chem. Soc.* **95,** 1497.
Ramqvist, L. (1966). *Powder Met.* **17,** 1–46.
Rassaerts, H., Nowotny, H., Vinek, G., and Benesovsky, F. (1967). *Mh. Chem.* **98,** 460–468.
Reed, T. B. (1969). *In* "High Temperature Technology" (N. K. Hiester, ed.), pp. 655–664. Butterworth, London.
Rexer, J., and Peterson, D. T. (1964). IMD Spec. Rep. No. 13, Amer. Inst. Min. Met. and Petr. Eng.
Rüdiger, O., and Röttger, H. (1964). *In* "Symposium sur la Metallurgie des Poudres," pp. 259–269. Ed. Metaux, Saint Germain-en-Laye.
Rudy, E. (1965). AFML-TR-65-2, Pt. I, Vol. 4. Air Force Materials Lab., Res. and Tech. Div., Air Force Systems Command, Wright-Patterson A.F.B., Ohio.
Rudy, E. (1969). AFML-TR-65-2 Pt 5, Compendium of Phase Diagram Data. Air Force Materials Lab., Res. and Tech. Div., Air Force Systems Command, Wright-Patterson A.F.B., Ohio.
Rudy, E., and Harmon, D. P. (1966). AFML-TR-65-2, Pt. I, Vol. 5. Air Force Materials Lab., Res. and Tech. Div., Air Force Systems Command, Wright-Patterson A.F.B., Ohio.

Rudy, E., and Windisch, S. (1967). *J. Amer. Ceram. Soc.* **50,** 272–273.
Rudy, E., Rudy, El., and Benesovsky, F. (1962). *Planseeber.* **10,** 42–64.
Rudy, E., Benesovsky, F., and Toth, L. (1963). *Z. Metkd.* **54,** 345–353.
Rudy, E., Harmon, D. P., and Brukl, C. E. (1965a). AFML-TR-65-2, Pt. I, Vol. 2. Air Force Materials Lab., Res. and Tech. Div., Air Force Systems Command, Wright-Patterson A.F.B., Ohio.
Rudy, E., Windisch, S., and Chang, Y. A. (1965b). AFML-TR-65-2, Pt. I, Vol. 1. Air Force Materials Lab., Res. and Tech. Div., Air Force Systems Command, Wright-Patterson A.F.B., Ohio.
Rudy, E., Windisch, S., and Hoffmann, J. R. (1966). AFML-TR-65-2, Pt. I, Vol. 6. Air Force Materials Lab., Res. and Tech. Div., Air Force Systems Command, Wright-Patterson A.F.B., Ohio.
Rudy, E., Windisch, S., Stoscik, A. J., and Hoffman, J. R. (1967). *Trans. AIME* **239,** 1247–1267.
Rudy, E., Windisch, S., and Brukl, C. E. (1968). *Planseeber.* **16,** 3–33.
Rundqvist, S., and Runnsiö, G. (1969). *Acta Chim. Scand.* **23,** 1191.
Russell, L. E., Bradbury, B. T., Harrison, J. D. L., Hedger, H. J., and Mardon, P. G., (eds.). (1964). "Carbides in Nuclear Energy" (*Proc. Symp. Harwell, England, 1963*). Macmillan, New York.
Smith, P. K., and Gilles, P. W. (1967). *J. Inorgan. Nucl. Chem.* **29,** 375–382.
Smith, G. V., Smith, R. G., and Thomas, A. G. (1964). *In* "Carbides in Nuclear Energy" (*Proc. Symp. Harwell, England, 1963*) (L. E. Russell *et al.*, eds.), pp. 261–265. Macmillan, New York.
Smith, G. S., Johnson, Qu., and Nordine, P. C. (1965). *Acta Crystallogr.* **19,** 668–673.
Stadelmaier, H. H., and Gregg, R. A. (1963). *Metall* **17,** 412.
Stanford Research Institute. (1969). "High Temperature Technology" (N. K. Hiester, ed.). Butterworth, London.
Storey, R. N., and Landise, R. A. (1970). *J. Crystal Growth* **6**(3), 261–265.
Storms, E. K. (1966). "Thermodynamics," Vol. 1. Int. At. Energy Agency, Vienna.
Storms, E. K. (1967). "The Refractory Carbides." Academic Press, New York.
Storms, E. K., and Krikorian, N. H. (1960). *J. Phys. Chem.* **64,** 1471.
Sugiyama, K., and Takahashi, T. (1970). *J. Chem. Soc. Japan. Ind. Chem. Sect.* **73,** 2105–2108.
Takahashi, T., and Freise, E. J. (1965). *Phil. Mag.* **12,** 1–8.
Toth, L., Benesovsky, F., Nowotny, H., and Rudy, E. (1961a). *Mh. Chem.* **92,** 956–960.
Toth, L., Nowotny, H., Benesovsky, F., and Rudy, E. (1961b). *Mh. Chem.* **92,** 794.
Troost, L. (1865). *C. R. Acad. Sci. Paris* **61,** 109.
Vahldiek, F. W., Mersol, S. A., and Lynch, C. T. (1966). *Trans. AIME* **236,** 1490.
van Arkel, A. E., and de Boer, J. H. (1925). *Z. Anorg. Allg. Chem.* **148,** 345.
Weiss, G. (1946). *Ann. Chim.* (*Paris*) **1**(12), 446.
Wilcox, W. R., Teviotdale, J. R., and Corley, R. A. (1968). *Trans. AIME* **242,** 588–591.
Williams, W. S. (1964). *Phys. Rev.* **135,** A505.
Wolfsgruber, H., Nowotny, H., and Benesovsky, F. (1967). *Mh. Chem.* **98,** 2403.
Yvon, K., and Parthé, E. (1970). *Acta Crystallogr.* **B26,** 149–153.
Zelikman, A. N., Loseva, S. S., and Tsetlina, N. J. (1947). *Tsvet. Metall.* **20,** 41.

Preparation of Chalcogenides and Pnyctides at Low Temperature

W. KWESTROO

Philips Research Laboratories
N. V. Philips' Gloeilampenfabrieken
Eindhoven, The Netherlands

I. Introduction

In preparative chemistry, purity has become a more and more important factor. In particular, compounds which are required for applications such as semiconducting, luminescent, or superconducting materials have to be prepared in a high state of purity. Pnyctides and chalcogenides of metallic elements are widely used for these applications.

A limited amount of these compounds can be made by precipitation from solution. For example, many sulfides and, to a lesser extent, some selenides are prepared by precipitation from solutions in water. Most chalcogenides and pnyctides, however, are synthesized directly from the elements (Ray, 1969; Hiscocks and Mullin, 1969). This seems to be an attractive preparation method, since many elements can be obtained in a very pure form. For most compounds, the procedure given in the literature is the heating of the elements together in evacuated, sealed ampoules. In many cases, this must be done at high temperatures (above 1000°C) and over extended reaction times (from days to weeks). Even then, inhomogeneous reaction products are the result. In order to accelerate the reaction and to obtain more homogeneous materials at lower temperatures, the component elements are often ground into small particles. Reisman and Berkenblit (1963), for instance, were able to prepare compounds like CdSe and ZnTe at 450–500°C by reducing the higher-melting or less volatile elements into the form of a fine powder and applying a very low heating rate (about 1°C min^{-1}). It is difficult, however, to avoid introduction of contaminants on reducing the materials to small-particle form.

It has been pointed out by Reisman and Berkenblit that in many cases, the reactions start at a temperature of 400–500°C, but stop immediately after a thin layer of the compound has formed on the surface of the higher-melting element. For instance, in the reaction of cadmium and selenium, at low temperature (about 400°C), an impermeable skin of CdSe is formed around the particles of cadmium metal. No reaction is observed up to about 1000°C. Then, the CdSe skin weakens (melting point of CdSe is 1230°C) and the vapor pressure of the metal becomes appreciable. Because of its exothermic character, the reaction starts with explosive intensity. The synthesis of ZnTe (Godau and Ormont, 1968) and of GaP (Miller, 1962) from the elements are other examples showing these difficulties.

If it were possible to prevent deposition of the inhibiting layer of the high-melting reaction product, the reaction would proceed gradually and at low temperature.

II. Preparation from the Elements Using Halides

The formation of an inhibiting layer can be prevented by carrying out the reactions in a suitable reaction medium in which the compound to be formed is more or less soluble. It is known that chalcogenides are soluble

TABLE I

MELTING POINTS OF IODIDES[a]

III A	IV B	V B	VI B	VII B	VIII	VIII	VIII	I B	II B	III B	IV A	V A
50 B 5												
191 Al 13											120 Si 14	61 P 15
Sc 21	600(II) Ti 22	750(II) V 23	856(II) Cr 24	638(II) Mn 25	502(II) Fe 26	515(II) Co 27	797 Ni 28	605(I) Cu 29	446 Zn 30	212 Ga 31	144 Ge 32	146 As 33
1004 Y 39	Zr 40	Nb 41	Mo 42	Tc 43	Ru 44	Rh 45	350 Pd 46	558(I) Ag 47	387 Cd 48	210 In 49	144 Sn 50	170 Sb 51
772 La 57	Hf 72	Ta 73	W 74	Re 75	Os 76	100 Ir 77	360(II) Pt 78	120(I) Au 79	259 Hg 80	440(I) Tl 81	402 Pb 82	408 Bi 83

Ce	Pr	Nd	Pm	Sm	Eu	Gd	Tb	Dy	Ho	Er	Tm	Yb	Lu
752	737	775		527(II)	527(II)	926	946	955	989	1020	1015	780(II)	1050
58	59	60	61	62	63	64	65	66	67	68	69	70	71

[a] According to Weast (1967). Chalcogenides and pnyctides of the elements, found in the part of the periodic system given in this table, were synthesized with the use of the iodides as a reaction medium.

in metal halides, since they can be recrystallized from these melt salts (Leverenz, 1950). Thus the reaction may proceed smoothly if a halide is used as reaction medium in the synthesis of chalcogenides starting with the elements. Experiments showed that synthesis from elements proceeds quickly at low temperatures in an evacuated silica tube if a metal halide is added (Kwestroo *et al.*, 1969). For reasons of chemical purity, the halide can be chosen containing the same metal as the chalcogenide to be synthesized. Thus ZnTe was prepared from the elements at 600°C using ZnI_2 (mp 446°C) as a "flux" in an evacuated silica tube.

Of the halides, iodides are especially suitable, since other halides often occur with water of crystallization, which results in decomposition into oxides or even explosion of the silica tubes. Another advantage of the use of iodides is the low melting point of these compounds as compared to the melting points of the other halides. A survey of the melting points of the metal iodides as given by Weast (1967) is given in Table I.

The use of iodides is also attractive from the point of view of their ease of formation from the elements at low temperature (Colton and Canterford, 1969; Canterford and Colton, 1968; Brown, 1968). This is especially useful when the metal iodide is easily hydrolyzed upon exposure to the atmosphere. Then, the iodide is made *in situ* from excess metal and the calculated amount of iodide. When the evacuated silica tube is heated gradually, the metal iodide is formed first at low temperature. When the temperature is raised, the remaining amount of metal then reacts with the other element, resulting in the formation of the compound required. An example is the synthesis of gallium phosphide, GaP. Appropriate amounts of gallium, red phosphorus, and iodine are heated up to 250°C in an evacuated tube. The temperature is kept at 250°C for 1 hr, during which period GaI_3 is formed. Then, the temperature is gradually increased to 500°C and kept there for 12 hr. The remaining gallium reacts smoothly with phosphorus, forming GaP. In this procedure, about 50% of the total weight is GaI_3. With small amounts of metal iodide, e.g., 10%, there was no smooth reaction, and indeed the course of the reaction was often found to be explosive.

Best results are generally obtained with a substantial amount of metal halide: 50% by weight of the compound formed is quite suitable. Another indication that the molten halide is the working constituent in this procedure is the relation between the temperature at which the reaction starts and the melting point of the halide used. Compounds of a metal having a halide with a high melting point start their formation at temperatures slightly above this point. For instance, chromium and selenium in CrI_2 did not react below about 850°C (mp of CrI_2 is 856°C), whereas the reaction of cadmium and selenium in CdI_2 is complete in a few hours at 500°C (mp of CdI_2 is 387°C). Compounds of the rare-earth metals also need rather high reaction temperatures, which is in accordance with the high melting points of the iodides of these elements (Table I).

Although application of the relatively high heating temperatures does introduce some of the difficulties mentioned earlier, the "halide method" remains attractive. The heating temperatures are still a few hundred degrees below those generally used in other procedures and the heating period is much shorter. Another advantage of the halide preparation method is the formation via the molten phase. After the metal halide is removed, a finely divided, homogeneous powder is the result.

III. Removal of the Halide

Generally, the removal of the halide poses no special problems. If the iodide is soluble in water and the compound is stable toward water, a simple washing procedure is used. This is applied to CdSe, ZnTe, and similar compounds. In special cases, aqueous solutions can be used. Thus

CuI can be removed with aqueous KI solution or with ammonia. In other cases, nonaqueous solvents, such as *N*-methyl pyrrolidone or dimethyl formamide, are suitable. Halides of the rare-earth elements, e.g., are soluble in dimethyl formamide. The more volatile halides can be sublimated in vacuum, or removed by heating the product in a stream of inert gas. The result of these halide-removing procedures is that about 0.1–1 wt% halide is still left in the compound formed. In some cases, such an amount does not affect the physical properties to be studied, or it may be tolerated in view of the further use of the compound, e.g., as a source in transport reactions. For instance, the iodine content of GaP transported by HCl was 0.1 ppm. With suitable aftertreatments (recrystallization), the halogen content of a compound may often be reduced to a very low value.

IV. Purity of the Products

Most compounds prepared by this method are synthesized at about 600°C. At this temperature, there is scarcely any contamination from the wall of the vessel. Thus the chemical purity is determined by the purity of the reacting elements and that of the reaction medium. These materials are available in a very pure state, so the required compounds can also be of high chemical purity. This is demonstrated in Tables II and III for CdTe and GaP, respectively. Details about preparation are given in Section VI. The purity of CdTe is determined by the purity of Cd, Te, and CdI_2.

GaP is prepared from Ga, P, and I. The iodine content is reduced to 500 ppm by heating the material at 250°C in a stream of H_2 gas. The purity of this product is given in Table III.

TABLE II

SPECTROCHEMICAL ANALYSIS OF STARTING MATERIALS AND OF THE RESULTING CdTe[a]

	Substance			
Impurity	Cd	Te	CdI_2	CdTe
Si	1	3	2	8
Mg	2	<0.9	1	<0.4
Fe	2	2	2	2
Ag	<0.4	<0.8	<0.8	<0.6
Cu	<0.07	<0.3	<0.5	<0.3

[a] Impurity concentrations in ppm by weight.

TABLE III

Spectrochemical Analysis of P and GaP, and Mass Spectrochemical Analysis of Ga and I (ppm)

Impurity	Substance[a]			
	Ga	P	I	GaP
Si	<0.04	1	nd	5
Mg	<0.07	<0.5	<0.05	<0.1
Fe	<0.02	<2	<0.05	<2
Al	<0.08	<0.5	nd	<0.7
Cu	nd	<0.1	<0.01	<0.1

[a] nd = not detected. Si in GaP is a contaminant from the Si ampoule.

For compounds for which the corresponding metal halides have melting points above 700°C, contamination by material of the silica vessel may occur in an increasing amount when the compounds are synthesized at temperatures somewhat above these. It is known that above 800°C, air may diffuse into the evacuated quartz ampoules (Golubkov *et al.*, 1966). Metals like manganese and rare-earth elements readily react with oxygen, and where the metal is in contact with the wall of the ampoule, the silica is corroded by these oxides. This may result in explosions after a short heating time. If the reactions are carried out in graphite crucibles placed in quartz ampoules, some oxide may still be found, but explosions no longer occur.

V. Characteristics and Applicability of the Method

Summarizing, the characteristics of the procedure are:

1. Complete reaction in one run.
2. Relatively short reaction time.
3. Low reaction temperature.
4. Finely divided, homogeneous reaction products.
5. Very pure products.

Up to now, about 50 compounds have been prepared by means of this method. Table IV lists these compounds, together with such data as melting points of compound and halide, heating temperature and time, and technique of halide separation. The method described has proven to be

TABLE IV

COMPOUNDS PREPARED ACCORDING TO THE HALIDE METHOD

Compound	Melting point (°C)	Halide used	Melting point of halide (°C)	Temperature of preparation (°C)	Heating time (hr)[a]	Method of metal halide separation[b]
ZnS	1020	ZnI_2	446	600	18	A
ZnSe	>1100	$ZnCl_2$	275	500	18	A
		ZnI_2	446	600	18	A
ZnTe	1238	$ZnCl_2$	275	500	18	A
		ZnI_2	446	600	18	A
CdS	—	CdI_2	387	600	6	A
CdSe	>1350	CdI_2	387	600	6	A
CdTe	1041	CdI_2	387	600	6	A
HgS	584	HgI_2	259	400	6	B
HgSe	—	HgI_2	259	400	6	B
GaP	—	$Ga + I_2$ (GaI_3)	212	500	6	C
GaAs	1238	$Ga + I_2$ (GaI_3)	212	400	6	C
Ga_2As_3	—	$Ga + I_2$ (GaI_3)	212	400	6	C
GaS	965	$Ga + I_2$ (GaI_3)	212	600	12	C
Ga_2S_3	1256	$Ga + I_2$ (GaI_3)	212	600	12	C
GaSe	960	$Ga + I_2$ (GaI_3)	212	600	12	C
Ga_2Se_3	>1020	$Ga + I_2$ (GaI_3)	212	600	12	C
InP	1070	$In + I_2$ (InI_3)	210	600	6	B
InAs	943	$In + I_2$ (InI_3)	210	600	6	A
InSb	535	$In + I_2$ (InI_3)	210	600	6	A
InSe	890	$In + I_2$ (InI_3)	210	600	6	A
SiS_2	1090[c]	$Si + I_2$ (SiI_4)	121	600	6	D
GeAs	—	$Ge + I_2$ (GeI_4)	144	600	6	D
$GeAs_2$	—	$Ge + I_2$ (GeI_4)	144	600	6	D
GeS	530	$Ge + I_2$ (GeI_4)	144	600	6	D
GeS_2	~800	$Ge + I_2$ (GeI_4)	144	600	6	D

TABLE IV (continued)

Compound	Melting point (°C)	Halide used	Melting point of halide (°C)	Temperature of preparation (°C)	Heating time (hr)[a]	Method of metal halide separation[b]
TiSe	—	TiI_2	600	700	24[d]	E
VSe	—	VI_2	750–800	800	24[d]	E
$NbTe_2$	—	$NbCl_5$	205	600	8	F
CrS	1550	$CrCl_2$	824	900	18	A
		CrI_2	868	900	18	A
CrSe	—	$CrCl_2$	824	900	18	A
		CrI_2	868	900	18	A
Cr_2Se_3	—	$CrCl_2$	824	900	18	A
MoS_2	1185	$MoCl_5$	194	600	8	D
$MoSe_2$	—	$MoCl_5$	194	600	8	D
WS_2	~1250	WCl_5	275	600	8	B
WSe_2	—	WCl_5	275	600	8	B
MnSe	—	$MnCl_2$	650	800	18[d]	G
MnTe	—	MnI_2	638	700	18[d]	E
FeP	1290	FeI_2	502	600	8	H
FeAs	1020	FeI_2	502	600	8	H
FeSe	—	FeI_2	502	650	12	E
CoSe	*e*	CoI_2	515	600	18	A
NiSe	*e*	NiI_2	797	850	24	A
Cu_2S	1100	CuI	605	625	2	I
CuSe	*e*	CuI	605	800	18	I
GeSe	—	$Ge + I_2$ (GeI_4)	144	600	6	B
$GeSe_2$	707	$Ge + I_2$	144	600	6	B
SnS	—	$Sn + I_2$ (SnI_2)	320	400	8	K
SnSe	861	$Sn + I_2$ (SnI_2)	320	400	8	K
SnTe	780	$Sn + I_2$ (SnI_2)	320	400	8	K
PbSe	—	$Pb + I_2$ (PbI_2)	402	600	6	L

TABLE IV (continued)

Compound	Melting point (°C)	Halide used	Melting point of halide (°C)	Temperature of preparation (°C)	Heating time (hr)[a]	Method of metal halide separation[b]
PbTe	917	$Pb + I_2$ (PbI_2)	402	600	6	L
La_2Se_3	—	LaI_3	772	850	24[d]	E
Nd_2Se_3	—	NdI_3	775	850	24	E
Sm_2Se_3	—	SmI_3	850	900	24[d]	E
$CuCr_2Se_4$	—	CuI	605	700	18	J

[a] Heating was performed in a silica boat except where otherwise noted.
[b] A: Solution in water. B: Heating at 350°C in Ar. C: Heating at 350°C in H_2. D: Heating at 250°C in Ar. E: Solution in dimethyl formamide. F: Floating at 300°C in Ar. G: Solution in *N*-methyl-2-pyrrolidone. H: Heating at 600°C in Ar. I: Solution in KI. J: Solution in ammonia. K: Heating at 400°C in Ar. L: Heating at 500°C in Ar.
[c] Sublimes.
[d] Graphite crucible.
[e] Red heat.

applicable for the synthesis of compounds containing S, Se, Te, P, As, and Sb as "negative" element (chalcogenides and pnyctides). Most metals listed in Table I react readily. If two elements can form more than one compound of different composition, these different compounds can be obtained by heating the appropriate stoichiometric amounts of the elements together with the metal halide.

In the following section, the preparation of nine compounds which are representative examples of the iodide method will be given in detail. The identity of the compounds was checked by means of x-ray analysis.

VI. Experimental

A. CdTe

A small silica boat containing 7.02 g of pure cadmium (pellets, 99.9999%, Soc. de la Vieille Montagne, Belgium), 7.98 g of pure tellurium (pieces, 99.9999%, Soc. Gén. Métal. de Hoboken, Belgium), and 10 g of cadmium iodide (CdI_2) (99.999%, Noury Baker N.V., Deventer, Holland) is placed in a silica tube. The evacuated and sealed tube is placed in a tubular furnace and heated to 600°C for 1 hr and kept at that temperature for about 6 hr. After the furnace has cooled to room temperature, the tube is opened and the silica boat is placed in a 400-ml silica beaker partly filled with deionized water. After boiling for a few minutes, the solid becomes detached from the wall of the boat and CdI_2 is dissolved. The resulting very fine powder is washed several times with water by decantation and dried at 90°C. The yield is nearly 15 g of a black crystalline powder with a particle size of 1 μm. The purity of the product is given in Table II.

B. GaP

A small silica boat containing a mixture of 4.3 g of gallium (99.9999%, Alusuisse, Neuhausen, Switzerland), 1.5 g of red phosphorus (lumps, 99.9999%, Koch-Light, Colnbrook, England), and 4.5 g of iodine (99.9999%, Merck, Darmstadt, Germany) is placed in a silica tube. This tube is vacuum-sealed, placed in a furnace, and heated at 250°C for 1 hr. Then, the temperature is gradually raised to 500°C at a rate of 50°C hr^{-1}. This temperature is maintained for 12 hr.

The oven is cooled to room temperature, the tube is opened, and the gallium iodide GaI_3 is removed by heating the reaction mixture at 350°C in a stream of H_2 gas. The result is about 5 g of microcrystalline GaP. The purity is given in Table III. To purify this material from iodine, it is recrystallized from Ga. Five grams of GaP together with 30 g of pure gallium are heated in an evacuated quartz tube for 2 hr at 1050°C. The furnace is slowly cooled to 500°C and opened at room temperature. Liquid gallium is poured off from the GaP crystals. Traces of attached gallium are removed by boiling in very pure 6 *N* hydrochloric acid. The product is dried at 70°C. Before the recrystallization, the iodine content is about 500 ppm by weight. After recrystallization using a platinum crucible, the iodine content is 0.01 ppm.

C. GaS

A mixture of 2.5 g of gallium, 0.95 g of sulfur, and 2.54 g of iodine is placed in a silica boat. This boat is placed in a silica tube, the tube is evacuated and sealed. The tube is heated first at 300°C for 1 hr and then slowly heated to 600°C. After 8 hr of heating at 600°C, the tube is opened and the material is heated in a nitrogen stream at 300°C. GaI_3 is removed by sublimation and 3 g of pure, powdery GaS is left.

D. Ga_2S_3

The same preparation method is followed as for GaS, but now 2.22 g of gallium, 1.23 g of sulfur, and 2.54 g of iodine are placed in the silica tube. The result after removing the GaI_3 by heating at 300°C in nitrogen is 3 g of pure, powdery Ga_2S_3.

E. $MoSe_2$

A mixture of 1.44 g of molybdenum, 1.58 g of selenium, and 3.04 g of iodine is placed in a silica boat in a silica ampoule. The ampoule is evacuated and sealed. It is then heated in a furnace for 1 hr at 300°C. Next, the temperature is slowly increased to 600°C and the ampoule is heated at this temperature for 8 hr. After cooling to room temperature, the reacted material is heated at 300°C in a nitrogen stream. A total of 2.56 g of $MoSe_2$ is left.

F. $CuCr_2Se_4$

A mixture of 1.27 g of copper, 2.08 g of chromium, 6.3 g of selenium, and 9.59 g of copper iodide (CuI) is heated at 600°C for 6 hr in an evacuated, sealed silica tube. After this period, the tube is opened and the contents washed with water in order to remove the CuI. The residue is dried and consists of 5 g of pure, powdery $CuCr_2Se_4$.

G. MnTe

A mixture of 1.5 g of manganese, 3.5 g of tellurium, and 5 g of manganese iodide (MnI_2) is placed in a graphite crucible and this in turn in a silica tube. These manipulations are carried out in a dry room under argon atmosphere. Then, the tube is evacuated, sealed, and heated in vertical position from room temperature to 700°C with a heating rate of about 100°C hr^{-1}. The temperature is kept at 700°C for 16 hr. After cooling to room temperature, the tube is opened and the reaction product is washed with dimethyl formamide. The residue is dried and the result is 5 g of MnTe.

H. TiSe

A mixture of 4.02 g of titanium, 4.67 g of selenium, and 6.31 g of iodine is brought in a graphite crucible under argon atmosphere. This crucible is placed in a silica tube, and the tube is evacuated, sealed, and heated in the vertical position at 300°C for 2 hr. Then, the temperature is increased slowly to 700°C with a heating rate of 100°C hr^{-1}. The tube is heated at 700°C for 24 hr, then cooled to room temperature. The reaction product is washed with dimethyl formamide to remove titanium iodide and the residue is dried. The result is 7.5 g of finely powdered TiSe.

I. La_2Se_3

A mixture of 2.7 g of lanthanum, 2.3 g of selenium, and 5 g of lanthanum iodide (LaI_3) is placed in a graphite crucible in an argon atmosphere. The same procedure as described for the preparation of TiSe is followed, but after reaching 700°C, the temperature is increased further at a rate of 50°C hr^{-1} to 850°C. This temperature is maintained for 18 hr. After cooling and washing with dimethyl formamide, the result is 5 g of homogeneous, powdery La_2Se_3.

REFERENCES

Brown, D. (1968). "Halides of the Lanthanides and Actinides." Wiley (Interscience), New York.

Canterford, J. H., and Colton, R. (1968). "Halides of the Second and Third Row Transition Elements." Wiley (Interscience), New York.

Colton, R., and Canterford, J. H. (1969). "Halides of the First Row Transition Metals." Wiley (Interscience), New York.

Godau, I., and Ormont, B. F. (1968). *Zh. Neorg. Mater.* **4**, 149.

Golubkov, A. V., Zhukova, T. B., and Sergeeva, V. M. (1966). *Zh. Neorg. Mater.* **2**, 77.

Hiscocks, S. E. R., and Mullin, J. B. (1969). *J. Mater. Sci.* **4**, 962.

Kwestroo, W., Huizing, A., and de Jonge, J. (1969). *Mater. Res. Bull.* **4**, 817.

Leverenz, H. W. (1950). "An Introduction to Luminescence of Solids." Wiley, New York.

Miller, J. F. (1962). *In* "Compound Semiconductors" (R. K. Willardson and H. L. Goering, eds.), Vol. I, p. 194. Reinhold, New York.

Ray, B. (1969). "II–VI Compounds," Chapter 2. Pergamon Press, Oxford.

Reisman, A., and Berkenblit, M. (1963). *J. Phys. Chem.* **67**, 22.

Weast, R. C. (1967). "Handbook of Chemistry and Physics," 48th ed. The Chemical Rubber Co., Cleveland, Ohio.

Author Index

Numbers in italics refer to the pages on which the complete references are listed.

B

C

D

E

F

I

J

K

L

M

N

O

P

R

S

T

Y

Subject Index

T

U

V